Basic Construction Materials

EIGHTH EDITION

Theodore W. Marotta
Professor of Civil Engineering Technology
Hudson Valley Community College

John C. Coffey, AIA
Instructor of Civil Engineering Technology
Hudson Valley Community College

Cynthia Brown LaFleur
Instructor of Civil Engineering Technology
Hudson Valley Community College

Christine LaPlante, PE, Ph.D
Department Chair Engineering and Industrial Technologies
Hudson Valley Community College

Prentice Hall
Boston   Columbus   Indianapolis   New York   San Francisco   Upper Saddle River
Amsterdam   Cape Town   Dubai   London   Madrid   Milan   Munich   Paris   Montreal   Toronto
Delhi   Mexico   City   Sao Paulo   Sydney   Hong Kong   Seoul   Singapore   Taipei   Tokyo

Editor in Chief: Vernon R. Anthony
Acquisitions Editor: David Ploskonka
Editorial Assistant: Nancy Kesterson
Director of Marketing: David Gesell
Marketing Manager: Derril Trakalo
Senior Marketing Coordinator: Alicia Wozniak
Project Manager: Holly Shufeldt
Senior Art Director: Jayne Conte
Cover Art: iStockphoto
Full-Service Project Management and Composition: Shiny Rajesh,
    Integra Software Services Pvt. Ltd.
Printer/Binder: LSC Communications
Cover Printer: LSC Communications
Text Font: 10/12, Minion

Many of the designations by manufacturers and seller to distinguish their products are claimed as trademarks. Where those designations appear in this book, and the publisher was aware of a trademark claim, the designations have been printed in initial caps or all caps.

Library of Congress Cataloging-in-Publication Data

Cataloging-in-Publication Data for this title can be obtained from the Library of Congress.

Prentice Hall
is an imprint of

www.pearsonhighered.com

ISBN 10:   0-13-512969-9
ISBN 13: 978-0-13-512969-2

PREFACE

In *Basic Construction Materials*, Eighth Edition, we present some of the basic materials used in the construction industry. We introduce these materials to prepare the reader for further academic course work in construction and engineering programs, or for entering the construction industry.

The basic materials selected are as follows:

Aggregates	Masonry
Asphalt, asphalt concrete	Iron and Steel
Portland cement, Portland cement concrete	Wood

These materials are widely used in construction and represent those over which field people in the industry have the most control. Shaping these materials to final size, protecting them from the elements, and fitting them together are accomplished in the field to a greater extent than with most other materials.

The format of this book consists of text material as well as an appendix that includes industry standards from the American Concrete Institute (ACI) and the Engineered Wood Association, and student-directed laboratory experiments. Because the construction industry is undergoing metrication, the appendix to this book also includes valuable metric information.

We appreciate the continued assistance of the engineering, construction, and manufacturers' associations that have provided valuable information for this book. Comments and assistance from our teaching and construction industry colleagues, as well as from our students, have been very helpful and are gratefully acknowledged. In addition, particular thanks are due to Kelly Seitter, Ferris State University, Eugene H. Wright, University of Nebraska–Lincoln, and John A. D'Angelo, PhD, PE, D'Angelo Consulting, for their assistance with the eighth edition text review. We would also like to acknowledge the continued support and assistance afforded us by the editorial staff at Pearson-Prentice Hall.

As with any publication, any errors or omissions are the responsibility of the authors. Therefore, we would appreciate notification of such, as well as suggested improvements, at j.coffey@hvcc.edu.

T.W.M.
J.C.C.
C.B.L.
C.M.L.

NEW TO THIS EDITION

The manufacture and use of the construction materials described in this text account for a substantial portion of all construction-related activity in the United States and throughout the world. Virtually all building projects use concrete, masonry, metals (including steel and iron), wood, or derivatives of these products. Asphalt materials have an ever-increasing share of all new and re-paved highway surfaces worldwide.

While a basic understanding of the proper fabrication and use of these materials remains a focus of this text book, new concerns about sustainability, the environmental costs of materials production, and the ability to recycle construction materials have taken a renewed focus in this eight edition of *Basic Construction Materials*.

Significant revisions and updates to the text include:

- A section on student experiments that can be easily understood now replaces the ASTM standards in the appendix of the previous editions. These experiments allow students and instructors to follow accepted guidelines for testing materials and analyzing the results.

- Extensive new information regarding Superpave has been added in Chapter 3 on asphalt materials, and this chapter has also been revised and updated.

- A new section highlighting the concept of sustainability in materials manufacture and use has been added in Chapter 1. Chapters 3, 4, 5, and 6, concerning asphalt, concrete, iron and steel, and wood, have been revised to reflect concerns about sustainability and materials recycling and reuse.

- A revised Instructors Test Bank is available in an electronic format along with PowerPoint presentations focusing on each chapter and providing updated lecture materials and exercises for presentation to students.

- New example problems have been developed for Chapters 1, 2, 3, and 4.

- New data recording sheets and excel spread sheets have been developed which will make it easier for students to record, analyze, and report the results of material experiments.

- New tables, figures, charts, and photos have been added to the text where applicable.

- Tables and figures are now separately identified so that readers may more easily follow and coordinate between text and graphic information.

INSTRUCTOR RESOURCES

To access supplementary materials online, instructors need to request an instructor access code. Go to www.pearsonhighered. com/irc, where you can register for an instructor access code. Within 48 hours after registering, you will receive a confirming e-mail, including an instructor access code. Once you have received your code, go to the site and log on for full instructions on downloading the materials you wish to use.

CONTENTS

Chapter 4

PORTLAND CEMENT CONCRETE 88

CHAPTER ONE

INTRODUCTION

Every construction project is intended to result in a finished structure which will perform certain functions in conformance with the project design requirements. Whether or not satisfactory results are achieved depends upon the materials selected and how they are used. The designer, the builder, and the user must all understand construction materials to produce the finished facility and to use it to best advantage. Knowledge of design procedures, construction methods, and maintenance practices is needed. Underlying all of these qualifications is a knowledge of materials. In order to be completely satisfactory, each material used must perform its function well over a sufficiently long time, and both original cost and maintenance expense must be reasonable.

THE CONSTRUCTION PROCESS

The construction process is initiated when a person or organization, which may be public or private, decides to improve the land with permanent or semipermanent additions. The initiator of a construction project, hereafter called the *owner*, therefore has a need, as well as the required financing, to complete the process. Private organizations may obtain the necessary funds through construction loans and mortgages, whereas public groups may use tax revenues and bonds as well as user fees to fund construction projects.

After a need is established, and financing has been obtained, the owner contracts with a design professional. The architect or engineer prepares plans, called *working drawings*, showing details and how the completed project will look. The plans indicate and briefly explain the various materials required. The specific details related to materials are covered in the *specifications*. These documents explain in great detail what materials to use, the characteristics of the materials, and what methods of inspection and testing

the owner's representative will use to evaluate the selected materials. The designer also incorporates in the specifications and working drawings all of the necessary building code, zoning, wetland, and other governmental requirements, where applicable.

After completion of the contract documents, contractors use them to prepare their estimates for bidding or negotiating purposes. The contractor, selected by bid on public works projects, and by bid or negotiation on private projects, enters into a contract with the owner to provide a completed project in accordance with the project contract documents. The two most common contracts are *lump sum* and *unit price*, depending upon the type of construction project. The lump sum estimate and subsequent bid requires the contractor to estimate all of the material quantities, installation, labor, and equipment costs to complete the project. The contractor then adds an overhead and profit figure to the total estimated cost. The lump sum system is usually associated with building construction projects.

Public highway construction typically uses the unit price system, where the owner supplies all of the quantities to the contractor in the bid documents. The contractor calculates cost factors for each material quantity unit as well as overhead and profit values. Therefore, each item in the estimate includes a cost factor as well as overhead and profit. Unit price contracts therefore make provisions for quantity changes, because the contractor will have based the bid price on the owner's estimate of material quantities. If large changes occur in estimated quantities, the contractor loses money on decreases and makes more money on increases, in which case the owner loses money. To protect both parties, a *general conditions* section is included in the contract documents. The general conditions are a set of guidelines developed by the design and construction communities to ensure fair practices. Because the project cost is determined by the submitted bid or negotiated price, changes to required work are dealt with by using *change orders*, which cover additional work, deleted work, or changed conditions.

The construction delivery system selected requires the contractor to deliver to the owner a completed project on time as scheduled, within budget, and at the specified quality levels. It is a given in the industry that the construction process will be undertaken in a safe manner. Therefore, project safety requirements are based on the Occupational Safety and Health Administration (OSHA) regulations, as well as numerous other regulations, and are the contractor's responsibility. The contractor is also responsible for maintaining a current Manufacturer's Safety Data Sheet (MSDS) file for all materials used on the project. The MSDS sheets, which are obtained from suppliers, cover potential material hazards and required safe handling techniques, as well as what should be done in case an accident occurs while using the material.

As a management tool, the contractor may elect, and/or be required by the contract, to submit a *project schedule* to the designer for approval. The proper scheduling of a project enables the contractor to allocate resources such as labor, money, equipment, and materials appropriately. The schedule created is updated as the project progresses, thus enabling the contractor to reallocate job resources as required. The schedule may also assist the contractor with cash flow analysis. Figure 1–1 shows a typical schedule, which incorporates various construction activities, starting and ending dates, activity duration, and critical activities, which, if delayed, may impact the scheduled completion date.

Basic materials, such as wood, asphalt, stone, and manufactured products, such as plywood sheets, cast iron pipe, and concrete masonry units (CMUs), must all be specified. Combinations of materials are commonplace, such as trusses consisting of glued laminated timber members in combination with steel members, cast iron pipe with portland cement lining, concrete beams reinforced with prestressed steel wire, or window and frame units containing glass, several kinds of metal, and plastic all in one *assembly*.

An assembly is either fully built at the factory (*shop assembled*) or partially completed in the factory and assembled in the field (*jobsite assembled*). Some of the types of work performed in the field are also manufacturing processes; for example, the mixing and placing of concrete and the cutting and welding of steel.

The contractor also has the responsibility of selecting materials for the project that comply with the technical specifications, because nothing may be incorporated into the project without the designer's review and/or approval. This submittal process requires that the contractor obtain catalog cuts, material samples, and shop drawings from suppliers and subcontractors to submit to the designer for approval. Catalog cuts might be used for door hardware, material samples for carpet, and shop drawings for structural steel. Some materials and assemblies may require the use of more than one of the submittal forms. The designer reviews the submittals for specification requirements and approves or rejects the contractor's choices. The contractor places copies of the approved catalog cuts, shop drawings, and, in some cases, approved material samples at the jobsite for construction use as well as for inspection and testing.

The owner is represented during the construction stage by an agent, usually the designer, who administers the contract impartially, by approving or rejecting materials and workmanship, by approving final construction, and by determining the amount of payment due. *Inspectors* are present at the jobsite to inspect the work in progress and perform field tests as part of construction supervision. Laboratory testing and field testing may be performed by an *independent testing laboratory*. The testing laboratory reports whether or not materials comply with specifications.

The builder uses materials in his operation which may not become part of the finished construction and are not controlled by the designer. Examples are temporary sheeting to hold back the sides of an excavation, and removable forms to hold concrete in the desired shape until it cures. The builder, like the designer, must select, inspect, and test materials best suited for the purpose from among those available.

Those who supply materials and partially or fully assembled components to be used in construction are called *suppliers* or *vendors*; included are manufacturers, quarries, sawmills, and others.

NEED FOR MATERIALS WITH VARIOUS QUALITIES

The construction industry requires materials for a vast range of uses. The qualities these materials possess are as varied as the strength and flexibility required of an elevator cable or the warm, wood grain appearance and smooth finish of a birch or maple cabinet.

The construction of a simple building, such as a house, requires selection of materials to perform the following tasks:

1. Footing
   (a) Distribute the weight of the building to the soil
   (b) Resist cracking despite uneven soil settlement
   (c) Resist corrosive attack from soil and water

2. Basement floor
   (a) Provide a smooth surface
   (b) Resist wear
   (c) Resist cracking despite upward water pressure or uneven soil settlement
   (d) Keep moisture out
   (e) Resist corrosive attack from soil and water

3. Basement walls
   (a) Support the rest of the building
   (b) Resist lateral side pressure from the earth
   (c) Keep moisture out
   (d) Resist corrosive attack from soil and water

4. Other floors and ceilings
   (a) Provide a smooth surface
   (b) Resist wear
   (c) Support furniture and people without sagging excessively or breaking
   (d) Provide a satisfactory appearance

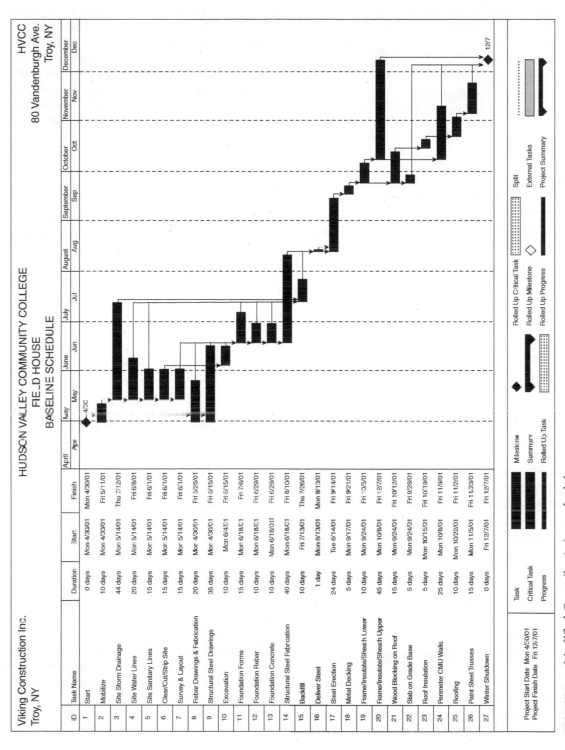

FIGURE 1–1. Modified Gantt (bar) chart schedule.

(Courtesy Eds Fleming)

    (e) Clean easily
    (f) Insulate against noise transmission

5. Outside walls
    (a) Support floors and roof
    (b) Resist lateral wind pressure
    (c) Provide a satisfactory appearance inside and out
    (d) Insulate against noise and heat transmission
    (e) Keep moisture out

6. Partitions
    (a) Support floors and roof
    (b) Provide a satisfactory appearance
    (c) Insulate against noise transmission

7. Roof
    (a) Keep moisture out
    (b) Support snow and other weights
    (c) Resist wind pressure and wind uplift
    (d) Provide a satisfactory appearance
    (e) Insulate against noise and heat transmission

The types of materials used in smaller buildings have become somewhat standardized. However, new materials are constantly being proposed, and the use of these materials results in lowered costs or improved living conditions. Their development and proper use require an understanding of materials.

A building is used to illustrate the points that a construction project includes many components that must perform various functions and that new materials must be constantly analyzed. However, the same is true of any other construction project. A project such as paving a street or laying a pipeline requires more kinds of material to perform several different functions than the casual observer would expect. New materials are continually available in these fields also.

As people's desires expand, the need is created for materials with new qualities. In order to explore space, lightweight materials were needed that could resist heat of a higher degree than ever before. Necessary qualities may be obtained by developing special treatments for common materials or by developing entirely new materials. For example, treatments have been developed to make wood highly fire resistant. Additionally, the development of steel allowed the construction of bridges with longer spans than had been possible with wood components.

SELECTING MATERIALS

We constantly encounter man-made objects built of materials carefully selected to be the most satisfactory ones for that particular use. Any satisfactory choice always requires a knowledge of construction materials and an adequate selection procedure.

A designer is selected who, among other things, is responsible for selection of all construction materials to achieve the desired performance within the budget cost. He considers the service each component must perform, appearance, original cost, maintenance expense, and useful life expectancy. Maintenance includes such operations as cleaning, preventing and repairing corrosion damage, and repairing or replacing damaged material.

Original cost and maintenance expense must be weighed together against useful life expectancy. Original cost and maintenance expense must also be balanced against each other. Often a low first cost means high maintenance expense and vice versa. However, this is not always so. Expensive material may be expensive to maintain. Even though it is the total cost that must be considered, it is important to remember that the original cost must be paid during construction and immediately thereafter, whereas the maintenance expense is paid through the life of the facility.

The designer may select the material or assembly needed, or prepare specifications describing the performance required and let the builder do the selecting within the requirements of the specifications subject to the approval of the person supervising construction.

If the designer specifies exactly what materials and assemblies are to be incorporated into the project, she knows, either from past experience or from investigation, that they will be satisfactory. She avoids the risk of using something new or unfamiliar. The designer also misses the opportunity of using something that is more economical or performs better. Specifications prepared this way are called *material specifications*. Material specifications may be either open or closed depending upon the wording. A closed or proprietary specification identifies specific products with no allowable substitutions. An open specification will name a proprietary product but allow substitutions, by adding the phrase "or approval equal." The open specification puts the burden on the designer to determine if the contractor's substitute is equal to or better than the named product.

If the designer specifies performance in terms of appearance, strength, corrosion resistance, and other features, he has the benefit of the builder's and vendor's experience in selecting the most economical materials. Specifications prepared this way are called *performance specifications*. Performance specifications may also include bonus and penalty clauses. In cases where highway pavement specifications stipulate pavement density and/or air void content, the contractor's work will be evaluated using established testing procedures and payments will be adjusted based on test results. They must be very carefully written to prevent any inferior products from satisfying the specification requirements; and the builder's selections must be carefully investigated to be sure they are acceptable according to the specifications. Both types of specification and various combinations of the two types are used.

The process of selection may include the following steps:

1. Analysis of the problem (e.g., performance required, useful life required, allowable cost, and maintenance expense).
2. Comparison of available materials or products with the criteria of step 1.

3. Design or selection of type of material, size, shape, finish, method of preserving, and method of fastening in place.

The method used to select construction materials based on these criteria is often referred to as a *life cycle–cost analysis.*

Though the method is not an exact science, if used correctly, the results will be of value when making material or system selections. The application of life cycle–cost analysis is usually limited to materials and systems that will be subjected to heavy, continuous use. For example, a highway engineer may analyze asphalt concrete pavements compared to portland cement concrete pavements, and a mechanical engineer may analyze different heating and cooling systems. The intent of the analysis in either case is to determine the most economical material or system, which will yield the lowest cost over its useful life.

SUSTAINABILITY

The concept of sustainability in building design and construction has grown out of concerns about better utilizing material and energy resources, enhancing the environment, and even creating jobs and lowering construction and operating costs. "Green building," "sustainability," and "environmentally friendly design" are all terms we may have heard, but what do these phrases actually mean and how do they relate to a textbook about construction materials?

For many years in the United States, when a building reached the end of its useful life and was no longer considered serviceable, it was demolished and the component parts of the building were removed and placed in a landfill. The material, energy, and human resources (labor) used to create the former building were considered expendable and a new facility was built using materials that were all newly manufactured. Jobs were created to produce these materials and construct the new building. Manufacturers, suppliers, designers, and builders all could make a profit on the venture. It seemed like a perfect system and was repeated over and over again.

What this process ignored, however, was that the material and energy resources needed to sustain this cycle were not limitless. Additionally, damage to our environment caused by air and water pollution in the manufacturing process and the accumulation of hazardous wastes from discarded building products was becoming more and more costly. We needed to find better ways to manage our built environment by conserving resources while limiting or reversing the impact on the environment and maintaining an economically viable construction industry.

Sustainability in building design and construction does not seek to prevent new building nor does it limit the choices we have in the materials we can use. The fundamentals of sustainability are formed around several related concepts:

First, we must design and construct new and renovated buildings through the most efficient use of finite material, energy, and labor resources. This includes choosing construction materials which will promote the maximum useful life of the building.

Second, we must seek to operate these facilities in a way that significantly reduces energy consumption and the production of green house gases which can contribute to climate change. Material choices which improve natural ventilation and help to control heat loss and gain are an example of this.

Third, we need to choose construction materials which can be recycled and reused when facilities must be reconstructed.

Let us look at each of these concepts as they relate to the construction materials discussed in this text.

Longevity. In central cities in Europe it is not unusual to find neighborhoods where the typical homes on a street may be 200 to 400 years old. Masonry and wood are the main components of these homes, yet they have remained functional and useful for centuries. It is noted in Chapter 4 that cast concrete portions of the Roman aqueduct system remain in use more than 2000 years after they were built.

In selecting materials today, designers must be aware of their appropriate use and inherent characteristics that will promote longevity of the buildings where they will be used. Proper detailing, protection, and compatibility with other materials are all elements the architect or design engineer must consider if new and renovated facilities are to remain useful over time. Builders must likewise be aware of the necessary preparation, fabrication, and placement of specified materials to insure they will continue to perform long after they are installed. As an example, concrete that is cast in place and properly finished and cured will perform over long periods of time with only normal maintenance. Concrete that is poorly fabricated or finished may remain serviceable only for short periods and can require replacement at significant cost to the building owner.

Energy Efficiency. All of the materials described in this text have properties relative to their energy footprint. Asphalt, concrete, and steel products use significant amounts of energy in their production. Wood products are, by contrast, a renewable resource, but must be carefully managed. Overproduction resulting from clear cutting contributes to global warming because the tree canopy is removed and depleted soils allow unrestrained runoff to harm rivers and streams and cause flooding.

Fortunately, new production technologies are emerging which are lowering the energy costs of producing concrete, steel, and asphalt paving materials. Many of these enhanced technologies are described in Chapters 3 and 4. Better forest management coupled with the expanded use of manufactured wood products is reducing lumber production costs. By-products such as wood strands and flakes that were once discarded in lumber mills can now be reused to make preengineered wood products, as described in Chapter 6.

Masonry (described in Chapter 7), which is one of our oldest building materials, can be used to aid in controlling heat loss and gain in buildings. Because of its mass and natural insulating properties, masonry products gain and lose heat slowly. The oldest known buildings in North America, the adobe masonry of the Southwest, utilized this concept to create structures which could resist solar heating of the day while allowing the slow loss of this heat to keep occupants warm at night. Masonry products continue to be used for this same purpose today.

Reusability and Recycling.   Virtually all of the construction products described in this textbook can be recycled to create new materials. Concrete removed from demolished structures can be crushed and stone aggregate removed for reuse. Reinforcing steel embedded in the concrete is isolated using magnets and can be reprocessed to make new iron and steel products. Asphalt paving can be remilled and used to create new paving. Most state and municipal highway projects require a percentage of all new paving to be remilled asphalt. Steel products are routinely recycled. In the past, the percentage of these recycled materials available for making construction grade steel was small, but this percentage is increasing as better separation technologies are being used to remove impurities from scrap steel. Composite materials manufactured from recycled wood and plastic waste are used as exterior decking and railing materials, as described in Chapter 6.

The Future of Sustainability

Sustainable design and construction principles are being widely adapted by governments, designers, and builders worldwide. This is true, not only because of the energy and material resource considerations, but also because green building has been proven to be a growing source of employment. Moreover, sustainable practices promote environmental health and safety of the public while significantly lowering energy costs.

In the United States, the U.S. Green Buildings Council (USGBC) is one of the primary forces in promoting sustainability in building design and construction. The Council has developed a program entitled "Leadership in Energy and Environmental Design (LEED)," which, through a point system, allows owners and designers of built facilities to apply for LEED certification. The advantage of certification is that it provides a measurable system for documenting sustainability and energy economies for new and renovated facilities. In the United States, many federal, state, and municipal agencies now require LEED certification for their projects. Private owners are increasingly seeking certification as well because the demonstrated savings and lowered operating costs of LEED-certified buildings are a sound business practice. Design and construction professionals may seek LEED accreditation as a means of demonstrating enhanced competence in energy- and environment-friendly design and construction practices.

PROPERTIES OF MATERIALS
Thermal Expansion

All building materials change size with a change in temperature, becoming smaller when colder and larger when hotter. A piece of material, if heated uniformly, expands, with each unit length becoming a certain percentage longer. This elongation takes place in all directions and is somewhat different for each material. In order to predict amounts of expansion and contraction to be expected, a *coefficient of expansion* is determined for each material. It is a decimal representing the increase in length per unit length per degree increase in temperature. The coefficient varies somewhat at different temperatures but is nearly constant for the range of temperatures involved in most cases so that one coefficient can be used for each material.

The coefficient of expansion for wrought iron is 0.0000067 in. per in. per °F or 0.0000121 cm per cm per °C. Each inch of length, width, or thickness becomes 1.0000067 in. if the temperature is increased 1°F and increases an additional 0.0000067 for each additional °F increase. As the temperature decreases, the dimensions decrease at the same rate. Table 1–1 shows the coefficients of expansion for some common building materials.

Materials to be used together in an assembly must have approximately the same coefficients of expansion, or else some provision must be made for their different expansions. The use of steel reinforcement in concrete is a good example of combining materials with approximately the same linear coefficients of thermal expansion. Long structural members may expand and contract so much that expansion room must be provided at the ends.

Example
Calculate the change in length of a 350-ft steel pipe that will be exposed to temperatures ranging from 55 to 200°F.

$$\delta = \alpha L(\Delta T)$$

where δ = total change in length (in., ft)

α = linear coefficient of thermal expansion

L = original length

ΔT = change in temperature (°F)(°C)

$$\delta = (0.0000065)(12 \times 350)(200 - 55)$$
$$\delta = 3.959 \text{ in.}$$

Calculate the change in length of a 120-m steel pipe that will be exposed to temperatures ranging from 12 to 95°C.

$$\delta = (0.0000117)120(95 - 12)$$
$$\delta = 0.1165 \text{ m}$$

Thermal Conductivity

A building used by people must be kept warmer than the surrounding air in cold climates and cooler than the surrounding air in hot climates. Heat flows to a cooler area much like water flows to a lower level. The flow continues

Table 1–1. Coefficients of expansion

Material	Coefficient of Linear Expansion	
	per °F	per °C
Asphalt	0.00034*	0.00061*
Portland cement concrete	Assumed to be 0.0000055 but varies from 0.000004 to 0.000007	Assumed to be 0.0000099 but varies from 0.000007 to 0.000013
Gray cast iron	0.0000059	0.0000106
Wrought iron	0.0000067	0.0000121
Structural steel	0.0000065	0.0000117
Stainless steel	0.0000055 to 0.0000096	0.0000099 to 0.0000173
	Parallel to grain	
Wood	0.000001 to 0.000003	0.000002 to 0.000005
	Perpendicular to grain	
	0.000015 to 0.000035	0.000027 to 0.000063
Mineral aggregate	0.000003 to 0.000007	0.000005 to 0.000013

*Coefficient of volumetric expansion; coefficient of linear expansion is not useful.

until outside and inside temperatures are equal. Heat movement takes place by conduction through any solid object separating areas of different temperatures.

It costs money to heat or cool a building, and the movement of heat in the wrong direction is expensive. The rate of movement varies with the material through which the heat passes. For large areas such as walls and roofs this rate is an important consideration. The rate is measured as thermal conductivity (U) in British thermal units (Btu) of heat transmitted per square foot of cross section per hour per °F difference in temperature between the two sides of the material.

Insulation, which is material with a very low U, is used to line large surfaces to lessen the rate of heat flow. The U of a material varies directly with its density. Dead air spaces in a material are effective in reducing the U factor. One of the better insulations, expanded plastic foam, consists of bubbles with the proportion of solid material less than 1 percent of the volume and the rest consisting of air or gas. Insulation is also made of fibers, ground particles, or other porous material. However, some structural materials have a low U factor and therefore serve as insulation also. Wood and certain types of lightweight concrete are two such materials that are covered in this book.

The resistance that construction materials offer to the flow of heat is called *thermal resistance* and is designated by the letter R. The reciprocal of the heat transfer coefficient U is R with a unit value of (hrft^2°F/Btu). The value of R is very useful for determining total heat flow of systems because individual R values for each component can be added together to determine the overall R value for the system. The higher the R value the better the insulating properties of the system or material are. Most of the materials used in construction have been tested and assigned R values based on the material's thickness as commonly used, or an R value rating per inch of material. For example, an expanded polystyrene

extruded panel with a cut surface and a density of 1.8 pounds per cubic foot (pcf) has an R value of 4.0; therefore, a 4-in.-thick panel would have a total R value of 16. CMUs are rated on overall thickness of the unit; therefore, an 8 in., 3-core unit has an R value of 1.11 for its 8-in. thickness. The calculations for heat loss and cooling loads of large buildings are normally completed by mechanical engineers who work with the architect to furnish a complete heating, ventilating, and air conditioning (HVAC) system for the structure.

Strength and Stress

All construction materials must resist loads or forces. A *force* is a push or pull that has a value and a direction. Loads on structures are separated into dead loads and live loads. *Dead loads* include the weight of the structural elements as well as permanent equipment such as boilers and air-conditioning units. *Live loads* are those imposed loads which may or may not be present and include occupants, furniture, wind, earthquake, and other variable load conditions.

A force exerted on the surface of an object is assumed to spread uniformly over the internal area of the object (Figure 1–2). A force does actually spread within the material, but not uniformly. *Stress* is force per unit area over which the force acts. It is obtained by dividing the force by the area on which it acts and is expressed as pounds per square inch (psi) or kips per square inch (Ksi). A kip is equivalent to 1000 lb.

Strength of a material, in general terms, is the ability to resist a force. That ability depends on the size and shape of the object as well as on the material of which it is made. An object with a large area is able to resist more force than an object of the same material but with a smaller resisting area.

In order that strength may be considered as a property of the material, it is necessary to relate strength to the material

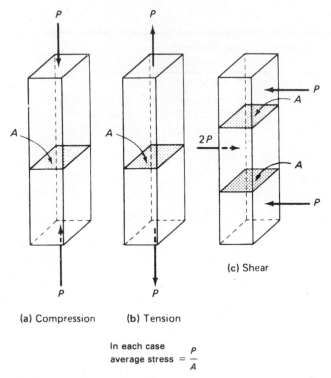

(a) Compression   (b) Tension

(c) Shear

In each case
average stress $= \dfrac{P}{A}$

FIGURE 1–2. Illustration of stresses.

itself regardless of its size or shape. Therefore, the *strength* of a material in technical terms is equal to the stress that the material can resist. Strength has the same units as stress, that is, psi, Ksi, or megapascals (MPa).

The useful strength of a material is equal to the stress at failure. Failure takes place when an object can no longer serve its purpose. The material may fail by breaking or by excessive deformation. *Deformation* means a change in the outside dimensions of an object caused by a force. The amount of deformation depends on the size and shape of the object as well as on the material of which it is made. As in the case of strength, it is desirable to relate deformation to the material itself regardless of its size or shape.

The term *strain* means the total change in dimension divided by the original dimension. Strain is the effect caused by stress. Strain is a ratio and therefore has no units. The amount of deformation and the original length must be measured in the same units to provide a correct ratio. They are usually measured in inches.

Strain can be shown by stretching a rubber band or compressing or twisting a piece of rubber hose. A sample of rubber subjected to a compressive force becomes substantially shorter and a little wider; one subjected to a tensile stress becomes substantially longer and a little narrower. The deformation and original length considered in computing strain are the ones in the direction of the stress.

Example

A metal bar with a rectangular cross section 1.5 in. by 2 in. breaks at a tensile load of 125,000 lb. Calculate the stress in the bar at failure.

$$s = \frac{P}{A}$$

where s = average computed stress (psi, Ksi)
P = external applied load (lb, kips)
A = cross-sectional area (sq in., sq ft)

$$s = \frac{125,000}{1.5 \times 2} = 41,667 \text{ psi}$$

The calculated stress may be converted directly to megapascals:

$$s = 41,667 \,\text{psi} \times 0.00689476 = 287.28 \text{ MPa}$$

or the problem can be solved using metric values:

$$1.5 \text{ in.} \times 0.0254 = 0.0381 \text{ m}$$
$$2.0 \text{ in.} \times 0.0254 = 0.0508 \text{ m}$$
$$125,000 \text{ lb} \times 4.44822 = 556,027 \text{ newtons (N)}$$

$$s = \frac{556\,027 \text{ N}}{0.0381 \text{ m} \times 0.0508 \text{ m}} = 287.29 \text{ Mpa}$$

If a lower force stretches the bar so far that it is no longer useful, its failure strength equals the stress found by dividing the lower force by the area of the original cross section. The deformation that can be allowed in the bar depends on what it is used for. Therefore, failure depends on the purpose for which the material is used. Failure could conceivably take place at a lower stress in one case with a particular material than in another case with the same material. Beams supporting a warehouse roof where appearance is not important can withstand any stress that does not break them; however, if the beams support a plaster ceiling, they fail at a stress that causes sufficient deflection to crack the plaster.

Through experience and the performance of tests, the stress that causes failure can be determined for various materials and uses. A knowledge of this stress is useful for designing purposes. However, nothing is designed to be stressed to the point where it is ready to fail. Instead, a lower stress called the *allowable stress* is selected, and this is the maximum stress allowed.

There are several reasons for not designing material to be stressed close to the failure stress:

1. The actual force (and therefore stress) on a structure may exceed expectations.

2. True failure stress may be somewhat less than that determined experimentally.

3. The simplified procedures used in design predict approximate stresses which may be exceeded somewhat in actuality.

4. Materials may be weakened by rusting (steel), rotting (wood), or spalling (concrete).

The failure stress is greater than the allowable stress by a factor called the *safety factor*. If failure stress is twice the value of the allowable stress, the safety factor is two. The safety factor equals the failure stress divided by the allowable stress.

Usually failure stress is determined experimentally. A safety factor and an allowable stress are selected by a committee of experts and the allowable stresses are published. Some organizations that publish allowable stresses are the American Institute of Steel Construction, the American Concrete Institute, and the National Forest Products Association. Designers select kinds of material and sizes and shapes of members to support loads that subject the member to stresses that are equal to or less than the allowable. Economy requires that the actual stress be near the allowable; if it is not, the material is being used inefficiently because less material would be adequate. The stress calculated according to load conditions is called the *working stress*.

Important factors considered in deciding on a safety factor are as follows:

1. How exactly loads can be predicted and calculated.
2. How exactly acting stresses can be calculated.
3. How exactly failure stresses can be determined.
4. How consistently the material conforms to the experimental strength.
5. How serious the consequences of a failure are.
6. How much warning the material gives before failing.
7. How much the material is likely to deteriorate under the conditions of use.

There are three kinds of stresses and corresponding strengths—compressive, tensile, and shearing. They depend on the position of the forces with respect to the object. The three types of stress are illustrated in Figure 1–2.

Stress is determined by dividing the acting force by the original area on which it acts. It is this area that resists displacement. Tensile and compressive stresses act on the cross-sectional area perpendicular to the direction of the force. Shear stresses act on the cross-sectional area parallel to the direction of the force. The force is not uniformly distributed across the area, but in computing tensile and compressive stresses it is assumed to be uniform with satisfactory accuracy.

Shearing stress acts unequally over an area, and the stress at the location of highest stress must be considered. The action of shearing stresses is complex compared to that of the axial stresses, tension, and compression.

Because the cross sections are changed in size by forces, they influence the stresses. If a force remains constant, the actual stress changes when the cross section changes. It is customary to compute stress on the basis of the area as it is before any force is applied. Computations are easier this way, and in all practical applications the area of interest is the original area, because any problem relating size to strength will be solved on the basis of original size, not a size distorted by a force.

Materials differ in their response to stress. A *ductile* material can be drawn into a thin, long wire by a tensile force. A *malleable* material can be flattened into a thin, wide sheet by a compressive force. A *brittle* material breaks with very little deformation—it appears to fail suddenly because there is no noticeable deformation to serve as a warning.

In this discussion, forces have been assumed to be applied once for a brief period of time. Ordinary tests made to determine strength consist of subjecting a sample of the material to a force that increases steadily until the material breaks. These tests take a few minutes and indicate static stress at failure. However, in a structure, forces may be applied for extended periods of time, they may be applied and removed many times, and they may be applied suddenly with impact or shock.

A material, even if brittle, deforms slowly when a force is applied to it for an extended period of years, even though the force is too small to cause failure in a short time. This deformation is called *creep*. The creep may be great enough to constitute failure.

Although a force of a certain amount may not cause breaking no matter how long it is applied, it may be large enough to cause breaking if it is applied and removed many times (tens of thousands of times), even if over a shorter time. For example, structural members of a bridge are subjected to application and removal of stress each time a vehicle crosses. Failure from this cause is called *fatigue*, and it occurs with very little deformation.

Because there is so little deformation, there is no warning and the break seems to be sudden. However, it begins as a tiny crack and becomes larger over many cycles until the structure fails by breaking. The smaller the stress, the more times it must be repeated to cause failure. There is a stress below which the material will not fail at any number of cycles, called the *endurance limit*. Any stress above this limit will cause failure if repeated enough times.

Specimens tested for endurance are generally subjected to bending first one way and then the opposite way. This reversal of stress produces failure at fewer cycles than the simple application and removal of force. The endurance limit found this way may be less than half the static stress at failure. For many steels, endurance limits (fatigue limit) range between 35 and 60 percent of the tensile strength.

Toughness is the capacity of a material to absorb energy while a force is applied to it. Energy is expended by a force acting over a distance and is absorbed by a material being forced to deform through a distance. Toughness is the product of stress and strain up to the point of fracture. It is computed by determining the area under the stress–strain curve (see Figure 1–3a), which is equivalent to multiplying the average stress by the total strain (force times distance). The result is called the *modulus of toughness*. Strength and ductility are both involved. The toughness of a material indicates its ability to withstand a sudden force, known as an *impact load* or *shock load*.

Resilience is the ability of a material to recover its original size and shape after being deformed by an impact load. The *modulus of resilience* is the product of stress and strain up to the elastic limit. It is a measure of the useful toughness, because beyond the elastic limit permanent deformation ordinarily renders the material unfit for further use. It is computed by determining the area under the stress–strain curve from zero to the elastic limit.

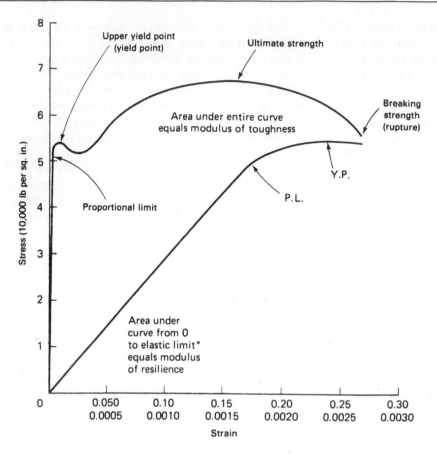

FIGURE 1–3A. Stress–strain diagram for ductile steel: upper curve (upper scale) shows relationship up to breaking point; lower curve (lower scale) shows curve with greater accuracy up to the yield point.

*Elastic limit is difficult to determine and is usually assumed to be at the proportional limit. This is very nearly correct and is accurate enough for ordinary use.

Modulus of Elasticity

Strain is directly proportional to the stress that causes it over a considerable range for many materials. At stresses higher than this range, each additional increment of stress causes greater strain than the previous increment of stress. The stress at which strain just begins to increase at a rate greater than in the proportional range is the *proportional limit*. An example is shown in Figure 1–3a. Stress in lb per sq in. (psi) is a very large number compared to the resulting strain for almost all construction materials.

The constant value of stress divided by strain is called the *modulus of elasticity*, or *Young's modulus*. The relationship is expressed as modulus of elasticity equals stress divided by strain, or $E = s/\varepsilon$. This relationship is known as Hooke's Law. Because ε is a ratio with no units, E has units of lb per sq in., the same as s, although E is not a stress. The modulus of elasticity indicates the stiffness or resistance to movement of a material. A stiff material deforms less under a given stress than does a material of less stiffness. A metal wire is very stiff compared to a rubber band of the same size, and the E of the metal wire is a much higher value. The modulus of elasticity is a characteristic which is different for each material. For example, A36 steel has a modulus of elasticity of 30×10^6 psi, gray cast iron's modulus is 15×10^6 psi, and concrete's

modulus of elasticity is about 3×10^6 psi with variations related to the concrete's compressive strength value.

Some materials do not have a range of constant relationship between stress and strain. The stress–strain relationship for this type of material is shown in Figure 1–3b. As stress is increased in a test specimen of this type of material, the strain increases at an increasing rate. There is no modulus of elasticity for such materials because there is no range of constant relationship between stress and strain. However, an E value is so convenient for design that an approximate value is sometimes used. Any ratio of stress–strain selected as the E value is correct at only one stress or possibly two. However, a ratio may be chosen which is reasonably close throughout the range of stresses that is encountered in use. This ratio is used for design. Methods used to determine approximate E ratios are shown in Figure 1–4.

Any of the values in the equation $E = s/\varepsilon$ can be determined if the other two are known. The modulus of elasticity has been determined by extensive testing for all commonly used construction materials, and the usual design problem is to find either stress or strain. In experimental work, a sample is tested under increasing stress with the strain being measured. Both are recorded at suitable intervals and the values used to plot their relationship. The complete test method is American Society for Testing and Materials (ASTM) E8

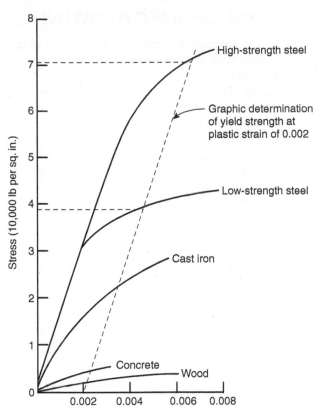

FIGURE 1–3B. Relative shapes of stress–strain diagrams for different materials.

standard method of tension testing of metallic materials and is often referred to as a *static-tensile test*.

The modulus E applies to compressive or tensile stresses and for most materials is nearly the same in compression and tension. The relationship of shearing stress to shearing

strain is designated E_s (the modulus of rigidity) and is a lower value.

Example

Calculate the change in length of a 50-ft steel rod subjected to a tensile load of 12,000 psi.

$$E = \frac{s}{\varepsilon}$$

where E = modulus of elasticity (psi)
  s = average computed stress (psi)
  ε = strain (in./in.)

$$30,000,000\,\text{psi} = \frac{12,000\ \text{psi}}{\varepsilon}$$

where ε = 0.0004 in./in

$$\varepsilon = \frac{\delta}{L}$$

where δ = total deformation or change in length (in.)
  L = original length (in.)

$$\delta = \varepsilon L$$

where δ = 0.0004 in./in. (50 ft × 12 in.)
  δ = 0.24 in.

Elastic and Plastic Properties

Elasticity is the property of a material that enables it to return to its original size and shape after a force is removed. Elasticity is not judged by the amount of strain caused by a given stress, but by the completeness with which the material

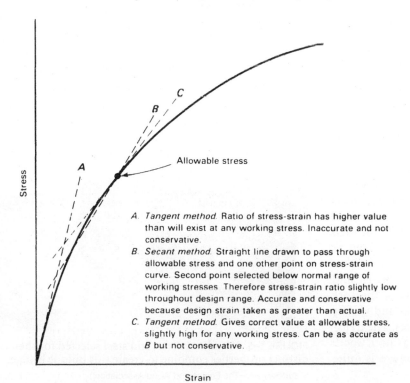

A. *Tangent method.* Ratio of stress-strain has higher value than will exist at any working stress. Inaccurate and not conservative.

B. *Secant method.* Straight line drawn to pass through allowable stress and one other point on stress-strain curve. Second point selected below normal range of working stresses. Therefore stress-strain ratio slightly low throughout design range. Accurate and conservative because design strain taken as greater than actual.

C. *Tangent method.* Gives correct value at allowable stress, slightly high for any working stress. Can be as accurate as B but not conservative.

FIGURE 1–4. Typical stress–strain diagram for material without a range of constant s/ε values, showing methods of determining a useable modulus of elasticity.

returns to its original size and shape when the force is removed. A metal wire does not stretch nearly as far as the same-size rubber band under the same force. However, it is just as elastic within its elastic range because it returns to its original size and shape when the force is removed, just as the rubber band does.

Plasticity is the property that enables a material changed in size or shape by a force to retain the new size and shape when the force is removed. Many materials are completely elastic (i.e., return exactly to original size and shape upon removal of a force) throughout a range of stress from zero to a stress called the *elastic limit*. At stresses greater than the elastic limit, the material takes a *permanent set* or a plastic deformation which remains when the force is removed. When a material is stressed beyond the elastic limit, the total strain is made up of recoverable elastic strain and permanent plastic strain.

Elastic strain takes place first, followed by the plastic strain; that is, as stress is increased starting from zero, the strain is entirely elastic until the elastic limit is reached, and then any additional strain is entirely plastic as stress greater than the elastic limit is imposed. For an elastic material, the total strain may be elastic or it may be elastic plus plastic, but it may not be plastic only. For each type of material, the elastic limit has a value in psi. It is the stress above which plastic deformation takes place and below which elastic deformation takes place. The elastic limit cannot be easily identified. Because it is at approximately the same location as the proportional limit, it may be taken as the same point for all but experimental work.

Yield point is the lowest stress at which an increase in strain occurs with no increase in stress. It is therefore at a point of zero slope in the stress–strain curve. In Figure 1–3a, the yield point is shown. Yield point is an important stress for steel (see Chapter 5). However, some steels have no yield point. As stress increases beyond the elastic limit, plastic deformation becomes unacceptable at some point and is considered failure for these steels. The stress at which the excessive plastic deformation is reached is the *yield strength*. In some materials a defined proportional limit and yield point are not obvious as shown in Figure 1–3b. Standard practice has been established wherein a straight line is constructed parallel to the elastic portion of the stress–strain curve at some specified strain offset, typically 0.002. The stress corresponding to a permanent deformation of 0.20 percent is indicated as the *offset yield strength*. This is demonstrated in Figure 1–3b as the point of intersection between the straight dashed line and the curves for each material tested. The term *yield stress* includes yield point or yield strength.

Materials of construction behave differently at high and low temperatures. Like many familiar objects, they are stronger and more brittle at low temperatures and weaker and more pliable (ductile) at high temperatures. The transformation is not noticeable in materials such as steel, wood, and concrete unless the change in temperature is quite extreme.

SOURCES OF INFORMATION

To use materials properly, it is necessary to understand the natural characteristics of the basic materials, the variations in these characteristics made possible through special techniques, and the ways in which materials can be used in combination with one another (Figure 1–5). This information is obtained from past performance of materials in use and from experimental investigation and tests performed on materials. Much of the information is published in technical reports or in advertising material prepared by suppliers. Some sources of information are described here.

Sweets Catalog File is a compilation of technical advertising literature published by suppliers. The file includes approximately a dozen categories, with each supplier's literature included in the appropriate category. Literature from about 2000 suppliers describing many thousands of products is included. Each supplier pays the publisher of *Sweets Catalog File*, the F. W. Dodge Company, a fee for the inclusion of its literature, and the file is available for nominal cost to designers who have a sufficient volume of business to justify receiving it.

Groups of suppliers producing the same product often set up *manufacturers' associations* to promote the use of their product. The association, which is financed by the suppliers, is a separate organization functioning to increase the usage of the product. The association does not sell the product or represent any one of the associated suppliers.

A manufacturers' association seeks to increase sales by finding new and better ways to use the product and by utilizing advertising campaigns. It conducts research and provides the latest findings to designers and builders, often at no charge. It provides technical assistance to designers and builders by means of published material including standard specifications and by personal visits from staff members to the office or jobsite to assist in solving unusual problems. It is to the association's advantage that its product be used successfully so that it will be used again. Some well-known manufacturers' associations are the Portland

FIGURE 1–5. Wood, concrete, and steel selected for their distinct properties combine to create this unique bridge.
(Courtesy APA—The Engineered Wood Association.)

Cement Association (PCA), National Clay Pipe Institute (NCPI), American Iron and Steel Institute (AISI), the Asphalt Institute, National Sand and Gravel Association (NSGA), American Plywood Association (APA), and National Ready Mix Concrete Association (NRMCA).

The *American Society for Testing and Materials* (ASTM) is an organization engaged in the standardization of specifications and testing methods and in the improvement of materials. It is made up of suppliers, designers, builders, and others interested in engineering materials. The organization publishes the ASTM Standards, containing more than 11,000 standard specifications and testing methods covering design, manufacture, construction, and maintenance for practically every type of construction material. The ASTM Standards consist of 75 volumes, each covering one field of interest and each under the jurisdiction of a standing committee which continually reviews and improves standards.

Each committee has members representing suppliers and users. New standards and revised standards are published as tentative for a time so that discussion can be considered before adoption of them is final. The ASTM Standards may be purchased one standard at a time; by section, each of which includes several related parts; or as a complete set. The standards are also available on CD-ROM, or online.

The *American Standards Association* (ASA) develops national industrial standards through the work of committees representing manufacturers, technical organizations, and government departments. The final standards are determined in much the same way as the standards of ASTM. The ASA also adopts the standards of other organizations and has adopted many of the ASTM standards.

There are many other national organizations with memberships and purposes similar to those of ASTM or ASA, but with narrower interests. These organizations develop standard specifications, inspection methods, and test procedures and also adopt ASTM or ASA standards. Some of these organizations are the American Association of State Highway and Transportation Officials (AASHTO), American Institute of Steel Construction (AISC), and American Concrete Institute (ACI).

Underwriters Laboratories (UL) is a nonprofit organization which investigates and tests materials, products, equipment, construction methods, and construction systems in its laboratories. A supplier may have its product tested for a fee, and, if approved, it will be included in the UL-approved list and the UL seal of approval may be displayed on the product. This approval is widely recognized as a safeguard against hazards to life and property, and specifications often require UL approval. The UL is particularly well known for evaluation of fire resistance of building components.

Professional organizations such as the American Society of Civil Engineers (ASCE) and the American Institute of Architects (AIA) devote much of their effort to improving design and construction practices. Valuable information concerning materials is published in their magazines and technical reports.

INSPECTION AND TESTING

Inspection means examining a product or observing an operation to determine whether or not it is satisfactory. The inspection may include scaling the dimensions, weighing, tapping with a hammer, sifting through the fingers, or scratching with a knife, as well as many other operations, some of which could conceivably be called tests. However, the results of the inspection and minor tests are not generally measurable. Often an inspection raises questions which are then resolved by testing.

A *test* consists of applying some measurable influence to the material and measuring the effect on the material. A common type of test consists of subjecting a sample of material to a measured force which is increased steadily until the material breaks or is deformed beyond a specified amount. This type of test measures the strength of the material directly by determining how strong the test specimen is. Some tests predict one characteristic by measuring another. For example, the resistance of aggregate to the destructive influence of freezing and thawing weather is predicted by soaking the aggregate in sodium sulfate or magnesium sulfate, drying in an oven, and determining the weight loss.

Inspection and tests can be categorized according to purpose as follows:

1. *Quality assurance or acceptance:* Inspection and tests performed to determine whether or not a material or product meets specific requirements in order to decide whether or not to accept or reject the material or product. A manufacturer performs such inspections and tests on raw materials he intends to use. A builder performs these inspections and tests on manufactured products and raw materials that he intends to use; and the owner's representative performs them on the builder's finished product.

2. *Quality control:* Inspection and tests performed periodically on selected samples to ensure that the product is acceptable. A supplier or manufacturer monitors his own operation by periodic checks of his product. The builder may check his product similarly. If control measures show the product to be below standards, the reason is determined and corrective measures taken.

3. *Research and development:* Inspection and tests performed to determine the characteristics of new products and also to determine the usefulness of particular inspection procedures and tests to judge characteristics or predict behavior of materials. A reputable manufacturing company tests a new product extensively before putting it on the market. Before adopting a new, simpler type of inspection or test procedure for acceptance or control, a highway department compares results obtained from the new procedure with results from the old procedure over a large range of conditions and over an extended period of time.

Tests for acceptance or control must usually be performed quickly. For reasons of economy, the tests cannot interfere with the manufacturing process. For the same reason, tests at a construction site cannot unduly delay the construction work. Because these tests are performed so many times, their cost is an important factor. Therefore, quick, inexpensive tests proven to be good indicators of actual performance are used extensively for these purposes. The type of test used in development of a product must give more exact results and is generally more time consuming and requires more expensive equipment.

An example will illustrate inspection and testing for the different purposes. A company making CMUs tries to reduce cost by using an industrial waste material as an aggregate. The proposed aggregate is examined and tested extensively before being used. It is then used in various combinations to make batches of CMUs, which are compared with each other by inspection and testing. The combination that proves to be most satisfactory is used to manufacture CMUs.

While blocks are in production, a continuous program is carried on to check the finished CMUs by inspection and tests. A certain percentage of the blocks are checked as a matter of routine to determine whether or not the quality changes. If there are indications of a change in quality, the cause of the change is determined and action taken to return to production of uniform quality.

The builder who purchases the CMU or the owner's representative then inspects and tests a certain percentage of the CMUs before accepting them. Each CMU is inspected for damage before being put into place in the structure. A final inspection is given to the entire project as a whole when it is completed.

Tests performed on samples of material from the same source do not yield exactly the same results for each sample. There are two reasons for this:

1. No material is perfectly homogeneous. There are slight differences in the composition of any substance from one point to another. In addition, there are always minute flaws which, though unimportant in a large mass of material, have a great effect if one is included in a small sample to be tested. Thus, one sample is not completely representative of the whole.

2. The testing methods, although performed according to standard procedures, cannot be duplicated exactly each time.

Some tests give nearly the same results when performed by different operators. These tests are said to have a high degree of *repeatability*. Tests have varying degrees of repeatability, which should be taken into account when interpreting the results. An indication of degree of repeatability is included in some test procedures. A supplier can be required to meet very exacting specifications if a test method is available that provides very accurate results. If there is no such method, more accuracy can be obtained by taking the average of several tests. If it is not feasible to do this, the specifications must be written to permit more variation in the product.

Variations in test results can be kept small by selecting large enough samples, by employing proper procedures for random sample selection, and by running enough tests to get a meaningful average and eliminating those results that are erroneous because of a faulty sample or faulty test performance.

A certain minimum size sample or minimum number of samples is needed to be truly representative of the whole. The more variable a material is, the larger or more numerous the samples must be. The less precise the test methods are, or the lower the correlation between test results and the property actually being investigated, the greater the number of times the test must be run.

The size and number of samples should be determined on a statistical basis. Size and number must be large enough to include all the characteristics to be tested, the least common characteristic once, and the more common ones in their proper proportion. Many testing procedures include minimum sizes or numbers of samples to accomplish this.

STANDARDS

The designer or builder may desire any number of properties in the material she is going to use. She must be able to specify the degree of each property in terms that she and the supplier understand, and she and the supplier must have some mutually acceptable means of determining whether or not the materials possess each property in sufficient amounts. The supplier must be able to prove to the buyer that the material possesses the properties desired to the degree desired.

In unusual cases a measurement or test may be devised for one specific application. Fortunately, this is not usually necessary. Standard measuring and testing methods are available and so are standard definitions of terms. Both buyer and supplier understand what is meant when standard terminology is used or reference is made to standard specifications, and both can use the same reproducible methods to determine whether or not the materials possess the required properties in sufficient quantity.

When standard specifications and standard testing methods are available, it makes no more sense to devise special, nonstandard specifications or test methods than it does to measure lengths with a yardstick or meter stick of nonstandard size. Material specifications consist largely of explanations of what properties a material must possess and the allowable limits for those properties.

A *testing method* is a specification explaining how to perform a test and how to measure the results. When the material is tested, if it possesses all the required properties to a sufficient degree, it is said to meet the standards or specifications.

Inspections often require measurements. A measurement may be as simple as determining the diameter of a piece of pipe by measuring it with a 6-ft rule to be sure it is of the size specified, or it may involve a more time-consuming and accurate measurement such as the determination of the percentage of air entrained in portland cement concrete.

A measurement may consist of determining the size of a crack in the end of a piece of wood by measuring it with a carpenter's rule. The piece of wood is considered to lose a certain percentage of its strength according to the size of the crack, and it must be discarded if it has a crack larger than a certain size. However, there are three types of cracks and each is measured in a different way. Therefore, the cracks must be measured according to standard specifications if all pieces are to be graded on an equitable basis. Tests can be effective only if they measure the appropriate characteristic the same way each time so that the results can be evaluated according to their relationship to past results. For the same reason, inspection should be performed in an identical way each time as much as possible.

Review Questions

1. Prepare an organizational chart depicting the construction process. Identify the parties involved, their functions, responsibilities, and connections with each other.

2. Explain the functions of independent testing laboratories in the construction industry.

3. Why must a builder/contractor understand materials?

4. Explain how the material used for a basement floor and the material used for a roof must be different.

5. As a research project, make a list of the kinds of materials used in (a) a water distribution system, (b) a city street pavement, (c) a sewage collection system, and (d) a roofing system.

6. Discuss the advantages of material specifications versus performance specifications. Research an actual project specification and identify the types of specification formats used and the reasons they were used.

7. A mechanical device rests in a level position on two supports, each 42 in. high, at 60°F. The support at one end is gray cast iron; at the other end, structural steel. What is the greatest amount the device can be out of level due to differences in expansion if the temperature rises to 165°F?

8. Calculate the change in length of a 100-ft-long steel bridge girder that will be exposed to a temperature range of −25 to 110°F.

9. What is the stress in a steel rod with cross section of 2.4 sq in. if it is subjected to a tension of 120,000 lb? If a concrete cylinder 6 in. in diameter is subjected to a compressive force of 122,700 lb?

10. Calculate the stress in a 65-ft steel cable that has a total deformation of 0.33 in.

11. Calculate the total deformation in a 15-ft, 2-sq-in. steel bar hanger that supports an 18-ton load.

12. A material is expected to fail at 50,000 psi. A safety factor of 2 is to be used. What is the allowable stress?

13. Calculate the compressive stress in a 6-in.-diameter by 12-in.-high concrete test specimen if the total deformation under load is 0.015. Calculate the total load on the test cylinder.

14. A material may fail by breaking or deforming excessively. Which is the case for creep? For fatigue? For impact failure?

15. What purpose does specifying organizations such as ASTM and ACI serve in the construction industry and why are these organizations important?

16. What is the purpose of manufacturers' associations and how do they accomplish their purpose?

17. What is the difference between inspection and testing?

18. Calculate the stress at failure of a concrete test cylinder with a measured diameter of 152.4 mm and a recorded load of 371 426 N.

19. Calculate the stress at failure of a 50-mm mortar cube that failed at 35 586 N.

20. Calculate the change in length of a 110-m-long steel pipe exposed to temperatures ranging from 5 to 90°C.

CHAPTER **TWO**

AGGREGATES

Aggregates are particles of random shape. They are found in nature as sand, gravel, stones, or rock that can be crushed into particles. They may also be by-products or waste material from an industrial process or mining operation. The term *aggregates* generally refers to mineral particles which have rock as their origin unless otherwise specified. These include sand, gravel, fieldstone, boulders, and crushed rock, because all are derived from rock by the forces of nature or, in the case of crushed rock, by a manufacturing process. *Rock* includes any large solid mass of mineral matter which is part of the earth's crust. Some other materials used as aggregates are blast-furnace slag, boiler slag, building rubble, refuse incinerator residue, and mine refuse.

Aggregate sizes vary from several inches to the size of the smallest grain of sand. In special cases aggregate larger than several inches may be used. Particles smaller than the size of a grain of sand are considered as impurities even if they are of mineral composition. Depending on the amount of impurities and the use to be made of the aggregate, these impurities may be tolerated or removed. In some cases these small particles are deliberately mixed with aggregate and are then considered as an additive.

The road building industry is the greatest consumer of aggregates. Aggregates are used as bases or cushions between the soil and traffic wheels or between the soil and pavement, and are also used in bituminous pavement and portland cement concrete pavement. They are used as leveling and supporting bases between the soil and all types of structures. They are used as *ballast,* which is the base for railroad tracks. They are used as protective and decorative coatings on roofs and floors. Another major use of aggregates is to filter water, which requires holding back suspended solids while allowing water to pass through.

There is not always a clear distinction between aggregate and the engineering material, soil, which also comes originally from rock. Some naturally occurring sand and gravel soils are usable as aggregate without processing. Most aggregate is soil that has been processed. Soil which remains in place and serves as the ultimate foundation of all construction is not considered as aggregate, no matter what its composition.

DEFINITIONS

Terms related to concrete aggregates are defined in ASTM C125, Terms Relating to Concrete and Concrete Aggregates. Many of the definitions are applicable to any type of aggregate and are given here to facilitate the discussions in this chapter.

coarse aggregate*: (1) Aggregate predominantly retained on the No. 4 (4.76-mm) sieve; or (2) that portion of an aggregate retained on the No. 4 (4.76-mm) sieve.

fine aggregate*: (1) Aggregate passing the $\frac{3}{8}$ in. sieve and almost entirely passing the No. 4 (4.76-mm) sieve and predominantly retained on the No. 200 (74-micron) sieve; or (2) that portion of an aggregate passing the No. 4 (4.76-mm) sieve and retained on the No. 200 (74-micron) sieve.

gravel*: (1) Granular material predominantly retained on the No. 4 (4.76-mm) sieve and resulting from natural disintegration and abrasion of rock or processing of weakly bound conglomerate; or (2) that portion of an aggregate retained on the No. 4 (4.76-mm) sieve and resulting from natural disintegration and abrasion of rock or processing of weakly bound conglomerate.

sand*: (1) Granular material passing the $\frac{3}{8}$-in. sieve and almost entirely passing the No. 4 (4.76-mm) sieve and

*The definitions are alternatives to be applied under differing circumstances. Definition (1) is applied to an entire aggregate, either in a natural condition or after processing. Definition (2) is applied to a portion of an aggregate. Requirements for properties and grading should be stated in the specifications.

16

predominantly retained on the No. 200 (74-micron) sieve, and resulting from natural disintegration and abrasion of rock or processing of completely friable sandstone; or (2) that portion of an aggregate passing the No. 4 (4.76-mm) sieve and predominantly retained on the No. 200 (74-micron) sieve, and resulting from natural disintegration and abrasion of rock or processing of completely friable sandstone.

bank gravel[†]: Gravel found in natural deposits, usually more or less intermixed with fine material, such as sand or clay, or combinations thereof; gravelly clay, gravelly sand, clayey gravel, and sandy gravel indicate the varying proportions of the materials in the mixture.

crushed gravel: The product resulting from the artificial crushing of gravel with substantially all fragments having at least one face resulting from fracture.

crushed stone: The product resulting from the artificial crushing of rocks, boulders, or large cobblestones, substantially all faces of which have resulted from the crushing operation.

crushed rock[‡]: The product resulting from the artificial crushing of all rock, all faces of which have resulted from the crushing operation or from blasting.

blast-furnace slag: The nonmetallic product, consisting essentially of silicates and aluminosilicates of lime and of other bases, which is developed in a molten condition simultaneously with iron in a blast furnace.

SOURCES

The earth's crust is solid rock called *bedrock,* and much of the crust is covered with soil particles originally derived from the bedrock. This soil is classified according to size, as gravel, sand, silt, and clay, with gravel being the largest and clay the smallest. Natural aggregates include sand, gravel, or larger stones, and bedrock reduced to particle size by manufacturing methods.

The sand and gravel occurring in nature were at one time broken from massive parent rock, transported by nature, and left in various types of deposits called *sand or gravel banks.* The processes that cause breaking, transporting, and depositing have operated continuously throughout the past and continue to operate now (Figure 2–1).

[†]Definition from ASTM D8.

[‡]Definition used in this book.

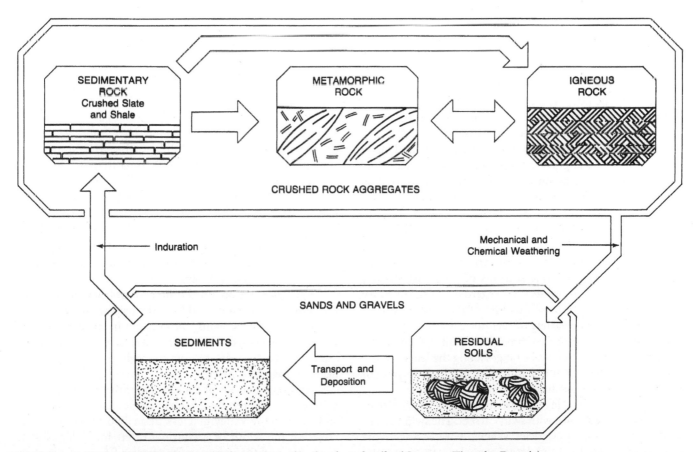

FIGURE 2–1. Geologic processes for the formation of bedrock and soils. *(Courtesy Timothy Dennis)*

Cracking of rock and eventual fracturing occurs mainly through expansion and contraction of the rock due to changing temperatures. The process is hastened if temperatures are sometimes low enough so that water within the cracks freezes and expands. The broken particles fall or roll short distances downhill under the force of gravity and may be carried farther by flowing water or the slowly flowing ice of a glacier. Many particles are broken loose originally by flowing water or glacial ice.

A freshly broken particle has a rough surface and an angular shape. The more it travels, the smoother its surface becomes and the more rounded its shape becomes. Rolling down a mountainside has a rounding effect. Sliding farther along, helped by rainwater runoff, and eventually tumbling along in a stream causes further rounding and smoothing. Being scraped along the ground first on one side and then on another while being carried in glacial ice has a particular effect on the shape and texture of aggregate particles. Glacial deposits consist of particles of widely varying sizes with some of the coarseness and angularity worn away. Many of the glacier-carried particles reach the streams of meltwater at the melting end of the glacier. They are then carried the same way with the same results as in any other stream. The deposits simply dropped by a glacier are called *till* and those carried farther by meltwater are called *outwash*. Outwash deposits are smoother, rounder, and more uniform in size than till.

The most perfectly rounded particles are those that are carried to the edge of a large body of water where they are washed back and forth incessantly by ocean or lake waves and become smooth and nearly spherical.

Rock particles, especially in dry climates or flat areas, may lie where they fall—next to a steep cliff, for example. Centuries of such deposition may provide sufficient quantities for commercial extraction and use of the aggregate which is rough-surfaced and of angular shape.

Many particles fall or roll into flowing water, and many deposits of sand and gravel are found in the beds of streams and rivers or where streams and rivers formerly were. The size of particles a stream carries is approximately proportional to the sixth power of the velocity of flow. The velocity depends on the slope, being greater with a greater slope, and on the quantity of water, being greater with a greater quantity. Greater quantities of water flow at certain times of the year, and steeper slopes in some sections of a flowing stream, cause faster flow in those sections regardless of the quantity of flow. These variations of quantity of flow and in-stream slope result in a wide variety of carrying capacities throughout a stream over a period of time. The result is a separation of aggregates into various size ranges along the length of the stream. In some cases gravel with remarkable uniformity of size is found in one location.

An example is a level, slow-flowing section of river following a steep section with swift current. The swift current carries large and small particles. When the water reaches the slow-flowing section, the large particles settle and the

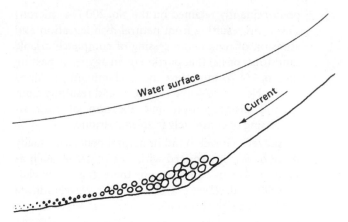

FIGURE 2–2. Section through river showing aggregates deposited because of reduction in water velocity due to flatter slope.

smaller ones are carried past. The result is a deposit of clean gravel of a fairly uniform size, containing no silt, clay, or trash. If the current continues to flow more and more slowly, smaller and smaller particles will gradually be dropped in succession as the water flows downstream. The entire reach of the river is then an aggregate deposit graded from larger particles upstream to smaller particles downstream (Figure 2–2). The very finest particles are carried to a lake or ocean and settle there where the movement of water is negligible. Lake deposits often contain too many fine particles to be good aggregate sources.

Glaciers are formed in high altitudes and pushed slowly down valleys by the weight of ice and snow piled up behind them. They scrape and gouge pieces of rock, large and small, from the sides of the valley and carry them slowly to the lower, melting edge of the glacier, where the larger particles are dropped and the smaller ones are carried away by the stream of meltwater. The melting edge moves downhill in winter and recedes uphill in the heat of summer. The flow of water is greater in the daytime than at night and greater in the summer than in winter. The variation in flow causes a wide variety of particle sizes to be deposited helter-skelter as the stream changes channels and cuts through previous deposits. This area around the variable melting edge, which is known as a *moraine,* contains many large boulders, and the area downstream has the aggregate deposits expected in a stream.

Aggregates are obtained from beds of lakes and streams, but more often are obtained from deposits where there were formerly lakes or streams. Movements in the earth's crust result in relocation of streams and lakes, so that some former stream and lake beds are now on high ground with their aggregate deposits intact.

At several periods of time much of the northern hemisphere was covered by glaciers. These glaciers, behaving similarly to the valley glaciers described previously, but covering the width of a continent, left huge deposits of aggregate, and the streams of meltwater flowing from their southern extremities also deposited huge quantities. The glaciers have receded and the streams and rivers they caused have disappeared. The deposits they left provide many of the best gravel banks now to be found.

METHODS OF EXTRACTION AND PROCESSING

Aggregate is recovered from deposits laid down in geologic times and from deposits still being laid down. The deposits are found on the ground surface and below the surface of the ground or water. Some aggregate is suitable for a specific use just as extracted, and some must be processed before being used.

Underwater Sources

Aggregate is brought up from lake and river bottoms by barge-mounted dredges with a single scoop or an endless chain of scoops and by dragline. The disturbance to the bottom and the motion of the scoop or bucket through the water cause some of the undesirable fine particles and lightweight material to be washed away as the load is being brought up. Barges are loaded and transported to shore, where their cargoes of aggregate are unloaded and stockpiled (Figure 2–3).

Aggregate is also pumped with pumps similar to those used for pumping concrete described in Chapter 4. Aggregate of sizes up to 6 in. and larger can be pumped and forced through a pipe to a barge or directly to the shore.

A knowledge of the characteristics of stream flow and deposition is required to locate the most likely places to find worthwhile aggregates. Samples of aggregates are brought to the surface and examined for desired characteristics before equipment is moved to a site to begin recovering them.

Unsuitable material may have to be removed first to reach the kind of aggregate that is wanted.

The operating area is generally controlled by government regulations to prevent interference with the natural flow of water and to preserve navigation channels. Many times when channels or harbors must be deepened for ships, the aggregate brought up from the bottom has commercial value.

Land Sources

Aggregates are excavated from natural banks, pits, or mines on land by bucket loaders, power shovels, draglines, and power scrapers. Unsuitable soil and vegetation, called *overburden*, must be removed to reach the deposits. Removal, which is accomplished with bulldozers and power scrapers, is called *stripping*. Once overburden has been removed (stripped), the surface is ready for drilling (Figure 2–4).

The landform indicates what type of deposit is below the surface. Landforms are studied by means of aerial photographs or field trips, and the best locations are pinpointed. Holes are bored or test pits dug, samples are brought up, and examinations and tests are made to determine the suitability of the aggregates for the intended purpose.

The landforms containing natural deposits can be recognized by one who understands the processes of nature by which they were formed. Certain types of glacial deposits consist of much silt and clay or particles of a wide range of sizes, including large boulders. Neither type is as valuable as glacial outwash, which contains cleaner aggregates of more uniform size.

If crushed rock is to be used as aggregate, it must be blasted loose with explosives (Figure 2–5) and then crushed by machinery to the size desired. Crushing provides a finished product of uniform size, and by proper blasting,

FIGURE 2–3. Aggregate barge processing and loading: (a) dragline excavating from the bottom of a bay; (b) dredging from a river bottom with chain of scoops. *(Courtesy of Carmeuse Line & Stone)*

FIGURE 2–4. Track drills used in a quarry to drill the holes required to blast out bedrock for aggregate production. *(Courtesy Callanan Industries, Inc.)*

crushing, and screening, the size can be controlled to suit the market (Figures 2-6a and 2-6b).

Particles of crushed rock have angular shapes and rough surfaces which are better suited for some uses than the more rounded shapes and smoother surfaces of naturally formed particles. These characteristics will be discussed later in this chapter. A rock formation has similar characteristics throughout so that when a good formation is found, a large supply of consistently good-quality aggregate is likely. This is usually not true of sand and gravel deposits which are likely to have inferior particles mixed with acceptable ones. A mixture is inevitable if particles were

transported through a geological time period from many rock formations to form the sand or gravel deposits. As a result, bank sand and gravel are often of poorer quality than crushed rock, and additional processing is needed to remove unacceptable particles. The removal of out-of-specification material is accomplished by screening and washing the aggregate.

Bank-run aggregate contains a range of sizes in any one deposit, and the finished product is normally screened to obtain separation of sizes and may be crushed to a smaller size. However, if a size larger than most of the available bank-run aggregate is needed, then crushed rock must be used. Generally, a crushed rock source has more versatility, and a natural bank is likely to be more economical for one particular type of aggregate (Figure 2–7).

ROCK TYPES

The constituents of the more common, or more important, natural mineral aggregates derived from rock are described in ASTM C294 and briefly summarized here (Table 2–1). Rock, from which most aggregate is derived, is of three types according to origin—igneous, sedimentary, and metamorphic. *Igneous rock* was at one time molten and cooled to its present form. *Sedimentary rock* at one time consisted of particles deposited as sediment by water, wind, or glacier. Most were deposited at the bottom of lakes or seas. The pressure of overlying deposits together with the presence of cementing materials combined to form rock. *Metamorphic rock* is either igneous or sedimentary rock that has been changed in texture, structure, and mineral composition, or in one or two of these characteristics, by intense geologic heat or pressure or both.

The natural mineral aggregates of whatever sizes and wherever found came from one of the three types of rock. Any given particle of aggregate may have been through the

FIGURE 2–5. Blasting bedrock. *(Courtesy of Carmeuse Line & Stone)*

FIGURE 2–6A. Aggregate production requires rock crushers, screens, and conveyors to crush, size, and transport the material. *(Courtesy Callanan Industries, Inc.)*

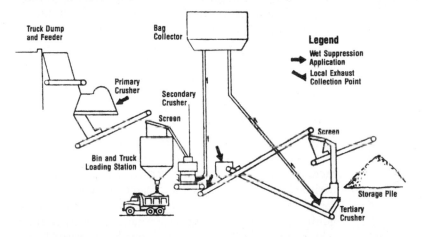

FIGURE 2–6B. Typical combination dust control system for crushing plants. *(Courtesy National Stone, Sand, and Gravel Association)*

FIGURE 2–7. Bucket loader filling truck with crushed rock aggregate from sized stock pile. *(Courtesy Callanan Industries, Inc.)*

Table 2–1. Rock and mineral constituents in aggregates. *(Courtesy Portland Cement Association)*

Minerals	Igneous rocks	Metamorphic rocks
Silica	Granite	Marble
Quartz	Syenite	Metaquartzite
Opal	Diorite	Slate
Chalcedony	Gabbro	Phyllite
Tridymite	Peridotite	Schist
Cristobalite	Pegmatite	Amphibolite
Silicates	Volcanic glass	Hornfels
Feldspars	Obsidian	Gneiss
Ferromagnesian	Pumice	Serpentinite
Hornblende	Tuff	
Augite	Scoria	
Clay	Perlite	
Illites	Pitchstone	
Kaolins	Felsite	
Chlorites	Basalt	
Montmorillonites		
Mica	*Sedimentary rocks*	
Zeolite	Conglomerate	
Carbonate	Sandstone	
Calcite	Quartzite	
Dolomite	Graywacke	
Sulfate	Subgraywacke	
Gypsum	Arkose	
Anhydrite	Claystone, siltstone,	
Iron sulfide	argillite, and shale	
Pyrite	Carbonates	
Marcasite	Limestone	
Pyrrhotite	Dolomite	
Iron oxide	Marl	
Magnetite	Chalk	
Hematite	Chert	
Goethite		
Ilmenite		
Limonite		

Note: For brief descriptions, see Standard Descriptive Nomenclature of Constituents of Natural Mineral Aggregates (ASTM C294).

cycle of rock formation, breaking, transporting, and depositing more than one time (Figure 2–1).

Igneous rock varies in texture from coarse grains to glasslike smoothness, depending on how quickly it cooled. Slower cooling creates a coarser texture. Some volcanic rock cooled as foam, resulting in very light weight because of the hollow bubbles.

Granite is a common, coarse-grained, light-colored igneous rock. *Gabbro* is a common, coarse-grained, dark-colored igneous rock. Both are extensively used in the construction industry. *Basalt* is a fine-grained equivalent of gabbro. *Diabase* is intermediate in grain size between gabbro and basalt. Diabase and basalt are known as *trap rock*. Trap rock provides excellent aggregate due to its very hard and low abrasive properties. Light-colored *pumice* and dark-colored *scoria* are two types of igneous rock filled with bubbles. They are used to produce lightweight aggregate.

Sedimentary rock generally shows stratification indicating the way it was laid down; it generally breaks more easily along the lines of stratification. Rock formed of gravel is called *conglomerate* and, if formed of sand, is either *sandstone* or *quartzite*. *Siltstone* and *claystone* are soft rock formed of silt or clay. *Shale* is hard claystone. All three break along planes or stratification to form flat particles. Therefore, they do not make the best aggregate, although shale is better because of its greater hardness.

Limestone (mainly calcium carbonate) and *dolomite* (mainly magnesium carbonate and calcium carbonate) were formed under salt water and are largely the remains of sea creatures. They do not break into flat particles. They are both rather soft but generally make satisfactory aggregates. *Chert,* which is formed from fine sand, is hard but often is not resistant to weathering.

It is difficult to generalize about metamorphic rock because the change, or metamorphosis, takes so many different forms. Metamorphic rock is dense but often forms platy particles. Generally, aggregate is hard and strong, but its platy shape is undesirable because during crushing, particles may fracture into flat and elongated pieces. *Marble* is a recrystallized limestone or dolomite. *Slate* is a harder form of shale. *Gneiss* is a very common metamorphic rock often derived from granite, but also derived from other rock. It is laminated but does not necessarily break along the laminations. *Schist* is more finely laminated than gneiss but of similar character. Granite, schist, and gneiss are often found together, separated from one another by gradual gradations.

It should be noted that gradations from one type of rock to another are common, and much rock does not fit into any definite category. Aggregate from a natural aggregate deposit may be of various kinds, and it is not then necessary to identify the parent rock types. Even when aggregate is to be obtained from a rock formation, identifying the rock types gives only a general indication of its characteristics which must be checked by testing samples. However, once characteristics are known for part of a geologic formation, only spot checks are necessary to verify characteristics of the entire formation because it was all formed over the same time by the same process.

PROPERTIES AND USES

The usefulness of aggregates to the engineering and construction fields depends on a variety of properties. Performance can be predicted from these properties, and, therefore, the selection of aggregate for a particular task is based on examination and tests. Rather than writing rigid specifications defining properties absolutely required, the specification writer must take into account the types of aggregate readily available and design the specifications to obtain the most suitable aggregate from local sources. A comparison of the specifications devised by various states for highway construction aggregate shows this to be the practice. States containing an abundance of high-quality natural aggregate have more demanding specifications.

Qualities that indicate the usefulness of aggregate particles to the construction industry are as follows:

1. Weight.
2. Strength of the particles to resist weathering, especially repetitive freezing and thawing.
3. Strength as demonstrated by the ability of the mass to transmit a compressive force.
4. Strength as demonstrated by the ability of the individual particles to resist being broken, crushed, or pulled apart.
5. Strength of the particles to resist wear by rubbing or abrasion.
6. Adhesion or the ability to stick to a cementing agent.
7. Permeability of the mass, or the ability to allow water to flow through, without the loss of strength or the displacement of particles.

Weight is of primary importance for large-size stone called *riprap* placed along the edge of a body of water to protect the bank or shore from eroding; for a *blanket* of stones placed to prevent erosion of sloping ground; and for stone-retaining walls held in place by wire baskets called *gabions*.

Resistance to weathering is necessary for long life of any aggregate unless it is used only indoors. The quality of resisting weathering is called *soundness*.

Strength of the mass is needed if the aggregate is to be used as a base to support the weight of a building, pipeline, or road, or if it is to be used in portland cement concrete or bituminous concrete. The strength of individual particles to resist being broken, crushed, or pulled apart is important when the aggregate is to be subjected to a load. Pressure on a particle can crush or break it, allowing movement of adjacent particles. Failure of too many particles causes enough movement to constitute failure. Aggregate particles embedded in portland cement are often subject to tension. Concrete pavement and other concrete structures are subject to tension, although not as severe as the compression they receive. The tension and compression must be carried through the aggregate particles as well as the cement paste.

Aggregate may be subject to rubbing and abrasion during processing and handling, and also in service. Aggregate may be chipped or ground by loading equipment, screening equipment, or conveyor belts. Aggregate having insufficient resistance to abrasion produces some additional small broken particles, and the original particles become somewhat smaller and more rounded. The result is that size, gradation, and shape are all changed from what was originally intended. Aggregate particles at the surface of all types of roadways are subject to abrasion from vehicle wheels. The particles throughout asphalt pavement are subject to abrasion because the pavement continuously shifts under the weight of traffic, causing the particles to rub each other.

Miscellaneous Uses

Various sizes of stone are used for riprap to protect natural or man-made earthwork. The individual pieces must be large enough so that the force of the water will not move them. Along a reservoir shore or the water's edge at a dam, protective aggregate consisting of particles from baseball size to basketball size might be dumped in a belt extending from low water to high water along the waterline location. On the banks of swiftly flowing streams, it may be considered necessary to place much larger sizes, with each one being fitted into place much like a stone wall, but lying against the bank in a belt extending from low-water to high-water level. Broken rock of irregular, slablike shapes is often used and put into place with a crane or backhoe. The chief requirements for riprap are high weight and low cost (Figure 2–8a).

FIGURE 2–8A. Riprap protection lining a retention pond.

Use of rock materials in these applications helps to prevent erosion of the fine soil particles along the edges of the shore and banks, toe of dams, and piers within waterways.

When the bank is too steep or the current too violent for riprap, gabions may be used to hold stones in place. A *gabion* is a basketlike container for stones that is made of heavy steel mesh, forming the shape of a block with level top and bottom and four vertical sides. Gabions are set in place and filled with stones the size of a fist or larger to act as riprap or to form a retaining wall to hold back an earth bank. They may be piled one on top of another as shown in Figure 2-8b. Stones are placed by hand or with machinery. The gabions are wired to each other to form a continuous structure. However, the structure is flexible and adjusts without breaking to uneven soil settlement or undermining by water current. The stone-filled gabions are permeable so that soil water flows through readily without building up pressure behind the gabion wall, thus preventing wall collapse. If stones are available nearby, there is little cost for production or transportation of materials. The finished gabion structure looks rustic, making it more suitable for some settings than concrete or steel work.

River and lake currents and tide movement are often diverted from their natural paths to control the deposition of waterborne particles—in order to fill a beach or to keep a ship-docking area from being filled, for example. This is done by building long, narrow obstructions to guide the flow of water in a new direction that will deposit suspended material where it is wanted or remove deposits from where they are not wanted. These low wall-like structures are called *training walls, breakwaters, groins,* or *jetties.* They are built of loose stones piled into the shape of a low, wide wall or of gabion construction where greater forces must be resisted.

High unit weight and reasonably good resistance to weathering are all that is required of the material for the uses described so far. Even though some substances are heavier and more weather resistant than natural stones, no other material approaches the advantages in low cost and ready availability in nearly every possible location. The useful life of the stones in any of these structures is very long, although in some cases abrasion, breakage, or washing away of stones can be expected to shorten life. Wire gabions should be inspected regularly for corrosion or wear. When replacement is needed, the entire structure is replaced, or a new one is built right over the original one.

Aggregate and Strength

Aggregate obviously cannot transmit a tensile force from one particle to another. Cementing agents, which are considered later in this chapter and in Chapters 3 and 4, combine with the particles to form a mass which can resist a small amount of tension. It may appear that particles transmit compressive forces from one to another. However, they do not. If particles with flat surfaces were piled vertically, as shown in Figure 2–9a, a compressive force could be transmitted through the pile just as it is in a structural column made of stone.

Aggregate cannot be piled in this way. It appears as shown in Figure 2–9b when in use. Figure 2–9b illustrates a cross section through a container of aggregate with a concentrated weight or force acting downward on one particle of

FIGURE 2–8B. Wire gabion retaining wall. *(Courtesy of Maccaferri Inc.)*

(a) Aggregate piled to transmit a compressive force. This is not a practical arrangement.

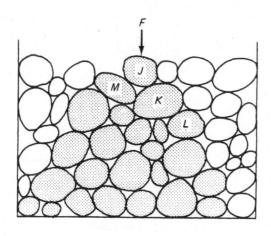

(b) Aggregate in a container. A compressive force on particle J is transmitted by shear through the shaded particles.

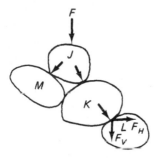

(c) Particle J pushes down and out on K and M like a wedge. Particle K transmits part of the force to L with horizontal and vertical components.

FIGURE 2–9. An "aggregate column" contrasted with a container of aggregate which transmits force by shear.

aggregate. Because of the random arrangement of particles, the concentrated load is necessarily transmitted to more particles as the force is transmitted deeper into the container and thereby is spread over most of the bottom of the container.

In order for the load to spread horizontally, there must be a horizontal force. The vertical load from the top is transmitted through the points of contact, as indicated in Figure 2–9c, over an ever larger area with ever smaller forces. The originally vertical force has a horizontal component at each point of contact below the point of original application. At the points of contact, if the surfaces are not perpendicular to the line of force, there is a tendency for the upper particle to slide across the lower particle or push the lower particle aside so that the lower one slides across the particle below it. The tendency to slide transversely is a shearing stress, and the strength to resist the sliding is the shearing strength of the aggregate (see Figure 1–2). This strength is the result of friction and interlocking between adjacent particles. Failure to resist the shearing stress results in some particles being forced closer together. Often other particles are pushed aside.

The surface of aggregate settles when particles are pushed closer together. The settlement could be considered as a strain caused by the imposed stress. Aggregate should be compacted so that all the significant settlement takes place before the aggregate is put to use. In that way, harmful settlement is eliminated. The only other way in which aggregate can settle is for particles to be crushed. They will not be crushed unless they are soft or very brittle. Soft or brittle particles should not be used.

Although aggregate particles can be crushed if a great enough force is exerted on them, in use, good aggregate will not hold still to be crushed; it will move before it can be crushed. Horizontal displacement takes place under a smaller force than is needed to crush or break the aggregate. Therefore, the controlling strength is shearing strength which is indicated by the load that can be carried without horizontal movement sufficient to be considered failure.

The tendency to move horizontally is resisted by friction and interlocking between particles. Both are illustrated in Figure 2–10. The friction that can be developed between two particles depends on the roughness or smoothness of the particle surfaces. The rougher the surface is, the more

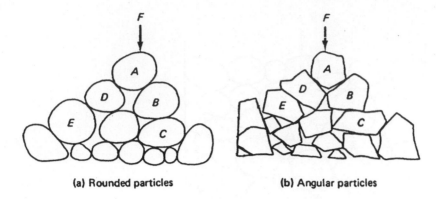

(a) Rounded particles   (b) Angular particles

Particles labeled *B* resist horizontal movement by friction.
Particles labeled *C, D,* and *E* resist by friction and interlocking.

FIGURE 2–10. Friction and interlocking.

resistance there is to sliding. The resistance developed by interlocking depends on the shape of the particles and is greatest for angular particles such as crushed rock, and least for well-rounded particles such as beach sand.

In Figure 2–10, particle *A* pushes down and to the right on particle *B*. Particle *A* cannot move downward without pushing particle *B* down or to the right. To go down or to the right, *B* must push *C* down or slide across it. In this example, *C* cannot be pushed down because it is held by particles under it. Particle *B* must overcome friction to slide over *C*. The force exerted by *A* on *B* may be enough to push *B* right over *C*. Theoretically, *C* could also push lower particles aside, but such a progression cannot continue without particles sliding over particles at some level.

Particle *D* is in a situation similar to particle *B,* but must move upward to overcome interlocking in order to move horizontally. A greater force is needed on *D* to push it up and over *E*. If the forces are great enough to cause horizontal movement at enough points of contact, the aggregate fails.

No force acts on a single particle, but some loads, such as the forces from the wheels of a vehicle, act on a small area at a time. Even wheel loads, however, act over a number of particles, which causes the actual transmittal of forces from particle to particle to be more complicated than indicated in Figure 2–9. Figure 2–9c shows a partial sketch of forces in one plane. The load is actually transmitted outward in all directions, forming an ever larger circle as it proceeds downward.

Figure 2–11 shows what an aggregate road looks like after failure. Overloading causes particles to be pushed aside and forced to ride up over adjacent particles.

Because of the way in which a concentrated force is spread out through a thickness of aggregate and converted to a lower pressure distributed over a larger area, aggregate is often used as a *base* to support a weight which is too heavy to be applied directly to the soil. A geotextile fabric placed between soil and aggregate prevents mixing the two, which would weaken the aggregate base. The aggregate base spreads the weight over a larger area of soil at a lower pressure. Generally, soil is not as strong as aggregate, but it consists of separate particles as aggregate does and so behaves in much the same way under a load. It fails in shear if overloaded. It is capable of settling more than aggregate and may also fail by settling excessively.

Aggregate is obtained from the best source and brought to the construction site, but the construction takes place on whatever soil is there. Therefore, the soil may be much weaker than available aggregate. Figure 2–12a shows how a concentrated wheel load is spread out over the soil by an aggregate base. Stress on the soil is reduced in proportion to the square of the depth of the aggregate base, because the area of the circle over which stress is spread is proportional to the square of the depth.

The radius of the circle of pressure equals the depth of aggregate multiplied by the tangent of the angle θ shown in Figure 2–12a. With the load applied over an area small enough to be considered a point, the forces are distributed to the soil over a circular area with a radius of $h \tan \theta$. The area of the circle equals the radius squared multiplied by pi (π). The mathematical equation ranges from $R = h \tan 30°$ to $R 5 = \tan 45°$, depending on the strength of the aggregate.

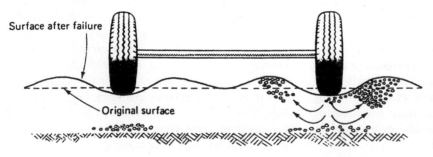

Surface after failure

Original surface

FIGURE 2–11. Aggregate failure under wheel loads.

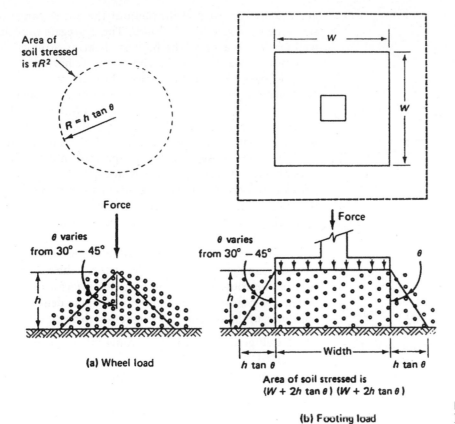

FIGURE 2–12. Wheel and footing loads transmitted to soil.

Aggregate with greater shearing strength spreads the load over a greater area, and the angle θ in Figure 2–12a is greater. The forces are not distributed uniformly within this circle, but for ordinary cases the pressure may be considered uniform.

Example

A wheel load (force) of 3000 lb is applied directly to a crushed rock base 8 in. deep, as illustrated in Figure 2–12a. Compute the pressure transmitted to the soil if the base material is of high quality, and angle θ can be considered to be 45°.

$$\text{Pressure} = \frac{\text{Force}}{\text{Area}} = \frac{\text{Force}}{\pi R^2}$$

$$\text{Area} = \pi(h\tan\theta)^2 = 3.14\left(\frac{8}{12} \times 1\right)^2$$

$$\text{Area} = 1.4 \text{ sq ft}$$

$$\text{Pressure} = \frac{3000 \text{ lb}}{1.4 \text{ sq ft}} = 2100 \text{ psf}$$

Figure 2–12b shows a concrete spread footing transmitting a structural column load through a crushed stone base to the soil. The pressure from the bottom of the footing spreads in all directions as it passes through the base to the soil. It may be assumed to spread in the shape of the footing (normally square or rectangular), becoming wider in all directions as the depth increases.

The highly concentrated load carried by the column is spread out within the footing until it is transmitted to the base over the entire bottom of the footing. The footing is one solid unit and does not transmit force in the same way as the aggregate particles. However, its purpose is the same. It reduces the pressure on the base as the base reduces the pressure on the soil. The footing could be built directly on soil. Whether or not to use a cushion of aggregate is decided by comparing the cost of a larger footing with the cost of the smaller footing plus the aggregate.

Example

A column load (force) of 33 kips acts on a 3 ft 0 in. × 3 ft 0 in. spread footing. Using Figure 2–12b, calculate the pressure on the soil if the depth of aggregate base is 8 in., and determine what depth of base is needed to reduce the pressure on the soil to 1.0 kip per sq ft. Assume angle θ is 40°. What is the maximum pressure on the aggregate?

$$\begin{array}{l}\text{Pressure 8 in.} \\ \text{below footing}\end{array} = \frac{\text{Force}}{\text{Area}} = \frac{\text{Force}}{(w + 2h\tan\theta)^2}$$

$$\text{Area} = \left(3 + 2 \times \frac{8}{12} \times 0.839\right)^2$$

$$\text{Area} = 17.0 \text{ sq ft}$$

$$\text{Pressure} = \frac{33 \text{ kips}}{17.0 \text{ sq ft}} = 1.94 \text{ Ksf}$$

$$\begin{array}{l}\text{Area required for} \\ \text{1 Ksf pressure}\end{array} = \frac{\text{Force}}{\text{Pressure}}$$

$$\text{Area} = \frac{33\,K}{1\,Ksf} = 33 \text{ sq ft}$$

$$\text{Area} = (W + 2h\tan\theta)^2$$

$$33 = (3 + 2h \times 0.839)^2$$

$$= (1.678h + 3)^2\, 2.82h^2 + 10.07h - 24 = 0$$

$$h = \frac{-10.07 + \sqrt{(10.07)^2 - 4 \times 2.82(-24)}}{2 \times 2.82}$$

$$h = 1.63 \text{ or, say, 1 ft 8 in.}$$

$$\frac{\text{Maximum Pressure}}{\text{(at bottom of footing)}} = \frac{\text{Force}}{\text{Area}}$$

$$\text{Pressure} = \frac{33}{3 \times 3} = 3.67\,Ksf$$

Loosely piled aggregate particles may be easily pushed aside. In other words, they lack shearing strength (Figure 2–13). Aggregates have a loose arrangement or structure somewhat like Figure 2–13a after handling. They are normally compacted into a tighter structure to increase friction and interlocking before a permanent load is placed on them. The reasons for compacting aggregate base material are to reduce its compressibility and to increase its shear strength.

Vibrations jar the particles into a close structure more effectively than simply rolling with a heavy weight. This can be demonstrated by filling a box with sand or gravel and compacting it by applying weight with a roller or some other means which does not jar or vibrate the material. A small amount of compaction takes place. Then shake the box lightly or rap it on the sides, and a substantial lowering of the level of the material takes place caused by compaction due to vibration.

Compaction results in an increase in density, and the density or unit weight can be used as an indication of the strength of an aggregate base. In Figure 2–13, example b is denser than example a.

Another way to increase density is to mix a variety of sizes. The smaller particles occupy spaces that would be voids if all the particles were large (Figure 2–14a). If the sizes are equally represented throughout the entire range of sizes, the aggregate is *well graded*. The aggregate must be well graded to achieve the highest shearing strength. A very strong base can be built with aggregate having a wide range of sizes of proportions selected to achieve the greatest density, as in Figure 2–14b. It is customary to obtain a high strength with a mixture of coarse and fine aggregates, which achieves a strength close to that of the ideal mixture with substantial economy in the cost of handling and mixing materials. Aggregates are separated into standard sizes by screening, and each size is stockpiled. If the mixed sizes were stockpiled, it would be difficult to prevent segregation in the stockpile. A desirable mixture is obtained by mixing two, and sometimes more, sizes of aggregate Figure 2–14c.

Factors which increase the shearing strength of aggregates are summarized here. All the factors, with the exception of the roughness of the particle surfaces, cause an increase in density. The amount of increase in density is an indication of the amount of strength gain to be expected.

1. A well-graded aggregate is stronger than one not well graded.

2. The larger the maximum size of aggregate is, the greater its strength. Larger particles provide greater interlocking, because particles must move upward for greater distances to override them.

3. The more flat, broken faces the particles have, the greater the strength developed through interlocking. Flat faces fit together more compactly with more contact between faces than if the particles are rounded. This does not mean the particles themselves should be flat. Flat particles slide readily over each other and result in lack of strength.

4. Compaction, especially by vibration, increases the shearing strength of aggregate of any size, shape, and gradation.

5. Rough particle surfaces increase strength because of greater friction between them.

Compaction

Compaction is the densification of a material resulting in an increase in weight per unit volume. The increase in density

(a) Aggregate dumped and spread. Strength due to friction only. Particles readily move downward and horizontally under a vertical load as indicated by arrows.

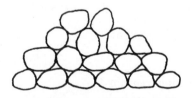

(b) Compacted aggregate. Strength due to friction and interlocking. Particles are pushed downward and horizontally as far as they can go and are now stable.

FIGURE 2–13. Increases in shear strength due to compaction.

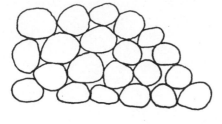
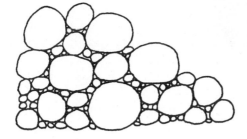
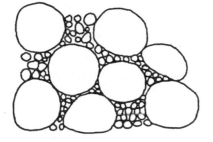

(a) Uniform size aggregate:
 1. Friction at few points of contact
 2. Poor interlocking
 3. High percentage of voids

(b) Well-graded aggregate:
 1. Friction at many points of contact
 2. Excellent interlocking
 3. Very few voids

(c) Mixture of coarse and fine aggregate:
 1. Friction at many points of contact
 2. Good interlocking
 3. Few voids
 4. Economical preparation

FIGURE 2–14. Increased density (and shearing strength) of well-graded aggregate.

of a material is related to the material's gradation, moisture content, and compactive energy used to densify the material. Well-graded aggregates compact more easily than poorly graded ones. Aggregates at *optimum moisture content* (omc) compact more easily than those with below or above omc because water acts as a lubricant during compaction. Therefore, aggregates below omc lack sufficient lubricant, and aggregates well above omc have voids filled with water, which is not a compressible material. The omc for a well-graded run-of-bank gravel might be about 7 percent, whereas a sand clay might require an omc of about 13 percent. The actual omc and maximum dry densities are determined by laboratory tests performed on aggregate samples obtained from the project.

The compactive effort applied to an aggregate entails one, or a combination, of the following: static weight, kneading, impact, or vibration. The equipment used to compact materials is grouped by classification, including sheepsfoot, vibratory, pneumatic, and the high-speed tamping foot. The equipment selected for compaction depends on the materials to be used; therefore, gravels are usually compacted with vibratory equipment (Figure 2–15), whereas sand-clay mixtures require sheepsfoot or high-speed tamping-foot equipment.

Compaction generally increases the strength of aggregates, reduces settlement, and improves bearing capacity. Therefore, when aggregates are used as subbase and base

course materials, specifications stipulate the maximum dry density or percentage compaction required. The dry density in the field can be determined by using the sand cone, balloon, or nuclear density gauge methods. When the field dry density is compared to a laboratory-determined maximum dry density, the percentage compaction can be calculated.

Contractors usually set up a test strip to select various compaction techniques to determine specification compliance. When the method of compaction and number of passes of the compaction equipment are determined, the contractor then calculates production rates.

The formula for determining the compacted cubic yards per hour is as follows:

$$\text{Cubic yards compacted per hour} = \frac{W \times S \times L \times 16.3}{P}$$

where W = compacted width per pass as recommended by the equipment manufacturer (ft)

S = speed (mph)

L = thickness of compacted lift (in.)

P = number of passes required to meet specification requirements

16.3 = constant which converts mixed units

FIGURE 2–15. Vibratory roller compacting aggregate. *(Courtesy Liz Elvin, NYS AGC)*

Example

Calculate the compacted cubic yards per hour, if a compactor with a 6-ft drum width travels at 7 miles per hour (mph) over an 8-in. lift of aggregate base and the test strip indicates four passes will be required to meet the density requirement of the specification.

Compacted cubic yards per hour

$$= \frac{6 \text{ ft} \times 7 \text{ mph} \times 8 \text{ in.} \times 16.3}{4} = 1369 \text{ cu yd}$$

This production estimate is based on a 60-minute hour and should be modified by a job efficiency factor based on anticipated field conditions:

Excellent = 90% = 54 = minute hour

Average = 83% = 50 = minute hour

Poor = 75% = 45 = minute hour

Therefore, a reasonably accurate production rate can be determined by multiplying the efficiency factor by the hourly production calculated:

Compacted cubic yard production = 1369 × 0.83

= 1136 cu yd

There are three volumetric measurements used for aggregates. *Bank measure* is the in situ, 1 cu yd of material before it is excavated and/or processed; *loose measure* is the material carried by hauling units; and *compacted measure* is used after compaction has been completed. For example, 1 cu yd bank measure may equal 1.30 cu yd loose and 0.8 cu yd compacted. These values indicate the swell and shrinkage characteristics of the material and are determined by field tests. If the contract payment is based on bank measure, the contractor divides the calculated compacted yards per hour by the 0.80 factor to determine bank cubic yards.

$$\text{Bank cubic yards} = \frac{1136}{0.80} = 1420 \text{ bank cu yd}$$

$$\text{Compacted cubic meter} = \frac{W \times S \times L}{P}$$

Where W = compacted width per pass as recommended by the equipment manufacturer (meters, m)

S = average speed (kilometers per hour, km/h)

L = thickness of compacted lift (millimeters, mm)

P = number of passes required to meet specification requirements

Calculate the compact cubic meters per hour, if a compactor with a 1.270-m drum travels at an average speed of 6.5 km/h over a 150-mm-thick fill, and test strip data indicate that three passes will be required to meet specification requirements:

Compacted cubic meters

$$= \frac{1.270 \text{ m} \times 6.5 \text{ km/h} \times 150 \text{ mm}}{3} \cong 413 \text{ m}^3/\text{h}$$

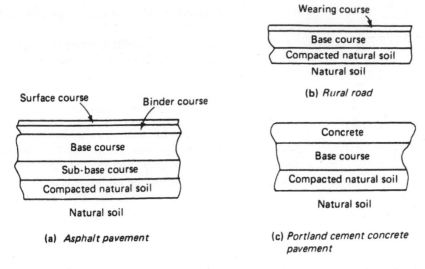

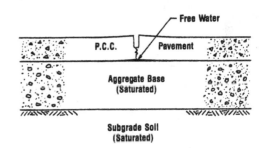

The load spreads over an area that increases with depth. Therefore the unit stress is less at greater depths. The material is arranged to be weaker and less expensive at greater depths.

FIGURE 2–16A. Typical pavement cross sections.

Pavement Base

Typical pavement construction consists of several layers or *courses* which reduce the pressure of concentrated wheel loads so that the underlying soil or *foundation* is not overloaded (Figure 2–16a). Wearing surfaces of asphalt mixtures and portland cement concrete are discussed in Chapters 3 and 4. The underlying soil may be that which is there naturally or may be hauled in to build a fill or embankment. It is not considered to be aggregate. It is strengthened by compaction with heavy construction equipment before a base or subbase is placed on it. Subbase material is usually unprocessed, run-of-bank material selected to meet specifications which have proven through performance to provide satisfactory material. Base material is more carefully selected.

In typical asphalt pavement, called *flexible* pavement, the base and subbase carry the load and distribute it to the soil under two or more thin layers of asphalt concrete. The concrete slab is the chief load-bearing element of portland cement concrete pavement, which is known as *rigid* pavement (Figure 2–16a).

The base under the rigid pavement slab spreads the load somewhat over the foundation, but is designed mainly to protect the pavement from the detrimental effects of too much moisture. These effects are frost action (heaving up against the slab when freezing and losing support by liquifying when thawing); standing water, which lowers the strength of the base; and *pumping,* which occurs when the pressure of passing traffic forces water out through pavement joints or at pavement edges, carrying small particles out with it (Figure 2–16b and 2–16c). Eventually, a hollow space is formed under the pavement by the removal of soil particles and the pavement fails.

Water does not accumulate if voids are large enough so the base drains freely and capillary water cannot rise into the base. A well-graded coarse aggregate with no appreciable

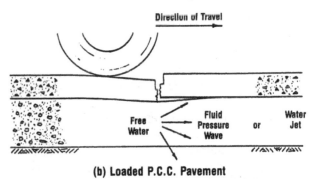

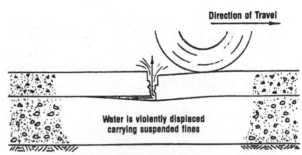

Note: Vertical dimensions of deformations are exaggerated for clarity.

FIGURE 2–16B. Pumping phenomena under portland cement concrete pavements. *(Courtesy National Stone, Sand, and Gravel Association)*

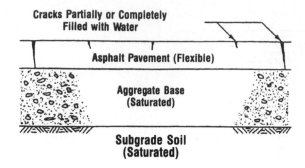

(a) Unloaded A.C. Pavement

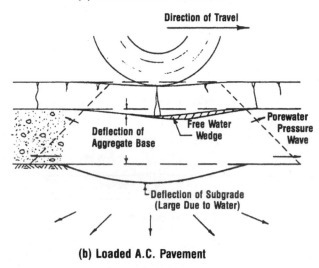

(b) Loaded A.C. Pavement

Note: Vertical dimensions of deformations are exaggerated for clarity.

FIGURE 2–16C. Action of free water in asphalt concrete pavement structural setting under dynamic loading. *(Courtesy National Stone, Sand, and Gravel Association)*

amount of fines is needed. Water may also be kept out if the aggregate has all voids filled to make it watertight. A well-graded aggregate with sufficient fine material to fill the voids is needed for watertightness.

The base under the wearing surface is the chief load-bearing element of rural or secondary roads (Figures 2-16a and 2-16c). The asphalt surface course is there to resist traffic abrasion and protect the base from rain. Therefore, the strength of the base aggregate to transfer the load to the foundation is of utmost importance. With a tight, water-repellant surface, aggregate must be well graded and compacted for strength but have enough open voids so that capillary water will not rise to be trapped under the surface course. Often the base is strengthened by mixing in just enough asphalt cement to coat the particles so they stick together but leave the voids open.

A base course with no surfacing must resist traffic and rain as well as support the load and transfer it satisfactorily to the foundation. The aggregate must be tightly bound for strength and watertightness, but should allow the rise of capillary moisture to replace moisture lost to the air. A lightly

traveled road does not need a protective covering over a properly constructed base.

The rise of water due to capillary action in an aggregate may be calculated using the following equation:

$$Z_c \cong \frac{4T_s}{d \times \gamma_{Water}}$$

Z_c equals the height of capillary rise. The term T_s indicates the order of magnitude of surface tension at the air–water interface and is equal to 0.75 g/cm. The theoretical tube diameter of the voids in the aggregate is represented by d and is nominally the effective size D_{10} in centimeters divided by 5. The density of water is 1 g/cm^3.

Example
Calculate the approximate capillary rise of water in sand with an effective size (D_{10}) of 0.6 mm, which is equal to 0.06 cm.

$$Z_c \cong \frac{4 \times 0.75 \text{ g/cm}}{0.06 \text{ cm/5} \times 1 \text{ g/cm}^3}$$

$$Z_c \cong \frac{0.3}{0.012} \cong 25 \text{ cm}$$

Freezing of water in the base or subbase does not cause much expansion. If water is drawn up by capillary action to replace the water removed by freezing and it in turn freezes, larger masses of ice called *ice lenses* are formed. These cause disruptive heaving of the surface. Coarse soils have such large voids that water cannot rise by capillarity. Fine soils with sufficient clay allow capillary water to rise very high, but movement is so slow that the quantity of water rising is insufficient to form ice lenses. The voids of silt-size particles are small enough to permit a capillary rise of several feet in quantities sufficient to form ice lenses. Aggregate sizes must be such that voids are either too large or too small to form ice lenses.

Springtime thawing of ice lenses results in a quantity of water being held under the surface course by frozen soil below it, which keeps it from draining. This removes the solid support of the base, causing the surface to break up under traffic.

The seepage of rainwater through the base carries fine particles out of the base suspended in the water. The more the particles are carried out, the faster the water flows and the larger the particles it carries are, weakening the base more and more.

Water rises by capillarity in all aggregate unless the voids are too large. Since a small amount of moisture gives added strength in cohesion to the base, it is desirable for capillary water to rise in sufficient quantity to replace evaporated water. If there is no surface over the base, capillarity should be encouraged. If the base is sealed by a watertight surface, the capillary water will accumulate under the surface course, weakening it by depriving it of solid support from the base. Therefore, large-size aggregate with large voids is used.

Road and airplane runway bases, whether covered with pavement or not, are subject to moving loads and are

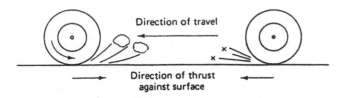

(a) Acceleration **(b) Deceleration or braking**

FIGURE 2–17. Horizontal traffic loads on pavement.

FIGURE 2–18. A road reclaimer/stabilizer cuts and blends up to a depth of 16 in. and can add stabilizing agents to the reclaimed material. *(Courtesy CMI Corporation)*

exposed to weather and running water—conditions which do not ordinarily affect aggregate bases for pipelines or footings. The moving traffic loads push horizontally. When accelerating or decelerating, tires change the vehicle velocity by pushing (backward to accelerate, forward to decelerate) against the surface below them (Figure 2–17). The vibration of traffic movement assists in loosening the bond and pushing particles to the sides. Freezing and thawing of the moisture in the aggregate, rainwater seepage downward, and movement of capillary water upward all tend to loosen the aggregate bond and remove fine particles, which further loosens and weakens the bond between particles.

Aggregate bases subject only to static loads and protected from weather and moving water need not meet the rigid standards required of pavement bases. Often the gradation of such bases is not critical. The gradation of base material for roads and airplane runways is more important in order to achieve maximum contact between particles and maximum watertightness.

Sufficient fine material must be used to ensure filling the voids without separating the larger particles from contact with each other. The larger particles are the load-bearing structure, and the fine particles hold or bind the coarser ones by preventing movement between them. The fine material is called *binder*. Base material may be a mixture of several sizes of processed aggregate or a mixture of soil and aggregate. The most efficient procedure is to mix the proper quantity and sizes of aggregate with the natural soil occurring on the site of the road or runway to produce suitable base material. Sometimes only a small percentage of aggregate is needed, and sometimes the soil is not usable at all so that the entire base must be aggregate. However, it is desirable to use the maximum amount of soil from as close to the finished construction as possible.

Stabilizing Aggregate

Aggregate strength can be improved by the addition of measured quantities of clay, which is a soil with very fine particles having properties unlike any of the larger soil particles. One of these properties is *cohesion* or the tendency to stick together. The strength due to cohesion is added to the shearing strength possessed by the aggregate. The clay, therefore, acts as a cement or paste. Other substances are also used for the same purpose. These include salts, lime, portland cement,

and bituminous cement. The use of these other substances to increase strength is called *stabilization* (Figure 2–18).

Calcium chloride and sodium chloride are the two salts mixed with aggregate to increase its strength. Coarse aggregate, fine aggregate, and binder must all be in the proper proportion, and the salt is mixed with them in quantities of 1 to $2\frac{1}{2}$ lb per sq yd of surface or 2 percent of the weight of aggregate. The salt, either calcium chloride or sodium chloride, is well mixed with the aggregate and the proper amount of moisture. The salt forms a brine with the water. This brine forms a film around each particle, which increases the strength of the aggregate in two ways.

The brine film is stronger in surface tension than ordinary moisture, and because of this it increases the cohesion of the particles. The brine film allows the particles to be forced closer together under compaction than they could be with ordinary moisture. It may be thought of as a better lubricant than plain water. Increased cohesion and increased density each cause an increase in strength.

Stabilizing with salt improves the roadway in two additional ways. These improvements result in greater strength, although in themselves they do not increase strength. Abrasion of the roadway surface by traffic causes fine particles to be lost as dust. The loss of these particles leaves voids which loosen larger particles, with the result that much of the aggregate is thrown to the sides of the road. In addition to damaging the road, the loss of the dust causes dust clouds, which are a nuisance and even a health hazard.

Brine does not evaporate as readily as untreated water, and therefore it holds fine particles in place much better despite abrasion by traffic, thereby lessening damage to the road and to nearby properties. Calcium chloride, in addition, is a hygroscopic substance, meaning that it absorbs moisture from the air. It therefore maintains a damp surface that prevents dusting.

Water percolating through aggregate removes some of the fine particles as it flows. The result is damage to the roadway similar to that caused by wheel abrasion on a dry roadway. Salt stabilization fills the voids with stationary moisture, closing the voids to the passage of water and preventing the washing through of fine particles.

Lime is also used to stabilize aggregate base material. The lime used is burned limestone in either of two forms—quicklime, which is calcium oxide containing magnesium oxide in an amount as high as 40 percent or as low as 0.5 percent $(CaO \cdot MgO)$; or hydrated lime, which is quicklime combined with enough water to produce $Ca(OH)_2 \cdot MgO$ or $Ca(OH)_2 \cdot Mg(OH)_2$. Unburned limestone cannot be used. Hydrated lime is more stable and therefore easier to store than quicklime, which hardens upon contact with air.

Lime is mixed with the aggregate in quantities of 2 to 4 percent of the aggregate weight. Water is used to accomplish thorough mixing and to combine chemically with the lime to form the final product, which is limestone.

Lime stabilizes aggregate in two ways. It reacts with clay, causing the particles to combine to form larger particles, giving a better gradation to aggregate containing too much clay. The new, larger particles will not swell excessively when moist as some clay does. Lime also causes a solidifying of the mass by reacting chemically with silica and alumina in the clay and aggregate. Calcium silicates and calcium aluminates are formed. These are cementing agents which act to hold the particles together. These cementing agents are also contained in portland cement and cause its cementing ability. A pozzolan (siliceous rock of volcanic origin) may be added to provide silica and alumina when the aggregate does not contain adequate amounts of these. These readily combine with the lime to form a bonding cement.

Portland cement or asphalt cement may be added to an aggregate base to increase its strength. Stabilization is not the same as manufacturing portland cement concrete or asphalt concrete. It refers to the addition and mixing of a small amount of cement to improve the strength of aggregate.

Concrete is aggregate stabilized with portland cement paste or asphalt cement so that a different material is formed which is no longer made up of particles. The material formed is continuous and rigid in the case of portland cement concrete, and semirigid in the case of asphalt concrete. Concrete strength depends on the strength of the aggregates, the strength of the cementing agent, and the strength of the adhesion between the two.

Permeability and Filters

Permeability is a measure of the ease with which a fluid, most commonly water, will flow through a material. Gravels have relatively high permeability, whereas sands and silts have lower permeability. The approximate permeability of a clean, uniform filter sand can be determined using Hazen's formula. The coefficient of permeability is denoted as k with units of cm/s.

Example

$$k \cong (D_{10})^2$$

where k = coefficient of permeability (cm/s)

$\quad\quad D_{10}$ = effective size based on gradation curve (mm)

Determine k for a clean sand that has an effective size D_{10} of 0.5 mm.

$$K \cong (0.5)^2 \cong 0.25\,cm/s$$

When more accurate permeability values are required, the material is tested in accordance with ASTM, AASHTO, or other standard procedures. Laboratory tests for the permeability of coarse-grained soils utilize constant head tests. For sands and silts, the falling head permeability test method is used, and for clays, a flexible wall permeameter consolidation-like test is used.

When a field permeability value is required for coarser grained soils, a percolation test is performed. When dealing with fine grained soils (low permeability), several test methods are used, including borehole tests, porous probes, and infiltrometers. The level and type of test depend on the data required and the use of the soil materials. The borehole and porous probe testing can measure in situ permeability in the horizontal and vertical direction relatively inexpensively. However, the permeability tested is predominantly the horizontal value. Similarly, soil permeability can be obtained at large depths; however, the volume of soil tested is small and unsaturated soil conditions are not taken into account. Infiltrometers test somewhat larger areas of soil and the vertical permeability of the saturated soil is obtained. Infiltrometer testing may be expensive; however, depending on the use of the soil, the data obtained may outweigh the costs associated with the testing.

The best permeability is obtained by using aggregate as large as possible and as uniform in size as possible. Both properties cause large voids with the result that water flows through easily. These voids would have to be filled with smaller particles for the aggregate to achieve its highest strength. Therefore, good permeability and high strength cannot be obtained together.

A filter consists of aggregate designed and installed for the purpose of holding back particles larger than a certain size while letting water flow through with a minimum of interference. A filter works as shown in Figure 2–19 with each layer being held in place by larger particles and in turn holding back smaller particles. Size and gradation are of primary importance for a filter.

Size must be such that the voids, which are smaller than the filter particles, are also smaller than the particles to be held back. However, they must be as large as feasible to permit water to flow through readily.

The filter catches all particles larger than the voids. If the filter particles are of uniform gradation, the void size is also uniform and all particles above the void size will be caught. Because of this fact, a filter can be designed to hold particles larger than a certain size.

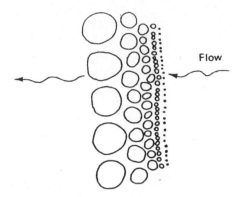

FIGURE 2-19. Filter particles positioned by flowing water.

and just too large to flow through the voids. This layer of particles, having smaller voids, catches smaller particles, and a layer of these smaller particles is formed. Layer after layer builds up until voids become so small that no particles flow through and water flow is somewhat restricted. Eventually, the filter is plugged. Several important types of filter are shown in Figure 2-20.

An underdrain system must be removed and replaced before it becomes completely plugged. A water well filter preventing sand from entering the well must be cleaned periodically by surging water back and forth through the filter. This removes fine particles by jarring them loose. They enter the well and are removed by pumping. This process is a major operation.

A filter for cleaning a drinking water supply requires cleaning daily or several times a day. It is a routine operation and is performed by forcing water backward rapidly through the filter, forcing out the finer layers which have built up. These are collected and disposed of as waste material.

A sewage filter requires occasional removal of a thin layer from the top as it becomes plugged, and eventual replacement after a sufficient thickness is removed. If the filter is below the ground surface, as shown in Figure 2-20, and the sewage filtered is not excessive, the filter operates

If the filter material is not uniform, some areas will allow particles of a certain size to go through and other areas will hold the same-size particles. Where filter particles are too small, the filter becomes plugged, and where filter particles are too large, particles that should be caught are allowed to pass through.

In a properly designed filter, the filter material originally lets all particles below the void size flow through, gradually forming a layer of particles smaller than the filter particles

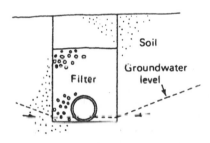

(a) *Underdrain*—lowers groundwater by carrying it away through a perforated pipe. The pipe is surrounded by a filter to prevent small particles from entering the pipe through the perforations, settling in the pipe, and restricting pipe flow.

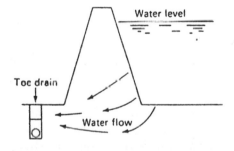

(b) *Toe drain*—underdrain that intercepts water seeping through and under an earth dam. Its purpose is to prevent undermining by piping which is the removal of soil particles by the flowing water until the dam is undermined. The filter prevents piping by holding the soil in place while allowing water to pass.

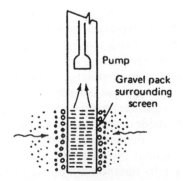

(c) *Gravel pack*—filter that prevents particles from entering the well as ground water flows through the soil, through the screen, through the pump, and to a water system at the surface.

FIGURE 2-20. Five types of filter.

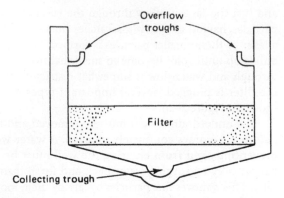

(d) *Water filter*—holds particles as water seeps downward through filter to collecting trough. When filter becomes nearly plugged, a small amount of water is pumped rapidly upward through the filter, overflowing into the troughs and carrying the filtered particles out with it. The particles that collect over many hours are removed in a few minutes.

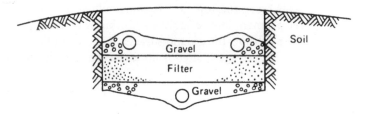

(e) *Sand filter*—sewage, with most of the solids removed by settling, flows through the upper pipes, into the filter through perforations, through the filter, and through perforations into the lower pipes which carry it to a stream or lake. The particles caught in the filter are organic and are consumed by bacteria. If properly designed, the filter will last indefinitely without cleaning.

FIGURE 2–20. Continued

indefinitely without maintenance because the sewage particles are consumed by microbes.

Filters may be divided into two categories. One type holds a mass of soil in place so that water can flow through the soil and then through the filter without carrying particles of soil with it. Mineral aggregates are always used for this type. The first three examples in Figure 2–20 are of this type. Another type of filter removes suspended particles from water as it flows through. Drinking water is cleaned this way before use, and waste water is cleaned this way before being returned to the ground or to a body of water. The last two examples in Figure 2–20 are of this type. Sand is used for this type of filter, but so are many other materials including coal, charcoal, and diatomaceous earth.

Similarly, under the new storm water management criteria, runoff waters are collected and treated through retention devices such as pond systems to ensure that the quality and quantity of water discharged to the environment meet regulatory requirements. To meet the quality requirements, sand filters in conjunction with pond systems are often used. Storm water management has become extremely important in the construction industry. Regulations require that all projects that disturb soils on site must maintain an erosion control plan and storm water management procedures. The extent of soil disturbance dictates the required storm water practices. These practices may include a series of pond systems as mentioned above, or simple drainage swale systems designed to filter out the fine particles from the runoff waters.

TESTS

The behavior of aggregates in use depends on the interrelationship of many properties. Many of these properties have been identified and defined. Standard tests have been devised to evaluate the properties. Performance can be predicted from test results based on past performances.

Cost is of great importance when large quantities of aggregates are being selected. Aggregate is always available at low cost, but the cost rises significantly if the aggregate must be handled one additional time or transported a great distance. It is often preferable to use the best aggregate available nearby rather than to improve it by processing or to obtain better aggregate from a greater distance. It must sometimes be decided whether to accept aggregate of a satisfactory quality or to pay for more processing to obtain aggregate of a better quality. The value of a particular property is often a matter of engineering judgment. Aggregate routinely used in some areas might not be acceptable where better aggregate is readily available. Because of this there is no absolute value required for many of the properties and even ASTM standards do not specify exact requirements for acceptability.

Size and Gradation

Particle sizes are important for all applications. The concept of aggregate size is difficult to express because the particles have odd shapes that cannot be measured easily and the shapes and sizes vary greatly in any one sample. The important features are *range of sizes,* or smallest and largest particles, and *gradation,* or distribution of sizes within the range covered. A few very large particles or a few very small particles do not ordinarily affect the performance of the mass of aggregate. Therefore, what is usually important is the range from the smallest particles that are contained in a significant amount to the largest particles that are contained in a significant amount.

A set of sieves fitting tightly one on top of the other is used to determine size and gradation of aggregate. A sample of the aggregate to be analyzed is placed in the top sieve, which has the largest holes. The second sieve has smaller holes, and each succeeding sieve has holes smaller than the sieve above it. At the bottom is a solid pan. The pan collects all particles smaller than the openings in the finest sieve, which is chosen to collect particles of the smallest significant size (Figure 2–21).

The *nest* of sieves is shaken, and each particle settles as far as it can. If no particle remains on the top sieve, then the top sieve size represents one kind of maximum size for the sample. The absolute maximum particle size is somewhere between this top sieve size and the size of the next sieve. If even a few particles remain on the top sieve, it is not known how large the maximum-size particles are. Usually it is not significant if only a small percentage remains on the top sieve.

Some particles fall through to the pan, and their size is not known either. If the lowest sieve has small enough holes to catch the smallest significant size, the size of particles in the pan is not important. However, an excessive quantity of material fine enough to reach the pan is of interest, just as an excessive quantity of aggregate larger than the largest sieve is of interest.

A sieve consists of a circular frame holding wires strung in such a way as to form square holes of a designated size. Particles are considered to be the size of the holes in the sieve on which they are caught. Whether or not flat particles and long, narrow ones go through a screen may depend on how they land on the screen while being shaken. Thus, chance may play a minor part in the number of particles retained on or passing through a sieve. Statistically, results are the same over a large number of tests of the same material. Extremely flat or elongated particles, which could cause the greatest variations, are unacceptable in nearly all cases and so are not ordinarily tested by sieve analysis.

Some standard sieve sizes commonly used for aggregates in construction and actual dimensions of the sides of

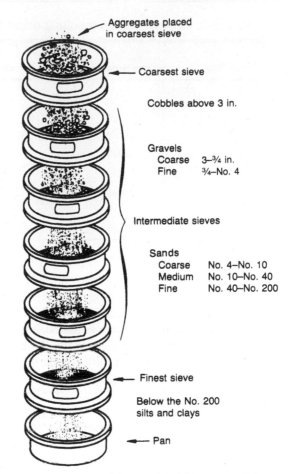

FIGURE 2–21. A nest of sieves with ASTM material designations by size. *(Courtesy The Asphalt Institute)*

the square openings are listed in Table 2–2. Nominal dimensions and permissible variations for openings in all standard sieves, as well as other specifications for their construction, are listed in ASTM E11, Standard Specifications for Wire-Cloth Sieves for Testing Purposes.

Table 2–2. Sieve sizes commonly used in construction

Sieve Designation		Nominal Opening (in.)
75 mm	3 in.	3.0
37.5 mm	1½ in.	1.5
19.0 mm	¾ in.	0.75
12.5 mm	½ in.	0.5
6.3 mm	¼ in.	0.25
4.76 mm	No. 4	0.187
2.36 mm	No. 8	0.0937
1.18 mm	No. 16	0.0469
0.6 mm	No. 30	0.0234
0.3 mm	No. 50	0.0117
0.15 mm	No. 100	0.0059
0.074 mm	No. 200	0.0029

Table 2–3. Sieve analysis results (coarse aggregate)

Sieve Size	Weight Retained (g)	Percent Retained	Cumulative Percent Retained	Percent Passing
4 in.	0	0	0	100
3 in.	540	11	11	89
$1\frac{1}{2}$ in.	1090	21	32	68
$\frac{3}{4}$ in.	1908	37	69	31
$\frac{1}{2}$ in.	892	17	86	14
No. 4	495	10	96	4
Pan	211	4	100.0	0.0
Total	5136	100.0		

Sizes designated in millimeters and inches or fractions of an inch indicate that clear openings between wires are squares with the given dimension as the length of the sides of the square. When the size is a number, such as No. 50, it means there are that number of openings in a lineal inch. The No. 50 sieve has a total of 50 openings per lineal inch or 2500 openings in a square inch. The openings are not $\frac{1}{50}$ of an inch in width because wire takes up much of the space. Therefore, these openings are smaller. Table 2–2 shows the length of the sides of each square opening, in millimeters and inches.

The results of a sieve analysis are tabulated and percentages computed as shown in Table 2–3. All relationships are by weight. The title "percent retained" refers to the percentage of the total that is retained on each sieve. The title "cumulative percent retained" refers to the percentage of the total that is larger than each sieve. It is, therefore, the sum of the percentage retained on the sieve being considered plus the percentage retained on each sieve coarser than the one being considered. The title "percent passing" means the percentage of the total weight that passes through the sieve under consideration. It is, therefore, the difference between 100 percent and the cumulative percentage retained for that sieve. A small error should be expected. Usually some dust is lost when the sieves are shaken. There is a gain in weight if particles left in the sieves from previous tests are shaken loose. However, these differences should be small if proper procedures are followed.

The weight of the original sample before sieving should agree with the sample weight after sieving. When the material is being tested for acceptance purposes, the difference between the two weights cannot differ by more than 0.3 percent based on the original dry-sample weight. The loss or gain due to sieving for general gradations is usually set at 2 percent and the total weight of the material after sieving is used as the divisor. The loss or gain of material due to sieving is usually a function of how clean the sieves were at the start of the test and how well the technician cleans each sieve during the test.

Example

Complete the sieve analysis calculations for the coarse aggregate data shown in Table 2-3.

Step one: Calculate percent retained.

$$\text{Percent retained} = \frac{\text{Weight of material retained on sieve}}{\text{Total sample weight}} \times 100$$

$$\% \text{ retained (4 in.)} = \frac{0}{5136} \times 100 = 0 = 0\%$$

$$\% \text{ retained (3 in.)} = \frac{540}{5136} \times 100 = 10.5 = 11\%$$

$$\% \text{ retained } (1\tfrac{1}{2} \text{ in.}) = \frac{1090}{5136} \times 100 = 21.2 = 21\%$$

$$\% \text{ retained } (\tfrac{3}{4} \text{ in.}) = \frac{1908}{5136} \times 100 = 37.1 = 37\%$$

$$\% \text{ retained } (\tfrac{1}{2} \text{ in.}) = \frac{892}{5136} \times 100 = 17.4 = 17\%$$

$$\% \text{ retained (No. 4)} = \frac{495}{5136} \times 100 = 9.6 = 10\%$$

$$\% \text{ retained (pan)} = \frac{211}{5136} \times 100 = 4.1 = \frac{4\%}{100\%}$$

The sum of the individual percents retained must equal 100 percent.

Step two: Calculate cumulative percent retained.

Cumulative percent retained (4 in.) = 0 = 0%

Cumulative percent retained (4 in., 3 in.) = 0 + 11% = 11%

Cumulative percent retained (4 in., 3 in., $1\frac{1}{2}$ in.) = 0 + 11% + 21% = 32%

Cumulative percent retained (4 in., 3 in., $1\frac{1}{2}$ in., $\frac{3}{4}$ in.) = 0 + 11% + 21% + 37% = 69%

Cumulative percent retained (4 in., 3 in., $1\frac{1}{2}$ in., $\frac{3}{4}$ in., $\frac{1}{2}$ in.) = 0 + 11% + 21% + 37% + 17% = 86%

Cumulative percent retained (4 in., 3 in., $1\frac{1}{2}$ in., $\frac{3}{4}$ in., $\frac{1}{2}$ in., No. 4) = 0% + 11% + 21% + 37% + 17% + 10% = 96%

Cumulative percent retained (4 in., 3 in., $1\frac{1}{2}$ in., $\frac{3}{4}$ in., $\frac{1}{4}$ in., No. 4, pan) = 0% + 11% + 21% + 37% + 17% + 10% + 4% = 100%

Step three: Calculate percent passing.

Percent passing = 100% − Cumulative percent retained

Percent passing (4 in.,) = 100% − 0 = 100%

Percent passing (3 in.,) = 100% − 11% = 89%

Percent passing (1 $\frac{1}{2}$ in.,) = 100% − 32% = 68%

Percent passing ($\frac{3}{4}$ in.,) = 100% − 69% = 31%

Percent passing ($\frac{1}{2}$ in.,) = 100% − 86% = 14%

Percent passing (No. 4) = 100% − 96% = 4%

Percent passing (pan) = 100% − 100% = 0%

Step four: Plot the percent passing versus sieve size on a gradation chart (Figure 2–22).

ASTM C136, Sieve or Screen Analysis of Fine and Coarse Aggregates, describes standard procedures for performing a sieve analysis and specifies the amount of error allowed. A brief description of the proper procedure follows. A representative sample is placed in the top sieve after the entire nest including the pan has been put together. The cover is placed on top. The nest of sieves is shaken by hand or by mechanical shaker long enough so that additional shaking cannot appreciably change the quantities retained on each sieve. The quantity retained on each sieve is removed from the sieve, using a brush to collect particles caught in the wire mesh, and weighed. Each quantity weighed should be kept on a separate sheet of paper until the sum of individual weights has been compared with the total weight of the sample. Then, if there is a discrepancy, the individual quantities can be reweighed. Percentages should be reported to the nearest whole number.

It is customary in the aggregate industry to process and stockpile aggregates in several size ranges, designated either as fine and coarse or by the size of the largest sieve retaining an appreciable percentage of the total particles. There are several advantages in this type of handling rather than stockpiling aggregate of the entire range of sizes in one pile. Segregation is more difficult to prevent in a quantity with a wide range of sizes. Also, a mixture of any desired range and gradation can be prepared with the proper proportions from several stockpiles. Therefore, the supplier with aggregate separated according to size is prepared to supply whatever the market requires. The mixture proportions can even be adjusted slightly during a project when conditions require it.

Other ASTM standards deal with specialized types of sieve analysis. ASTM C117, Standard Method of Test for Materials Finer than No. 200 Sieve in Mineral Aggregate by Washing, provides a method for washing clay particles through the sieves when the clay is stuck together in chunks or adheres to larger particles. ASTM D451, Sieve Analysis of Granular Mineral Surfacing for Asphalt Roofing and Shingles; ASTM D452, Sieve Analysis of Nongranular Mineral Surfacing for Asphalt Roofing and Shingles; and ASTM D546, Sieve Analysis of Mineral Filler, set forth other specialized sieve analysis procedures. Mineral filler is very fine, dustlike aggregate used in bituminous concrete to fill voids between fine aggregate particles.

The results of a sieve analysis are often plotted on graph paper with sieve sizes on the horizontal axis as the abscissa and percent coarser (retained) and finer (passing) on the vertical axis as the ordinates. Sample plots are shown in Figure 2–22. The horizontal axis is divided according to a logarithmic scale because of the wide range of sizes to be plotted. The largest sieve openings may be several hundred

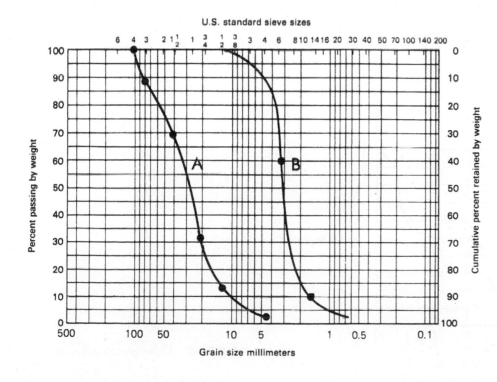

FIGURE 2–22. (a) Curve representing the example problem gradation; (b) typical gradation curve—uniform aggregate.

times the size of the smallest sieve openings. The smaller sizes must be spaced far enough apart for clarity; but if the same scale is used for the larger sizes, an excessively long sheet of paper is needed. Linear scales are therefore not used. The semilogarithmic scale is much more satisfactory.

A graph of percent by weight versus sieve sizes is called a *gradation chart.* It is a better presentation in some ways than a tabulation. Size, range, and gradation can be seen on the graph. The range of sizes can be obtained from graph or tabulation with equal ease. However, finding the size and gradation may require the use of a gradation chart. There are several ways in which the size of aggregate is defined.

It has been found that the filtering performance of an aggregate can be predicted by the particle size that has a certain percentage by weight finer than its own size. This size is designated by the letter D with a subscript denoting the percentage finer. In Figure 2–22 curve B, D_{15} is 2 mm and D_{85} is 4 mm.

The *effective size,* used to designate size of aggregate to be used as a filter for sewage or drinking water, is that diameter or size on the graph which has 10 percent of the total finer than its size. It is not necessarily a sieve size and cannot be found accurately without a gradation curve. In Figure 2–22 curve B, it is 1.4 mm. Sieve sizes may increase to the left as in Figure 2–22 or increase to the right as in Figure 2–23.

The *maximum size* of aggregate, when used in the design of portland cement concrete mixes, is taken for that purpose to be the size of the sieve next above the largest sieve

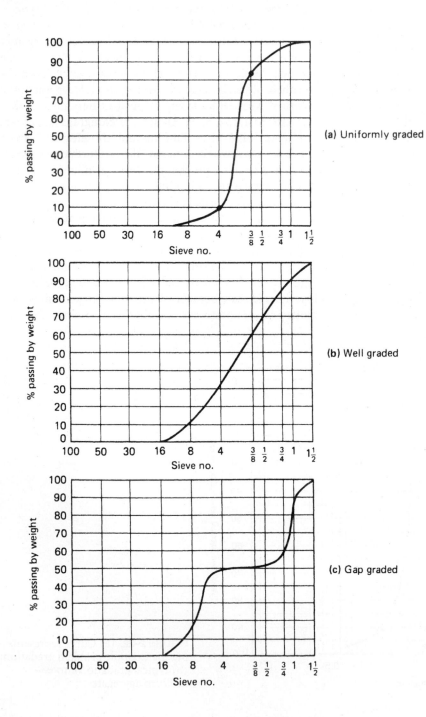

(a) Uniformly graded

(b) Well graded

(c) Gap graded

FIGURE 2-23. Three gradation types.

that has 15 percent of the total sample coarser than it (cumulative percentage retained). Therefore, in Table 2–3 the maximum aggregate size would be 3 in.

Fineness modulus is a value used in the design of portland cement concrete mixes to indicate the average size of fine aggregate. It may also be used to indicate the average size of coarse aggregate. It is determined by adding the cumulative percentages retained on specified sieves and dividing by 100. Specified sieves include 6 in., 3 in., $1\frac{1}{2}$ in., $\frac{3}{4}$ in., $\frac{3}{8}$ in., No. 4, No. 8, No. 16, No. 30, No. 50, and No. 100. The sample must be run through all these sieves, omitting those too large to retain any aggregate, but not any that are smaller than the smallest aggregate. Calculating the fineness modulus requires the inclusion of cumulative percentages retained on each sieve.

The fineness modulus should be determined to two decimal places. The whole number indicates the sieve on which an average-size particle would be retained. To locate this sieve, count upward from the pan a number of sieves corresponding to the whole number of the fineness modulus. The average size is further defined by considering it to be located between the designated sieve and the sieve above it, a portion of the distance upward, as indicated by the decimal. The decimal 0.75 would locate the average size at 75 percent of the way up from the lower sieve toward the upper one.

Gradation, meaning the distribution of particle sizes within the total range of sizes, can be identified on a graph as well graded, uniform, or gap graded (sometimes called skip graded). *Well graded* means sizes within the entire range are in approximately equal amounts, although there will be very small amounts of the largest and smallest particles. *Uniform* gradation means that a large percentage of the particles are of approximately the same size. *Gap graded* or *skip graded* means that most particles are of a large size or a small size with very few particles of an intermediate size. Typical curves for the three types of gradation are shown in Figure 2–23.

The shape of the curve aids in identifying the type of gradation. A line nearly vertical indicates that a large quantity of material is retained on one or possibly two sieves. In Figure 2–23a, 85 percent of the particles are finer than the $\frac{3}{8}$ in. sieve,

and only 10 percent are finer than the No. 4 sieve. Therefore, 75 percent of the material is caught on the No. 4 sieve. In other words, much of the aggregate is the same size, and the material is uniform.

A line with a constant slope, as in Figure 2–23b, changes the same amount in the vertical direction with each equal increment in the horizontal direction. This indicates that approximately the same quantity of material is retained on each successive sieve and therefore that the aggregate being tested is well graded.

A horizontal or nearly horizontal line, as in Figure 2–23c, indicates that there is no change or little change in percent finer through several successive sieves. Therefore, no material or very little material is retained on these sieves, and there is a gap in the gradation. Aggregate can be processed or mixed to provide any of these gradations if desired. In nature aggregate seldom occurs in these ways, and gradation charts usually take less idealized shapes.

The gradation curve does not give a precise indication of uniformity, although if curves for two aggregates are plotted at the same scale, one could tell which is more uniform. The *uniformity coefficient* is a mathematical indication of how uniform the aggregate is. It is determined by dividing the diameter or size of the D_{60} by the diameter or size of the D_{10}. The smaller the answer, the more uniform in size the aggregate. The uniformity coefficient cannot be determined from a tabulation of sieve analysis results because the D_{60} and D_{10} are usually between sieve sizes, and a plot must be made to locate them.

Aggregate size and gradation are often specified by listing sieve sizes and a range of "percent passing" for each size. The sample being tested is sieved with the specified sieves and is acceptable if, for each sieve, the sample's percentage falls within the specified range. A plot of the upper and lower specified limits is called an *envelope*. An aggregate meets the specifications if its gradation curve plots entirely within the envelope (Figure 2–24). A gradation curve is not needed but it shows the aggregate's relationship to the envelope in detail that cannot be obtained from a tabulation.

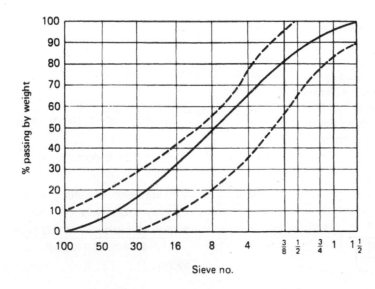

FIGURE 2–24. Gradation curve that fits within an envelope, thereby meeting the specification.

Another way to specify size and gradation is with a range for the effective size and a range for the uniformity coefficient.

Example

The gradation chart (particle size distribution curves) can be used for the comparison of different soil materials. Using Figure 2–22, calculate the effective size and the uniformity coefficient for sample A. Determine the filter compatibility between soils A and B with B as the soil and A as the filter material.

Effective size = D_{10} − diameter corresponding to 10% finer.

$$D_{10} = 11\,mm$$

Uniformity coefficient = C_u

$$C_u = \frac{D_{6c}}{D_{1c}}$$
$$= \frac{34\,mm}{11\,mm}$$
$$= 3.09$$

A well-graded soil will have a uniformity coefficient greater than 4 for gravels and 6 for sands typically.

Filter performance criteria   A = filter

B = soil

$$\frac{D_{15}(\text{of filter})}{D_{85}(\text{of soil})} < 4\text{ to }5 < \frac{D_{15}(\text{of filter})}{D_{15}(\text{of soil})}$$
$$\frac{15\,mm}{4\,mm} < 4\text{ to }5 < \frac{15\,mm}{2\,mm}$$
$$3.75 < 4\text{ to }5 < 7.5$$

Results meet the filter criteria for preventing piping of the B soil through Soil A (grading of drainage aggregates to control piping).

Surface Area

The surface area of a quantity of aggregate is sometimes important. A ratio of surface area to volume or surface area to weight is determined and used for computations dealing with surface area. Of all possible particle shapes, a sphere has the lowest ratio of surface area to volume or weight. Particles that roughly approximate spheres in shape also have roughly the same surface to volume ratio as spheres, and all other shapes have greater ratios, with the ratio being greater as the particle differs more from a spherical shape. The following calculation shows the ratio of surface area to volume for a sphere:

$$\text{Surface area} = 4\pi r^2$$
$$\text{Volume} = \left(\frac{4\pi r^3}{3}\right)$$
$$\text{Ratio} = \frac{\text{Surface area}}{\text{Volume}} = \frac{4\pi r^2}{\left(\frac{4\pi r^3}{3}\right)} = \frac{3}{r}$$

The ratio is therefore 3 divided by the radius of the sphere. The ratio is greater for small particles than for large particles because the ratio becomes greater as the radius becomes smaller.

In waste water filters, organic matter is consumed by bacteria that live on aggregate surfaces. Consumption, and therefore the quantity of waste water that can be treated, is proportional to the numbers of bacteria which are proportional to the total surface of the particles. It is, therefore, advantageous to have a large total surface per volume of aggregate to support more bacteria.

It is also of importance if aggregate particles are to be bound together for strength. The design of an asphalt paving mix requires enough liquid asphalt to form a coat of a certain thickness completely over each particle. An estimate is made of the square feet of surface per pound of aggregate, and the proportion by weight of asphalt material to aggregate is determined according to the amount of surface to be coated. The particles are always bulky in shape for strength so that the surface area is small (slightly larger than for spheres), and the result is that a small quantity of asphalt material is sufficient to coat the particles.

Weight–Volume Relationships

The total volume of an aggregate consists of solid particles and the voids between the particles. The total volume is important because aggregate must be ordered to fill a certain volume. Aggregate for a filter must cover a certain number of square feet to a particular depth. Aggregate for a roadbed must be placed in a certain width and thickness for a specified number of miles.

The volume of solid matter is also of importance. The volume of one aggregate particle consists of a mass of solid material. However, all particles used as aggregates contain some holes or pores. The pores range in size from a large open crack that can hold small particles to holes that cannot be seen, but can absorb water. For some uses the volume of these pores is important and for some uses it is not.

Example

Calculate the solid volume and percent of voids in a fine aggregate if it has a specific gravity of 2.65 and a bulk unit weight of 111.3 pcf (1782.85 kg/m³).

$$\text{Solid volume} = \frac{\text{Weight (lb)}}{\text{Specific gravity} \times \text{Unit weight of water}}$$
$$SV = \frac{111.3\,lb}{2.65 \times 62.4\,lb\,per\,cu\,ft} = 0.673\,cu\,ft$$

$$\%\,\text{Voids} =$$
$$\frac{\text{Specific gravity} \times \text{Unit weight of water} - \text{Unit weight}}{\text{Specific gravity} \times \text{Unit weight of water}} \times 100$$
$$= \frac{2.65 \times 62.4\,lb/cu\,ft - 111.3\,lb/cu\,ft}{2.65 \times 62.4\,lb/cu\,ft} \times 100$$
$$= 32.7\%$$

$$\text{Solid volume} = \frac{1782.85 \text{ kg}}{2.65 \times 1000 \text{ kg/m}^3} = 0.673 \text{ m}$$

$$\%\text{Voids} = \frac{2.65 \times 1000 \text{ kg/m}^3 - 1782.85 \text{ kg}}{2.65 \times 1000 \text{ kg/m}^3} \times 100$$

$$= 32.7\%$$

In asphalt concrete, a percentage of the volume of the pores is filled with liquid asphalt material. Therefore, the correct quantity of asphalt cement for the mix includes enough to coat the particles plus enough to partially fill the pores.

In portland cement concrete, the mixing water completely fills the pores in the aggregate. Lightweight aggregate, because of its very porous structure, may absorb so much water that there is not enough remaining to combine with the cement satisfactorily.

The volume of the pores in aggregate used as a base or as a filter is of little importance except that freezing and thawing cause the breaking of porous particles more so than of more solid particles. A few fine particles may enter the larger pores in coarse aggregate, but their total volume is negligible. In base material the voids between coarse aggregate particles are filled with fine aggregate. The volume of voids must be known to obtain the correct volume of fine aggregate but the volume of pores is not needed.

It is usually desirable to know the volume of aggregate in relationship to its weight. This is so because the quantity needed is determined according to the volume it must occupy, but that quantity is ordered and measured for payment by weight.

Various combinations are used to relate weight and volume, depending on how the aggregate is to be used. The possibilities include using total volume (solids and voids), volume of solids including pores, or volume of solids less volume of pores; and using wet weight, saturated, surface-dry weight, or oven-dry weight. These alternatives are illustrated in Figure 2–25.

1. The volume of aggregate may include solid matter, plus pores in the particles, plus voids. This is called *bulk volume* of aggregate.

2. The volume may include solid matter, plus pores in the particles but not voids. This is called the *saturated, surface-dry volume.*

3. The volume may include solid matter only, not pores or voids. This is called *solid volume.*

4. The weight may include solid matter, plus enough water to fill the pores, plus free water on the particle surface. This is called *wet weight.*

5. The weight may include solid matter, plus enough water to fill the pores. This is called *saturated, surface-dry weight.*

6. The weight may include solid matter only. This is called *oven-dry weight.*

If aggregate particles are soaked in water until all pores are filled and then removed and wiped dry, the pores will remain filled with water for a time. If that aggregate is placed

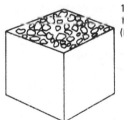

1 cu. ft. box holds 1 cu. ft. of aggregate (bulk volume)

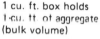

(a) *Bulk volume*—volume of solid particles including their pores plus volume of voids.

(b) *Saturated, surface—dry volume*—volume of solids including pores.

(c) *Solid volume*—volume of solid material not including pores.

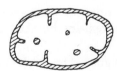

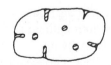

(d) *Wet weight*—weight of solid material plus absorbed water filling the pores plus some free water on the particle surfaces. The amount of free water is variable.

(e) *Saturated, surface-dry weight*—weight of solid material plus enough absorbed water to fill the pores.

(f) *Oven dry weight*—weight of solid material only. This is the most consistent weight because no water is included.

FIGURE 2–25. Kinds of volume and weight for aggregate particles.

into a calibrated container of water, the rise in water level indicates the saturated, surface-dry volume of aggregate because the volume added to the container of water consists of solids plus pores saturated with water.

If aggregate is dried in an oven until all moisture is driven from the particle pores and then placed into a calibrated container of water, the rise in water level indicates the solid volume because the volume added to the container of water consists only of solid material. Some time is needed for the water to enter the pores, and the solid volume cannot be determined at once. In soaking particles to fill their pores with water, a 24-hour soaking period is considered sufficient.

The simplest weight-volume relationship is expressed as unit weight of aggregate. The procedure for determining it is contained in ASTM C29, Unit Weight of Aggregate. The sample of aggregate is oven-dried, and a cylindrical metal container of known volume is filled and weighed. The procedure for filling the container is specified in detail to ensure consistent results. The container is filled in three equal layers, and each layer is compacted with 25 strokes of a rod of standard dimensions.

Density or unit weight values for aggregates are expressed as total density γ or dry density γ_d. When total density is determined as well as water content for an aggregate, the dry density can then be calculated as follows.

Example

1. Calculate the moisture content of a soil that has a volume of 0.4987 cu ft, total weight of 41.34 lb, and an oven-dry weight of 32.85 lb.

$$mc = \left(\frac{W - W_s}{W_s} \right) \times 100$$

$$= \left(\frac{41.34 \, lb. - 32.85 \, lb}{32.05 \, lb} \right) \times 100$$

$$= 25.8\%$$

where W = total weight

W_s = weight of solids

mc = moisture content

2. Calculate γ_d (dry density) if an aggregate weighs 47.72 lb, has a moisture content of 6.3 percent, and occupies a volume of 0.4987 cu ft.

$$\gamma = \frac{W}{V} = \frac{47.72 \, lb}{0.4987 \, cu \, ft} = 95.69 \, pcf$$

$$\gamma_d = \frac{\gamma}{1 + mc} = \frac{95.69 \, pcf}{1 + 0.063} = 90.02 \, pcf$$

where γ = total density or unit weight (pcf)

W = weight of aggregate, including water (lb)

V = volume (cu ft)

mc = moisture content expressed as a decimal

The oven-dry weight of aggregate divided by its bulk volume is its unit weight. This is not the same as solid unit weight of aggregate particles, which is the oven-dry weight divided by the actual volume of the particles not including voids.

Specific gravity (SG) of a substance is the ratio of the solid unit weight of that substance to the unit weight of water. The specific gravity of aggregate particles is useful in calculations, particularly those to convert weight of the irregularly shaped particles to saturated, surface-dry volume or to solid volume.

There are two kinds of specific gravity used with aggregate particles. *Bulk specific gravity* is based on oven-dry weight and saturated, surface-dry volume of the aggregate particles. Pores in the particles are considered part of the volume. *Apparent specific gravity* is based on oven-dry weight and solid volume of the particles. Either one of these can be considered as a true specific gravity, and each has its own use.

A type of specific gravity called *effective specific gravity* is used in the design of asphalt concrete. It is derived by dividing oven-dry weight by the weight of a volume of water equal to the solid volume plus the volume of pores that is not filled with asphalt cement when aggregate and cement are mixed.

Dividing the weight of a certain volume of any substance by the weight of the same volume of water is equivalent to dividing unit weight of the substance by unit weight of water, and so the result is specific gravity of the substance. Methods for determining bulk and apparent specific gravity and absorption for aggregate are contained in ASTM C127, Specific Gravity and Absorption of Coarse Aggregate, and ASTM C128, Specific Gravity and Absorption of Fine Aggregate. Bulk specific gravity is sometimes determined using saturated, surface-dry weight, and that method is included in these two ASTM standards.

The methods are briefly described here.

Specific Gravity

Specific Gravity of Coarse Aggregate

1. A representative sample of coarse aggregate weighing about 5 kg is dried by heating and weighed at intervals until two successive weighings show no loss of weight. The final weight is recorded as the oven-dry weight.

2. The oven-dried sample is soaked in water for 24 hours.

3. The aggregate is removed from the water and dried with a cloth until no film of water remains, but the particle surfaces are damp.

4. The aggregate is then weighed. The result is saturated, surface-dry weight in air.

5. The aggregate is submerged in water at 23 ± 1.7°C (73.4 ± 3°F) in a wire basket, and the submerged weight is obtained. The submerged weight of the basket must be deducted. The submerged particles displace a volume of water equal to their own volume including all pores because the pores are filled with water before being submerged.

There is a weight loss when submerged which is equal to the weight of water displaced.

6. Oven-dry weight of aggregate in air divided by the difference between saturated, surface-dry weight in air and weight submerged equals bulk specific gravity. The weight loss in water is the weight of a quantity of water equal to the saturated, surface-dry volume of aggregate. Therefore, this calculation determines the ratio of weight of a volume of aggregate (solid matter plus pores) to the weight of an equal volume of water (Refer Equation 1 example problem).

7. Oven-dry weight of aggregate in air divided by the difference between oven-dry weight of aggregate in air and weight submerged equals apparent specific gravity. In this case the weight loss in water is the weight of a quantity of water equal to the volume of solid matter of aggregate. Therefore, this calculation determines the ratio of weight of aggregate solid matter without pores to weight of an equal volume of water. It should be understood that oven-dry weight of solid matter alone is the same as oven-dry weight of solid matter plus pores since the empty pores have no weight.

The only difference between calculating bulk specific gravity and apparent specific gravity is that the saturated, surface-dry volume is used for bulk specific gravity and the volume of solid matter alone is used for apparent specific gravity. Using saturated, surface-dry particles means that saturated, surface-dry volume is the basis for determining the weight of an equal volume of water to find bulk specific gravity. Using oven-dry particles means that the solid volume is the basis for determining the weight of an equal volume of water to find apparent specific gravity (Equation 2).

8. Bulk specific gravity is based on oven-dry weight unless specified otherwise. It may be determined on a saturated, surface-dry basis by using the saturated, surface-dry (SSD) weight in air as aggregate weight rather than the oven-dry (OD) weight and becomes bulk SSD (Equation 3). This variation is covered in ASTM C127.

9. *Absorption*, which is the percentage of the weight of water needed to fill the pores compared to the oven-dry weight of aggregate, is computed by dividing the difference between SSD weight and OD weight in air by the OD weight in air (Equation 4).

Example
Calculate the specific gravities and absorption given the following data:

$$\text{Given: SSD weight in air} = 5480\,\text{g}$$

$$\text{Submerged weight} = 3450\,\text{g}$$

$$\text{OD weight} = 5290\,\text{g}$$

$$\text{Bulk SG} = \frac{\text{OD weight}}{\text{SSD weight} - \text{Submerged weight}}$$

$$= \frac{5290}{5480 - 3450} = 2.61 \text{ (Equation 1)}$$

$$\text{Apparent SG} = \frac{\text{OD weight}}{\text{OD weight} - \text{Submerged weight}}$$

$$= \frac{5290}{5290 - 3450} = 2.88 \text{ (Equation 2)}$$

$$\text{Bulk SSD} = \frac{\text{SSD weight in air}}{\text{SSD Weight in air} - \text{submerged weight}}$$
$$\text{(Equation 3)}$$

$$= \frac{5480\,\text{g}}{5480\,\text{g} - 3450\,\text{g}}$$

$$= \frac{5480\,\text{g}}{2030\,\text{g}}$$

$$= 2.70$$

$$\%\,\text{Absorption} = \frac{\text{SSD weight} - \text{OD weight}}{\text{OD weight}}$$

$$= \frac{5480 - 5290}{5290} \times 100 \text{ (Equation 4)}$$

$$= 3.6\%$$

Specific Gravity of Fine Aggregate

1. A representative sample of fine aggregate weighing about 1000 g is dried to a constant weight, the oven-dry weight.

2. The oven-dried sample is soaked in water for 24 hours.

3. The wet aggregate is dried until it reaches a saturated, surface-dry condition. This condition cannot be easily recognized with small-size aggregate. To determine the point at which it reaches this condition, the drying must be interrupted frequently to test it. The test consists of putting the aggregate into a standard metal mold shaped as the frustum of a cone and tamping it 25 times with a standard tamper. The cone is removed vertically and the fine aggregate retains the mold shape if sufficient moisture is on the particle surfaces to cause cohesion. The aggregate is considered saturated, surface-dry the first time the molded shape slumps upon removal of the mold.

4. A representative sample consisting of 500 g of aggregate (saturated, surface-dry weight) is put into a 500-cm^3 container, and the container is filled with water at $23 \pm 1.7°C$ ($73.4 \pm 3°F$).

5. The full container is weighed. The total weight consists of the sum of the following:
   (a) known weight of flask,
   (b) known weight of aggregate (SSD), and
   (c) unknown weight of water. The weight of water can be determined by subtracting flask and aggregate weights from the total weight.

6. The entire contents is removed from the container, with additional rinsing as required to remove all particles, and the oven-dry weight of aggregate is determined.

7. The container is weighed full of water.

8. The foregoing procedures provide all the data needed to compute SG and absorption. Saturated, surface-dry volume of aggregate is used for bulk SG, and solid volume is used for apparent SG. Oven-dry weight of aggregate is used for either type of SG, and SSD weight may be used for bulk SG if so stated. Either SG equals oven-dry weight of aggregate divided by weight of container filled with water plus weight of aggregate in air, minus weight of container filled with aggregate and water. For bulk SG, the SSD weight of aggregate in air is used in the denominator, and for apparent SG the oven-dry weight of aggregate in air is used in the denominator. The procedure is illustrated in Figure 2–26.

9. Absorption is the weight of water needed to fill the particle holes, divided by the weight of solid matter and expressed as a percentage. It is computed by dividing SSD weight minus oven-dry weight by oven-dry weight.

Example

Calculate the specific gravities and absorption given in the following data:

$$\text{Given: SSD weight} = 500\text{ g}$$
$$\text{OD weight} = 492.6\text{ g}$$
$$\text{Flask} + \text{water weight} = 537.6\text{ g}$$
$$\text{Flask} + \text{Water} + \text{Fine aggregate weight} = 846.2$$

Steps 1 through 4
Known weight of aggregate SSD (W_A) plus an unknown weight of water to fill flask of known volume.

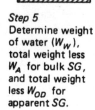

Step 5
Determine weight of water (W_W), total weight less W_A for bulk *SG*, and total weight less W_{OD} for apparent *SG*.

Step 6
Determine oven dry weight of aggregate (W_{OD}).

Bulk specific gravity =

$$\frac{\text{OD weight}}{[(\text{Flask} + \text{Water}) + (\text{SSD wt})] - [\text{Flask} + \text{Water} = \text{Fine aggregate}]}$$

$$= \frac{492.6}{[537.6 + 500] - [846.2]} = 2.57$$

Apparent specific gravity =

$$\frac{\text{OD weight}}{[(\text{Flask} + \text{Water}) + (\text{OD weight})] - [\text{Flask} + \text{Water} + \text{Fine aggregate}]}$$

$$= \frac{492.6}{[537.6 + 492.6] - [846.20]} = 2.68$$

$$\% \text{ Absorption} = \frac{\text{SSD weight} - \text{OD weight}}{\text{OD weight}} \times 100$$

$$= \frac{500\text{ g} - 492.6\text{ g}}{492.6\text{ g}} \times 100$$

$$= 1.50\%$$

Deleterious Matter

Excessive amounts of foreign material are detrimental in aggregate used for any purpose. What constitutes an excessive amount depends on the usage of the aggregate. Very little

Step 8

$W_F - W_W$ = Weight of water of same volume as aggregate

Aggregate volume is SSD volume if W_A used in step 5, and it is solid volume if W_{OD} is used.

$$SG = \frac{W_{OD}}{W_F - W_W}$$

Bulk *SG* is obtained if W_A used in step 5, and apparent *SG* is obtained if W_{OD} is used.

Using the weights obtained through step 7 including weight of container (W_C)

Step 7
Determine weight of water to fill flask (W_F).

$$SG = \frac{W_{OD}}{W_C + W_F + W_A - (W_C + W_W + W_A)} = \frac{\text{OD agg. wt}}{\text{wt water same vol. as agg. } SSD}$$

FIGURE 2–26. Specific gravity of fine aggregate.

foreign matter can be permitted in aggregate for portland cement concrete, asphalt concrete, or filters. Permissible amounts are greater for aggregate used as a base.

ASTM C33 contains allowable limits for seven types of deleterious substances which must be controlled in aggregates to be used for portland cement concrete. Maximum permissible quantities are listed according to a percentage of the weight of the entire sample for fine aggregates and coarse aggregates. The seven categories are friable particles, material finer than No. 200 sieve, soft particles, coal and lignite, chert, organic impurities, and materials reactive with the alkalis in cement.

Friable particles are those which are easily crumbled, such as clay lumps, weak sandstone, or oxidized ores. An excessive amount of these causes a change to a gradation with more fine particles when the friable ones are broken in use. The end result is similar to using an aggregate with excessive fine material in it. The method of testing for friable particles is described in ASTM C142, Test for Friable Particles in Aggregates. Friable particles are described as those that can be crushed between thumb and forefinger without using fingernails. Material is separated on sieves before the test, and after all friable particles are broken, each fraction is sieved on a sieve finer than the one it was retained on. The weight of crushed particles that goes through the finer sieves divided by the total weight of the test sample gives the percentage of friable particles. ASTM C33 limits friable particles to 3 percent for fine aggregates and 2–10 percent for coarse aggregates when either is to be used for portland cement concrete.

Material finer than No. 200 sieve is that material which passes through the No. 200 sieve in a washed sieve analysis performed according to ASTM C117, Test for Materials Finer than No. 200 Sieve in Mineral Aggregates by Washing. The material must be washed through the sieves because much of it may be stuck to larger particles. One reason the fine material is objectionable is that it coats larger particles. The coating is a hindrance to the adherence of portland cement paste or asphalt cement to the aggregates. In portland cement concrete, the fine material, whether loose or coating a particle, absorbs water before the water can combine with cement to form a paste.

ASTM C33 limits material finer than No. 200 sieve in fine aggregate to 3 percent for portland cement concrete subject to abrasion and 5 percent for other portland cement concrete, and in coarse aggregate to 1 percent for all concrete. If the finer material is stone dust which has a bulky shape, it is not as harmful as clay particles which have a flat shape, and these limits may be increased to 5 and 7 percent, respectively.

Fine material may be objectionable in filter material because it will either be washed through the filter or partially plug the filter, depending on the relative sizes of the fine material and the filter material. Plugging to any extent is always undesirable, as it lowers the quantity of water that can pass through the filter. Fine material flowing through is objectionable if it settles to the bottom and impedes the flow of water in a conduit following the filter or if the filter's purpose is to provide clear water.

Soft particles are those that are marked with a groove after being scratched on a freshly broken surface by a pointed brass rod under a force of 2 lb in accordance with ASTM C235, Test for Scratch Hardness of Coarse Aggregate Particles. The test is simple and suitable for field investigation of a possible aggregate source. Soft particles are detrimental when the aggregate is subject to abrasion, such as in a gravel road, bituminous concrete road, or portland cement concrete floor subject to steel wheel traffic. The main concern is that soft particles will be crushed or rubbed into powder, thereby interrupting the continuity of the aggregate structure by removing some of the aggregate particles that are needed either to transmit shear or to maintain a continuous surface. Therefore, soft particles are of little concern in fine aggregate. ASTM C33 limits soft particles in coarse aggregate to 5 percent for concrete in which surface hardness is important and has no limit for soft particles in fine aggregate.

Lightweight pieces are particles in coarse or fine aggregate that have an SG substantially less than that of the aggregate as a whole. They are objectionable for several reasons. They are often soft or weak. If they consist of coal or lignite, they cause unsightly pitting and black staining at the surface of a portland cement concrete structure and have an SG of about 2.0. Chert particles with an SG of 2.35 expand because of their porous particle structure and cause pitting in concrete.

The test for determining the percentage of lightweight pieces in aggregate is described in ASTM C123, Lightweight Pieces in Aggregate. It consists of placing the aggregate sample into a mixture of liquids proportioned to have an SG between that of the acceptable aggregate and that of the lightweight pieces. The SG of the liquid is designed according to the SGs of particles to be separated. The lightweight particles float and are skimmed or poured out. Zinc chloride in water may be used, or carbon tetrachloride or kerosene may be blended with a heavy liquid to produce the desired SG. ASTM C33 limits coal and lignite in fine and coarse aggregate to 0.5 percent when surface appearance of concrete is important and to 1 percent when surface appearance is not important. Chert is limited in coarse aggregate to 1 percent for concrete exposed to severe weather and 5 percent for concrete with mild exposure.

Organic impurities are nonmineral material of an organic type, mainly tannic acid, sometimes found in fine aggregate. Organic material hinders the hardening of portland cement paste and so must be limited in fine aggregate to be used in portland cement concrete. A method of testing for excessive organic impurities is contained in ASTM C40, Organic Impurities in Sands for Concrete. The principal value of the test is to furnish a warning that further tests are necessary before the aggregate can be approved.

The method consists of preparing a reference solution of standard brown color and comparing it with a solution containing a sample of the aggregate being tested. A measured quantity of the fine aggregate is mixed with a specified solution of sodium hydroxide in water. After 24 hours the solution containing aggregate is compared to the standard color solution. The more organic material there is in the aggregate, the darker the solution is. If it is darker than the standard color, it presumably contains excessive organic material.

Suspect aggregates should not be used for portland cement concrete unless it can be proven that mortar made from it is as strong, or nearly as strong, as mortar made from the same aggregate with the organic impurities washed out. The method for making this comparison is described in ASTM C87, Effect of Organic Impurities in Fine Aggregate on Strength of Mortar. Cubes of mortar are made from both washed and unwashed aggregate, and the average crushing strengths compared. The mortar made from the unwashed aggregate should be at least 95 percent as strong as the mortar made with clean aggregate. This test takes one week for the mortar cubes to cure, compared to one day for the simpler presumptive test.

Reactive aggregates are those which contain minerals which react with alkalis in portland cement, causing excessive expansion of mortar or concrete. The expansion causes disintegration which may not be apparent in use for several years. The reaction is either alkali–silica or alkali–carbonate, depending on the type of aggregate. The reaction takes place when mortar or concrete is subject to wetting, extended exposure to humid atmosphere, or contact with moist ground. It can be prevented by using cement with a low alkali content or with an additive which has been proven to prevent the harmful expansion. The additives either combine with alkalis while the cement paste is still in a semiliquid state, thus removing the alkalis, or prevent the reaction between the deleterious substance and the alkali. The alternative is to check aggregates for reactivity whenever it is suspected. The tests are difficult and usually time consuming, and no one test is entirely satisfactory for all cases. Often aggregates are accepted based on past experience in similar cases and in doubtful cases the aggregate is tested.

A standardized microscope examination which is useful to determine the quantity of reactive material in aggregate is described in ASTM C295, Petrographic Examination of Aggregates for Concrete. However, the actual results caused by the reactive substance are a better criterion than the quantity of reactive substance in the aggregate.

A chemical method, described in ASTM C289, Test for Potential Reactivity of Aggregates (Chemical Method), indicates the potential reactivity by the amount of reaction between a sodium hydroxide solution and the aggregate submerged in it. This test is not completely reliable but serves as an indicator before undertaking a longer test.

Another method, described in ASTM C227, Test for Potential Alkali Reactivity of Cement-Aggregate Combination (Mortar Bar Method), is to make mortar specimens and measure them for possible expansion while stored at uniform temperature and moisture over a period of at least 3 months and preferably 6 months. This method is recommended to detect only alkali-silica reactions, because carbonate aggregates of substantial reaction potential give very little indication of it during this test.

ASTM C342, Test for Potential Volume Change of Cement-Aggregate Combinations, describes a similar test to determine volume change of reactive aggregates in mortar specimens exposed to wide variations in temperature and moisture over a period of 1 year. This test applies particularly to certain cement-aggregate combinations found in parts of the central United States.

ASTM C586, Potential Alkali Reactivity of Carbonate Rocks for Concrete Aggregates (Rock Cylinder Method), indicates the potential alkali-carbonate reactivity between cement and limestone or dolomite. Small cylinders cut from aggregate are immersed in a sodium hydroxide solution, and change in length is determined over a period of about a year. The method is meant for research or screening of a possible source rather than to check conformance to specifications.

Miscellaneous Properties

Toughness, which means resistance to abrasion and impact, is indicated either by the Deval test described in ASTM D2, Abrasion of Rock by Use of the Deval Machine, and ASTM D289, Abrasion of Coarse Aggregate by Use of the Deval Machine; or by the Los Angeles abrasion test described in ASTM C131, Resistance to Abrasion of Small Size Coarse Aggregate by Use of the Los Angeles Machine, and ASTM C535, Resistance to Abrasion of Large Size Coarse Aggregate by Use of the Los Angeles Machine. ASTM D2 describes the Deval test for crushed rock, and ASTM D289 describes the Deval test for other aggregates. The two Los Angeles tests are for any coarse aggregate—one for smaller and one for larger sizes. The Los Angeles tests are of more recent origin. They can be run in less time and provide a greater range of results so that differences in toughness are more apparent.

Toughness is an important quality for aggregate subjected to mixing in a portland cement concrete mixer or an asphalt concrete pugmill (actions similar to that of the Deval and Los Angeles tests); to compaction with heavyweight or vibratory compaction equipment as roadbeds and asphalt pavement are; or to steel-wheeled or hard-rubber-tired traffic as some industrial floors are. In addition, the particles in gravel roads and asphalt concrete roads rub against each other throughout the lifetime of the road each time traffic passes by. All of these processes can break and abrade particles, thus changing the design gradation and opening holes in the surface.

In the Deval test, aggregate is rotated 10,000 times in a cylinder with six steel spheres for ASTM D289 and no spheres for ASTM D2. Coarse aggregate of various sizes may be tested but none smaller than the No. 4 sieve. Material broken fine enough to pass the No. 12 sieve is expressed as a percentage of the total sample weight to indicate the susceptibility to abrasion and breakage.

In the Los Angeles test (Figure 2–27), aggregate of the type to be used is combined with 12 steel spheres or fewer for smaller aggregates and rotated in a cylinder 500 times for small-size coarse aggregate and 1000 times for large-size coarse aggregate. A shelf inside the cylinder carries the aggregate and steel balls to a point near the top where they fall once each revolution. This fall plus the addition of more spheres for larger aggregates causes greater abrasion and breakage than in the Deval tests. Material broken fine enough to pass the No. 12 sieve is expressed as a percentage of the total sample weight to indicate the susceptibility to abrasion and breakage.

If the aggregate particles are all of equal toughness, the loss of weight in fine particles increases in direct proportion to the number of revolutions. If there are some very weak

FIGURE 2–27. Los Angeles Abrasion Machine. *(Courtesy The Asphalt Institute)*

particles, they will be completely crushed in the early part of the test, while the tougher particles will continue to lose weight at a constant rate throughout the test. In this case there will be a rapid loss of weight at first, tapering off to a uniform rate of loss near the end.

A very tough aggregate with a small but significant percentage of very weak particles is not as valuable as an aggregate that is consistently moderately tough. Yet each type could produce the same results at the end of a test. A method of checking on the uniformity of toughness throughout the sample consists of comparing the weight of material passing the No. 12 sieve after one-fifth of the revolutions with the weight passing after all the revolutions. The ratio should be approximately one-fifth if the sample is of uniform toughness, and will be higher if the sample is not uniform.

Soundness of aggregates means resistance to disintegration under weathering including alternate heating and cooling, wetting and drying, and freezing and thawing. Expansion and contraction strains caused by temperature changes impose a stress on aggregate (as well as on any other material) which may eventually cause breaking. Chemical changes that slowly disintegrate some types of aggregate are brought about by atmospheric moisture which contains dissolved gases. Drying and rewetting renews the chemical attack, which is generally the dissolving of a constituent of the aggregate. By far the most destructive effect of weathering is caused by freezing and thawing. Pores in the particles become filled with water, which freezes and expands within

the pores, exerting great pressure that tends to break the particle open at the pores.

ASTM C88, Soundness of Aggregates by Use of Sodium Sulfate or Magnesium Sulfate, provides a method for measuring soundness by immersing aggregate in a sodium sulfate or magnesium sulfate solution and removing and oven-drying it. Each cycle requires a day's time. This procedure causes an effect similar to weathering in that particles are broken away from the aggregate, but at an accelerated rate. It is believed that salt crystals accumulating in the aggregate pores exert an expanding pressure similar to that caused by the formation of ice. After the specified number of cycles, which varies with the intensity of weathering that must be resisted, the weight in material that passes a sieve is determined as an indication of the soundness of the aggregate. The sieve is slightly smaller than the size that retained all the aggregate originally for coarse aggregate and the same size as the one that retained all the aggregate for fine aggregate.

The procedure is designed to test resistance to the freeze–thaw cycle because this is the most destructive type of weathering, and aggregate with high resistance to freezing and thawing can resist any weather. An alternative test may be made by freezing and thawing the wet aggregate, but many cycles and a long period of time are necessary if the freezing and thawing during the useful life of the aggregate are to be simulated.

Hydrophilic aggregate is that aggregate which does not maintain adhesion to asphalt when it becomes wet. The word *hydrophilic* means "loves water." The implication is that hydrophilic aggregate prefers water to asphalt. Some silicious aggregates, for example, quartzite, are hydrophilic and therefore cannot be used satisfactorily with asphalt cement without special preparation of the asphalt cement. ASTM D1664, Coating and Stripping of Bitumen-Aggregate Mixtures, provides a method for testing aggregates for adhesion to asphalt. Aggregate is completely coated with the asphalt and submerged in water for 16–18 hours. The amount of asphalt coating that strips away from the aggregate is determined by a visual estimate of whether the aggregate surface left coated is more or less than 95 percent of the total surface.

Sampling

Aggregate tests and inspection must be performed on representative samples. Ideally, a representative sample is a small quantity with exactly the same characteristics as the entire quantity. Expressed practically, a representative sample closely reproduces those characteristics of the entire mass that is to be tested. Methods of taking samples must avoid *segregation* which is any separation of particles on the basis of some property. The most common segregation is by particle size, with smaller particles tending to become separated from larger ones. ASTM D75, Sampling Stone, Slag, Gravel, Sand and Stone Block for Use as Highway Materials, provides methods for proper sampling of aggregate.

Samples are required for the following:

1. Preliminary investigation of a possible source of supply, whether a rock formation, an aggregate deposit, or an industrial by-product. The supplier makes this investigation before investing the money to extract and process aggregate.

2. Acceptance or rejection of a source of supply by the buyer. This is a preliminary determination. An inspection and tests are made for this purpose by a prospective buyer who intends to buy large quantities for one project or a series of projects. An example is a state public works department which approves or disapproves gravel or sand pits for state projects for the coming year or other period of time.

3. Acceptance or rejection by the buyer of specified material from the supplier. Inspection and tests are performed as a final check for conformance to the agreement at the time of delivery.

4. Control of removal and processing operations. The supplier assures himself that his product remains of consistent quality by testing it.

A natural deposit is investigated by making test holes and examining or testing their entire contents. A layered deposit with somewhat different characteristics in each layer requires a sample cutting through a large number of layers. Aggregate sources frequently include pockets or areas of nontypical aggregate. Separate samples may be needed from each of these pockets or areas whenever they are observed (Figure 2–28).

A representative sample from a stockpile that may be segregated requires one large sample made up of samples from top, middle, and bottom of the pile. A representative sample from a bin should be taken in several increments while aggregate is being discharged by intercepting the entire cross section of the stream of particles each time but not the very first or very last particles being discharged.

The samples collected in any of the ways noted are generally combined in appropriate proportions to make one large sample representing the entire quantity of aggregate. That is, the size of each separate sample has the same relationship to the size of the combined sample as the quantity of aggregate represented by each separate sample has to the entire quantity of aggregate. When variations in characteristics or amount of

FIGURE 2–28. Natural deposit of mineral aggregates. *(Courtesy Spectra Environmental Group Inc.)*

segregation are of importance, each uncombined sample is inspected and tested separately.

The combined samples are often too large to use and must be reduced in size for handling and testing without changing characteristics that are to be investigated. Sample size is reduced by quartering or by dividing in a sampler splitter. Either method provides a representative sample. The two methods are illustrated in Figure 2–29.

SPECIAL AGGREGATES

Lightweight aggregates are those that have a unit weight of no more than 70 lb/cu ft (1120 kg/m^3) for fine aggregate, 55 lb/cu ft (880 kg/m^3) for coarse aggregate, and 65 lb/cu ft (1040 kg/m^3) for combined fine and coarse aggregate. These weight limitations and other specifications for lightweight aggregate for structural use are found in ASTM C330,

Lightweight Aggregates for Structural Concrete, and ASTM C331, Lightweight Aggregates for Concrete Masonry Units. The purpose of using lightweight aggregate in concrete structures is usually to reduce the weight of upper parts of a structure so that the lower supporting parts (foundations, walls, columns, and beams) may be smaller and therefore cost less. Lightweight aggregates are also used in insulating concrete. Transportation costs are less for lightweight aggregates and for lightweight concrete or masonry products than for their traditional, heavier counterparts.

Three types of lightweight aggregate are used for concrete in which strength is of major importance. Volcanic rock such as pumice, scoria, or tuff, all of which contain numerous air bubbles, and man-made particles prepared by expanding blast-furnace slag, clay, diatomite, fly ash, perlite, shale, slate, or vermiculite, are used in portland cement concrete structural members or for concrete masonry units.

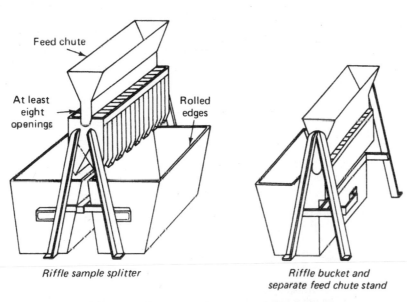

Riffle sample splitter

Riffle bucket and separate feed chute stand

(a) Large riffle samplers for coarse aggregate

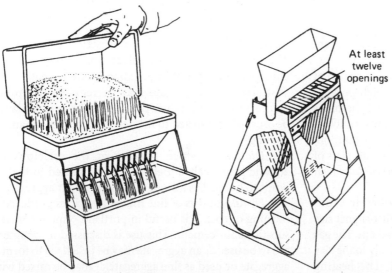

Note – May be constructed as either closed or open type. Closed type is preferred.

(b) Small riffle sampler for fine aggregate

FIGURE 2–29. Reducing sample size. *(Copyright ASTM INTERNATIONAL. Reprinted with permission.)*

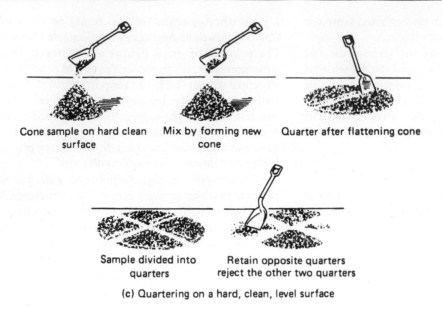

Cone sample on hard clean surface

Mix by forming new cone

Quarter after flattening cone

Sample divided into quarters

Retain opposite quarters reject the other two quarters

(c) Quartering on a hard, clean, level surface

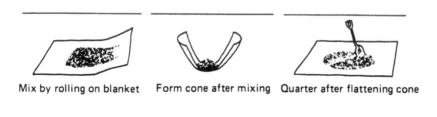

Mix by rolling on blanket

Form cone after mixing

Quarter after flattening cone

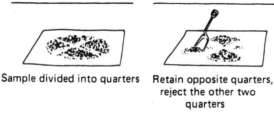

Sample divided into quarters

Retain opposite quarters, reject the other two quarters

(d) Quartering on a canvas blanket

FIGURE 2-29. Continued

In addition, cinders from the combustion of coal or coke are used for masonry units only.

As slag flows from a blast furnace in a molten stream at temperatures of 1400 to 1600°C, it is chilled with high-pressure water spray forced into the molten mass where it becomes steam as it cools the slag. The expanding steam causes bubbles so that the slag is frothy by the time it cools. The operation, which is completed in a few minutes, results in a hard mass of lightweight material called *expanded slag* or *foamed slag* which must be crushed into aggregate size.

Certain slags, shales, and slates when heated to about 2000°F expand to as much as seven times their original size because of the expansion of gas formed within them. The gas may be formed from minerals occurring naturally in the clay or rock or from a chemical that is added. The heating process, which usually takes place in a rotary kiln, must be rapid so that the gas expands with explosive force sufficient

to expand the particle and create a series of discrete air cells. The expansion must take place when the particle is soft enough from the heat to be expanded by the gas rather than shattered. The aggregate may be reduced to proper size before heating to produce particles of the desired size or may be crushed and sorted to the desired size after expansion and cooling.

Diatomite, which consists of the skeletons of tiny aquatic plants called diatoms, can be heated to the melting point to be used as a cinderlike, lightweight aggregate.

Fly ash consists of fine mineral particles produced by the burning of coal. It is useful in portland cement concrete as a substitute for cement. This use is discussed in Chapter 4. Fly ash to be used as an aggregate may be pelletized to form coarse aggregate or used as fine aggregate. It may be mixed with coal mine wastes for aggregate manufacture. In fact, the production of aggregates made of fly ash in combination with other materials for road bases, railroad ballast, and asphalt or portland cement concrete occupies many large industries.

Perlite is volcanic glass in spherical particles of concentric layers. It contains water which, if heated rapidly enough, becomes steam with enough force to shatter the spheres into particles and expand the particles.

Vermiculite includes a variety of water-bearing minerals derived from mica. These expand perpendicularly to the layers when steam is formed by rapid heating.

Cinders used as aggregates are fused into lumps by combustion of coal or coke and are not the softer ashes formed by lower-temperature combustion. Cinders contain some unburned material which is undesirable. Sulfur compounds found in cinders corrode steel, and cinders or concrete containing cinders should not be placed in contact with steel.

Manufactured sands are fine aggregates produced by the crushing and screening of quality rock. They are used as fine aggregates in areas where the natural sands are of poor quality or scarce.

Lightweight aggregate is sometimes incorporated into concrete primarily for heat insulation. Any of the aggregates used for lightweight concrete designed primarily for strength except cinders may also be used for insulating concrete. However, aggregates prepared by expanding natural minerals such as perlite and vermiculite are the lightest and generally the best insulators of all these. ASTM C332, Lightweight Aggregates for Insulating Concrete, specifies a maximum weight of 12 lb/cu ft (dry loose weight) for perlite and 10 lb/cu ft for vermiculite.

Lightweight aggregate may have a tendency to stain concrete surfaces because of iron compounds that are washed out by rain. ASTM C641, Staining Materials in Lightweight Concrete Aggregates, describes procedures for testing this tendency.

Heavy aggregates are those with higher specific gravities than those of aggregates in general use, although there is no definite line separating them from ordinary aggregates. They are used primarily to make heavy concrete. Heavy concrete is needed in special cases to resist the force of flowing water or to counterbalance a large weight, on a bascule bridge, for instance. Heavy aggregate is also used in nuclear radiation–shielding concrete where greater density provides greater shielding. Natural minerals used are iron minerals with specific gravities as high as 4 to 5.5 and barium minerals with specific gravities up to 4.5. Steel punchings, iron shot, and a by-product of the production of phosphorus called ferrophosphorus (SG of 5.7 to 6.5) are used as heavy aggregates.

Natural heavy mineral aggregates and ferrophosphorus are described in ASTM C638, Constituents of Aggregates for Radiation-Shielding Concrete, and general requirements are specified in ASTM C637, Aggregates for Radiation-Shielding Concrete.

Radiation shielding is also accomplished by the inclusion of natural boron minerals, or boron minerals heated to partial fusion, as aggregates in the concrete shield. These are also described in ASTM C638.

Review Questions

1. How is bedrock reduced to aggregate particles in nature?

2. How would you differentiate between a gravel deposit of glacial till and one of glacial outwash?

3. A 2000-lb wheel load is to be supported by aggregate over soil that can withstand a pressure of 1000 lb/sq ft. What depth of aggregate is needed if $\theta = 40°$?

4. A pipe is to be installed under the ground on a bed of aggregate 6 in. thick. The entire weight on the aggregate from pipe and soil above the pipe is 1800 lb/lineal foot of pipe. What is the pressure on the soil if $\theta = 45°$? Assume the pipe load on the aggregate acts on a line.

5. How do all stabilizing techniques increase the strength of aggregate?

6. Why should a filter be made of uniformly graded aggregate? Explain.

7. Using the following data, determine the percentage retained, cumulative percentage retained, and percentage passing for each sieve. Plot the gradation curve. Determine the effective size and uniformity coefficient if appropriate. Determine the fineness modulus and check ASTM C33 gradation requirements.

a.

Sieve Size (in.)	Weight Retained (g)
3	736
2	984
$1\frac{1}{2}$	1642
$\frac{3}{4}$	1030
$\frac{3}{8}$	625
Pan	96

b.

Sieve Size	Weight Retained (g)
No. 4	59.5
No. 8	86.5
No. 16	138.0
No. 30	127.8
No. 50	97.0
No. 100	66.8
Pan	6.3

c.

Sieve Size (in.)	Weight Retained (g)
3	1.62
2	2.17
$1\frac{1}{2}$	3.62
$\frac{3}{4}$	2.27
$\frac{3}{8}$	1.38
Pan	0.21

d.

Sieve Size	Weight Retained (g)
No. 4	10.2
No. 8	81.6
No. 16	92.3
No. 30	122.8
No. 50	116.2
No. 100	46.3
Pan	19.1

8. How many cubic yards of aggregate must be ordered for a road base 16 in. thick and 2 miles long with a top width of 30 ft, if the side slopes are one on one, or 45°?

9. How many tons of coarse aggregate will be required to fill a trench 5 ft deep, 3 ft wide, and 300 ft long? The coarse aggregate unit weight equals 93.6 lb per cu ft.

10. A sample of coarse aggregate weighs 1072 g when oven-dried, 1091 g when saturated surface-dried, and 667.6 g when submerged. Calculate the bulk specific gravity, apparent specific gravity, and absorption.

11. A sample of lightweight coarse aggregate weighs 532.3 g when oven-dried, 615.7 g when saturated surface-dried, and 191.3 g when submerged. Calculate the bulk specific gravity, apparent specific gravity, and percent absorption.

12. A sample of fine aggregate weighs 501.2 g when SSD and 491.6 g when OD. The flask weighs 540.6 g when filled with water and 843.1 g when filled with the aggregate sample and water. Calculate the bulk specific gravity, apparent specific gravity, and percent absorption.

13. A sample of fine aggregate weighs 544 g when SSD and 530 g when OD. The flask weighs 654 g when filled with water and 985 g when filled with fine aggregate and water. Calculate the bulk specific gravity, apparent specific gravity, and absorption.

14. Calculate the solid volume and percent of voids of an aggregate that weighs 111.2 pcf and has a specific gravity of 2.62.

15. What is the difference in detrimental effects of clay lumps and clay particles in aggregate to be used for portland cement concrete?

16. Why is a comparison made between the percent of the sample passing the No. 12 sieve after one-fifth of the Los Angeles Abrasion test and the percent passing at the completion of the test? Explain.

17. What are the four situations in which aggregate must be sampled and tested?

18. Describe the structure of lightweight aggregate.

19. Research local Department of Transportation coarse and fine aggregate specifications for concrete. Compare them to the ASTM C33 requirements located in the Appendix.

20. Calculate the solid volume and percent of voids of an aggregate that has a density of 1581 kg/m^3 and a specific gravity of 2.59.

21. Calculate the metric tons of gravel required to fill an excavated area 70 m by 50 m by 2 m deep if the gravel has a density of 1890 kg/m^3.

22. Calculate the compacted cubic yards per hour if a compactor with a 5-ft drum width travels at 3 miles per hour over a 10-in. lift of aggregate base course with 4 passes required to meet density specifications under average operating conditions.

23. Calculate in centimeters and in inches the approximate capillary rise of water in a material with an effective size (D_{10}) of 0.09 mm.

CHAPTER THREE

ASPHALT

Bituminous materials are important construction materials. They are strong cements, durable, highly waterproof, and readily adhesive. Bituminous materials are also highly resistant to the action of most acids, alkalis, and salts. They can be found on all types of construction projects, from buildings to highway and heavy construction. They are used in roofing systems, sealants and coatings, and in pavements. Asphalt combined with 95 to 97 percent aggregate constitutes asphalt mixtures, which is the material most commonly used in constructing the nation's roadways.

HISTORY

The use of asphalt by humans can be traced back to approximately 6000 B.C. Asphalts were used as cements to hold stonework together in boat building and as waterproofing in pools and baths. Some asphalt was mixed with sand and used to pave streets and palace floors.

The Egyptians made use of asphalt in the mummification process and as a building material. The Greeks and Romans not only used asphalt as a building material but also used burning asphalt as a military weapon.

The asphalt used by these ancient civilizations was *natural asphalt* formed when crude petroleum oils rose to the earth's surface and formed pools (Figure 3–1). The action

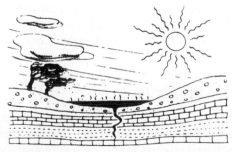

FIGURE 3–1. Formation of natural asphalt. *(Courtesy of Caterpillar Inc.)*

of the sun and wind drove off the lighter oils and gases, leaving a heavy residue. The residue was asphalt with impurities such as water and soil present. Using crude distillation processes, cementing and waterproofing materials were obtained from this residue.

Bituminous Materials

The American Society for Testing and Materials (ASTM) defines bitumens, asphalt, and tar as follows:

Bitumens: mixtures of hydrocarbons of natural or pyrogenous origin or combinations of both, frequently accompanied by their nonmetallic derivatives, which may be gaseous, liquid, semisolid, or solid, and which are completely soluble in carbon disulfide.

Asphalt: a dark brown to black cementitious material, solid or semisolid in consistency, in which the predominating constituents are bitumens which occur in nature as such or are obtained as residue in refining petroleum.

Tar: brown or black bituminous material, liquid or semisolid in consistency, in which the predominating constituents are bitumens obtained as condensates in the destructive distillation of coal, petroleum, oil shale, wood, or other organic materials, and which yields substantial quantities of pitch when distilled.

Natural Asphalt

Many pools of natural asphalt still exist; the largest are the Bermudez deposit in Venezuela and the asphalt lake on the island of Trinidad. Sir Walter Raleigh obtained Trinidad lake asphalt to caulk his ships during a voyage to the New World. Until the development of distillation processes to produce asphalt from crude petroleum, the deposits were a major source of asphalt. In the United States, the LaBrea "tar" pits in Los Angeles, California, are of interest because of the fossil remains and skeletons of prehistoric animals found in the source of naturally occurring asphalt.

Tar

Asphalt should not be confused with coal tar because asphalt is readily soluble in most petroleum products and tar is resistant to petroleum-based solvents. Asphalt is composed almost entirely of bitumens, whereas tar has low bitumen content.

Tar is generally produced as a by-product of coke or of heating coal. As coal is heated, the gases generated are refined to produce road tars, roofing tars, waterproofing pitches, creosote oils, and other tar chemicals. Coal tars have high viscosities and good adhesive properties. As sealers, emulsified in water-based products, coal tars are highly effective in resisting the degradation to pavements from the contamination caused by petroleum-based (oil or gasoline) products.

Asphalt Binder

Even though natural asphalt does occur, the majority of asphalt used in construction today is distilled from petroleum crude oil. Depending on its use and crude oil source, the asphalt can be produced in a variety of types and grades ranging from hard, brittle materials to thin liquids, to be used in a variety of applications. Approximately 80 percent of the asphalt produced is used in paving and related industries, while the remaining 20 percent is used in the manufacture of roofing materials and systems, and miscellaneous areas such as metal coatings and waterproofing.

Petroleum crude oils are generally classified on the basis of their crude oil content.

1. Asphaltic base crude (almost entirely asphalt)
2. Paraffin base crude (contains paraffin but no asphalt)
3. Mixed base crude (contains both paraffin and asphalt)

The amount of asphalt obtained from a crude oil is based on its America Petroleum Institute (API) gravity; the higher the gravity, the lower the asphalt content, and the lower the API crude gravity, the higher the asphalt content. Figure 3–3 illustrates the production of asphalt, and other petroleum products, from a barrel of crude oil.

Asphalt "binders," sometimes called "cements," are the bituminous glue that has been used for many years. In the modern world, asphalt is used extensively for paving and roofing applications. Its ability to change properties with temperature is what makes asphalt a good construction material.

At cold temperatures asphalt is a very hard brittle material. As the temperature of the asphalt is increased, it softens, and at its higher storage temperature, approximately 300°F, it becomes a liquid. This is shown schematically in Figure 3–2. These are the properties that make asphalt a good construction material. At high temperatures it is a liquid that flows well to coat aggregate and allows it to be workable during placement as a paving material. At normal climatic temperatures it is a strong durable material that has strong binding characteristics.

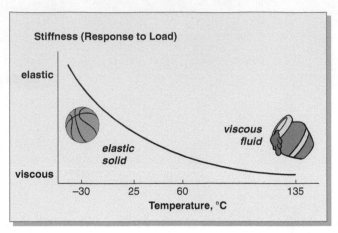

FIGURE 3–2. Relation between temperature and stiffness of asphalt binder. *(Courtesy John D'Angelo Ph.D, P.E., D'Angelo Consulting, LLC)*

Refining of Asphalt Binders

Paving grade asphalt binders are typically produced from petroleum oil. The asphalt binder is produced through a distillation process at a petroleum refinery. Not all petroleum or crude oil is suitable for the production of asphalt binder. Crude oil designated as heavy or sour, such as those originating from Venezuela, produces the highest percentage of asphalt binder. Light or sweet crude oil, such as those originating from the Middle East, will produce little to no asphalt binder, but a higher percentage of light gas oils or gasoline.

In a typical refining operation, the crude oil is processed through a distillation tower. The primary unit in the refining process is usually an atmospheric distillation tower. The crude oil is heated to between 650 and 700°F and fed into the atmospheric tower. Because of their various molecular weights, materials will rise in the tower based on their boiling points and are pulled off at various levels to produce gasoline, kerosene, and fuel oils. The residual or heaviest portion is the asphalt binder. This material may be further refined by the use of a vacuum distillation tower. By refining at pressures below atmospheric pressure, additional lighter fractions can be removed from the crude oil. A diagram of the process is shown in Figure 3–3.

The type or grade of the asphalt binder is governed by the source of the crude oil used and any additional processing done following distillation. The production of asphalt binders that meet specifications for areas with warmer climates and/or high traffic demand involves additional processing such as the addition of rubber or polymers to improve properties.

Superpave Binder Tests

For many years, asphalt was characterized simply by penetration and viscosity. The penetration test is an empirical measure of the relative hardness of asphalt cement. Empirical tests yield information about the relative behavior of

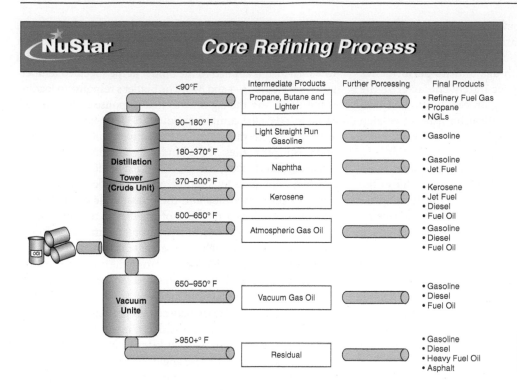

FIGURE 3–3. Petroleum refining process and output. *(Courtesy of NuStar Energy, LP)*

materials based on experience; however, they do not provide any information about fundamental properties, and in the case of asphalt, its elasticity or strength.

Penetration was measured as the actual depth of penetration (measured in mm) a loaded pin of 100 g would make in a small sample of asphalt binder at a stated temperature, for a specified time period, typically 5 seconds. The penetration test was popular because it yielded information about the "stiffness" of asphalt, with predictable outcome in terms of performance. Before an actual "test procedure" was adopted, in the early years, asphalt binder was evaluated by forming a small ball and "chewing" the sample to determine the material's hardness.

The measurement of viscosity of the asphalt binder, or the binder's resistance to flow, grew in prevalence in the 1960s. However, while considered "fundamental" in that the test measured a behavior of the asphalt binder, the test was not reliable over a range of temperatures that asphalt binders are in service; neither did it distinguish a fundamental property of asphalt binder—elasticity.

As asphalt refining processes changed, and as a result of widespread Hot Mix Asphalt (HMA) pavement failures across the country, which were largely attributed to increased traffic volumes, and in particular heavy truck traffic, the Strategic Highway Research Program (SHRP) produced testing protocols to characterize asphalt binders. The Superpave binder tests developed under SHRP are just one part of the overall Superpave (Superior Performing Asphalt Pavements) design practice, which was implemented in the mid-1990s and is the predominant HMA pavement design system in use today.

The Superpave binder (asphalt) is selected based on the climate of the pavement, the expected traffic, and the location in the pavement structure. The binders are evaluated at the expected highest and lowest pavement temperatures. These pavement temperatures are used to determine the Superpave binder grade. For example, a PG 64-22 represents PG for performance grade; and 64 is the highest pavement temperature in degrees Celsius (147°F) and −22 the lowest pavement temperature in degrees Celsius (−8°F) at which the binder can be expected to perform without failure. Generally, if the absolute values of the highest and lowest temperature ends of the PG specified binder are added and exceed 92, the asphalt binder will need to be modified (typically with polymers or rubber). This is also done to enhance the desired properties of a neat (unmodified) asphalt binder. Most often, asphalt binders are modified to achieve additional "stiffness" to resist permanent deformation (rutting) at higher pavement temperatures subjected to heavy traffic.

The Superpave binder specification is based on the "rheological" (the study of deformation and flow) properties of the asphalt binder, and is measured over the wide range of temperatures and aging conditions that the asphalt will be exposed to during its service life. Various pieces of equipment are used to measure stress–strain relationships in the binder at the specified test temperatures. This equipment includes the dynamic shear rheometer (DSR), bending beam rheometer (BBR), and the direct tension tester (DTT). Measuring the binders' rheological properties over a wide range of temperatures, loading conditions, and aging conditions allows performance relationships to be established between the test results and the pavement. The asphalt binder is tested using new equipment and test procedures as described in the American Association of State Highway and Transportation Officials (AASHTO) standards.

Over the design life of a pavement, the asphalt binder changes with time. This is primarily through oxidative aging. As the binder ages, the rheolgical properties change. The binder will become stiffer with age and its performance characteristics will change. To accurately characterize the asphalt binder, the pavement aging process has to be simulated. The Superpave binder specification uses a two-step process to simulate aging of the binder—the rolling thin film oven test (RTFOT) and the pressure aging vessel (PAV).

The RTFOT is used to simulate the aging that takes place during production and up to the first year of life of the pavement. The binder is poured into cylindrical glass bottles, placed horizontally in a convection oven, and rotated at 163°C for 85 minutes. This process creates a thin film of asphalt on the inside of the bottles which ages due to heat and the injection of air into the bottle. The RTFOT is used as a standardized process for the aging of the binder, not for duplicating the actual aging that will take place in the field. A single laboratory test cannot duplicate the complicated process that actually takes place during HMA production where the asphalt undergoes an extreme aging due to high temperatures. The "production" aging process is affected by aggregate type, mix plant type, rate of production, haul time, lay-down process, and density in the pavement. The intent of the RTFOT is only to establish a standardized process that can be used in a purchase specification that will simulate early aging of the binder during production.

The next step in the simulation of the aging process for the binder is the PAV. The RTFO aged binder is placed in shallow pans approximately 3 mm thick. The pans are placed in a pressure vessel at 2.1 MPa at 100°C for 20 hours. During the PAV process, oxygen is driven into the binder, simulating long-term aging of the binder in the pavement. Again as in the RTFO, the PAV only simulates aging in a standardized manner for a purchase specification and does not purport to duplicate the actual aging that takes place in the pavement. The actual aging is affected by pavement density, film thickness, and climate.

The DSR is used to measure the high temperature and intermediate temperature properties of the asphalt binder. An asphalt binder sample is placed between metal plates and a torsional load is applied to the binder at the specified temperature, see Figures 3–4 and 3–5. The binder's response to loading is measured and the lag time to that response is determined. From this data the complex modulus G* ("G star") and phase angle δ are determined and are used to calculate the viscous and elastic properties of the binder. This is measured on the unaged, RTFOT-aged, and the PAV-aged binder.

The Superpave tests on the original binder measures the complex shear modulus and the phase angle of the binder. The phase angle is a measure of the elastic response of the binder and a good indicator of polymer modification as shown in Figure 3–5. The phase angle is a measure of how much out of phase the load is from the response on the binder. If a material is elastic when loaded, there is an immediate response or resistance to movement. However, for viscous materials there is a time lag between the loading and response. Neat asphalt binders when tested at temperatures where the modulus is approximately 1.0 kPa are almost completely viscous and will have a phase angle of almost 90 degrees.

Research has linked the binder phase angle to the molecular weight of the binder. Polymer modification changes the properties of the binder and this can be seen in the phase angle. Polymer modification will make the binder much more elastic in its response to loading. It is typical to measure phase angles of 77 degrees or less at the temperatures where the modulus of the binder is 1.0 kPa for polymer modified binders. The phase angle is a direct measure of the elastic response of the binder.

There has been extensive study indicating that the current AASHTO M-320 Table 1 specification did not address the true rutting potential of polymer-modified asphalt binders. To address this issue, the Multi-Stress Creep and Recovery Test (MSCR) was developed as a replacement for the existing AASHTO M320 high temperature binder test run on the RTFOT.

FIGURE 3–4. Asphalt sample set on the dynamic shear rheometer. *(Photo courtesy of John D'Angelo, Ph.D, P.E., D'Angelo Consulting, LLC)*

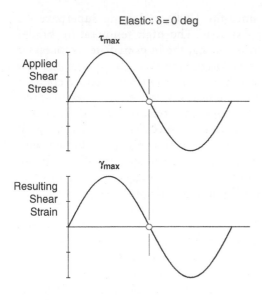

Elastic: δ = 0 deg

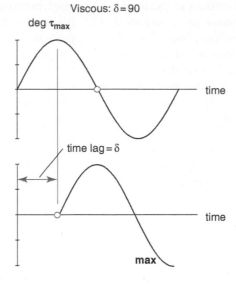

Viscous: δ = 90 deg

FIGURE 3–5. Schematic showing phase angle and its relation to elastic and viscous behavior. *(Courtesy of John D'Angelo, Ph.D, P.E., D'Angelo Consulting, LLC)*

Multiple binders, both neat and polymer modified, were evaluated in the development of a new binder test to determine high temperature rutting property for binders. Equipment for testing of the binders was focused on the existing DSR. This equipment has been widely accepted by highway agencies for use in determining the rheological properties of binders in specifications. The new MSCR test has potential to differentiate the effects of various polymer modification processes on asphalt binders and is anticipated by owners (state highway agencies), producers, and contractors in the industry.

The MSCR test applies 1-second creep loading with a 9-second recovery over multiple stresses (0.1 and 3.2 kPa) at 10 cycles for each stress level. The test is started at the lower stress level and increased to the next stress level at the end of every 10 cycles, with no time lags between cycles. The average nonrecovered strain for the 10 creep and recovery cycles is then divided by the applied stress for those cycles yielding the nonrecoverable compliance J_{nr}.

$$J_{nr} = avg. \ \gamma_u \ / \ \tau$$

where γ_u = unrecovered strain from the end of the 9-second recovery portion of the creep and recovery test.

   τ = shear stress applied during the 1-second creep portion of the creep and recovery test.

The BBR is used to measure the low temperature stiffness properties of the binder. In the test, a beam of asphalt is subjected to three-point loading in a low temperature bath at 10°C above the expected low pavement temperature (see Figure 3–6). The deflection of the beam is measured with time. From this measurement the low temperature creep compliance and slope of the creep compliance curve are

FIGURE 3–6. Beam of asphalt binder loaded in the BBR rheometer load frame. *(Courtesy of John D'Angelo, Ph.D, P.E., D'Angelo Consulting, LLC)*

determined. The low temperature stiffness of the binder is determined from the creep compliance curve measured at 60 seconds loading time. The slope of the curve or "m value" at 60 seconds loading time is also included in the specification as an indication of the binder's ability to dissipate stress buildup during contraction of the pavement as it cools. The AASHTO M320 low temperature specification requires that the BBR-measured stiffness be 300 MPa and the m value be 0.30 at the test temperature.

The DTT measures the fracture strength and strain at failure of the binder. The binder is tested in a low temperature bath set at 10°C above the expected low pavement temperature (see Figure 3–7). A small binder sample is placed on pins and tested by pulling in direct tension at a constant strain rate of 1 mm per min. The testing is conducted in the temperature range where the binder fails in a brittle mode. The current DTT test is intended to evaluate if the binder will exhibit some ductile properties at the expected low pavement temperatures and therefore be able to dissipate energy and avoid cracking. The current AASHTO M320 specification requires the DTT only when the BBR stiffness is between 300 and 600 MPa and the m value is above 0.30. For these cases the DTT strain at failure must be greater than 1.0 percent for the binder to meet specification requirements.

These tests are combined to create a performance-related specification for the asphalt binder and have been implemented since the 1990s under the Superpave PG Binder Grading system. The high temperature binder properties address rutting, the intermediate temperature properties address fatigue, and durability and the low temperature properties address low temperature cracking in the pavement. The test procedures and specifications are based on the stress–strain relationships of the binder under loading. The tests and specification also attempt to predict the changing properties of asphalt binder with time (e.g., stiffness and oxidation). This allows for an asphalt binder to be specified to expected performance properties that will minimize the potential for pavement failure.

Additional tests, not part of Superpave characterization but still in use today, are the flash point test and ductility test. The flash point test is a safety test. Because asphalt binders must be heated to be used in construction, the flash point of asphalt binder tells the user the maximum temperature to which the material may be heated before an instantaneous flash will occur in the presence of an open flame. The flash point is usually well above the normal heating ranges of asphalt binders. Though usually not specified, the fire point of asphalt binder is the highest temperature at which the material will support combustion.

The Cleveland Open Cup (COC) flash point test (Figure 3–8) is usually used to determine the flash point of asphalt binder. A brass cup is filled with the proper sample

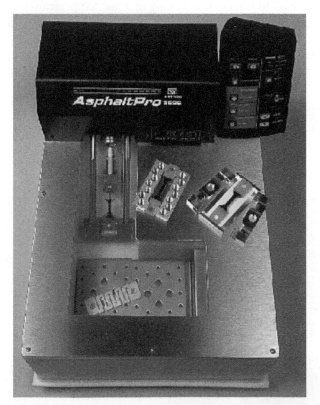

FIGURE 3–7. Direct tension test bath and load frame. (*Courtesy of John D'Angelo, Ph.D, P.E., D'Angelo Consulting, LLC*)

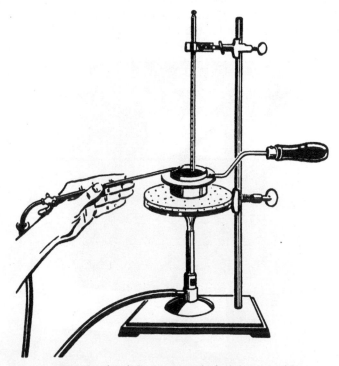

FIGURE 3–8. Cleveland Open-Cup Flash Point test. (*Courtesy The Asphalt Institute*)

amount of asphalt binder and heated at a specified rate of temperature increase. A small flame is passed over the surface of the asphalt cement being heated, and the temperature at which an instantaneous flash occurs is called the flash point.

Ductility is considered an important property of asphalt binders. Asphalt cements possessing ductility are more adhesive to aggregates than asphalt cements lacking ductility. However, some asphalt binders having a high degree of ductility are also more temperature susceptible.

That is, their consistency will change more in response to a temperature change.

Figure 3–9 is a typical "Certificate of Analysis" that is generated routinely to ensure asphalt binders meet all of the specifications of AASHTO M320. The certificate provides the user with the information (e.g., mixing, compaction temperatures, and flash point) they will need to handle the asphalt binder at the HMA plant and in densification of the HMA produced with a particular asphalt binder.

Avery Lane Terminal:Newington, NH
PG ASPHALT BINDER CERTIFICATION

Sample Source: Avery Lane   PG Classification: 64-28   Certification: ☑

Lot #:   64-28/09/25   PII Sample #: 500-916   Process Control Test: ☐

Sample Date: 11/18/2009   Tank #:   5

TEST	RESULT	UNIT/TEMP.	SPECIFICATION	METHOD
Specific Gravity:	1.0408	60°F (15.6°C)	Report Only	AASHTO T 228
Specific Gravity:	1.0347	77°F (25°C)	Report Only	AASHTO T 228
API Gravity:	4.45	API	Report Only	
LBS/GAL @ 60°F:	8.677	LBS	Report Only	ASTM TABLE 8
Smoke Point:		°C	Report Only	FM 5-519
Flash Point (C.O.C.):	273		Min. 230°C	AASHTO T 48
Rotational Viscosity:	0.474	Pa-s @ 135°C	Max. 3.0 Pa-s	AASHTO T 316
Rotational Viscosity:	0.131	Pa-s @ 165°C	Max. 3.0 Pa-s	AASHTO T 316
ORIGINAL BINDER				
DSR: G*/Sin Delta:	1.303	kPa @ 64°C	Min. 1.00 kPa	AASHTO T 315
RTFO RESIDUE				AASHTO T 240
Mass Loss:	-0.836	% @ 163°C	Max. 1.00%	
DSR: G*/Sin Delta:	3.776	kPa @ 64°C	Min. 2.2 kPa	AASHTO T 315
PAV RESIDUE				AASHTO R 28
DSR: G*Sin Delta:	3671	kPa @ 22°C	Max. 5000 kPa	AASHTO T 315
BBR Creep Stiffness				AASHTO T 313
Stiffness (S):	268	MPa @ -18C	Max. 300.0 MPa	
m-Value:	0.316	@ -18C	Min. 0.300	
Critical Cracking Temperature		°C	Report Only	CCT

Recommended laboratory mixing temp., Min/Max, °C: 155/161

Recommended laboratory compaction temp, Min/Max, °C: 144/149

Data obtained from manufacturer's certification or independent laboratory testing

For: Pike Industries, Inc.

Pete M Moore

Certifying Agent

Technician Peter Moore NETTCP #225   This material is modified using PPA. Do not use with any liquid anti-strip without consulting the supplier

FIGURE 3–9.

Asphalt Emulsions

With the increasing realization that the world supply of petroleum is limited and expensive, highway engineers are constantly looking at alternatives to reduce cost and extend natural resources. The use of asphalt emulsions, in addition to the use of recycled asphalt pavement (RAP) and warm mix asphalt (WMA), helps preserve natural resources and limit emissions.

Emulsified asphalts are asphalts that are suspended in water. They are produced by shearing the hot asphalt binder into minute globules and dispersing them in water that is treated with an emulsifying agent. The asphalt is called the discontinuous phase and the water the continuous phase. The asphalt emulsion is produced in a colloidal mill which applies shearing stress to the asphalt and water as it passes between a stationary plate and a rotating plate to ensure uniform interaction.

An inverted emulsion may be formed with asphalt as the continuous phase and the water in minute globule size as the discontinuous phase. This inverted emulsion is usually produced with asphalt binder that has been cut with a small amount of an MC-type (petroleum) solvent.

If the asphalt globule has a negative charge, the emulsion produced is classified as anionic. When the asphalt globule has a positive charge, the emulsion is classified as cationic. By varying the materials and manufacturing processes, three emulsion grades are produced in either the anionic or cationic state.

Because like charges repel, the asphalt globules are kept apart until the material comes in contact with aggregate particles and the charges are neutralized or the water evaporates. The process of the asphalt globules coming together in rapid and medium curing emulsions is called the break or set. The slow-setting emulsions depend primarily upon evaporation of water to set.

Specifications for emulsified asphalts are included in the ASTM standards. Asphalt emulsions, which have a variety of viscosities, asphalt cement bases, and setting properties, are available.

Saybolt Furol Viscosity Test. The consistency properties of anionic and cationic emulsions are measured by the Saybolt Furol viscosity test. As a matter of testing convenience and to achieve suitable testing accuracy, two testing temperatures are used (77 and 122°F), depending on the viscosity characteristics of the specific type and grade of asphalt emulsion. The test procedure is basically the same as that used to test asphalt cements. The unit of measure is the poise.

Distillation Test. The *distillation test* (Figure 3–10) is used to determine the relative proportions of asphalt cement and water in the asphalt emulsion. Some grades of emulsified asphalt also contain an oil distillate. The distillation test provides information on the amount of this material in the emulsion. Also, the distillation test provides an asphalt cement residue, on which additional tests (e.g., penetration, solubility, and ductility) may be made as previously described for asphalt cement.

The test procedure is substantially the same as that described for liquid asphalt. A 200-g sample of emulsion is distilled to 500°F. The principal difference in the emulsion distillation test is that the end point of distillation is 500°F rather than 680°F, and an iron or aluminum alloy still and ring burners are used instead of a glass flask and Bunsen burner. This equipment is designed to prevent trouble that may result from foaming of emulsified asphalt as it is being heated. The end point of distillation is carried to 500°F, and this temperature is held for 15 minutes in order to produce a smooth, homogeneous residue.

Settlement Test. The *settlement test* detects the tendency of asphalt globules to "settle out" during storage of emulsified asphalt. It provides the user with an element of protection against separation of asphalt and water in unstable emulsions that may be stored for a period of time.

A 500-ml sample is placed in each of two graduated cylinders, stoppered, and allowed to stand undisturbed for 5 days. Small samples are taken from the top and bottom parts of each cylinder. Each sample is placed in a beaker and

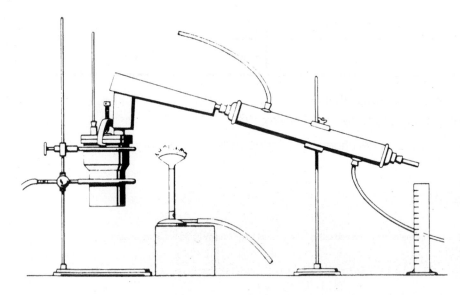

FIGURE 3–10. Distillation test for emulsified asphalts. *(Courtesy The Asphalt Institute)*

weighed. The samples are then heated until water evaporates; residues are then weighed. The weights obtained provide the basis for determining the difference, if any, between asphalt cement content in the upper and lower portions of the graduated cylinder, thus providing a measure of settlement.

Sieve Test.   The *sieve test* complements the settlement test and has a somewhat similar purpose. It is used to determine quantitatively the percentage of asphalt cement present in the form of pieces, strings, or relatively large globules. Such nondispersed particles of asphalt might clog equipment and would tend to provide nonuniform coatings of asphalt on aggregate particles. This nonuniformity might not be detected by the settlement test, which is of value in this regard only when there is a sufficient difference in the specific gravities of asphalt and water to allow settlement.

In the sieve test, 1000 g of asphalt emulsion is poured through a U.S. standard No. 20 sieve. For anionic emulsified asphalts, the sieve and retained asphalt are then rinsed with a mild sodium oleate solution. For cationic emulsified asphalts, rinsing is with distilled water. After rinsing, the sieve and asphalt are dried in an oven, and the relative amount of asphalt retained on the sieve is determined.

Demulsibility Test.   The *demulsibility test* is used only for rapid- and medium-setting grades of anionic asphalt emulsions. It indicates the relative rate at which colloidal asphalt globules coalesce (or break) when spread in thin films on soil or aggregate particles.

Calcium chloride coagulates, or flocculates, are the minute globules present in anionic emulsified asphalts. To make the test, a 100-g sample is thoroughly mixed with a calcium chloride solution. The mixture is then poured over a No. 14 sieve and washed. The degree of coalescence is determined from the amount of asphalt residue remaining on the sieve.

A high degree of demulsibility is required for the rapid-setting grade of anionic emulsified asphalt because it is expected to break almost immediately upon contact with the aggregate surface. Therefore, a very weak calcium chloride solution is used for the demulsibility test on these products. A somewhat more concentrated solution is used when testing medium-setting grades, because they are formulated to break more slowly.

Slow-setting grades are often used in mixes containing fine aggregates or in other applications where rapid coalescence of asphalt particles is undesirable. The cement mixing test is therefore used in lieu of the demulsibility test as a control for the setting rate of these products.

Cement Mixing Test.   The *cement mixing test* is performed by adding 100 ml of emulsion diluted to 55 percent residue with water to 50 g of high, early strength portland cement with stirring for thorough mixing. Additional water is stirred in. The mixture is then washed over a No. 14 sieve, and the percentage of coagulated material retained on the sieve is determined.

As noted, the cement mixing test is used instead of the demulsibility test for slow-setting grades of emulsified asphalt. It is specified for both the anionic and cationic types to assure products substantially immune from rapid coalescence of asphalt particles in contact with fine-grained soils or aggregates.

Coating Ability and Water Resistance Test.   The *coating ability and water resistance* test determines the ability of an emulsified asphalt to do the following:

1. Coat an aggregate thoroughly.
2. Withstand mixing action while remaining as a film on the aggregate.
3. Resist the washing action of water after mixing is completed.

The test is primarily intended to aid in identifying asphalt emulsions that are suitable for mixing with coarse-graded aggregate intended for job use. For specification purposes, the test is required only for cationic, medium-setting asphalt emulsions.

A 465-g air-dried sample of aggregate that is to be used on a project is mixed with 35 g of emulsified asphalt for 5 minutes. One-half of the mixture is removed from the pan and placed on absorbent paper, and the percentage of coated particles is determined.

FIGURE 3–11. Saybolt Furol Viscosity. *(Courtesy Suit-Kote Corporation)*

The remaining mixture in the pan is carefully washed with a gentle spray of tap water and drained until the water runs clear. This mixture is then placed on absorbent paper and the percentage of coated aggregate particles is determined.

This procedure is repeated for wet aggregate (9.3 ml of water mixed with the air-dried aggregate) before mixing with emulsified asphalt.

ASPHALT PAVEMENTS

The basic idea in building a road or parking area for all-weather use by vehicles is to prepare a suitable subgrade or foundation, provide necessary drainage, and construct a pavement that will do the following:

1. Have sufficient total thickness and internal strength to carry expected traffic loads.
2. Prevent the penetration or internal accumulation of moisture.
3. Have a top surface that is smooth and resistant to wear, distortion, skidding, and deterioration by weather and de-icing chemicals.

The subgrade ultimately carries all traffic loads. Therefore, the structural function of a pavement is to support a wheel load on the pavement surface and transfer and spread that load to the subgrade without overtaxing either the strength of the subgrade or the internal strength of the pavement itself.

Figure 3–12 shows wheel load W being transmitted to the pavement surface through the tire at an approximately uniform vertical pressure P_0. The pavement then spreads the wheel load to the subgrade so that the maximum pressure on the subgrade is only P_1. By proper selection of pavement materials and with adequate pavement thickness, P_1 will be small enough to be easily supported by the subgrade.

Asphalt pavement is a general term applied to any pavement that has a surface constructed with asphalt. Normally, it consists of a surface course (layer) of mineral aggregate coated and cemented with asphalt and one or more supporting courses, which may be of the following types:

1. Asphalt base, consisting of asphalt–aggregate mixtures
2. Crushed stone (rock), slag, or gravel
3. Portland cement concrete
4. Old brick or stone block pavements

Asphalt pavement structure consists of all courses above the prepared subgrade or foundation. The upper or top layer is the *asphalt wearing surface*. It may range from less than 1 in. to several inches in thickness, depending on several design factors.

Though a variety of bases and subbases may be used in asphalt pavement structures, more commonly they consist of compacted granular materials (such as crushed rock, slag, gravel, sand, or a combination of these) or stabilized soil. One of the main advantages of asphalt pavements is that a variety of materials may be used; thus, economy is achieved by using locally available materials.

Generally, it is preferable to treat the granular material used in bases. The most common treatment is to mix asphalt with the granular material, thus producing an asphalt base. A 1-in. thickness of asphalt base is about equal in load-carrying performance to a 2- or 3-in. thickness of granular base materials not treated with asphalt.

Untreated bases and subbases have been widely used in the past. However, as modern traffic increases in weight and volume, these bases show performance limitations. Consequently, it is now common practice to limit use of untreated bases to pavements designed for lower volumes of traffic.

When the entire pavement structure above the subgrade consists of asphalt mixtures, it is called a *full-depth asphalt pavement* or *perpetual pavement*. This is generally considered the most modern and dependable type of pavement for present-day traffic. A perpetual pavement typically consists of subgrade, asphalt base mixture, an intermediate course, and a surface course, totaling approximately 12 in. The pavement is perpetual because as the wearing or surface course ages and degrades, the surface is milled and replaced, while the lower courses remain intact.

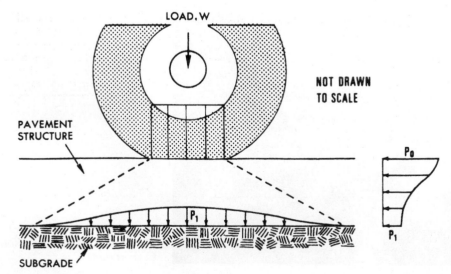

FIGURE 3–12. Spread of wheel load through pavement structure. *(Courtesy The Asphalt Institute)*

DETERMINING REQUIRED PAVEMENT THICKNESS

A significant advance in highway engineering is the realization and demonstration that structural design of asphalt pavement is similar to the problem of designing any other complex engineering structure. When asphalt pavement was first being introduced, determining the proper thickness was a matter of guesswork, rule of thumb, and opinion based on experience. Almost the same situation once prevailed in determining the dimensions of masonry arches and iron and steel structures. However, these early techniques have long since yielded to engineering analysis. Similarly, based on comprehensive analysis of vast volumes of accumulated data, the structural design of asphalt pavements has now been developed into a reliable engineering procedure. Research aimed at further refinements and a fully rational design procedure is continuing.

There is no standard thickness for a pavement. However, the Asphalt Institute has published general guidelines to be used for parking areas and driveways (Table 3–1). Required total thickness for roadways and airfields is

Table 3–1. Suggested pavement thickness for parking areas and driveways.

PAVEMENT THICKNESS FOR DRIVEWAYS FOR PASSENGER CARS

THICKNESS REQUIREMENTS IN INCHES

	FULL DEPTH ASPHALT CONCRETE		ASPHALT CONCRETE SURFACE		PLANT-MIX SURFACE USING LIQUID ASPHALT		ASPHALT SURFACE TREATMENT	
	Asphalt Concrete Surface	Asphalt Concrete Base[1]	Asphalt Concrete Surface	Crushed Rock Base[2]	Asphalt Plant-Mix Surface	Crushed Rock Base[2]	Asphalt Surface Treatment	Crushed Rock Base[2]
Gravelly or sandy soils, well drained	1	2–3	3	2	4.5*	2	1**	6–8
Average clay loam soils, not plastic	1	3–4	3	2–4	4.5*	2–4	1**	8–10
Soft clay soils, plastic when wet	1	4–5	3	4–6***	4.5*	4–6***	1**	10–12***

[1] Prime required on subgrade.
[2] Prime required on base.
* Must be spread and compacted in layers not exceeding 1½ inches in depth and the volatiles (petroleum solvents or water) allowed to evaporate before the next layer is placed. Also, a seal coat may be required as a final surfacing.
** Economical but relatively limited service life. Usually less than 1 inch thick.
*** Two inches of coarse sand or stone screenings recommended between subgrade and base as an insulation course.

PAVEMENT THICKNESS FOR PARKING AREAS FOR PASSENGER CARS

THICKNESS REQUIREMENTS IN INCHES

	FULL DEPTH ASPHALT CONCRETE		ASPHALT CONCRETE SURFACE		PLANT-MIX SURFACE USING LIQUID ASPHALT		ASPHALT SURFACE TREATMENT	
	Asphalt Concrete Surface	Asphalt Concrete Base[1]	Asphalt Concrete Surface	Crushed Rock Base[2]	Asphalt Plant-Mix Surface	Crushed Rock Base[2]	Asphalt Surface Treatment	Crushed Rock Base[2]
Gravelly or sandy soils, well drained	1	2–3	3	2	4.5*	2	1**	6–8
Average clay loam soils, not plastic	1	3–4	3	2–4	4.5*	2–4	1**	8–10
Soft clay soils, plastic when wet	1	4–5	3	4–6***	4.5*	4–6***	1**	10–12***

[1] Prime required on subgrade.
[2] Prime required on base.
* Must be spread and compacted in layers not exceeding 1½ inches in depth and the volatiles (petroleum solvents or water) allowed to evaporate before the next layer is placed. Also, a seal coat may be required as a final surfacing.
** Economical but relatively limited service life. Usually less than 1 inch thick.
*** Two inches of coarse sand or stone screenings recommended between subgrade and base as an insulation course.

Table 3–1. (continued)

PAVEMENT THICKNESS FOR PARKING AREAS FOR HEAVY TRUCKS							
THICKNESS REQUIREMENTS IN INCHES							
FULL DEPTH ASPHALT CONCRETE		ASPHALT CONCRETE SURFACE		SURFACE TREATMENT ON PENETRATION MACADAM		PLANT-MIX SURFACE USING LIQUID ASPHALT	
Asphalt Concrete Surface	Asphalt Concrete Base[1]	Asphalt Concrete Surface	Crushed Rock Base[2]	Asphalt Surface Treatment	Asphalt Penetration Macadam Base[1]	Asphalt Plant-Mix Surface	Crushed Rock Base[2]
Gravelly or sandy soils, well drained	1.5	3–5	4.5	0–4	1*	7–10	7**	2–4
Average clay loam soils, not plastic	1.5	5–6	4.5	4–6	1*	10–11.5	7**	4–6
Soft clay soils, plastic when wet	1.5	6–8	4.5	6–10***	1*	11.5–14.5	7**	6–10***

[1] Prime required on subgrade.

[2] Prime required on base.

* Usually less than 1 inch thick.

** Must be spread and compacted in layers not exceeding $1\frac{1}{2}$ inches in depth and the volatiles (petroleum solvents or water) allowed to evaporate before the next layer is placed. Also, seal coat may be required as a final surfacing.

*** Two inches of coarse sand or stone screenings recommended between subgrade and base as an insulation course.

(*Courtesy The Asphalt Institute*)

determined by engineering design procedure, which considers the following factors:

1. Traffic to be served initially and over the design service life of the pavement.

2. Strength and other pertinent properties of the prepared subgrade.

3. Strength and other influencing characteristics of the materials available or chosen for the layers or courses in the total asphalt pavement structure.

4. Any special factors peculiar to the road being designed.

Traffic Analysis

The weight and volume of traffic that a road is expected to carry initially and throughout its design life influence the required thickness of asphalt pavement structure, as well as mix selection and mix design. Traffic for many years was evaluated by using annual average daily traffic (AADT) data. The Superpave system introduced evaluation of traffic in terms of equivalent single axle loads (ESALs), which is a component in the specification of a mix design. Using ESALs distinguishes the stresses the heavy loads of trucks and buses impart on a pavement more accurately than AADT.

Subgrade Evaluation

There are several methods for evaluating or estimating the strength and supporting power of a subgrade, including the following:

1. Loading tests in the field on the subgrade itself; for example, the plate bearing test uses large circular plates loaded to produce critical amounts of deformation on the subgrade in place.

2. Loading tests in a laboratory using representative samples of the subgrade soil. Some commonly used tests are (a) California bearing ratio (CBR) test, which is sometimes used on the subgrade in place in the field; (b) Hveem stabilometer test; and (c) triaxial test.

3. Evaluations based on classification of soil by identifying and testing the constituent particles of the soil. Four well-known classification systems are (a) American Association of State Highway and Transportation Officials (AASHTO) Classification System; (b) Unified Soil Classification System; (c) Corps of Engineers, U.S. Army; and (d) U.S. Federal Aviation Administration (FAA) method.

Asphalt Paving-Mix Design

The design of asphalt paving mixes, as with other engineering materials designs, is largely a matter of selecting and proportioning materials to obtain the desired qualities and properties in the finished construction. The overall objective for the design of asphalt paving mixes is to determine an economical blend and gradation of aggregates and asphalt that yields a mix having the following:

1. Sufficient asphalt to ensure a durable pavement.

2. Sufficient mix stability to satisfy the demands of traffic without distortion or displacement.

3. Sufficient voids in the total compacted mix to allow for a slight amount of additional compaction under traffic loading without flushing, bleeding, and loss of stability, yet low enough to keep out harmful air and moisture.

4. Sufficient workability to permit efficient placement of the mix.

There are currently three methods used to design asphalt paving mixes, and the selection and use of these mix design methods is principally a matter of engineering requirements. Each of the methods has unique features and advantages for particular design problems.

There are currently three methods used to design asphalt paving mixtures:

1. Marshall method
2. Hveem
3. Superpave

All methods use as the primary criterion the level of air voids at production of HMA, utilizing AASHTO T-209, Theoretical Maximum Specific Gravity and Density of Bituminous Paving Mixtures, and AASHTO T-166 Standard Method of Test for Bulk Specific Gravity of Compacted HMA Mixtures Using Saturated Surface Dry Specimens, or their ASTM counterparts. All three methods target an air void range, at plant production, of 3 to 5 percent. The differences in the systems lie in their tests to predict performance and the method for compacting specimens for analysis by AASHTO T-166 (Bulk Specific Gravity of Compacted Mixtures).

The Marshall mix design method was developed by Bruce Marshall of the Department of Transportation in Mississippi in 1939. The complete Marshall method may be found in the Asphalt Institute Manual MS-2, "Mix Design Methods." The Marshall method was the prevalent form of mix design and analysis of HMA mixtures until the mid-1990s. Over time, it has generally been replaced by the Superpave system by most state agencies.

The Marshall method utilizes a "hammer" to produce bulk density (unit weight) specimens for analysis of air voids. The method also utilizes analysis of "stability and flow" as a predictive indicator of a mixture's performance. Stability is measured as the maximum load measured in pounds when a load is applied to a 4″ temperature-conditioned specimen at a rate of 50.8 mm per minute or 2″ per minute. Flow is measured as the degree of deformation at maximum load, expressed in 0.25 mm or 0.01 in.

Because of the long successful history of the Marshall system, some agencies still use the method today, including the Federal Aviation Administration and many local municipalities that are comfortable with its known performance. However, after significant and widespread pavement failures, largely attributed to increased traffic volumes, truck traffic, and tire pressures, the Federal Highway Administration (FHWA) invested years and millions of dollars of research to improve the design of HMA mixtures to better withstand current demands. While the evaluation of air voids was done under the Marshall system, it was believed the stability and flow aspect of the system was inadequate for the demands of today's traffic, especially in the predication of permanent deformation, or rutting.

The Superpave (superior performing asphalt pavements) was the outcome of the above-mentioned research

conducted in the early 1990s. A majority of state highway agencies use the Superpave system today.

The Superpave system integrated the following three considerations for mix design selection:

1. Performance-graded binder specifications [based on climatic temperature, with engineering consideration for "bumping" if traffic loads or conditions (static intersections) warrant].

2. Aggregate "consensus" properties, which are minimal standards for the acceptance of aggregates, include coarse aggregate angularity (crush count or fractured faces) to ensure a stable aggregate structure, sand equivalent (a measure of clay content), and fine aggregate angularity. In some cases, materials will need to be imported for the most stringent specifications of Superpave at the highest traffic levels. However, the Superpave system has evolved to generally accept historically well-performing aggregates that are available locally.

3. Superpave mix design system. The Superpave mix design system utilizes a Superpave gyratory compactor (SGC) to produce the bulk specimens for evaluation of air voids. The SGC is capable of applying consolidation pressure of 600 kPa (80 psi), at an internal angle of 1.16°, and speed of 30 revolutions per minute. The number of gyrations used to compact a specimen is a function of the amount of ESALs the pavement will encounter over a design life of 20 years. The gyratory compactor was envisioned to better predict in situ air voids over the design life of a pavement (20 years), using a mechanical system with pressure, angle, and revolutions which simulated traffic better than the Marshall "hammer."

The Superpave system did not have a "performance" component until very recently. Three tests were identified by work completed at Arizona State for the National Cooperative Highway Research Program, NCHRP 9-19. Those tests were flow time, flow number, and dynamic modulus. The equipment for these tests was developed under NCHRP 9-29, and the Superpave Performance Tester (SPT), also known as Asphalt Mix Performance Tester (AMPT), is the result. As of 2009, initial procurement of equipment for state highway agencies is underway, and the implementation of these tests as an evaluation of mix designs will become the standard.

The Hveem method is still in use in a few states, mostly in the West. The Hveem system also evaluates asphalt mixtures in terms of air voids, and like the Marshall method measures stability, using the Hveem stabilometer.

TYPES OF ASPHALT PAVEMENT CONSTRUCTION
Plant Mix

Asphalt concrete paving mixtures prepared in a hot mix plant are known as hot mix asphalt or HMA. HMA is considered the highest quality type of asphalt mixture because the processing at the plant removes all ambient moisture from aggregates, which ensures the best bond between the aggregates and the asphalt binder. The aggregates are heated from 275 to 325°F and are measured and proportioned based on the mix design. The asphalt is also carefully measured, and the materials are mixed. The HMA is hauled to the construction site, where it is spread on the roadway by a "spreader" or "paver." The smooth layer of HMA following the paver is compacted by rollers to proper density before the HMA cools.

The production of HMA requires abundant energy. Heating of aggregates to produce HMA typically consumes 2.5 gallons no. 2 fuel oil for 1 ton of HMA. Because of limited natural resources, the industry is currently exploring warm mix asphalt (WMA) technologies to reduce fuel consumption and emissions. WMA technologies are either mechanical (foaming) or chemical (additive) and permit reduction in temperatures in the range of 50 to 75°F.

Asphalt mixes containing liquid asphalt also may be prepared in central mixing plants. The aggregate may be partially dried and heated or mixed as it is withdrawn from the stockpile. These mixes are usually referred to as *cold mixes*, even though heated aggregate may have been used in the mixing process.

Asphalt mixtures made with emulsified asphalt and some cutback asphalts can be spread and compacted on the roadway while quite cool. Such mixtures are called *cold-laid asphalt plant mixes*. They are hauled and placed in normal warm-weather temperatures. To hasten evaporation of emulsification water or cutback solvents, these mixtures, after being placed on the roadway, are sometimes processed or worked back and forth laterally with a motor grader before being spread and compacted.

Mixed-in-Place (Road Mix)

Emulsified asphalt and many cutback asphalts are fluid enough to be sprayed onto and mixed into aggregate at moderate- to warm-weather temperatures. When this is done on the area to be paved, it is called *mixed-in-place construction*. Although *mixed-in-place* is the more general term, and is applicable whether the construction is on a roadway, parking area, or airfield, the term *road mix* is often used when construction is on a roadway.

Mixed-in-place construction can be used for surface, base, or subbase courses. As a surface or wearing course, it usually is satisfactory for light and medium traffic rather than heavy traffic. However, mixed-in-place layers covered by a high-quality asphalt plant-mix surface course make a pavement suitable for heavy traffic service. Advantages of mixing in place include the following:

1. Utilization of aggregate already on the roadbed or available from nearby sources and usable without extensive processing.
2. Elimination of the need for a central mixing plant. Construction can be accomplished with a variety of machinery often more readily available, such as motor graders, rotary mixers with revolving tines, and traveling mixing plants.

Slurry Seal

A *slurry seal* is a thin asphalt overlay applied by a continuous process machine to worn pavements to seal them and provide a new wearing surface. Slurry seals are produced with emulsified asphalts. Aggregates used for slurry seals must be hard, angular, free of expansive clays, and uniformly graded from a particle size about the thickness of the finished overlay down to No. 200. Crushed limestone or granite, slag, expanded clays, and other lightweight materials are typical aggregates used for slurry seals. The truck-mounted equipment (Figure 3–13) transports, proportions, mixes, and applies the slurry seal.

FIGURE 3–13. Continuous process slurry seal machine. (*Courtesy Slurry Seal, Inc.*)

RECYCLED ASPHALT CONCRETE

The HMA industry is the largest recycling group, in terms of tons, nationally. *Asphalt pavement recycling* combines reclaimed pavement materials with new materials to produce asphalt mixtures that meet normal specification requirements. The most common methods of pavement removal include ripping and crushing and cold milling or planing of the existing pavement. One of the milling techniques is called *mill and fill* and is used on high-traffic roads. The wearing surface is milled off and trucked back to the hot mix plant, while trucks are leaving the plant with HMA concrete to be used as the wearing surface replacement. The operation is a continuous removal and replacement process, thereby reducing some of the traffic delays normally associated with roadway reconstruction projects.

The materials produced by these methods are classified as reclaimed asphalt pavement (RAP) or reclaimed aggregate material (RAM). Because RAM contains no asphalt, it is handled and processed using the same techniques as virgin aggregates. The RAP contains aged asphalt, and therefore is subject to crushing limitations and stockpile height requirements no greater than 10 ft. Also, if possible, it must be protected from the weather (Figure 3–14).

The use of RAP to produce new paving materials is based on the heat-transfer method. The virgin and RAM materials are superheated and fed into the pugmill where the RAP, new asphalt, and, if required, a recycling agent are introduced. The mixing process allows for some heat transfer to take place, and complete temperature equilibrium is usually reached sometime later, after the mix is discharged into trucks or a storage silo.

The amount of reclaimed asphalt pavement that can be used in the recycled mix depends on (1) the moisture content and stockpile temperature of the reclaimed material, (2) the required temperature of the recycled mix, and (3) the temperature of the superheated aggregate. If optimum conditions exist, as much as 40 percent of the new mixture can be RAP. However, a RAP utilization of 20 to 30 percent is a realistic range for normal operations.

ASPHALT SPRAY APPLICATIONS

Many necessary and useful purposes are served when paving asphalt, temporarily in a fluid form, can be sprayed in uniform and controlled amounts onto a surface.

Surface Treatments and Seal Coats

A sprayed-on application of asphalt to a wearing surface, with or without a thin layer of covering aggregate, is called an *asphalt surface treatment.* By definition, such surface treatments are 1 in. or less in thickness. Sometimes these surface treatments are included in the original construction. More often they are applied to old pavements after a period of service and before surface deterioration from traffic wear and weathering proceeds too far.

The sprayed-on asphalt serves to improve or restore the waterproof condition of the old pavement surface. Also, it serves to arrest any scuffing or raveling of the wearing surface. The addition of a cover of aggregate over the sprayed-on asphalt restores and improves the skid resistance of the wearing surface.

Multiple-surface treatments consist of two or more alternate layers of sprayed-on asphalt and aggregate cover.

Surface treatments that have waterproofing or texture improvement, or both, as their main purpose are called *seal coats.*

Single- or multiple-surface treatments with aggregate cover also may be placed on granular-surfaced roads to upgrade them for traffic. The treatment eliminates dust, protects the road by shedding water, and provides a smoother riding surface. It is a useful, low-cost, all-weather improvement of a granular-surfaced road, but it has limited traffic capacity and should be used only where traffic is light or where the period of expected service is limited.

Tack Coats and Prime Coats

Each layer in an asphalt pavement should be bonded to the layer beneath. This is accomplished by spraying onto the surface of the underlying layer a thin coating of asphalt to bind the layers together. This thin spread of asphalt is called a *tack*

FIGURE 3–14. Milling machines remove asphalt concrete pavements to line and grade. The material removed will be recycled (RAP). *(Courtesy Astec Industries, Inc.)*

coat. Tack coats are used to bond asphalt layers to a portland cement concrete base or old brick and stone pavements.

When an asphalt pavement or asphalt surface treatment is to be placed on a granular base, it is desirable to treat the top surface of the base by spraying on a liquid asphalt that will seep into or penetrate the base. This is called *priming*, and the treatment is called *prime coat.* Its purpose is to serve as a transition from the granular material to the asphalt layer and bind them together. A prime coat is different from a tack coat as to type and quantity of asphalt used. However, both are spray applications.

ASPHALT PLANTS

Asphalt paving mixes made with asphalt cement are prepared at an asphalt mixing plant. Here, aggregates are blended, heated, and dried, and mixed with asphalt cement to produce a hot-asphalt paving mixture. The mixing plant may be small and simple, or it may be large and complex, depending on the type and quantity of asphalt mixture being produced.

As shown in Figure 3–15, components of an asphalt plant are as follows:

1. Cold aggregate storage
2. Drying
3. Screening
4. Hot storage
5. Measuring and mixing

Aggregate is removed from storage, or stockpiles, in controlled amounts and passed through a dryer where it is dried and heated. The aggregate then passes over a screening unit that separates the material into different size fractions and deposits them into bins for hot storage. The aggregate and mineral filler, when used, are then withdrawn in controlled amounts, combined with asphalt, and thoroughly mixed. This mix is hauled to the paving site.

During the production of HMA, various tests are utilized to measure the quality of mix being produced. Constant monitoring or quality control ensures compliance to specifications.

Deviations from established job mix formula (JMF) gradation targets established during the mix design process can cause significant swing in the mixture's air voids because of changes in surface area that affect the required asphalt binder content. A typical JMF is shown in Figure 3–16, and although a JMF is a blueprint for what an asphalt plant will actually produce, its gradation targets are important production parameters.

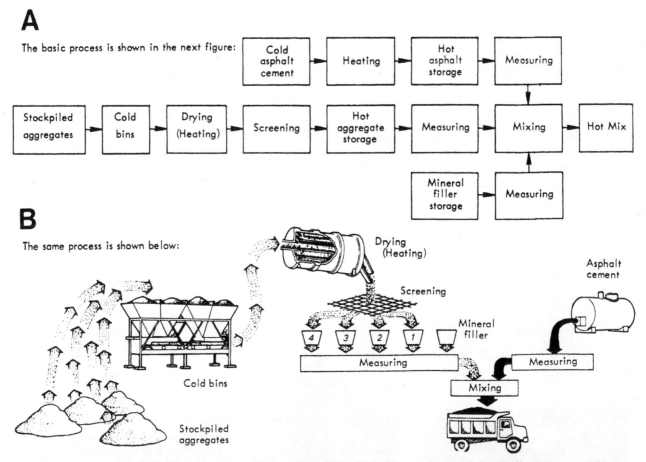

FIGURE 3–15. Basic batch facility operations shown (A) in flow chart form and (B) schematically. (*Courtesy The Asphalt Institute*)

DESIGN ESAL LEVEL
3.0 to <30

SUPERPAVE Job Mix Formula
12.5 mm Nominal Maximum Size

12.5 mm Nominal Maximum Size

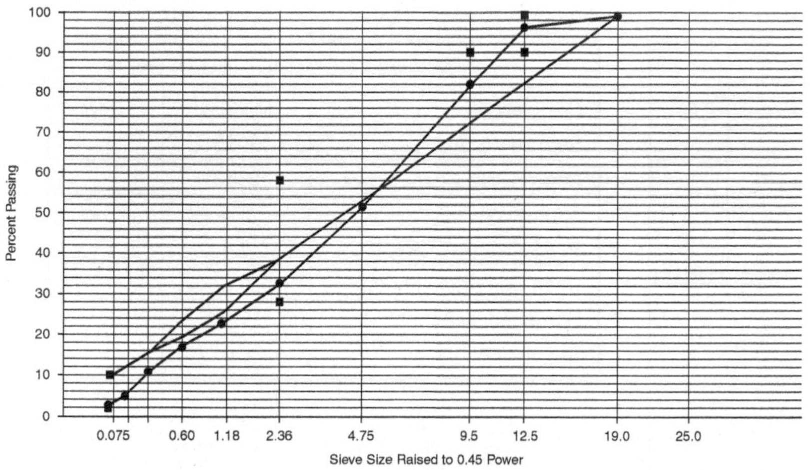

REGION	JMF No.
8	99D4003cV
PRODUCER	Callanan Industries. Inc.
LOCATION	East Kingston, NY Plant #19

Aggregates		Source Number	Blend %	
Coarse	No. 1 Stone	8–13RS	32	1 Fric. Blend 100
	No. 1 Non-Carbonate Stone			
	No. 1A stone	8–13RS	25	1A Fric. Blend
	No. 1A Non-Carbonate Stone			
Fine	Manufactured	8–13R	23	100
	Natural Sand	8–100F		
	Manuf.			
	Mineral Filler			Sand Blend
	RAP	Tway	20	

Aggregates		Blend Limits to Maintain C.A. P's		
		C.A.A.	F.A.A.	Flat / El.
Coarse	No. 1 Stone	100		3.8
	No. 1 Non-Carbonate Stone			
	No. 1A stone	100		3.6
	No. 1A Non-Carbonate Stone			
Fine	Manufactured		47.4	
	Manufactured Natural Sand			
	RAP		44.8	

Sieve Size (mm)		0.075	0.15	0.3	0.6	0.18	2.36	4.75	9.5	12.5	19.0	25	% PGB	Binder Grade
% Passing	General Limits	2–10					28–58		<90	90–100	100			PG 64–28
	JMF Range	0.0–8.0	0–10	6–16	12–22	18–28	28–38	47–57	78–88	92–100	95–100		n/a	
	Target Value	3.0	5	11	17	23	33	52	83	97	100		5.0	

Submitted for Review by: David Dachehausen Date: Jan. 25, 2009

Accepted for Verification / Production by: NYSDOT Date:

Revised

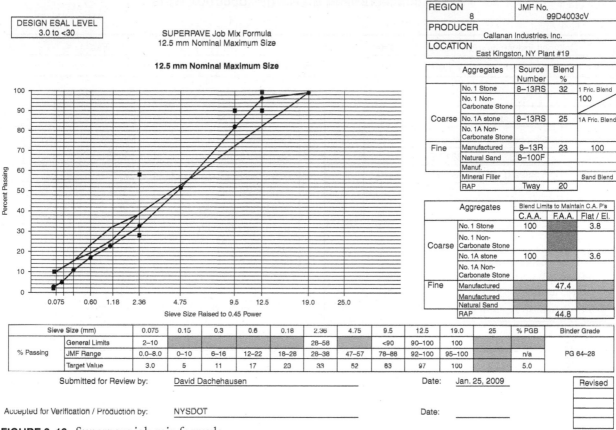

FIGURE 3–16. Superpave job mix formula.

HMA stockpile samples are routinely monitored to predict when a change in proportions might be necessary because of a change in the constituent aggregate's gradation. An HMA production composite sample is also routinely evaluated as quality control and is often a specification requirement. Gradation of the HMA composite sample is typically prepared in an ignition oven. An HMA sample is dried to a constant weight, then weighed, and placed in the ignition oven and is heated to 1000–1200°F. This extreme heat burns the asphalt binder and what is left, once cooled, is evaluated for gradation and compared to the JMF targets. The ignition process is also used to verify the asphalt binder content by comparing the initial weight, and the weight of the sample after the asphalt binder has been consumed through ignition. A correction factor is applied because the intense heat will also break down the aggregates, resulting in overstated asphalt binder content (Figure 3–17).

Measuring the HMA mixture's air voids is a common specification requirement. An air void content between 3 and 5 percent at production is sufficient to cause binder film coating of aggregates to ensure durability and allow for air void structure to permit space for the asphalt binder to flow under traffic load or temperature stress.

The air voids, and other important derived volumetric properties, such as voids in mineral aggregate and voids filled with binder (Figure 3–18), are determined by evaluating the mixtures, "Rice"(voidless density), or Maximum Specific Gravity and Density of Bituminous Paving Mixtures per AASHTO T-209. Bulk specimens are also necessary and are produced either by Marshall hammer or Superpave gyratory compactor methods, and then evaluated for their Bulk Specific gravity per AASTHO T-166, Standard Method of Test for Bulk Specific Gravity of Compacted Bituminous Mixtures Using Saturated Surface Dry Specimens. A simple calculation will yield the mixture's air voids.

$$\% \text{ Air Voids} = \frac{(\text{Maximum Theoretical} - \text{Bulk Specific Gravity})}{\text{Maximum Theoretical}} \times 100$$

Other routine tests in the quality control of HMA include RAP binder content, aggregate moisture, mixture moisture, and temperature.

Types of Asphalt Plants

Asphalt plants are classified as *stationary* or *portable*—both *batch* and *continuous-mix type.* The stationary plant is permanently situated and is not normally dismantled and moved. The portable plant can be easily disassembled, moved by rail or highway, and reassembled with a minimum of time and energy.

ASPHALT CONCRETE DRUM MIX PLANT PRODUCTION TESTS

Plant	Callanan Industries Inc.	Location	East Kingston			Facility #		10282
Mix Type	12.5mm	Lot#	18	Sublot# A	JMF #	99D4003cV		

Moisture Content				
Calculation	Composite	Mixture	Rap	QA
Wet Wt (A)	4023	3017.6		
Dry Wt. (B)	3899.0	3015		
Diff Wt (A-B)	127	2.6		
(A-B/B)/100	3.3	0.09		

ASPHALT CONTENT – BURN OVEN			
Calculation	Mixture	Rap	QA
Sample Wt (A)	3015		
Agg Wt. (B)	2852		
Asphalt Wt A-B	163		
% Asphalt (A-B)/A x 100	5.4		
Oven Correction Factor	0.3		
Corrected % Asphalt	5.1		
Target % Asphalt	5.0		

Sample Times	
Moisture	Asphalt Content
Composite:	Mixture:
Mixture:	Rap:
Rap:	

Sieve Size	QC Gradation			% Pass Rap	% Agg 100.0	Total Percent Passing	QA Gradation			% Pass Rap	% Agg	Total Percent Passing	JMF Target	JMF Range	
	Weight	% Ret.	% Pass.				Weight	% Ret.	% Pass.						
50 mm			100.00			100.00							100	95	100
37.5 mm			100.00			100.00							100	95	100
25.0 mm			100.00			100.00							100	95	100
19.0 mm			100.00			100.00							100	95	100
12.5 mm	85.5	3.00	97.0			97.0							97	92	100
9.5 mm	399.2	14.01	83.0			83.0							83	78	88
4.75 mm	884	31.03	51.9			51.9							52	47	57
2.36 mm	541.9	19.02	32.9			32.9							33	28	38
1.18 mm	285.2	10.01	22.9			22.9							23	18	28
.600 mm	173	6.07	16.8			16.8							17	12	22
.300 mm	168.3	5.91	10.9			10.9							11	6	16
.150 mm	140.5	4.93	6.0			6.0							5	0	10
.075 mm	90	3.16	2.8			2.8							3	0	6
Pan	80.8	2.84													
Total	2848.4														

Quality Control By: _____   Date: ___6/25/2009___
David Dachenhausen

Quality Assurance By: _____   Date: _____

FIGURE 3–17. Gradation by ignition method.

Producer: Callanan Industries Inc.
Location: East Kingston
Facility No. 10282
Date: 6/25/2009

Mix Type 12.5mm
Lot No. 18 Sublot A
JMF Number 99D4003cV

Computation of Volumetric Mix Properties

Superpave

Specimen ID	Weight – Grams: In Air	Weight – Grams: In Water	Weight – Grams: S.S.D.	Volume CC	Maximum Specific Gravity G_{mm}	@ N-design: Bulk Specific Gravity	@ N-design: % G_{mm}	@ N-design: Percent Air Voids	@ N-design: Specimen Height mm	@ N-initial: Specimen Height mm	@ N-initial: Bulk Specific Gravity G_{mb}	@ N-initial: % G_{mm}
a	c	d	e	f	g	i	j	k	h	l	m	n
				e-d		c/f	100 (i/g)	100-n			(l/h) X g	100 (m/g)
A	4798.5	2792.3	4804.9	2012.6		2.384	95.47	4.525	117.3	131.7	2.124	85.0
B	4789.4	2786.7	4793.8	2007.1		2.336	95.55	4.445	116.7	130.9	2.127	85.2
QC Avg.					2.497	2.385	95.51	4.49	117.0	131.3	2.125	85.1
QA Avg.												

Sample	Quality Control 1A	Quality Control 1B	Quality Assurance 1A	Quality Assurance 1B
A	2009.8	2009.0		
D	1209.8	1381.4		
E	2414.5	2586.2		
A+D-E	805.1	804.2		
Gmm	2.496	2.498		
Average Gmm	2.497			

Quality Control By: (signature)	Date:
David Dachenhausen	6/25/2009
Quality Assurance By: (signature)	Date:
	6/25/2009

Combined Aggregate Bulk Specific Gravity Gsb

Aggregate	Bulk S.G.	% Blend
3A'S	2.683	0
2'S	2.683	0
3/8'S	2.683	32
Screnings	2.690	23
Sand	2.602	0
Sand	2.632	0
1A's	2.683	25
	0.000	0
	0.000	0
RAP	2.699	20

Gsb	2.684
% Asphalt	5.0
VMA	15.83
VFA	71.66
Dust to Effective Binder	
% Pass. #200	4.6
Gb	1.037
Pbe	4.93
F/Pbe	0.93

Gsb = Combined Agg. Bulk S.G.
Gb = Specific Gravity of Binder
Pbe = effective binder Content
F/Pbe = Dust to Eff. Asphalt Content

Gyrations at N_{Des}	75
Gyrations at N_{int}	8

FIGURE 3–18. Computation of volumetric mix properties.

FIGURE 3–19. Asphalt concrete batch plant. *(Courtesy Astec Industries, Inc.)*

In the batch-type mixing plant (Figure 3–19), different size fractions of hot aggregate in storage bins are withdrawn in desired amounts to make up one batch for mixing. The entire combination of aggregate is then dumped into a mixing chamber called a *pugmill*. The pugmill may be set for a 15-second dry mix of the aggregates and an additional 45 seconds of wet mixing time after the asphalt cement is introduced into the pugmill. The asphalt, which has also been weighed, is thoroughly mixed with the aggregate. After mixing, the material is emptied from the pugmill into one batch.

In the continuous-type mixing plant, aggregate and asphalt are withdrawn, combined, mixed, and discharged in one uninterrupted flow. The combining of materials is generally done by volumetric measurements based on unit weight. Interlocked devices, feeding the aggregate and asphalt into one end of the pugmill mixer, automatically maintain the correct proportions. While being mixed, the materials are propelled by stirring paddles to the discharge end.

Figure 3–20 illustrates the operation of a dryer drum or continuous-type mixing plant. Aggregates with controlled gradations are placed in the cold feed bins (1). The mix design (JMF) determines the proportions of the aggregate drawn from the bins and moved to the drum mixer by the cold feed conveyor (2). The exact weight of the aggregate feed is monitored by the automatic weighing system (3), which is interlocked with the asphalt pump (5). The asphalt pump draws the correct quantity of asphalt cement from the storage tank (6) and feeds the asphalt into the drum mixer where the aggregates and asphalt are blended together (4). Dust escaping from the mixing system is captured in the dust collector (7). The completely mixed asphalt concrete is continuously fed into a surge storage silo (9) by the hot mix conveyor (8). The entire production system is controlled and monitored by the plant operator stationed in the control van (10).

Storage of Hot-Mix Asphalt

Should paving operations be temporarily interrupted, rather than stopping production at the plant, a *surge bin* may be installed and used for temporarily storing the hot mix. This is usually a round, silo-type structure, the lower end of which is cone shaped. Hot mix is dumped into the top of the silo so as to fall vertically along the vertical axis of the structure. The bin is designed so that segregation of the mix is held to a minimum. As it is withdrawn from the bottom, its uniformity is maintained.

Surge bins also speed the loading of trucks with hot mix. The trucks can be filled in a matter of seconds, while a truck at the plant has to wait for the production of several individual batches before it is loaded.

Surge bins are insulated and can store 50 to 100 tons of mix. They can usually store hot mix for up to 12 hours with no significant loss of heat or quality.

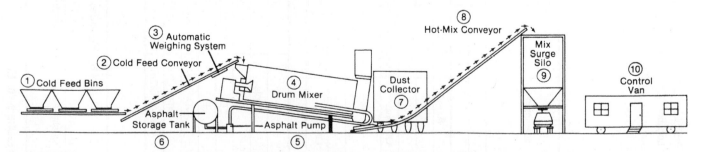

FIGURE 3–20. Basic drum mix facility. *(Courtesy The Asphalt Institute)*

FIGURE 3–21. Asphalt concrete drum mix plant with surge/storage silos. *(Courtesy Astec Industries, Inc.)*

Where paving operations can lay the hot mix at a rate faster than the plant can produce it, full surge bins at the beginning of the day will increase the plant's effective daily output.

Sometimes it is necessary or desirable to store hot-asphalt mixes for more than 12 hours. Storage silos, similar to surge bins, are used for this purpose. The capacity of these heated silos may be as much as 350 tons. Hot-mix asphalt ages quite rapidly during exposure to air. The oxidizing process hardens the asphalt. For long-term storage, an atmosphere free of oxygen is used to fill the silo to prevent age-hardening of the hot mix.

Storage silos may be remotely located. This makes it possible for the asphalt plant to serve a larger area and provide paving mix at times when the plant would normally not be operating (Figure 3-21).

ESTIMATING ASPHALT CONCRETE

The unit of measure for the purchase of asphalt concrete is the ton (2000 lb). Therefore, the number of tons of asphalt concrete required to complete a paving job must be determined. Some paving contractors utilize quantity charts or slide-rule-type paving calculators to determine required quantities of material, and others produce their own tables based on local materials.

On small paving jobs the quantities are usually determined using a rule of thumb which states that 1 ton of asphalt concrete will cover 80 sq ft, 2 in. thick.

Example

$$\text{Area to be paved} = 2600 \text{ sq ft}$$
$$\text{Thickness} = 2 \text{ in.}$$
$$\frac{2600 \text{ sq ft}}{80 \text{ sq ft/ton}} = 32.5 \text{ tons}$$

Larger paving estimates are based on the unit weight of the asphalt concrete. Asphalt concrete's unit weight will range from 140 to 150 lb per cu ft. The following formula is used:

$$T = \frac{A \times t \times uw}{2000 \text{ lb}}$$

where T = tons of asphalt concrete required

A = area to be paved in sq ft

t = thickness of pavement in ft

uw = unit weight of asphalt concrete

Example

$$\text{Area to be paved} = 180,000 \text{ sq ft}$$
$$\text{Thickness} = 3 \text{ in} = 0.25 \text{ ft}$$
$$\text{Unit weight of asphalt concrete} = 150 \text{ lb/cu ft}$$
$$T = \frac{180,000 \text{ sq ft} \times 0.25 \text{ ft} \times 150 \text{ lb/cu ft}}{2000 \text{ lb/ton}}$$
$$T = 3375 \text{ tons}$$

A rule of thumb for quickly calculating quantities of HMA uses a conversion factor of 110 pounds per square yard per 1″ of HMA.
For example:
For an area a total of 100 sq yd at 3″ deep, the quantity of HMA to be reordered would be determined as follows:

$$\frac{(100 \times 330)}{2000} = 16.5 \text{ tons}$$

PREPARATION OF UNPAVED SURFACES

The following roadway surfaces are generally considered unpaved:

1. Compacted subgrade
2. Improved subgrade
3. Untreated base
4. Nonsurfaced aggregate roadway

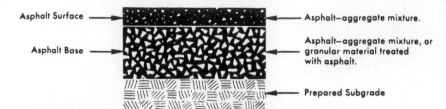

Asphalt Surface ——————— Asphalt–aggregate mixture.

Asphalt Base ——————— Asphalt–aggregate mixture, or granular material treated with asphalt.

←——— Prepared Subgrade

FULL DEPTH ASPHALT PAVEMENT

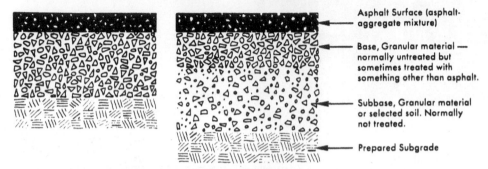

Asphalt Surface (asphalt-aggregate mixture)

Base, Granular material — normally untreated but sometimes treated with something other than asphalt.

Subbase, Granular material or selected soil. Normally not treated.

←——— Prepared Subgrade

ASPHALT PAVEMENT WITH UNTREATED BASE (AND SUBBASE)

Asphalt Surface, asphalt–aggregate mixture.

Base—portland cement concrete.

Asphalt–aggregate mixture.

←——— Prepared Subgrade.

ASPHALT PAVEMENT WITH PORTLAND CEMENT CONCRETE OR COMBINED PORTLAND CEMENT CONCRETE AND ASPHALT BASE

FIGURE 3–22. Asphalt pavement cross sections showing typical asphalt pavement structures. (*Courtesy The Asphalt Institute*)

Certain treated or stabilized granular bases are considered unpaved surfaces when being prepared for asphalt paving. Bases that have been treated or stabilized with either asphalt or portland cement are considered paved surfaces and are excluded from this classification.

Typical asphalt pavement structures are illustrated in Figure 3–22.

Prepared Subgrade

A *prepared subgrade* is one that has been worked and compacted. This may be the foundation soil or a layer of stabilized soil, select soil, or otherwise improved subgrade material.

The riding quality of the pavement surface depends largely on proper construction and preparation of the foundation material. The roadway should be shaped and proof-rolled so that the paving equipment has no difficulty in placing the material at a uniform *thickness* to a smooth grade.

Weather conditions should be suitable and the roadway surface should be firm, dust free, and dry, or just slightly damp, when paving operations are started.

Untreated Base

Untreated aggregate bases and some chemically stabilized bases other than portland cement and asphalt-treated bases should be properly shaped and proof-rolled.

When asphalt pavement courses are to be placed on an untreated aggregate base, loose aggregate particles should be swept from the surface of the roadway using power brooms. Care should be taken, however, not to dislodge or otherwise disturb the bond of the aggregate in the surface of the base. When the loose material has been properly removed, only the tops of the aggregate in the surface will be exposed, and the pieces solidly embedded in the base. When the surface has been cleaned, it is ready to be primed with asphalt.

For priming, an asphalt distributor sprays about 0.2 to 0.5 gal per sq yd liquid asphalt, usually MC-30 or MC-70, over the surface. The asphalt then penetrates or soaks into the surface. If it is not absorbed within 24 hours after application, too much prime has been used. To correct this condition, sand should be spread over the surface to blot the excess asphalt. Care should be taken to prevent overpriming. The prime should be fully set and cured before placing the asphalt mixture on the base.

Asphalt sometimes is mixed into the top 2 or 3 in. of base material in lieu of priming when it is difficult to obtain uniform and thorough penetration. This provides a tough working surface for equipment, a waterproof protective layer, and a bond to the superimposed construction.

Nonsurfaced Aggregate Roadways

Nonsurfaced aggregate roadways are similar to untreated aggregate bases. They differ primarily in that they have been used as a roadway by traffic and, typically, have been in use for a considerable time.

Again, it cannot be stressed too strongly that the riding quality of the surface depends to a great degree on conditioning and preparing the underlying pavement structure. It is advisable, where the aggregate-surfaced roadway is rough and uneven, to bring the surface to grade by scarifying the top few inches of material, blending in more aggregate as may be necessary, compacting, and priming with asphalt. If the aggregate roadway has had a previous asphalt treatment and is rough and uneven, it is advisable to place a *leveling course* of hot-asphalt plant mix ahead of the first layer of the surface course.

Placing a leveling course is an operation employed when the road surface is so irregular that it exceeds the leveling capabilities of the paver. The paver is usually effective on irregularities of lengths not longer than $1\frac{1}{2}$ to 2 times the wheel base of the machine. Beyond these lengths, outside help is needed. Generally, irregularities no longer than 40 or 50 ft can be handled with a traveling stringline device and automatic screed control on the paver.

PAVED SURFACES
Flexible-Type Pavements

Structural distress occurring in old asphalt pavements is usually the result of inadequate design, inadequate execution of the design including compaction, or both. Poor mix design can also cause several types of distress, and excess asphalt may cause corrugation or rutting. Cracking may be caused by excessive pavement deflection under traffic due to an inadequate pavement structure or a spongy foundation. Insufficient or oxidized asphalt may also cause cracking because the mix may be brittle.

Before the old pavement receives an overlaying course of asphalt, it should first be inspected. The remedy for whatever failures exist depends on the type and extent of the distress. If failure is extensive, reconstruction will probably be necessary. In any case, all needed repairs to the old surface should be made before paving.

A slick surface on an old pavement may be caused by aggregate polishing under traffic or by too much asphalt in the mix. If the cause is excess asphalt, it should be either burned off or removed with a *heater-planer* before placing the overlay.

Old pavements having small cracks should be given a fog seal before overlaying. A *fog seal* is a light application of emulsified asphalt diluted with water and sprayed over the surface. Larger cracks warrant having the surface treated

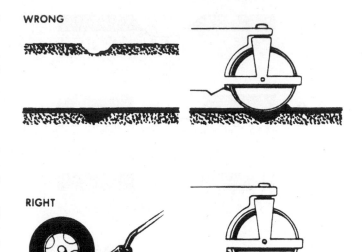

FIGURE 3–23. Potholes should be filled and compacted before spreading the first course. *(Courtesy The Asphalt Institute)*

with an emulsion slurry seal. The *slurry seal* is a mixture of emulsified asphalt, fine aggregate, and mineral filler, with water added to produce a slurry consistency.

Potholes in an old pavement that is otherwise strong should be filled with asphalt concrete and compacted before the roadway surface is paved (Figure 3–23). If the base beneath the old pavement has failed, the damaged areas will eventually show in the new paved surface. Patching, of course, must be done before paving, and the patch should be deep enough to strengthen the base.

All bleeding and unsuitable patches, excess asphaltic crack or joint filler, loose scale, and any surplus bituminous material should be removed from the surface of the existing pavement.

All depressions of 1 in. or more should be overlaid with a truing and leveling (T and L) course and compacted ahead of the surfacing operation. All surfaces, both horizontal and vertical, that will be in contact with the new asphalt surface must be thoroughly cleaned. Cleaning flat surfaces is usually done with rotary brooms, but washing or flushing may be necessary to remove clay or dirt. A tack coat should be applied to the existing pavement and to all vertical faces. These include curbs, gutters, drainage gratings, manholes, and other contact surfaces. A uniform coating of liquid asphalt or asphalt emulsion will provide a closely bonded waterproof joint. For repairs to pavements that will not be resurfaced, the procedure illustrated in Figure 3–24 should be used.

Rigid-Type Pavements

Distress in rigid-type pavements also must be corrected before resurfacing. Additional layers of asphalt mixtures must be thick enough not only to provide required additional strength but also to minimize the reflection of cracks from old pavement in the new surface.

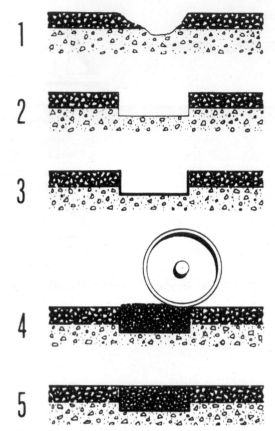

FIGURE 3–24. Pothole permanent repair: (1) untreated pothole; (2) surface and base removed to firm support; (3) tack coat applied; (4) full-depth asphalt mixture placed and compacted; (5) finish patch compacted to level of surrounding pavement. *(Courtesy The Asphalt Institute)*

Preparation of damaged or distressed rigid pavements differs from that for asphalt pavements. Preparation may include one or more of the following:

1. Breaking large cracked slabs into smaller pieces and seating them firmly with heavy rollers (rubblelizing)
2. Cracking slabs that rock under traffic and seating them with heavy rollers
3. Undersealing to provide uniform support
4. Patching disintegrated and spalled areas
5. Sealing cracks to prevent the intrusion of water from below

Pumping at the joints is caused by the rocking of slabs under traffic. The wet subgrade is removed from beneath the slab, and this causes cracking and breaking in the joint area. Usually a pavement can be stabilized by undersealing with asphalt specially prepared for this purpose. Otherwise, the rocking slabs should be cracked into smaller pieces and seated.

In some cases, portions of the pavement may be disintegrated or so badly broken up that the pavement fragments will need to be removed. In such cases, these areas should be prepared and patched with asphalt concrete prior to the resurfacing operation.

Filler material is removed from joints and cracks in the slabs to a depth of at least $1/4$ in. The joints are refilled with an asphalt joint-filling material. Any asphalt patches having excess asphalt, as well as any excess crack filler that may have accumulated next to cracks and joints, are removed.

As a final step, before any asphalt paving is placed, the pavement surface should be swept clean and given a tack coat. Where preparation measures have caused settlement or pavement roughness, a leveling course of asphalt concrete should be placed prior to the resurfacing operation.

INSPECTION OF MIX

Close cooperation between the paving crew and the asphalt plant is essential in securing a satisfactory and uniform job. A fast means of communication should be established between the paving operation and the asphalt plant so that any change in the mixture can be made promptly. When possible, the paving inspector and the plant inspector should frequently exchange visits. When the paving inspector is familiar with plant operations, he can easily determine if changes at the plant are necessary to improve the mix. The plant inspector, on the other hand, by being familiar with the paving operation can better understand problems attendant to it.

Every truckload of material should be observed as it arrives. The mix temperature should be checked regularly. If it is not within specified tolerance, the mix should not be used.

Mistakes in batching, mixing, and temperature control can and do occur, and these errors may sometimes go unnoticed by the plant inspector. Consequently, loads arriving at the spreader may be unsatisfactory. When the paving inspector rejects a load, he should record his action, with the reason for rejection, both on the ticket and in his diary so that the proper deduction can be made from the pay quantities. If appropriate, a sample should be obtained for laboratory analysis. A record should also be kept of the loads accepted and placed. These records should be checked daily, or more frequently, with those of the plant inspector so that no discrepancies exist when work is completed.

Mix Deficiencies

Some mix deficiencies that may justify discarding the mix are as follows:

1. *Too hot:* Blue smoke rising from the mix usually indicates an overheated batch. The temperature should be checked immediately. If the batch exceeds maximum specification limits, it should be discarded. If it exceeds optimum placing temperature but does not exceed the specification limit, the batch is usually not discarded, but immediate steps should be taken to correct the condition.

2. *Too cold:* A generally stiff appearance, or improper coating of the larger aggregate particles, indicates a cold

mixture. Again, the temperature should be checked immediately. If it is below the specification limit, it should be discarded. If it is within the specification limit but below optimum placing temperature, steps should be taken immediately to correct the situation.

3. *Too much asphalt:* When loads have been arriving at the spreader with the material domed up or peaked and suddenly a load appears lying flat, it may contain too much asphalt. Excessive asphalt may be detected under the screed by the way the mix slacks off.

4. *Too little asphalt:* A mix containing too little asphalt generally can be detected immediately if the asphalt deficiency is severe. It has a lean, granular appearance and improper coating, and lacks the typical shiny, black luster. The pavement surface has a dull, brown appearance, and the roller does not compact it satisfactorily. A less severe deficiency is difficult to detect by appearance; suspicions should be checked by testing.

5. *Nonuniform mixing:* Nonuniform mixing shows up as spots of lean, brown, dull-appearing material within areas having a rich, shiny appearance.

6. *Excess coarse aggregate:* A mix with excess coarse aggregate can be detected by the poor workability of the mix and by its coarse appearance when it is on the road. Otherwise, it resembles an overrich mix.

7. *Excess fine aggregate:* A mix with an excess of fine aggregate has a different texture from a properly graded mix after it has been rolled. Otherwise, it resembles a lean mix.

8. *Excess moisture:* Steam rising from the mix as it is dumped into the hopper of the spreader indicates moisture in the mix. It may be bubbling or popping as if it were boiling. The mix may also foam so that it appears to have too much asphalt.

9. *Miscellaneous:* Segregation of the aggregates in the mix may occur because of improper handling and may be serious enough to warrant rejection. Loads that have become contaminated because of spilled gasoline, kerosene, oil, and the like should not be used in the roadway.

THE PAVING OPERATION

Spreading and compacting asphalt mixtures is the operation to which all the other processes are directed. Aggregates have been selected and combined; the mix designed; the plant and its auxiliary equipment set up, calibrated, and inspected; and the materials mixed together and delivered to the paver.

Asphalt mix is brought to the paving site in trucks and deposited directly into the paver or in windrows in front of the paver. The paver then spreads the mix at a set width and thickness as it moves forward. In doing so, the paver partially compacts the material. Immediately or shortly thereafter and while the mix is still hot, steel-wheeled and rubber-tired rollers are driven over the freshly paved strip, further compacting the mix. Rolling is usually continued until the pavement is compacted to the required density, or until the temperature has dropped to a point where further compaction may produce detrimental results.

Figure 3–25 shows a paver with an extended screed placing and compacting asphalt pavement. The flow of materials through the paver is illustrated in Figure 3–26.

After the pavement course has been compacted and allowed to cool, it is ready for additional paving courses or ready to support traffic loads.

The Asphalt Paver

The *asphalt paver* spreads the mixture in a uniform layer of desired thickness and shape, or finishes the layer to the desired elevation and cross section, ready for compaction. Modern pavers are supported on crawler treads or wheels.

FIGURE 3–25. Paver with extended screed placing asphalt concrete. *(Courtesy Advance Testing Company)*

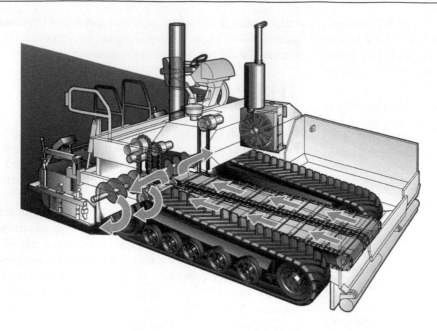

FIGURE 3–26. Flow of materials through a typical asphalt paver. (*Courtesy of Caterpillar Inc.*)

These machines can place a paved layer of less than 1 in. to about 10 in. in thickness over a width of 6 to 32 ft. Working speeds generally range from 10 to 70 ft per minute.

On small paving jobs it is sometimes more convenient and economical to use a tow-type paver than to use the larger, self-powered pavers. Tow-type pavers (Figure 3–27) are attached to the rear of the dump truck that hauls the asphalt mix from the plant.

Hand-Spreading Operations

The increasing use of asphalt in construction has resulted in an increased number of related projects in which asphalt mixes are used. For example, asphalt paving mixes are used for constructing driveways, parking areas, shoulders, and sidewalks. Such incidental construction is becoming more and more a part of the paving contract.

Normally, mix requirements for these jobs are the same as for roadway mixes. However, where it is anticipated that the major portion of placing will be by hand, mixes should be designed for good workability. This is easily achieved by decreasing the amount of coarse aggregate in the mix. Placing and compacting methods for hand-placed asphalt mixes are the same as for machine-spread paving.

Certain differences in construction methods are required, especially where special equipment such as sidewalk pavers are not available. In addition, there might be places on a regular road-paving job where spreading with a paver is either impractical or impossible. In these cases, hand spreading may be permitted.

Placing and spreading by hand should be done very carefully and the material distributed uniformly to avoid segregation.

Material should not be broadcast or spread from shovels as this causes segregation. Rather, it should be deposited by shovels or wheelbarrows into small piles and spread with asphalt rakes or wide-blade lutes. Any part of the mix that has formed into lumps and does not break down easily should be discarded.

The Roller Operation

Rolling should start as soon as possible after material has been spread. Rolling consists of three consecutive phases: *breakdown* or *initial rolling, intermediate rolling,* and *finish rolling.* Estimated asphalt concrete compaction production may be calculated using the compaction equation introduced in Chapter 2, and converting compacted cubic yards per hour into tons per hour of compacted hot-mix asphalt concrete.

Breakdown rolling compacts the material beyond that imparted by the paver, to obtain practically all of the density

FIGURE 3–27. Flow of material through a tow-type paver. (*Courtesy The Asphalt Institute*)

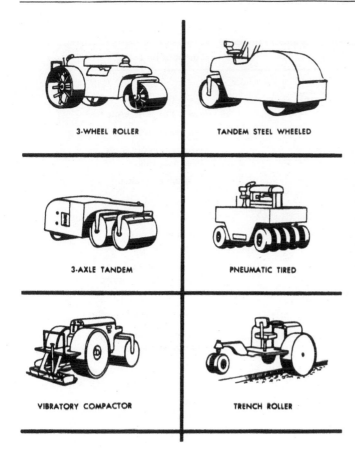

FIGURE 3-28. Types of compactors. *(Courtesy The Asphalt Institute)*

it needs. Intermediate rolling densifies and seals the surface. Finish rolling removes roller marks and other blemishes left from previous rolling. Rollers available for these operations are illustrated in Figure 3–28 and include the following:

1. Steel wheeled
2. Pneumatic tired
3. Vibrating
4. Combination steel wheeled and pneumatic tired

Steel-wheeled rollers have been and may be used for all three rolling phases. A pneumatic-tired roller sometimes is used for breakdown rolling, but is generally preferred for intermediate rolling. Vibrating rollers (Figure 3–29) are also used primarily for intermediate rolling. During rolling, the roller wheels should be kept moist with only enough water to avoid picking up material. Rollers should move at a slow, uniform speed with the drive roll nearest the paver (Figure 3–30). Rollers should be in good mechanical condition. The line of rolling should not change suddenly, nor should the roller be reversed quickly, thereby displacing the mix. Any major change in roller direction should be done on stable material. If rolling causes displacement of the material, the area affected should be loosened and restored to original grade with loose new material before being rolled again.

FIGURE 3-29. Vibratory rollers compacting asphalt concrete. *(Courtesy Advance Testing Company)*

Two important areas of rolling are the transverse and longitudinal joints. The transverse joints are perpendicular to the centerline of the pavement, and the longitudinal joints are parallel to the centerline.

The roller should be placed on the previously compacted material with about 6 in. of the roller wheel on the uncompacted mix. The roller should work successive passes, each covering 6 to 8 in., until the entire width of the drive roll is on the new mix.

If the specified density of asphalt pavement mix is not obtained during construction, subsequent traffic will further consolidate the pavement. This consolidation occurs principally in the wheel paths and appears as channels in the pavement surface.

Most mixtures compact quite readily if spread and rolled at temperatures that assure proper asphalt density. Refer to Table 3–2 for items that influence compaction and suggested corrective measures.

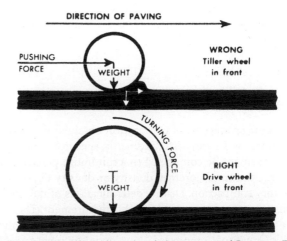

FIGURE 3-30. Rolling direction is important. *(Courtesy The Asphalt Institute)*

Table 3–2. Summary of influences on compaction.

Item	Effect	Corrections*
Aggregate		
• Smooth Surfaced	Low interparticle friction	Use light rollers Lower mix temperature
• Rough Surfaced	High interparticle friction	Use heavy rollers
• Unsound	Breaks under steel-wheeled rollers	Use sound aggregate Use pneumatic rollers
• Absorptive	Dries mix—difficult to compact	Increase asphalt in mix
Asphalt		
• Viscosity		
—High	Particle movement restricted	Use heavy rollers Increase temperature
—Low	Particles move easily during compaction	Use light rollers Decrease temperature
• Quantity		
—High	Unstable & plastic under roller	Decrease asphalt in mix
—Low	Reduced lubrication—difficult compaction	Increase asphalt in mix Use heavy rollers
Mix		
• Excess Coarse Aggregate	Harsh mix—difficult to compact	Use heavy rollers
• Oversanded	Too workable—difficult to compact	Reduce sand in mix Use light rollers
• Too Much Filler	Stiffens mix—difficult to compact	Reduce filler in mix Use heavy rollers
• Too Little Filler	Low cohesion—mix may come apart	Increase filler in mix
Mix temperature		
• High	Difficult to compact—mix lacks cohesion	Decrease mixing temperature
• Low	Difficult to compact—mix too stiff	Increase mixing temperature
Course Thickness		
• Thick Lifts	Hold heat—more time to compact	Roll normally
• Thin Lifts	Lose heat—less time to compact	Roll before mix cools Increase mix temperature
Weather Conditions		
• Low Air Temperature	Cools mix rapidly	Roll before mix cools
• Low Surface Temperature	Cools mix rapidly	Increase mix temperature
• Wind	Cools mix—crusts surface	Increase lift thickness

*Corrections may be made on a trial basis at the plant or job site. Additional remedies may be derived from changes in mix design.

Pavement Density

The degree or amount of compaction obtained by rolling is determined by *Density tests*. Ordinarily, specifications require that a pavement be compacted to a minimum percentage of either maximum theoretical density or density obtained by laboratory compaction. Density determinations of the finished pavement are necessary to check this requirement. These are made in the laboratory in accordance with the test method, Specific Gravity of Compressed Bituminous Mixtures (AASHTO T166), on samples submitted by the paving inspector.

Care should be exercised in obtaining and transporting these samples to the field laboratory to ensure a minimum of disturbance. Sampling of compacted mixes from the roadway should be done in accordance with AASHTO T168, Sampling Bituminous Paving Mixtures. A pavement power saw or coring machine used for taking samples provides the least disturbance of samples and compacted pavement (Figure 3–31).

Example

The density of an asphalt concrete mix is determined by laboratory tests to be 148.9 pcf. The density of a field

FIGURE 3–31. Cores being taken for laboratory density tests. *(Courtesy Advance Testing Company)*

core taken from a section of compacted pavement by test is 146.2 pcf.

$$PC = \frac{\gamma_F \, (pcf)}{\gamma_L \, (pcf)} \times 100$$

where PC = percent compaction (%)

γ_F = density of field core or nuclear density test value (pcf)

γ_L = density of laboratory specimen (pcf)

$$PC = \frac{146.2}{148.9} \times 100 = 98.2\%$$

Currently Air Permeability tests and Nuclear Density tests have progressed to the point where compaction may be controlled without having to obtain a sample of compacted pavement. These nondestructive tests also make possible the rapid determination of density so that additional compaction may be given, if required, while the pavement is still hot (Figure 3–32).

AUXILIARY EQUIPMENT

Asphalt Distributor

The asphalt distributor is used to apply either a prime coat or a tack coat to the surface to be paved. Prime coats are applications of liquid asphalt to an absorbent surface such as granular base. Tack coats are very light applications of liquid asphalt to an existing paved surface which is thoroughly cleaned.

The *distributor* (Figure 3–33) consists of a truck or trailer on which is mounted an insulated tank with a heating system, usually oil burning. The distributor has a power-driven pump and a system of spray bars and nozzles through which the asphalt is forced, under pressure, onto the construction surface.

It is important that asphalt sprayed from a distributor be spread over the surface uniformly at the desired rate of application. This requires a pump in good working order and free-flowing spray bars and nozzles.

To obtain a desired rate of application, the speed of the distributor must be determined for a given pumping rate and coverage width. Distributors usually have a bitumeter which indicates the forward speed of the distributor in feet per minute. This speed must be held constant if the asphalt coverage is to be uniform.

Motor Grader

In some situations the *motor grader* may be used to spread asphalt plant mixes, for example, in the placement of leveling courses. A leveling course is a thin layer of asphalt plant mix placed under a pavement's wearing surface.

FIGURE 3–32. Nuclear Density test on compacted pavement. *(Courtesy Advance Testing Company)*

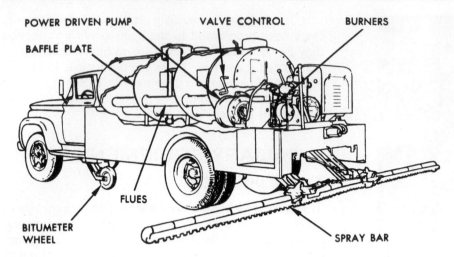

POWER DRIVEN PUMP   VALVE CONTROL   BURNERS

BAFFLE PLATE

FLUES

BITUMETER WHEEL

SPRAY BAR

FIGURE 3–33. Bituminous distributor truck. *(Courtesy The Asphalt Institute)*

The leveling course helps the paver lay a smooth, uniform-wearing course by removing irregularities in the old pavement.

Windrowing Equipment

Windrows of asphalt plant mixes are sometimes placed in the roadway and in front of the asphalt paver. An elevator attachment fixed to the front end of the paver picks up the asphalt mix and discharges it into the hopper. This eliminates the necessity of having trucks backing up and discharging their loads as the paver moves forward.

Windrowing equipment may also be used for controlling the amount of material for leveling courses spread by a motor grader.

Incidental Tools

Adequate hand tools and proper equipment for cleaning and heating them should be available for the paving operation. Incidental tools include the following:

1. Rakes
2. Shovels
3. Lutes
4. Tool heating torch
5. Cleaning equipment
6. Hand tampers
7. Small mechanical vibrating compactors
8. Blocks and shims for supporting the screed of the paver when beginning operations
9. Rope, canvas, or timbers for construction of joints at ends of runs
10. Joint cutting and painting tools
11. Straightedge

Not all of these tools are necessarily used on every paving project or every day of a particular job. Rakes, shovels, and lutes are frequently used by personnel around the paver. While the paver is operating, the laborers work or rework a portion of the mix to fit the paving around objects or fill in areas that the paver does not pave adequately or cannot reach during the paving operation. Ramps, intersections, and areas broken up by bridge columns and piers are typical areas where hand paving will be required.

ASPHALT ROOFING PRODUCTS

The preservative and waterproofing characteristics of asphalt make it an ideal material for roofing systems. The unit of measure for roofing materials is 100 sq ft and is referred to as 1 square. For example, a 15 square roof would be 1500 sq ft.

Asphalts used in the production of roofing products include *saturant*, or oil-rich, asphalts, and a harder, more viscous asphalt known as a *"coating asphalt."* The softening point of saturants varies from 100 to 160°F, and the softening point of coating asphalt may run as high as 260°F.

The sheet material used in the production of roofing products may be either an *organic felt sheet* or an *inorganic glass fiber mat*. Fiberglass shingles are governed by ASTM D3462 and organic felt shingles by ASTM D225; both standards require the shingles to pass tear-strength and nail withdrawal tests. In general, organic shingles have higher tear and nail-withdrawal strengths. Shingles are manufactured in many colors and profiles. One of the most popular types of shingle is the architectural. The architectural shingle has a random pattern and is easier to install than the 3-tab shingle because the alignment is not a problem. Cellulose fibers from rags, paper, and wood are processed into a dry felt which must have certain characteristics of strength, absorptive capacity, and flexibility. The glass fiber mat is composed of continuous or random thin glass fibers bonded with plastic binders. Glass mats are coated and impregnated with asphalt in one operation, while the felt sheets are impregnated with the saturant first and then

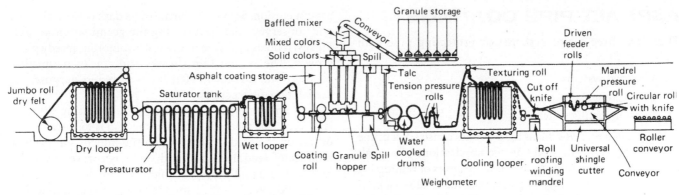

FIGURE 3–34. Flow diagram of a typical roofing plant. *(Courtesy Asphalt Roofing Manufacturers Association)*

coated with the coating asphalt. Coating asphalts generally contain a finely ground mineral stabilizer, giving the coating increased weathering resistance and increased resistance to shattering in cold weather. Some mineral stabilizers used are silica, slate dust, talc, micaceous materials, dolomite, and traprock.

Certain asphalt roofing products are coated with mineral aggregate granules. The granules protect the coating asphalt from light and offer a weathering protection. The mineral granules also increase the fire resistance of asphalt roofing. The mineral granules also impart the various colors or color blends that are available in roofing products.

The manufacturer distributes finely ground talc or mica powder on the surfaces of rolled roofing and shingles to prevent their sticking together before they are used. The powder has no other purpose and usually disappears from exposed surfaces soon after the roofing is installed.

The flow diagram in Figure 3–34 illustrates the production of asphalt roofing. During production, the materials are rigidly inspected to ensure conformance with standards. Some important items checked are the following:

1. Saturation of felt to determine quantity of saturant and efficiency of saturation
2. Thickness and distribution of coating asphalt
3. Adhesion and distribution of mineral granules
4. Weight, count, size, coloration, and other characteristics of finished product before and after it is packaged

The installation of asphalt roofing usually requires small amounts of accessory asphaltic materials. *Flashing cements* are used as part of a flashing system at points of vertical intersection such as where the roof meets a wall, chimney, or vent pipes. Flashing cements are processed to remain elastic through the normal temperature ranges of summer to winter. They are elastic after setting, and normal expansion and contraction of the roof deck should not cause separation to occur.

Lap cements are thinner than flashing cements and are used to make a watertight bond between laps of roll roofing. *Roof coatings* are thin liquids applied with brush or by spraying to resurface old roofs and may be either emulsified or cutback asphalts.

Manufactured asphalt roofing products are produced to ASTM standards. Asphalt roofing that is listed by the Underwriters Laboratories, Inc. (UL) as *A, B,* or *C* will not ignite easily, readily spread flames, or create flaming brands to endanger nearby buildings. UL also rates shingles for wind resistance. They must withstand a 60-mph wind for 2 hours to carry the UL label.

Prepared roofing products are those products which are manufactured and packaged ready to apply to the roof deck, usually by nailing. Asphalt shingles and certain roll roofings are considered prepared roofing products.

Another asphalt roofing system is the built-up roof. The *built-up roof* is used on flat or almost flat roofs and consists of alternate layers of roofing felt and asphalt covered with an aggregate coating.

Roofing asphalt is heated in an asphalt kettle, and is raised to the roof by pulley or pumped directly from the kettle to the roof. The hot asphalt is mopped or distributed by hot-asphalt applicators and covered with roofing felt in layers until the last coat of asphalt is applied, depending on the number of plies specified. The last asphalt application is a flood coat which is immediately covered with mineral aggregates before it cools. The aggregate is generally spread mechanically, but can be done by hand.

Asphalts used for built-up roofs are softer asphalts and have greater temperature susceptibility so that they soften and flow slightly during warm weather. This softening enables the asphalt to "heal" small cracks which may develop from contraction, expansion, or minor movements due to settlement of the building.

Roofing asphalt is available for use in any climate and on any slope on which built-up roofing can be used. Refer to ASTM D312 Section 4 for the types and general applications.

ASPHALT PIPE COATINGS

There are three major systems for protecting pipe with asphalt coating:

1. Wrapped systems
2. Mastic systems
3. Interior coating systems

Asphalt-wrapped systems for pipelines consist of a prime coat followed by either one or two applications of asphalt enamel in conjunction with one or more layers of reinforcing and protective wrapping. The wrapping material is asphalt-saturated felt or asphalt-saturated glass wrap.

Mastic systems for pipelines consist of a prime coat followed by a coating of a dense, impervious, essentially voidless mixture of asphalt, mineral aggregate, and mineral filler. The minimum thickness is usually $1/4$ in. The finished mastic coating is usually coated with whitewash.

Interior coating systems consist of a prime coat followed by a centrifugally cast layer of asphalt enamel about $1/32$ to $3/32$-in. thick.

ASPHALT MULCH TREATMENTS

Stabilizing slopes and flat areas on construction projects is a chronic problem. Soil erosion caused by wind and water can be prevented by the establishment of vegetation which anchors the soil in place.

However, during the germination of seed and early plant growth, mulch must be used to prevent erosion. There are two acceptable methods for using asphalt in the mulching process:

1. Asphalt spray mulch
2. Asphalt mulch tie-down

Asphalt spray mulch is usually an emulsified asphalt sprayed on the newly seeded area. The thin film of asphalt has three beneficial effects. First, it holds the seed in place against erosion. Second, by virtue of its dark color, it absorbs and conserves solar heat during the germination period. Finally, it holds moisture in the soil, promoting speedy plant growth. The asphalt film shrinks and cracks, permitting plant growth. Eventually, after it has served its purpose, the asphalt film disintegrates.

An *asphalt mulch tie-down* can be done using either of two acceptable methods. The first method is to spread straw or hay on the graded slope. When the mulch is in place, a mixture of seed, fertilizer, and water is sprayed over the mulch. The liquid passes through the mulch into the soil. The liquid asphalt is then sprayed over the mulch to lock it in place.

The second method requires spreading the seed and fertilizer on the prepared soil, followed by spraying the asphalt and mulching material simultaneously.

ASPHALT JOINT MATERIALS

Asphalt is used as a *joint* and *crack filler* in pavements and other structures. The joint between two concrete pavement slabs may be filled with hot asphalt. The asphalt's properties of adhesion and flexibility allow the two separate slabs to expand and contract without permitting moisture to penetrate the joint. The asphalt material used may also be premolded in strips. The strips are composed of asphalt mixed with fine mineral substances, fibrous materials, cork, or sawdust, manufactured in dimensions suitable for use in joints. The premolded strips are inserted at joint locations before placement of the portland cement concrete and eliminate the filling of the joint openings after the slabs have cured.

Asphalt is a widely used construction material with diverse applications. To ensure that asphalt and its products are used properly, organizations such as the Asphalt Institute, Asphalt Roofing Manufacturers' Association, and others produce technical literature concerning asphalt uses. For an in-depth treatment of the material, such technical literature is recommended.

Review Questions

1. How are natural asphalts formed?
2. What are the differences between asphalt and tar?
3. What are bituminous materials?
4. What are the classifications of petroleum crude oil, based on their asphalt content?
5. How is the percentage of asphalt in a crude oil determined?
6. What distillation processes are used to produce asphalt?
7. What do the basic tests on asphalt materials measure?
8. What are the basic requirements for an asphalt pavement?
9. What factors must be considered when designing an asphalt pavement's thickness?
10. What objectives must the design of an asphalt pavement mix meet?
11. What are tack coats and prime coats?
12. What is the suggested pavement thickness for a full-depth asphalt concrete driveway, if the in situ soils are nonplastic clayey loams?
13. What is surge storage, and how is it used in asphalt concrete production?

14. Calculate the number of tons of asphalt concrete required to pave a 300-sq-ft area 3-in. thick. The asphalt concrete has a density of 150 lb per cu ft.

15. How many tons of asphalt concrete will be required to overlay a 2.5-mile-long, 24-ft-wide road with 3 in. of wearing course asphalt concrete that has a density of 147.9 pcf?

16. Briefly detail the paving operation utilizing mechanized equipment.

17. The core densities listed were obtained from a compacted asphalt concrete pavement: core A = 142.9 pcf, core B = 149.9 pcf, and core C = 143.4 pcf. Determine the percentage of compaction if the laboratory density of the mix was 147.1 pcf. Do any of the cores indicate potential problems with the pavement?

18. What is an asphalt distributor truck, and what functions does it serve on a pavement project?

19. Describe briefly how asphalt roofing materials are produced.

20. How is asphalt used for slope stabilization in highway construction?

21. Calculate the number of tons of asphalt concrete required to pave a 500-sq-ft area 2-in. thick. The asphalt concrete has a density of 148.6 lb per cu ft.

22. How many tons of asphalt concrete will be required to overlay a 3-mile-long, 24-ft-wide road with $2\frac{1}{2}$ in. of wearing course asphalt concrete that has a density of 146.3 pcf?

23. What are the suggested thicknesses for an asphalt concrete surface parking area for heavy trucks on gravelly or sandy well-drained soils?

24. Calculate the metric tons of asphalt concrete required to pave 2 km of roadway 4 m wide and 80 mm thick. The density of the paving material is 2370 kg/m^3.

25. Calculate the compacted tons per hour of asphalt concrete if a 4.5-ft-wide roller traveling at 3 mph over a 3-in.-thick lift requires five passes to achieve a required density of 147.6 pcf. (The project field conditions are average.)

CHAPTER FOUR

PORTLAND CEMENT
CONCRETE

Concrete is the world's most widely used construction material, and the Portland Cement Association estimates its annual production in excess of 5 billion cu yd. Concrete has many characteristics that make it such a widely used construction material. Among them are raw material availability, the ability of concrete to take the shape of the form it is placed in, and the ease with which its properties can be modified. The ability of concrete to modify such properties as its strength, durability, economy, watertightness, and abrasion resistance is most important. As illustrated in Table 4–1, there are a great many variables that affect the properties of concrete. The ease with which concrete can be modified by its variables can often work to the disadvantage of the user if quality control is not maintained from the first to the last operations in concrete work.

Basically, *concrete* is 60 to 80 percent aggregates (i.e., sand and stone), which are considered "inert" ingredients, and 20 to 40 percent "paste" (i.e., water and portland cement), considered the active ingredient. These materials are combined, or mixed, and cured to develop the hardened properties of concrete.

During this production sequence, concrete is often produced by one firm with products supplied by three or four other firms and sold in an unfinished state to a contractor who will place, finish, and cure it. During this process, it will be subject to the weather and other variables. Therefore, in order to ensure that the concrete initially designed for a specific function is the same concrete that ends up in service, careful consideration must be given to all of the variables which may affect it.

HISTORY

The development of cementing materials can be traced back to the Egyptians and Romans and their use of masonry construction. The Egyptians used a cement produced by a heating process, and this may have been the start of the technology. Roman engineering upgraded simple lime mortars with the addition of volcanic ash which increased their durability, as evidenced by the sound structures that still stand.

A good example of the durability and longevity of Roman concrete is the Aqua Virgo built around 19 B.C. Today it still carries spring water over 11 miles to the Trevi Fountain in Rome.

The development of concrete or cement technology as we know it today probably can be traced back to England, where in 1824 Joseph Aspdin produced a portland cement from a heated mixture of limestone and clay. He was awarded a British patent, and the name "portland cement" was used because when the material hardened, it resembled a stone from the quarries of Portland, England. Several of Joseph Aspdin's contemporaries were involved in the same research, but evidence shows that he fired his product at approximately the clinkering temperature, thus producing a superior product.

The production of portland cement in the United States dates back to 1872, when the first portland cement plant was opened at Coplay, Pennsylvania. Today, the production of portland cement is considered a basic industry, and it occurs in almost all regions of the world.

MANUFACTURE OF PORTLAND CEMENT

The manufacture of portland cement requires raw materials which contain lime, silica, alumina, and iron. The sources of these elements vary from one manufacturing location to another, but once these materials are obtained, the process is rather uniform (Table 4–2).

As illustrated in Figure 4–1, the process begins with the acquisition of raw materials such as limestone, clay, and sand. The limestone is reduced to an approximately 5-in. size in the primary crusher and further reduced to $3/4$ in. the

Table 4–1. Chart showing the principal properties of good concrete, their relationships, and the elements which control them

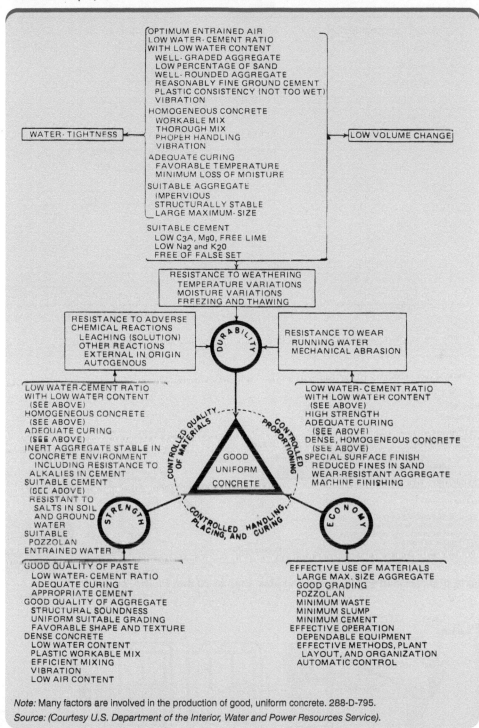

OPTIMUM ENTRAINED AIR
LOW WATER- CEMENT RATIO
WITH LOW WATER CONTENT
  WELL- GRADED AGGREGATE
  LOW PERCENTAGE OF SAND
  WELL- ROUNDED AGGREGATE
  REASONABLY FINE GROUND CEMENT
  PLASTIC CONSISTENCY (NOT TOO WET)
  VIBRATION
HOMOGENEOUS CONCRETE
  WORKABLE MIX
  THOROUGH MIX
  PROPER HANDLING
  VIBRATION
ADEQUATE CURING
  FAVORABLE TEMPERATURE
  MINIMUM LOSS OF MOISTURE
SUITABLE AGGREGATE
  IMPERVIOUS
  STRUCTURALLY STABLE
  LARGE MAXIMUM- SIZE

SUITABLE CEMENT
  LOW C_3A, MgO, FREE LIME
  LOW Na_2 and K_2O
  FREE OF FALSE SET

WATER- TIGHTNESS

LOW VOLUME CHANGE

RESISTANCE TO WEATHERING
  TEMPERATURE VARIATIONS
  MOISTURE VARIATIONS
  FREEZING AND THAWING

RESISTANCE TO ADVERSE
CHEMICAL REACTIONS
  LEACHING (SOLUTION)
  OTHER REACTIONS
    EXTERNAL IN ORIGIN
    AUTOGENOUS

RESISTANCE TO WEAR
  RUNNING WATER
  MECHANICAL ABRASION

DURABILITY

LOW WATER-CEMENT RATIO
WITH LOW WATER CONTENT
  (SEE ABOVE)
HOMOGENEOUS CONCRETE
  (SEE ABOVE)
ADEQUATE CURING
  (SEE ABOVE)
INERT AGGREGATE STABLE IN
CONCRETE ENVIRONMENT
  INCLUDING RESISTANCE TO
  ALKALIES IN CEMENT
SUITABLE CEMENT
  (SEE ABOVE)
  RESISTANT TO
    SALTS IN SOIL
    AND GROUND
    WATER
SUITABLE
  POZZOLAN
ENTRAINED WATER

LOW WATER- CEMENT RATIO
WITH LOW WATER CONTENT
  (SEE ABOVE)
HIGH STRENGTH
ADEQUATE CURING
  (SEE ABOVE)
DENSE, HOMOGENEOUS CONCRETE
  (SEE ABOVE)
SPECIAL SURFACE FINISH
  REDUCED FINES IN SAND
  WEAR-RESISTANT AGGREGATE
  MACHINE FINISHING

CONTROLLED QUALITY OF MATERIALS
CONTROLLED PROPORTIONING
CONTROLLED HANDLING, PLACING, AND CURING

GOOD UNIFORM CONCRETE

STRENGTH

ECONOMY

GOOD QUALITY OF PASTE
  LOW WATER- CEMENT RATIO
  ADEQUATE CURING
  APPROPRIATE CEMENT
GOOD QUALITY OF AGGREGATE
  STRUCTURAL SOUNDNESS
  UNIFORM SUITABLE GRADING
  FAVORABLE SHAPE AND TEXTURE
DENSE CONCRETE
  LOW WATER CONTENT
  PLASTIC WORKABLE MIX
  EFFICIENT MIXING
  VIBRATION
  LOW AIR CONTENT

EFFECTIVE USE OF MATERIALS
  LARGE MAX. SIZE AGGREGATE
  GOOD GRADING
  POZZOLAN
  MINIMUM WASTE
  MINIMUM SLUMP
  MINIMUM CEMENT
EFFECTIVE OPERATION
  DEPENDABLE EQUIPMENT
  EFFECTIVE METHODS, PLANT
    LAYOUT, AND ORGANIZATION
  AUTOMATIC CONTROL

Note: Many factors are involved in the production of good, uniform concrete. 288-D-795.

Source: (Courtesy U.S. Department of the Interior, Water and Power Resources Service).

secondary crusher. All of the raw materials are stored in the bins and proportioned prior to delivery to the grinding mill. The wet process results in a slurry, which is mixed and pumped to storage basins. The dry process produces a fine ground powder which is stored in bins.

Both processes feed rotary kilns where the actual chemical changes will take place. The material is fed into the upper end of the kiln, and as the kiln rotates, the material passes slowly from the upper to the lower end at a rate controlled by the slope and speed of rotation of the kiln. As the material passes through the kiln, its temperature is raised to the point of *incipient fusion*, or clinkering temperature, where the chemical reactions take place. Depending on the raw materials, this temperature is

Table 4–2. Sources of raw materials used in the manufacture of portland cement

Calcium	Iron	Silica	Alumina	Sulfate
Alkali waste	Blast-furnace flue dust	Calcium silicate	Aluminum-ore refuse*	Anhydrite
Aragonite*	Clay*	Cement rock	Bauxite	Calcium sulfate
Calcite*	Iron ore*	Clay*	Cement rock	Gypsum*
Cement-kiln dust	Mill scale*	Fly ash	Clay*	
Cement rock	Ore washings	Fuller's earth	Copper slag	
Chalk	Pyrite cinders	Limestone	Fly ash*	
Clay	Shale	Loess	Fuller's earth	
Fuller's earth		Marl*	Granodiorite	
Limestone*		Ore washings	Limestone	
Marble		Quartzite	Loess	
Marl*		Rice-hull ash	Ore washings	
Seashells		Sand*	Shale*	
Shale*		Sandstone	Slag	
Slag		Shale*	Staurolite	
		Slag		
		Traprock		

Note: As a generalization, probably 50 percent of all industrial by-products have potential as raw materials for the manufacture of portland cement.

*Most common sources.

Source: Portland Cement Association.

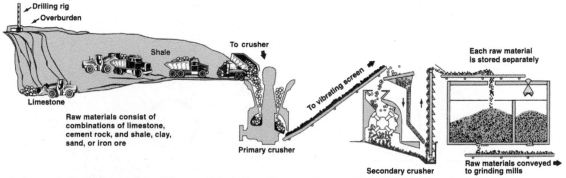

1. Stone is first reduced to 125 mm (5 in.) size, then to 20 mm (³⁄₄ in.), and stored.

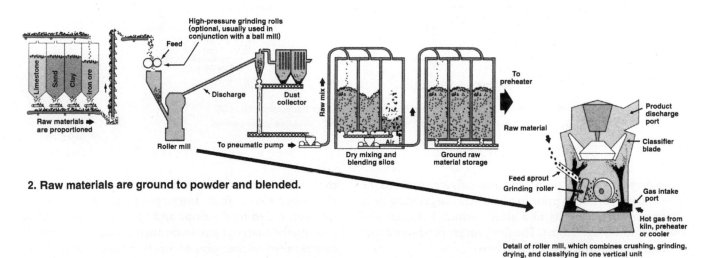

2. Raw materials are ground to powder and blended.

Detail of roller mill, which combines crushing, grinding, drying, and classifying in one vertical unit

FIGURE 4–1. Flow chart of the manufacture of portland cement. *(Courtesy Portland Cement Association)*

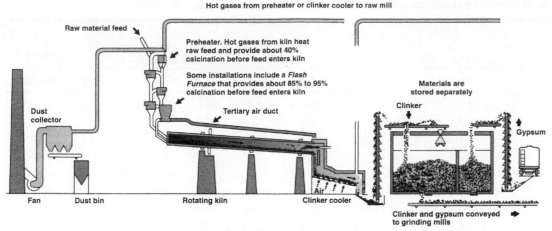

3. Burning changes raw mix chemically into cement clinker. Note four-stage preheater, flash furnaces, and shorter kiln.

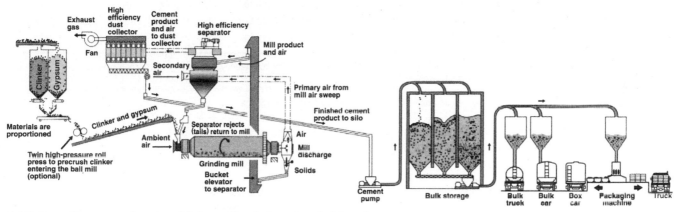

4. Clinker with gypsum is ground into portland cement and shipped.

FIGURE 4–1. Continued

usually between 2400°F (1316°C) and 2700°F (1482°C). Chemical recombinations of the raw ingredients take place in this temperature range to produce the basic chemical components of portland cement.

The *clinker* produced is black or greenish black in color and rough in texture. Its size makes it relatively inert in the presence of moisture. From clinker storage, the material is transported to final grinding where approximately 2 to 3 percent gypsum is added to control the setting time of the portland cement when it is mixed with water. The resulting concrete will have improved shrinkage and strength properties.

The portland cement produced is either distributed in bulk by rail, barge, or truck or packaged in bags. Bulk cement is sold by the barrel, which is the equivalent of four bags or 376 lb, or by the ton. Bag cement weighs 94 lb ± 3 percent and is considered to be 1 bulk cubic foot of cement.

The manufacture of portland cement involves the use of many technical skills to ensure a uniform product. Engineers, chemists, and technicians are all employed in the process of determining and controlling the various chemical and physical properties of the cements produced. For practical purposes, Type I portland cement contains the oxides illustrated in Table 4–3.

CHEMICAL COMPOSITION OF PORTLAND CEMENT

Portland cements are composed of four basic chemical compounds, shown with their names, chemical formulas, and abbreviations:

1. Tricalcium silicate: $3CaO \cdot SiO_2 = C_3S$
2. Dicalcium silicate: $2CaO \cdot SiO_2 = C_2S$
3. Tricalcium aluminate: $3CaO \cdot Al_2O_3 = C_3A$
4. Tetracalcium aluminoferrite: $4CaO \cdot Al_2O_3FeO_3 = C_4AFe$

The relative percentages of these compounds can be determined by chemical analysis. Each of the components exhibits a particular behavior, and it can be shown that by modifying the relative percentages of these compounds, the behavior of the cement can be altered.

Tricalcium silicate hardens rapidly and is largely responsible for initial set and early strength. In general, the early strength of portland cement concretes will be higher with increased percentages of C_3S. However, if moist curing is continued, the later strength after about 6 months will be greater for cements with a higher percentage of C_2S.

Table 4–3. Chemical and compound composition and fineness of some typical cements

Type of Portland Cement	Chemical Composition(%)							Potential Compound Composition (%)*				Blaine fineness m2/kg
	SiO$_2$	Al$_2$O$_3$	Fe$_2$O$_3$	CaO	MgO	SO$_3$	Na$_2$O eq	C$_3$S	C$_2$S	C$_3$A	C$_4$AF	
I (min-max)	18.7-22.0	4.7-6.3	1.6-4.4	60.6-66.3	0.7-4.2	1.8-4.6	0.11-1.20	40-63	9-31	6-14	5-13	300-421
I (mean)	**20.5**	**5.4**	**2.6**	**63.9**	**2.1**	**3.0**	**0.61**	**54**	**18**	**10**	**8**	**369**
II** (min-max)	20.0-23.2	3.4-5.5	2.4-4.8	60.2-65.9	0.6-4.8	2.1-4.0	0.05-1.12	37-68	6-32	2-8	7-15	318-480
II (mean)**	**21.2**	**4.6**	**3.5**	**63.8**	**2.1**	**2.7**	**0.51**	**55**	**19**	**6**	**11**	**377**
III (min-max)	18.6-22.2	2.8-6.3	1.3-4.9	60.6-65.9	0.6-4.6	2.5-4.6	0.14-1.20	46-71	4-27	0-13	4-14	390-644
III (mean)	**20.6**	**4.9**	**2.8**	**63.4**	**2.2**	**3.5**	**0.56**	**55**	**17**	**9**	**8**	**548**
IV (min-max)	21.5-22.8	3.5-5.3	3.7-5.9	62.0-63.4	1.0-3.8	1.7-2.5	0.29-0.42	37-49	27-36	3-4	11-18	319-362
IV (mean)	**22.2**	**4.6**	**5.0**	**62.5**	**1.9**	**2.2**	**0.36**	**42**	**32**	**4**	**15**	**340**
V (min-max)	20.3-23.4	2.4-5.5	3.2-6.1	61.8-66.3	0.6-4.6	1.8-3.6	0.24-0.76	43-70	11-31	0-5	10-19	275-430
V (mean)	**21.9**	**3.9**	**4.2**	**63.8**	**2.2**	**2.3**	**0.48**	**54**	**22**	**4**	**13**	**373**
White (min-max)	22.0-24.4	2.2-5.0	0.2-0.6	63.9-68.7	0.3-1.4	2.3-3.1	0.09-0.38	51-72	9-25	5-13	1-2	384-564
White (mean)	**22.7**	**4.1**	**0.3**	**66.7**	**0.9**	**2.7**	**0.18**	**63**	**18**	**10**	**1**	**482**

*Values represent a summary of combined statistics. Air-entraining cements are not included. For consistency in reporting, elements are reported in a standard oxide form. This does not mean that the oxide form is present in the cement. For example, sulfur is reported as SO$_3$, sulfur trioxide, but, portland cement does not have sulfur trioxide present. "Potential Compound Composition" refers to ASTM C 150 (AASHTO M 85) calculations using the chemical composition of the cement. The actual compound composition may be less due to incomplete or altered chemical reactions.

**Includes fine-grind cements.

Source: Adapted from Portland Cement Assoiaiton (1996) and Gebhardt, RF "Survey of North American Portland Cements" (1994).

Dicalcium silicate hardens slowly, and its effect on strength increases occurs at ages beyond one week.

Tricalcium aluminate contributes to strength development in the first few days because it is the first compound to hydrate. It is, however, the least desirable component because of its high heat generation and its reactiveness with soils and water containing moderate-to-high sulfate concentrations. Cements made with low C_3A contents usually generate less heat, develop higher strengths, and show greater resistance to sulfate attacks.

Tetracalcium aluminoferrite assists in the manufacture of portland cement by allowing lower clinkering temperature. C_4AFe contributes very little to the strength of concrete even though it hydrates very rapidly.

Most specifications for portland cements place limits on certain physical properties and chemical composition of the cements. Therefore, the study of this material requires an understanding of some of these basic properties.

PHYSICAL PROPERTIES OF PORTLAND CEMENT

One factor which affects the hydration of cement, regardless of its chemical composition, is its *fineness*. The finer a cement is ground, the higher the heat of hydration and resulting accelerated strength gain. The strength gain due to fineness is evident during the first 7 days.

For a given weight of cement, the surface area of the grains of a coarse-ground cement is less than for a fine-ground cement. Because the water is in contact with more surface area in a fine-ground cement, the hydration process occurs more rapidly in this cement. If the cement is ground too finely, however, there is a possibility of prehydration due to moisture vapor during manufacturing and storage, with the resulting loss in cementing properties of the material. There is evidence to show that in some cases very coarse-ground particles may never completely hydrate.

Table 4–3 shows the compound composition and fineness of each type of portland cement. Currently the fineness of cement is stated as *specific surface* (i.e., the calculated surface area of the particles in square meters per kilogram [m²/kg] of cement). Even though the specific surface is only an approximation of the true area, good correlations have been obtained between specific surfaces and those properties influenced by particle fineness. Higher specific surfaces indicate a finer-ground, and usually a more active, cement.

Soundness is the ability of a cement to maintain a stable volume after setting. An unsound cement will exhibit cracking, disruption, and eventual disintegration of the material mass. This delayed-destruction expansion is caused by excessive amounts of free lime or magnesium.

The free lime is enclosed in cement particles, and eventually the moisture reaches the lime after the cement has set. At that time the lime expands with considerable force, disrupting the set cement.

The current test for soundness in cement is the ASTM Autoclave test. Standard specimens of neat cement paste are subjected to high pressure and temperature for 3 hours.

After cooling, the length of bars is compared with the length before testing and cements that exhibit an expansion of no more than 0.50 percent are considered to be sound. Since the introduction of the Autoclave test, there have been almost no cases of delayed expansion due to unsound cements.

WATER–CEMENT REACTION

Hydration is the chemical reaction that takes place when portland cement and water are mixed together. The hydration reaction is considered complete at 28 days. The reaction depends on available moisture. Figure 4–2 indicates strength gains for different curing conditions.

When cement is mixed with water to form a fluid paste, the mixture will eventually become stiff and then hard. This process is called *setting*. A cement used in concrete must not set too fast, for then it would be unworkable, that is, it would stiffen and become hard before it could be placed or finished.

Table 4–4. Approximate relative strength of concrete as affected by type of cement

Type of Portland Cement		Compressive Strength (Percent of Strength of Type I or Normal Portland Cement Concrete)			
ASTM	CSA	1 day	7 days	28 days	3 mos.
I	Normal	100	100	100	100
II		75	85	90	100
III	High-Early-Strength	190	120	110	100
IV		55	55	75	100
V	Sulfate-Resisting	65	85	100	

Source: Portland Cement Association.

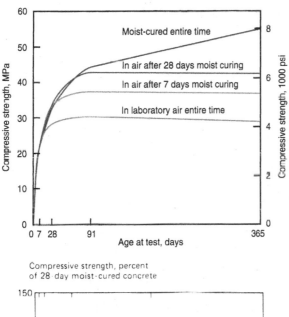

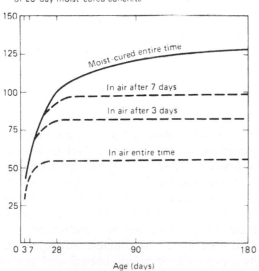

FIGURE 4–2. Strength of concrete continues to increase as long as moisture is present for hydration of cement. *(Courtesy Portland Cement Association)*

When it sets too slowly, valuable construction time is lost. Most portland cements exhibit initial set in about 3 hours and final set in about 7 hours. If gypsum were not added during final grinding of normal portland cement, the set would be very rapid and the material unworkable.

False set of portland cement is a stiffening of a concrete mixture with little evidence of significant heat generation. To restore plasticity, all that is required is further mixing without additional water. There are cases where a *flush set* is exhibited by a cement, and in this case the cement has *hydrated* and further remixing will do no good. The actual setting time of the concrete will vary from job to job depending on the temperature of the concrete and wind velocity, humidity, placing conditions, and other variables.

The ability of a cement to develop compressive strength in a concrete is an important property (Table 4–4). The compressive strengths of cements are usually determined on standard 2-in. (50.8-mm) cubes. The results of these tests are useful in comparing strengths of various cements in neat paste conditions. *Neat paste* is water, cement, and a standard laboratory sand used to standardize tests. The tests will not predict concrete strength values due to the variables in concrete mixtures that also influence strength.

The heat generated when water and cement chemically react is called the *heat of hydration*, and it can be a critical factor in concrete use. The total amount of heat generated depends on the chemical composition of the cement, and the rate is affected by the fineness, chemical composition, and curing temperatures (Figure 4–3).

Approximate amounts of heat generation during the first 7 days of curing using Type I cement as the base are as follows:

Type I	100%
Type II	80–85%
Type III	150%
Type IV	40–60%
Type V	60–75%

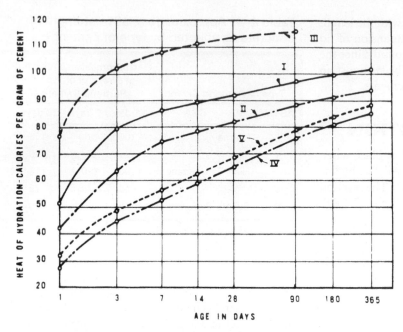

HEAT OF HYDRATION FOR VARIOUS TYPES OF CEMENT

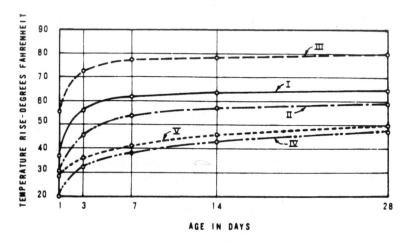

TEMPERATURE RISE OF CONCRETE

Tests of mass concrete with $4\frac{1}{2}$-in. maximum aggregate, containing 376 pounds of cement per cubic yard in 17-by 17-in. cylinders, sealed and cured in adiabatic calorimeter rooms

FIGURE 4–3. Heat of hydration and temperature rise for concretes made with various types of cement. 288-D-118 (*Courtesy U.S. Department of the Interior, Water and Power Resources Service*)

Concrete has a low tensile strength; it is approximately 11 percent of concrete's compressive strength. To allow the use of concrete in locations where tensile strength is important or increased compressive strength is required, steel is used to reinforce the concrete. (Refer to Table 4–8 for reinforcing steel sizes.) Steel used for reinforcing concrete can be welded wire mesh, deformed reinforcing bars, or cable tendons. Steel has a high strength-to-weight ratio, and its coefficient of thermal expansion is almost the same as concrete's.

Plain reinforced concrete can be used for most construction. The steel is positioned in the form and the concrete is placed around it. When the concrete has cured, the materials are bonded together and act as one.

Prestressed concrete requires the application of a load to the steel before concrete placement. When the concrete has cured, the load is removed from the steel and the concrete is placed in compression. Posttensioned concrete involves the application of a load to the steel after the concrete has cured. Open ducts are left in the concrete through which steel tendons are placed, or plastic-coated steel tendons are draped in the formwork before the concrete is placed. When the concrete has cured, loads are applied to the steel.

Posttensioned concrete is versatile because the loads applied to the steel can be changed according to actual conditions of structure loading.

TYPES OF PORTLAND CEMENT ASTM C150

ASTM Type I (Normal)

This type is a general concrete construction cement utilized when the special properties of the other types are not required. It is used where the concrete will not be subjected to sulfate attack from soil or water or be exposed to severe weathering conditions. It is generally not used in large masses because of the heat generated due to hydration. Its uses include pavements and sidewalks, reinforced concrete buildings, bridges, railway structures, tanks, reservoirs, culverts, water pipes, and masonry units.

ASTM Type II (Moderate Heat or Modified)

Type II cement is used where resistance to moderate sulfate attack is important, as in areas where sulfate concentration in groundwater is higher than normal but not severe. Type II cements produce less heat of hydration than Type I, hence their use in structures of mass such as piers, abutments, and retaining walls. They are used in warm-weather concreting because of their lower temperature rise than Type I. The use of Type II for highway pavements will give the contractor more time to saw control joints because of the lower heat generation and resulting slower setting and hardening.

ASTM Type III (High-Early-Strength)

Type III cements are used where an early strength gain is important and heat generation is not a critical factor. When forms have to be removed for reuse as soon as possible, Type III supplies the strength required in shorter periods of time than the other types. In cold-weather concreting, Type III allows a reduction in the heated curing time with no loss in strength.

ASTM Type IV (Low Heat)

Type IV cement is used where the rate and amount of heat generated must be minimized. The strength development for Type IV is at a slower rate than Type I. It is primarily used in large mass placements such as gravity dams where the amount of concrete at any given time is so large that the temperature rise resulting from heat generation during hardening becomes a critical factor.

ASTM Type V (Sulfate-Resisting)

Type V is primarily used where the soil or groundwater contains high sulfate concentrations and the structure would be exposed to severe sulfate attack.

Air-Entraining Portland Cements

ASTM C175 governs the air-entraining cements, Types IA, IIA, and IIIA. The three cements correspond to Types I, II, and III, with the addition of small quantities of air-entraining materials integrated with the clinker during the manufacturing process. These cements provide the concrete with improved resistance to freeze–thaw action and to scaling caused by chemicals and salts used for ice and snow removal. Concrete made with these cements contains microscopic air bubbles, separated, uniformly distributed, and so small that there are many billions in a cubic foot.

White Portland Cement

White portland cement is a true portland cement, its color being the principal difference between it and normal portland cement. The cement is manufactured to meet ASTM C150 and C175 specifications. The selected raw materials used in the manufacture of white cement have negligible amounts of iron and manganese oxide, and the process of manufacture is controlled to produce the white color. Its primary use is for architectural concrete products, cement paints, tile grouts, and decorative concrete. Its use is recommended wherever white or colored concrete or mortar is desired. Colored concretes are produced by using a coloring additive, and the white cement allows for more accurate control of colors desired.

Portland Blast-Furnace Slag Cements

In these cements, granulated blast-furnace slag of selected quality is interground with portland cement. The slag is obtained by rapidly chilling or quenching molten slag in water, steam, or air. Portland blast-furnace slag cements include two types, Type IS and Type IS-A, conforming to ASTM C595. These cements can be used in general concrete construction when the specific properties of the other types are not required. However, moderate heat of hydration (MH), moderate sulfate resistance (MS), or both are optional provisions. Type IS has about the same rate of strength development as Type I cement, and both have the same compressive strength requirements.

Portland-Pozzolan Cements

IP, IP-A, P, and P-A designate the portland-pozzolan cements with the A denoting air-entraining additives as specified in C595. They are used principally for large hydraulic structures such as bridge piers and dams. These cements are manufactured by intergrinding portland cement clinker with a suitable pozzolan such as volcanic ash, fly ash from power plants, or diatomaceous earth, or by blending the portland cement or portland blast-furnace slag cement and a pozzolan.

Masonry Cements

Type I and Type II masonry cements are manufactured to conform to ASTM C91 and contain portland cement, air-entraining additives, and materials selected for their ability to impart workability, plasticity, and water-retention properties to the masonry mortars.

Special Portland Cements

Oil Well Cement   Oil well cement is used for sealing oil wells. It is usually slow setting and resistant to high pressures and temperatures. The American Petroleum Institute Specifications for Oil Well Cements (API standard 10A) cover requirements for six classes of cements. Each class is applicable for use at a certain range of well depths. Conventional portland cements are also used with suitable set-modifying admixtures.

Waterproof Portland Cement   Waterproof portland cement is manufactured by the addition of a small amount of calcium, aluminum, or other stearate to the clinker during final grinding. It is manufactured in either white or gray color and is used to reduce water penetration through the concrete.

MIXING WATER

In the production of concrete, water plays an important role. It is used to wash aggregates, as mixing water, during the curing process and to wash out mixers.

The use of an impure water for aggregate washing may result in aggregate particles being coated with silt, salts, or organic materials. Aggregates that have been contaminated by such impure water may produce distressed concrete due to chemical reactions with the cement paste or poor aggregate bonding. In most cases, comparative tests should be run on possible contaminated aggregates.

It is generally accepted that any potable water can be used as mixing water in the manufacture of concrete. Duff Abrams found seawater having a 3.5 percent salt content adequate in producing concrete so that some waters used in concrete making are not potable.

Questionable water supplies may be used for concrete if mortar cubes made with the water meet C94 requirements as shown in Table 4–5. Impurities in water may also affect volume stability and cause *efflorescence* (the leaching of free lime), discoloration, and excessive reinforcement corrosion.

A water source similar in analysis to any of the water supplies in Table 4–6 is probably satisfactory for use in concrete. The waters represent the approximate compositions of water supplies for most cities with populations over 20,000 in the United States and Canada. The units used to designate foreign matter in water are parts per million (ppm) and designate the weight of foreign matter to the weight of water.

Table 4–5. Acceptance criteria for questionable water supplies (ASTM C94 or AASHTO M 157)

	Limits	Test Method
Compressive strength, minimum percentage of control at 7 days	90	C 109* or T 106
Time of set, deviation from control, hour:min	from 1:00 earlier to 1:30 later	C 191* or T 131

*Comparisons should be based on fixed proportions and the same volume of test water compared to control mix using city water or distilled water.
Source: Portland Cement Association.

Table 4–6. Typical analyses of city water supplies and seawater, parts per million

	Analysis Number						
Chemical	1	2	3	4	5	6	Seawater*
Silica (SiO_2)	2.4	0.0	6.5	9.4	22.0	3.0	—
Iron (Fe)	0.1	0.0	0.0	0.2	0.1	0.0	—
Calcium (Ca)	5.8	15.3	29.5	96.0	3.0	1.3	50 to 480
Magnesium (Mg)	1.4	5.5	7.6	27.0	2.4	0.3	260 to 1410
Sodium (Na)	1.7	16.1	2.3	183.0	215.0	1.4	2190 to 12,200
Potassium (K)	0.7	0.0	1.6	18.0	9.8	0.2	70 to 550
Bicarbonate (HCO_3)	14.0	35.8	122.0	334.0	549.0	4.1	—
Sulfate (SO_4)	9.7	59.9	5.3	121.0	11.0	2.6	580 to 2810
Chloride (Cl)	2.0	3.0	1.4	280.0	22.0	1.0	3960 to 20,000
Nitrate (NO_3)	0.5	0.0	1.6	0.2	0.5	0.0	—
Total dissolved solids	31.0	250.0	125.0	983.0	564.0	19.0	35,000

*Different seas contain different amounts of dissolved salts.
Source: Portland Cement Association.

The upper limit of total dissolved solids is usually 2000 ppm. Although higher concentrations are not always harmful, certain cements may react adversely. Therefore, possible higher concentrations should be avoided. The effects of many common impurities have been well documented in technical literature.

Carbonates and bicarbonates of sodium and potassium affect the setting times of concrete. Sodium carbonate may cause very rapid setting; bicarbonates may either accelerate or retard the set. In large concentrations, these salts can materially reduce concrete strength. When the sum of these dissolved salts exceeds 1000 ppm (0.1 percent), tests for setting time and 28-day strength should be made.

Sodium chloride or sodium sulfate can be tolerated in large quantities; waters having concentrations of 20,000 ppm of sodium chloride and 10,000 ppm of sodium sulfate have been used successfully. Carbonates of calcium and magnesium are not very soluble in water and are seldom found in high enough concentrations to affect concrete properties. Bicarbonates of calcium and magnesium are present in some municipal water supplies, and concentrations of the bicarbonate of up to 400 ppm are not considered harmful.

Concentrations of magnesium sulfate and magnesium chloride up to 40,000 ppm have been used without harmful effects on concrete strength. Calcium chloride is often used as an accelerator in concrete in quantities up to 2 percent by weight of the cement. It cannot be used in prestressed concrete or in concrete containing aluminum conduit or pipe.

Iron salts in concentrations of up to 40,000 ppm have been used successfully; however, natural groundwater usually contains no more than 20 to 30 ppm.

The salts of manganese, tin, zinc, copper, and lead may cause reductions in strength and variations in setting times. Salts that act as retarders include sodium iodate, sodium phosphate, sodium arsenate, and sodium borate, and when present in amounts as little as a few tenths of 1 percent by weight of cement, they can greatly retard set and strength development. Concentrations of sodium sulfide as low as 100 ppm warrant testing.

Generally, seawater containing 35,000 ppm of salt can be used in nonreinforced concrete, which will exhibit higher early strength with a slight reduction in 28-day strength. The reduction in 28-day strength is usually compensated for in the mix design. Seawater has been used in reinforced concrete; however, if the steel does not have sufficient cover or if the concrete is not watertight, the risk of corrosion is increased greatly. Seawater should never be used in prestressed concrete.

Aggregates from the sea may be used with fresh mixing water because the salt coating would amount to about 1 percent by weight of the mixing water.

Generally, mixing waters having common inorganic acid concentrations as high as 10,000 ppm have no adverse effects on concrete strength. The acceptance of mixing waters containing acid should be based on concentrations of acids in ppm rather than the pH value, as the latter is an intensity index.

Sodium hydroxide concentrations of 0.5 percent by weight of cement do not greatly affect the concrete strength, provided quick set does not occur. Potassium hydroxide in concentrations up to 1.2 percent by weight of cement can reduce strengths of certain cements while not materially affecting strengths of others.

Industrial waste water and sanitary sewage can be used in concretes. After sewage passes through a good disposal system, the concentration of solids is usually too low to have any significant effect on concrete. Waste waters from tanneries, paint factories, coke plants, chemical plants, galvanizing plants, and so on may contain harmful impurities. As with all questionable water sources, it pays to run the comparative strength tests before using such waters in concrete manufacturing.

Sugar in concentrations of as little as 0.03 to 0.15 percent by weight of cement will usually retard the setting time of cement. There may be a reduction in 7-day strength and an increase in 28-day strength. When the amount of sugar is raised to 0.20 percent by weight of cement, the set is accelerated. When the sugar exceeds 0.25 percent, there may be rapid setting and a reduction in 28-day strength. Water containing an excess of 500 ppm of sugar should be tested.

Clay or fine particles can be tolerated in concentrations of up to 2000 ppm. Though the clay may affect other properties of cement, the strength should not be affected at higher concentrations. Silty water should settle in basins before use to reduce the suspended silts and clays.

Mineral oils have less effect on strength development than vegetable or animal oils; however, when concentrations are greater than 2 percent by weight of cement, a strength loss of approximately 20 percent or more will occur.

Organic impurities such as algae in mixing water may cause excessive strength reductions by affecting bond or by excessive air entrainment.

As with all of the ingredients used in concrete production, if the water available is questionable, the comparative property tests should be run. Sometimes the concrete mix can be modified to compensate for water which produces low strength or exhibits other adverse characteristics.

The use of water containing acids or organic substances should be questioned because of the possibility of surface reactions or retardation. The other concern with curing water is the possibility of staining or discoloration due to impurities in the water.

AGGREGATES

The aggregate component of a concrete mix occupies 60 to 80 percent of the volume of concrete, and its characteristics influence the properties of the concrete. The selection of aggregates will determine the mix design proportion and the economy of the resulting concrete. It is therefore necessary to understand the importance of aggregate selection, testing, and handling as discussed in Chapter 2.

Aggregates selected for use should be clean, hard, strong, and durable particles, free of chemicals, coatings of clay, or other materials that will affect the bond of the cement paste. Aggregates containing shale or other soft and porous organic particles should be avoided because they have poor resistance to weathering. Coarse aggregates can usually be inspected visually for weaknesses. Any aggregates that do not have adequate service records should be tested for compliance with requirements. Most concrete aggregate sources are periodically checked to ensure that the aggregates being produced meet the concrete specifications.

The commonly used aggregates such as sand, gravel, and crushed stone produce normal-weight concrete weighing 135 to 160 lb/cu ft. Various expanded shales and clays produce structural lightweight concrete having unit weights ranging from 85 to 115 lb/cu ft. Much lighter concretes weighing 15 to 90 lb/cu ft, using vermiculite, pumice, and perlite as aggregates, are called insulating concretes. The use of materials such as barites, limonite, magnetic iron, and iron particles produces heavyweight concretes with weights going as high as 400 lb/cu ft.

Aggregates must possess certain characteristics to produce a workable, strong, durable, and economical concrete. These basic characteristics are shown in Table 4–7 with their significant ASTM test or practice designation and specification requirement.

The most common test used to measure abrasion resistance of an aggregate is the Los Angeles rattler. A quantity of aggregate is placed in a steel drum with steel balls, the drum is rotated for a preset number of revolutions, and the percentage of material worn away is determined. Though the test is a general index of aggregate quality, there is generally no direct correlation with concrete abrasion using the same

Table 4–7. Characteristics and tests of aggregates

Characteristic	Significance	Test Designation*		Requirement or Item Reported
Resistance to abrasion and degradation	Index of aggregate quality; wear resistance of floors, and pavements	ASTM C 131 ASTM C 535 ASTM C 779	(AASHTO T 96)	Maximum percentage of weight loss. Depth of wear and time
Resistance to freezing and thawing	Surface scaling, roughness, loss of section, and aesthetics	ASTM C 666 ASTM C 682	(AASHTO T 161) (AASHTO T 103)	Maximum number of cycles or period of frost immunity; durability factor
Resistance to disintegration by sulfates	Soundness against weathering action	ASTM C 88	(AASHTO T 104)	Weight loss, particles exhibiting distress
Particle shape and surface texture	Workability of fresh concrete	ASTM C 295 ASTM D 3398		Maximum percentage of flat and elongated particles
Grading	Workability of fresh concrete; economy	ASTM C 117 ASTM C 136	(AASHTO T 11) (AASHTO T 27)	Minimum and maximum percentage passing standard sieves
Fine aggregate degradation	Index of aggregate quality; Resistance to degradation during mixing	ASTM C 1137		Change in grading
Uncompacted void content of fine aggregate	Workability of fresh concrete	ASTM C 1252	(AASHTO T 304)	Uncompacted voids and specific gravity values
Bulk density (unit weight)	Mix design calculations; classification	ASTM C 29	(AASHTO T 19)	Compact weight and loose weight
Relative density (specific gravity)	Mix design calculations	ASTM C 127 fine aggregate ASTM C 128 coarse aggregate	(AASHTO T 85) (AASHTO T 84)	—
Absorption and surface moisture	Control of concrete quality (water–cement ratio)	ASTM C 70 ASTM C 127 ASTM C 128 ASTM C 566	(AASHTO T 85) (AASHTO T 84) (AASHTO T 255)	—
Compressive and flexural strength	Acceptability of fine aggregate failing other tests	ASTM C 39 ASTM C 78	(AASHTO T 22) (AASHTO T 97)	Strength to exceed 95% of strength achieved with purified sand
Definitions of constituents	Clear understanding and communication	ASTM C 125 ASTM C 294		—

(Continued)

Table 4–7. (continued)

Characteristic	Significance	Test Designation*		Requirement or Item Reported
Aggregate constituents	Determine amount of deleterious and organic materials	ASTM C 40 ASTM C 87 ASTM C 117 ASTM C 123 ASTM C 142 ASTM C 295	(AASHTO T 21) (AASHTO T 71) (AASHTO T 11) (AASHTO T 113) (AASHTO T 112)	Maximum percentage of individual constituents
Resistance to alkali reactivity and volume change	Soundness against volume change	ASTM C 227 ASTM C 289 ASTM C 295 ASTM C 342 ASTM C 586 ASTM C 1260 ASTM C 1293	(AASHTO T 303)	Maximum length change, constituents and amount of silica, and alkalinity

*The majority of the tests and characteristics listed are referenced in ASTM C33. Reference 4-22 presents additional test methods and properties of concrete influenced by aggregate properties.
Source: Portland Cement Association.

aggregate. If wear resistance is critical, it is more accurate to run abrasion tests on the concrete itself.

Porosity, absorption, and pore structure determine the freeze–thaw resistance of an aggregate. If an aggregate particle absorbs too much water, when it is exposed to freezing there will be little room for water expansion. At any freezing rate there may be a critical particle size above which the particle will fail if completely saturated.

There are two ways to determine the resistance to freezing–thawing of an aggregate: past performance records, if available, and the freeze–thaw test on concrete specimens containing the aggregates. Specimens are cast and cyclically exposed to freezing and thawing temperatures, with the deterioration measured by the reduction in the dynamic modulus of elasticity of the specimens.

Although aggregates are generally considered to be the "inert" component of a concrete mix, alkali-aggregate reactions do occur. Past performance records are usually adequate, but if no records are available or if a new aggregate source is being used, laboratory tests should be performed to determine the aggregate's alkali reactiveness.

Fresh concrete is affected by particle texture and shape more than hardened concrete. Rough-textured or flat aggregates require more water to produce a workable concrete than rounded or cubical, well-shaped aggregates. The National Ready Mix Concrete Association, based on the use of well-shaped cubical aggregate as the standard, suggests that flat, elongated, or sharply angular aggregates will require 25 lb of water more per cubic yard, thus requiring more cement to maintain the same water–cement ratio. The use of rounded gravel aggregates will usually allow a 15-lb reduction in mixing water, with the resulting reduction in cement, thus producing a savings to the concrete producer. It is recommended that long, flat particles not exceed 15 percent by weight of the total aggregate. This requirement is important when using manufactured sand because it contains more flat, elongated particles than natural sand.

As described in Chapter 2, the method of determining aggregate gradation and maximum aggregate size is by sieve analysis. The grading and maximum size of aggregates affect relative aggregate proportion as well as cement and water requirements, workability, economy, porosity, and shrinkage of concrete. Variations in grading may seriously affect the uniformity of concrete from one batch to another. Harsh sands often produce unworkable mixes and very fine sands often produce uneconomical concretes. Generally, aggregates that have smooth grading curves, that is, no excesses or deficiencies, produce the most satisfactory results. For workability in leaner mixes, a grading that approaches the maximum recommended percentage passing through each sieve is desirable. Coarse grading is required for economy in richer mixes. Generally, if the water–cement ratio is held constant and if the proper ratio of coarse-to-fine aggregate is chosen, a wide range in grading may be used with no effect on strength.

Usually more water is required for smaller aggregates than for larger maximum sizes. Conversely, the larger sizes require less water, and therefore less cement, to maintain a constant ratio with its resulting economy. The higher cost of obtaining or handling aggregates in the 2-in.-plus range usually offsets the saving in cement.

Generally, in higher-strength ranges, smaller aggregates will produce higher strengths than larger aggregates.

In certain cases, aggregates that have been gap graded may be used to produce higher strengths in stiff concrete mixes. Close control of gap-graded mixes is required because the variations may produce segregation.

RECYCLED CONCRETE MATERIALS

With the greening of the construction industry, the use of recycled construction materials has become more prevalent. The use of these materials reduces landfill waste disposal, increases aggregate supplies in areas of scarcity, and provides economic benefits through life cycle cost analysis.

Recycled concrete material (RCM) is produced by the crushing of existing concrete structures when they are demolished. It may consist of concrete curbs, sidewalks, pavements, as well as building concrete. The concrete may be crushed and processed on-site using portable crushers or trucked to a central facility for processing. The removal of reinforcing steel and embedments is performed by magnetic separators, mounted on the crushing system, as well as the manual removal of these items.

When portable crushing systems are used, the crushed concrete produced usually becomes backfill and base course material for on-site use, thus reducing material and transportation costs.

The concrete processed at a central facility will normally be graded, washed, and stockpiled as coarse and fine aggregate. These recycled aggregates will be tested and evaluated for their potential use as partial or complete substitutions for natural aggregates in portland cement concrete, asphalt paving mixes, and subgrade material. The quality control system used to monitor these recycled materials is critical, due to the variability of the initial concrete that was processed.

The use of recycled concrete aggregates may also help earn credit toward the U. S. Green Building Council's Leadership in Energy & Environmental Design (LEED) Green Building Rating System, increasing the chance to obtain LEED project certification.

The maximum sizes for coarse aggregates are usually based on the ACI 211.1 recommendations as shown in the following example (refer to Table 4–8 for rebar diameters).

Example

1. **Nonreinforced member**—One-fifth the minimum dimension of nonreinforced members. With a minimum dimension of 15 in.

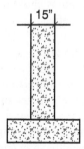

$\frac{1}{5} \times 15$ in. = 3-in. maximum aggregate size

2. **Reinforced members**—Three-fourths the spacing between reinforcing bars or between reinforcing bars and the forms.

Table 4–8. ASTM standard metric reinforcing bars

Bar Size	Diameter, mm [in.]	Cross-sectional Area, mm² (in.²)	Weight kg/m (lb/ft)
#10 [#3]	9.5 [0.375]	71 [0.11]	0.560 [0.376]
#13 [#4]	12.7 [0.500]	129 [0.20]	0.944 [0.668]
#16 [#5]	15.9 [0.625]	199 [0.31]	1.552 [1.043]
#19 [#6]	19.1 [0.750]	284 [0.44]	2.235 [1.502]
#22 [#7]	22.2 [0.875]	387 [0.60]	3.042 [2.044]
#25 [#8]	25.4 [1.000]	510 [0.79]	3.973 [2.670]
#29 [#9]	28.7 [1.128]	645 [1.00]	5.060 [3.400]
#32 [#10]	32.3 [1.270]	819 [1.27]	6.404 [4.303]
#36 [#11]	35.8 [1.410]	1006 [1.56]	7.907 [5.313]
#43 [#14]	43.0 [1.693]	1452 [2.25]	11.38 [7.65]
#57 [#18]	57.3 [2.257]	2581 [4.00]	20.24 [13.60]

Note: Soft metric conversions are now the standard sizes. Because many readers are familiar with the eleven inch-pound bars, the equivalent inch-pound bar sizes and equivalent nominal dimensions are shown within brackets.

Source: Printed with permission of the American Concrete Institute.

Subtract the distance occupied by stirrups, bars, and covers from the total dimension.

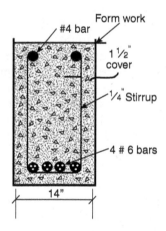

Width of member	14 in.
Two $\frac{1}{4}$ in. stirrups	$-\frac{1}{2}$ in.
Two $1\frac{1}{2}$ in. covers	–3 in.
Four $\frac{3}{4}$ in. bars	–3 in.
Clear space	$7\frac{1}{2}$ in.

$7\frac{1}{2}$ in. divided into three spaces = three $2\frac{1}{2}$ in. spaces

$$\frac{3}{4} \times 2\frac{1}{2} \text{ in.} = 1.875 \text{ in.} = 1\frac{7}{8} \text{ in.}$$

Check cover

$$\frac{3}{4} \times 1\frac{1}{2} \text{ in.} = 1\frac{1}{8} \text{ in.}$$

In this example, cover governs. Select the next lower commercially available aggregate (1 in.).

3. **Nonreinforced slabs**—One-third the depth of nonreinforced slabs on grade.

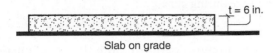

Slab on grade

6 in. $\times \frac{1}{3}$ = 2-in. maximum aggregate size

If, in the judgment of the engineer, the concrete is workable and can be placed without honeycomb or excessive voids, these requirements may be modified.

The weight of an aggregate per unit volume is called its *bulk unit weight* because the volume is occupied by aggregates and voids. A container of known volume is filled with aggregate following ASTM C29 methods and then weighed to determine the aggregate's bulk unit weight, or density in pcf or kg/m³.

Insulating lightweight aggregates	6–70 lb/cu ft
Structural lightweight aggregates	30–70 lb/cu ft
Normal-weight concrete aggregates	75–110 lb/cu ft
Heavyweight concrete aggregates	110–290 lb/cu ft

The specific gravity is not a measure of aggregate quality, but is used to design and control concrete mixes. The specific gravity is defined as the ratio of the solid unit weight of a substance to the weight of an equal volume of water. In the metric system under standard conditions, the unit weight of water may be taken as unity; therefore, the solid unit weight of a substance in grams per cubic centimeter is equal to its specific gravity.

For use in concrete mix design, the bulk specific gravity is determined on aggregates in a saturated, surface-dry state. The ASTM procedures for determining specific gravity are ASTM C127 and ASTM C128.

The moisture content of the aggregate is separated into internal moisture called absorbed moisture and external or surface moisture (Figure 4–4). The moisture content of an aggregate must be known so that the batch weights and water content of concrete may be controlled. All certified concrete production facilities have automatic equipment to measure moisture conditions of the aggregates and make adjustments in batch weights.

ADMIXTURES

The basic concrete mix design can be modified by the addition of an admixture. *Admixtures* are defined as any material other than portland cement, aggregates, and water added to a concrete or mortar mix before or during mixing. Usually good concrete-mixing practice will impart the qualities that will be required of the concrete when it is placed into service. Very often by using a different type of cement or modifying the mix design desired, property changes can be obtained for special concrete service.

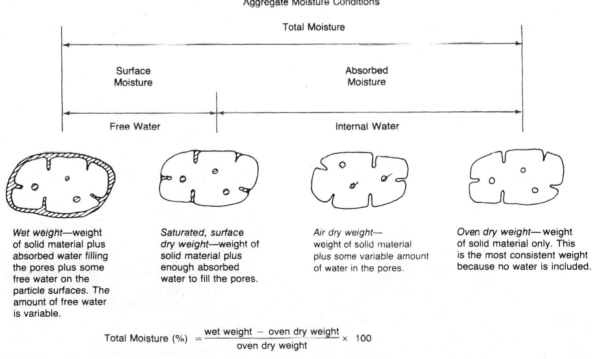

Aggregate Moisture Conditions

Total Moisture

Surface Moisture   Absorbed Moisture

Free Water   Internal Water

Wet weight—weight of solid material plus absorbed water filling the pores plus some free water on the particle surfaces. The amount of free water is variable.

Saturated, surface dry weight—weight of solid material plus enough absorbed water to fill the pores.

Air dry weight— weight of solid material plus some variable amount of water in the pores.

Oven dry weight—weight of solid material only. This is the most consistent weight because no water is included.

$$\text{Total Moisture (\%)} = \frac{\text{wet weight} - \text{oven dry weight}}{\text{oven dry weight}} \times 100$$

$$\text{Absorbed Moisture (\%)} = \frac{\text{saturated surface dry weight} - \text{oven dry weight}}{\text{oven dry weight}} \times 100$$

$$\text{Free Moisture (\%)} = \text{Total Moisture (\%)} - \text{Absorbed Moisture (\%)}$$

FIGURE 4–4. Aggregate moisture conditions.

There are times, however, when it is important to modify the design, and the only way of making a change is through the use of an admixture. Admixtures are generally used for one or more of the following reasons (Table 4–9):

1. To improve workability of the fresh concrete
2. To reduce water content, thereby increasing strength for a given water–cement ratio
3. To increase durability of the hardened cement
4. To retard setting time or increase it

5. To impart color to concrete
6. To maintain volume stability by reducing or offsetting shrinkage during curing
7. To increase concrete resistance to freezing and thawing

Most admixtures perform more than one function; for example, when an air-entraining admixture is used, increased resistance to freeze–thaw cycles in the hardened concrete, a reduction in bleed water, and increased workability in the fresh concrete can be expected. Sometimes the

Table 4–9. Concrete admixtures by classification

Type of Admixture	Desired Effect	Material
Accelerators (ASTM C 494 and AASHTO M 194, Type C)	Accelerate setting and early-strength development	Calcium chloride (ASTM D 98 and AASHTO M 144) Triethanolamine, sodium thiocyanate, calcium formate, calcium nitrite, calcium nitrate
Air detrainers	Decrease air content	Tributyl phosphate, dibutyl phthalate, octyl alcohol, water-insoluble esters of carbonic and boric acid, silicones
Air-entraining admixtures (ASTM C 260 and AASHTO M 154)	Improve durability in freeze–thaw, deicers, sulfate, and alkali-reactive environments. Improve workability	Salts of wood resins (Vinsol resin), some synthetic detergents, salts of sulfonated lignin, salts of petroleum acids, salts of proteinaceous material, fatty and resinous acids and their salts, alkylbenzene sulfonates, salts of sulfonated hydrocarbons
Alkali-aggregate reactivity inhibitors	Reduce alkali-aggregate reactivity expansion	Barium salts, lithium nitrate, lithium carbonate, lithium hydroxide
Antiwashout admixtures	Cohesive concrete for underwater placements	Cellulose, acrylic polymer
Bonding admixtures	Increase bond strength	Polyvinyl chloride, polyvinyl acetate, acrylics, butadiene-styrene copolymers
Coloring admixtures (ASTM C 979)	Colored concrete	Modified carbon black, iron oxide, phthalocyanine, umber, chromium oxide, titanium oxide, cobalt blue
Corrosion inhibitors	Reduce steel corrosion activity in a chloride-laden environment	Calcium nitrite, sodium nitrite, sodium benzoate, certain phosphates or fluosilicates, fluoaluminates, ester amines
Dampproofing admixtures	Retard moisture penetration into dry concrete	Soaps of calcium or ammonium stearate or oleate. Butyl stearate. Petroleum products
Foaming agents	Produce lightweight, foamed concrete with low density	Cationic and anionic surfactants. Hydrolized protein
Fungicides, germicides, and insecticides	Inhibit or control bacterial and fungal growth	Polyhalogenated phenols. Dieldrin emulsions. Copper compounds
Gas formers	Cause expansion before setting	Aluminum powder
Grouting admixtures	Adjust grout properties for specific applications	See Air-entraining admixtures, Accelerators, Retarders, and Water reducers
Hydration control admixtures	Suspend and reactivate cement hydration with stabilizer and activator	Carboxylic acids. Phosphorus-containing organic acid salts
Permeability reducers	Decrease permeability	Latex. Calcium stearate
Pumping aids	Improve pumpability	Organic and synthetic polymers. Organic flocculents. Organic emulsions of paraffin, coal tar, asphalt, acrylics. Bentonite and pyrogenic silicas. Hydrated lime (ASTM C 141)

Table 4–9. (continued)

Type of Admixture	Desired Effect	Material
Retarders (ASTM C 494 and AASHTO M 194, Type B)	Retard setting time	Lignin Borax Sugars Tartaric acid and salts
Shrinkage reducers	Reduce drying shrinkage	Polyoxyalkylene alkyl ether Propylene glycol
Superplasticizers* (ASTM C 1017, Type 1)	Increase flowability of concrete Reduce water-cement ratio	Sulfonated melamine formaldehyde condensates Sulfonated naphthalene formaldehyde condensates Lignosulfonates Polycarboxylates
Superplasticizer* and retarder (ASTM C 1017, Type 2)	Increase flowability with retarded set Reduce water–cement ratio	See superplasticizers and also water reducers
Water reducer (ASTM C 494, and AASHTO M 194, Type A)	Reduce water content at least 5%	Lignosulfonates Hydroxylated carboxylic acids Carbohydrates (Also tend to retard set so accelerator is often added)
Water reducer and accelerator (ASTM C 494 and AASHTO M 194, Type E)	Reduce water content (minimum 5%) and accelerate set	See water reducer, Type A (accelerator is added)
Water reducer and retarder (ASTM C 494 and AASHTO M 194, Type D)	Reduce water content (minimum 5%) and retard set	See water reducer, Type A (retarder is added)
Water reducer—high range (ASTM C 494 and AASHTO M 194, Type F)	Reduce water content (minimum 12%)	See superplasticizers
Water reducer—high range—and retarder (ASTM C 494 and AASHTO M 194, Type G)	Reduce water content (minimum 12%) and retard set	See superplasticizers and also water reducers
Water reducer—mid range	Reduce water content (between 6 and 12%) without retarding	Lignosulfonates Polycarboxylates

*Superplasticizers are also referred to as high-range water reducers or plasticizers. These admixtures often meet both ASTM C494 and C1017 specifications simultaneously.

Source: Portland Cement Association.

admixture produces an adverse reaction in conjunction with the desired one. For example, some finely powdered workability admixtures tend to increase drying shrinkage.

Because the effectiveness of an admixture varies with the type and amount of cement, aggregate shape, gradation, proportion, mixing time, water content, concrete, and air temperature, it is usually advisable to make trial mixes before using an admixture.

Trial mixes or small sample batches duplicate job conditions as accurately as possible so that the admixture dosages and results will be close to job expectations. Trial mixes also allow study of the admixtures' compatibility if more than one admixture is to be used in the concrete.

When deciding whether to use a certain admixture or not, the designer should consider a mix design modification, a comparison of costs between the modification, and the admixture cost including purchase, storage, and batching costs and any adverse reactions that can be expected.

It is advised that any admixture being considered meet or exceed appropriate specifications and that the manufacturer's recommendations be followed.

The most commonly used admixtures today are air-entraining agents, accelerators, retarders, and water reducers. The air-entraining admixtures are covered by ASTM C260; the others are covered by ASTM C494. Although both of the specifications set no limits on chemical composition and are based more on performance, they require the producer of the admixture to state the chloride content of the product.

The air-entraining admixture is used to increase the durability of concrete by protecting it against freeze–thaw cycle damage. When a concrete has been placed in service and it is subjected to water saturation followed by freezing temperatures, the resulting expansion of the water into ice usually will damage the concrete. By entraining air in concrete to form a microscopic air-void system, the expansion is provided a relief valve system. The air-void system in the hardened concrete paste allows water to freeze, with the empty air voids providing room for the expansion that occurs as water changes to ice.

The normal mixing of concrete will entrap a certain percentage of air, depending on aggregate size, temperature, and other variables. However, this air is usually in the form of widely spaced large bubbles, and under a microscope can be distinguished from entrained air. The ability to handle the expansion of the water is a function of the spacing factor or average distance from any point in the paste to the nearest air void. The recommended spacing factor is 0.008 in. (Table 4–10).

Table 4–10. Effect of production procedures, construction practices, and environment on control of air content in concrete

Procedure/Variable	Effects	Guidance
Batching sequence	Simultaneous batching lowers air content.	Add air-entraining admixture with initial water or on sand.
	Cement-first raises air content.	
Mixer capacity	Air increases as capacity is approached.	Run mixer close to full capacity. Avoid overloading.
Mixing time	Central mixers: air content increases up to 90 sec. of mixing.	Establish optimum mixing time for particular mixer.
	Truck mixers: air content increases with mixing.	Avoid overmixing.
	Short mixing periods (30 seconds) reduce air content and adversely affect air-void system.	Establish optimum mixing time (about 60 seconds).
Mixing speed	Air content gradually increases up to approx. 20 rpm.	Follow truck mixer manufacturer recommendations.
	Air may decrease at higher mixing speeds.	Maintain blades and clean truck mixer.
Admixture metering	Accuracy and reliability of metering system will affect uniformity of air content.	Avoid manual-dispensing or gravity-feed systems and timers. Positive-displacement pumps interlocked with batching system are preferred.
Transport and delivery	Some air (1 to 2%) normally lost during transport.	Normal retempering with water to restore slump will restore air.
	Loss of air in nonagitating equipment is slightly higher.	If necessary, retemper with air-entraining admixture to restore air.
		Dramatic loss in air may be due to factors other than transport.
Haul time and agitation	Long hauls, even without agitation reduce air, especially in hot weather.	Optimize delivery schedules. Maintain concrete temperature in recommended range.
Retempering	Regains some of the lost air.	Retemper only enough to restore workability. Avoid addition of excess water.
	Does not usually affect the air-void system.	
	Retempering with air-entraining ad mixtures restores the air-void system.	Higher admixture dosage is needed for jobsite admixture additions.
Belt conveyors	Reduces air content by an average of 1%.	Avoid long conveyed distance if possible.
	Reduce the free-falling effect at the end of conveyor.	
Pumping	Reduction in air content ranges from 2 to 3%.	Use of proper mix design provides a stable air-void system.
	Does not significantly affect air-void system.	Avoid high slump, higher air content concrete.
		Keep pumping pressure as low as possible.
	Minimum effect on freeze–thaw resistance.	Use loop in descending pump line.
Shotcrete	Generally reduces air content in wet-process shotcrete.	Air content of mix should be at high end of target zone.
Internal vibration	Air content decreases under prolonged vibration or at high frequencies.	Do not overvibrate. Avoid high-frequency vibrators (greater than 10,000 vpm). Avoid multiple passes of vibratory screeds.
	Proper vibration does not influence the air-void system.	Closely spaced vibrator insertion is recommended for better consolidation.
Finishing	Air content reduced in surface layer by excessive finishing.	Avoid finishing with bleed water still on surface.
		Avoid over finishing. Do not sprinkle water on surface prior to finishing. Do not steel trowel exterior slabs.
Temperature	Air content decreases with increase in temperature	Increase air-entraining admixture dosage as temperature increases.
	Changes in temperature do not significantly affect spacing factors.	

The size of the effective air voids is in the range of 50 to 500 microns, the largest and smallest voids having little effect in protecting the paste.

The actual dosages required to produce a given air content will vary. If excess fines are present, for example, a higher dose of admixture is required to produce the required air content.

Generally, if air-entraining admixtures are used, the water content may be decreased 0.3 to 4 lb per 1 percent of air with the same workability due to the ball bearing action of the air voids. If no reduction in water content is made and cement content is maintained, a decrease in compressive strength of 3 to 5 percent for each 1 percent of air entrained will result.

To accelerate the setting time of concrete, the admixtures used are usually soluble chlorides, carbonates, and silicates, the most widely used being calcium chloride. The general dosage of calcium chloride should not exceed 2 percent by weight of cement, and it should not be used in prestressed concrete because of corrosion. Calcium chloride may be used during the winter to speed up initial setting time to allow earlier finishing. Calcium chloride is not an antifreeze and does not substantially lower freezing temperatures of the concrete. It does, however, get the concrete through its early setting stage faster, and it is during this stage when freezing will damage the concrete the most. The effects of calcium chloride on the strength of concrete are shown in Figure 4–5.

For concrete placements during warm weather, a retarder is generally used. The retarders are primarily organic compounds such as lignosulfonic acid salts or hydroxylated carboxylic acid salts.

For normal conditions the setting time of concrete can be reduced 30 to 50 percent for normal dosages. The use of a retarder during warm weather helps to offset the decreased setting time due to higher placement temperatures. Retarders can also be used to reduce cold joints, allow smaller crews to finish flat work, and permit later control joint sawing. Water reducers are usually retarders that have been modified to suppress or modify the retarding effect.

Superplasticizers or high-range water reducers are chemical dispersants that, when added to a concrete mix with a 3- to 3 1/2-in. slump, can increase the slump to 8 to 10 in. depending on the dosage rate and other mix components. This increase is temporary and with the passage of time the mix will revert to the original slump. The admixtures work by increasing the dispersion of cement particles thereby reducing interparticle friction. The increased dispersion of the cement particles allows for a more complete hydration to take place and as a result higher compressive and flexural strengths are achieved.

Superplasticizers are used to solve difficult placement problems such as tight constricted formwork, dense rebar configurations, and situations where the concrete must be pumped, conveyed, or chuted over long distances.

Microsilica or silica fume is a by-product of the silicon and ferrosilicon industries. During the production of these materials a silica-based gas is given off. As it rises it cools rapidly into tiny, glassy spherical particles, which are collected in a bag house, a system used to filter the hot gases vented from the submerged electric-arc furnaces.

The chemical makeup of microsilica is similar to fly ash and portland cement in that all three contain the same basic chemical compounds. However, the relative distribution by percent of the chemical compounds is somewhat different. The physical characteristics of microsilica are quite different. Average diameters of particles of microsilica are 100 times finer than cement, the specific gravity of microsilica is 2.2 versus a common 3.15 for cement, and the bulk density is 9 to 25 pcf versus 94 pcf for portland cement.

During the hydration process, two products are formed: (1) calcium silicate hydrate, the glue or binder of the system, and (2) a nonbinder calcium hydroxide, which may occupy

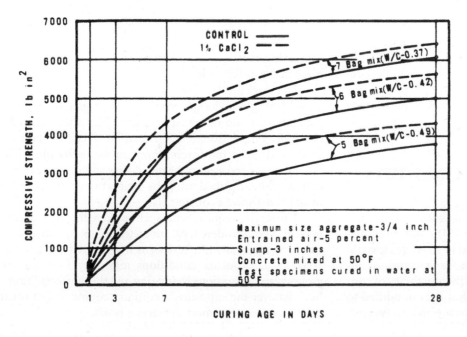

FIGURE 4–5. The effects of calcium chloride on the strength of concrete of different cement contents and at different ages with Type II cement. 288-D-1530. *(Courtesy U.S. Department of the Interior, Water and Power Resources Service)*

as much as a quarter of the volume of the hydration products. The calcium hydroxide, when present in large quantities, may make the concrete more vulnerable to chemical and sulfate attack as well as to adverse alkali-aggregate reactions. The calcium hydroxide can also combine with carbon dioxide to form a soluble salt, which causes efflorescence as the salt is leached out of the concrete.

Microsilica's pozzolanic reaction with calcium hydroxide and water produces more aggregate-binding calcium silicate gel, which improves bonding within the matrix, reduces permeability, and increases compressive and flexural strengths, abrasion resistance, and durability.

ASTM C618 is the standard specification that governs the use of fly ash, a by-product of the combustion of powdered coal as a mineral admixture. The fly ashes are classified as pozzolanic materials; that is, they possess little or no cementitious value but will, in the presence of moisture, chemically react with calcium hydroxide to form cementitious compounds. The standard describes Class F fly ash as the by-product of burning anthracite or bituminous coal and as having pozzolanic properties. Class C, produced from the burning of lignite or subbituminous coal, has both pozzolanic and some cementitious properties.

Concrete specifications have usually limited the fly ash substitution by weight for portland cement to a range of 15 to 25 percent maximum; with the easing of these restrictions, replacement values as high as 50 percent are successfully being used on concrete construction projects. These new concrete mixes are called high volume fly ash concrete (HVFAC) and have increased the environmental benefits of using the waste by-product of burning coal in a safe, economical manner, as well as enhancing the concretes' mix properties.

The use of fly ash as a mineral admixture or cement substitute is discussed in ACI 211 with examples of the methods used to determine the fly ash content in a properly designed concrete mix.

When concrete is manufactured with the addition of fly ash, its placeability, workability, and pumpability increase because of the ball bearing effect of the microscopic ash particles. Concrete with fly ash will normally take longer to set, thus extending the finishing time, and overall will take longer to reach the desired compressive strength. Because the required strengths are not usually reached at 28 days, it is advisable to prepare enough test specimens to allow for testing at later ages up to 56 days.

PROPORTIONING CONCRETE INGREDIENTS

Duff Abrams published his initial research on the water–cement ratio (w/c) concept in 1918, indicating that the ratio of water to cement was related to concrete strength. Currently it is also accepted that lower water–cement ratios reduce the permeability and improve the durability of concrete. The term *water–cement ratio* has been modified to *water–cementitious material ratio* (w/c + m) or (w/cm),

because pozzolans such as fly ash and microsilica are widely used today as part of the reactive ingredients in concrete. Because concrete strength is inversely proportional to the water–cement ratio, a reduction in water while maintaining cement content will give an increase in strength. A rule of thumb for good concrete of 0.45 to 0.58 water–cement ratio states that each 0.01 reduction in the water–cement ratio will increase 28-day strength by 100 psi. It also follows that for a given strength, if the water is reduced, then the cement may be reduced with resulting economy.

In higher-strength concretes, the use of a water reducer is required because of the low water–cement ratio being used (0.30 to 0.35). If concrete of this water–cement ratio is to be workable, a water reducer is required to raise the workability of such a mix.

The determination of the relative amounts of materials required to produce a concrete that will be economical and workable in the plastic state and that will have the required properties in the hardened state is called mix design or proportioning. Proportioning may vary from the simple 1:2:3 formula, which means 1 part cement, 2 parts fine aggregate, and 3 parts coarse aggregate, to the ACI 211.1 mix design procedure, which is included in the Appendix. Some of the proportioning is based on empirical information and some on tests and calculations. What is important is that the concrete to be produced from the initial proportioning satisfies its service requirements. When proportioning concrete, the technician or engineer must look at the entire job. Factors such as mixing apparatus, concrete handling and transportation, finishing, curing, and strength requirements must be considered.

Concrete economy is basically a matter of reducing cement content, as it is usually the most expensive ingredient, without sacrificing the service requirements.

Admixtures such as water reducers have played a part in reducing cement contents while still maintaining required strengths. Cement reductions are also possible by utilizing the stiffest mix placeable, the largest maximum aggregate size, and the proper ratio of fine-to-coarse aggregates. Generally, labor, handling, and forming costs are the same for varying concretes so that basic savings can only be recognized in the initial design.

The workability of the fresh concrete is an important factor during mix design. Maximum aggregate sizes, water contents, method of transporting, and finishing must be considered here. Concrete that can be placed in slabs on grade with little reinforcing may not be easily placed in heavily reinforced walls or beams. A very stiff concrete that is to be machine-finished would be useless to a contractor who must hand-trowel a concrete slab.

The service requirements such as strength and exposure must also be considered. All concretes do not have the same strength requirement; mix designs may range from 2000 to 10,000 psi. Exposure conditions may vary from sulfate groundwater conditions to freezing and thawing conditions, but whatever the exposure condition, the mix design must produce concrete to meet the service needs.

Trial batches are usually produced and tested before actual production of concrete begins. A materials testing laboratory usually performs this operation. The laboratory is supplied with the mix requirements and samples of materials to be used on the job. Its job is to design the most economical concrete mix that will satisfy the job requirements. One of the problems inherent in this system is the size of the trial mixes, and usually when production begins, adjustments are made in the design.

Concrete proportioning can be performed by trial batching. If the cement ratio is given, the technician can produce a paste (water and cement) for that water–concrete ratio and vary the fine aggregate and coarse aggregate to produce different aggregate ratios. The batches are usually produced based on past trial batches so that the analysis of all possible combinations is not required. The trial batch that meets the design requirements is then enlarged to job batch sizes, and if it requires further refinement, it can be made when production starts.

Another method used is to base designs on past experience. If the concrete producer has good records, it could possibly supply concrete based on past performance.

A widely used method for mix designs is ACI 211.1, Recommended Practice for Designing Normal and Heavyweight Concrete. The complete procedure can be found in the Appendix.

The proportioning of concrete mixes is not an exact science, and human judgment is an important factor to be considered. The technician or engineer will not find all of the mix design solutions in charts and tables, but in a blend of his skill and judgment and the technical information available.

CONCRETE ESTIMATING

Determining concrete quantities for a construction project requires volumetric calculations because concrete is estimated and purchased by the cubic yard or cubic meter. The contractor completes these volumetric calculations and adds an appropriate waste factor to the calculated quantities. Typical waste factors for concrete construction range from 3 to 8 percent, with lower values used for formed placements and higher values used for slab on grade projects. Waste factors may also include spillage and pump losses due to concrete that will remain in the pump lines and hopper. Waste factors should be based on prior field experience if the contractor has historical data available.

Example
Calculate the concrete required to cast a 40 ft × 60 ft by 5-in. thick slab on a prepared subgrade (1 cu yd = 27 cu ft).

$$\text{Volume (cu ft)} = 0.417 \text{ ft} \times 40 \text{ ft} \times 60 \text{ ft} = 1001 \text{ cu ft}$$

$$\text{Volume (cu yd)} = \frac{1001 \text{ cu ft}}{27 \text{ cu ft/cu yd}} = 37.1 \text{ cu yd}$$

If the subgrade for this slab has not been fine graded accurately, a waste factor of 7 percent might be appropriate. Therefore, the quantity calculated would be increased.

$$\text{Volume (cu yd)} = 37.1 \times 1.07 = 39.7 \cong 40 \text{ cu yd}$$

Calculate the concrete required to cast a wall 3 m high, 20 m long and 250 mm thick.

$$\text{Volume (m}^3\text{)} = 3 \text{ m} \times 20 \text{ m} \times 0.250 \text{ m} = 15 \text{ m}^3$$

When orders for concrete on large projects are made, contractors might place a hold on a specified quantity of concrete with a balance to be determined and ordered as the concrete placement nears completion. A contractor using hold-and-balance ordering is reducing the possibility, for example, of 11 cu yd of concrete arriving at the jobsite, when 3 cu yd is all that will be required to complete the placement.

The construction estimator will also be concerned with the specification requirements of the concrete for the project. The estimator might use a checklist with regard to concrete strength, reinforcement, inserts, and curing requirements and other project specification details.

CONCRETE MANUFACTURING

The production of concrete for the job covers a wide range of methods. Concrete can be mixed by hand in small portable mixers, in transit mix trucks, and in large stationary mixers. But no matter how the concrete is mixed, the end result desired is the same—a quality concrete meeting the design requirements. To produce quality concrete, the batching and measuring of ingredients must be done accurately. Therefore, most specifications require that materials be weighed and combined rather than combined by volume. Water is the one ingredient that is usually measured out either way, by weight or by volume. Weighing materials allows for adjustments in moisture conditions, especially in the fine aggregates where bulking can occur due to moisture.

All weighing equipment should be checked periodically and adjustments made when required. Admixture equipment should be checked daily since overdoses can be detrimental to quality concrete production. From the discussion on air-entraining admixtures, it can be seen that an overdose here could possibly reduce the concrete strength considerably.

Concrete Mixing

The actual mixing of concrete is performed for the most part by mixing equipment. The mixing equipment is usually rated for two functions: first, the actual mixing of the ingredients to produce concrete, and second, the agitating of the already-mixed concrete. The agitating capacity is higher than the mixing capacity. The mixing equipment can be stationary, mounted on wheels and towable, mounted on a truck for transit mix, or mounted on crawlers for paving operations.

Stationary mixing equipment can be found on jobs which require large amounts of concrete at steady rates. A good example is a large concrete dam; the amount of concrete required and the steady placement scheduled permit the semipermanent installation. Large highway paving jobs also may utilize stationary mixers. The concrete is produced at one central location and transported to the paving equipment by special agitator trucks or in ordinary dump trucks if conditions permit. The economy of such a system should be evident when comparing the cost of transit mix trucks to dump or agitator trucks. The actual mixing drums may vary from 2 to 12 cu yd and may be placed in tandem so that while one drum is discharging, the other drum is mixing a batch.

The required mixing time of a stationary mixer may vary, but generally 1 minute is required for the first cubic yard and 15 seconds for each additional cubic yard or a fraction of a cubic yard. The mixing time is measured from the time all of the solid ingredients are in the drum, provided that all of the water has been added before one-fourth of the mixing time has elapsed. Many of the stationary mixers have timing devices which can be set and locked to prevent the discharge of the concrete before proper mixing time has elapsed.

ASTM C94 specifies mixing time based on drum revolutions. Generally, 70 to 100 revolutions at a rotation rate designated by the manufacturer of the mixer is required to produce uniform concrete. No more than 100 revolutions at mixing speed can be made. All revolutions over 100 must be made at agitating speed. ASTM C94 also specifies that discharge of the concrete shall be completed within $1\frac{1}{2}$ hours or before the drum has revolved 300 revolutions—whichever comes first—after the introduction of the mixing water to the cement and aggregates, or after the introduction of cement to the aggregates.

Concrete-Mixing Methods

Stationary or central mixers are also used by some ready-mix producers. The concrete is mixed at the central yard and delivered to various jobsites in transit mix trucks (Figure 4–6). The trucks are able to deliver more concrete per trip because they are operating at agitating capacity which is higher than mixing capacity. The truck in Figure 4–6 is rated for 11 cu yd mixing capacity and 14.5 cu yd agitating capacity by the Truck Mixer Manufacturers Bureau (TMMB) as approved by the National Ready Mix Concrete Association (NRMCA).

The truck in Figure 4–7 is a mobile concrete production system sometimes referred to as a concrete mobile. Bins on the truck carry coarse and fine aggregates and cement, while tanks carry water and admixtures. The system produces concrete on-site to required specifications; for larger projects the bins and tanks are reloaded at the jobsite.

Many of the stationary mixers in use today are actually portable in that the components are maintained on trailers, and after the job's completion, the plant can be disassembled easily and transported to the next site.

For smaller concrete jobs many contractors rely on the portable mixer or construction mixer. The capacities of

FIGURE 4–6. Front-discharge transit mixer. *(Courtesy Clemente Latham Concrete Division of Callanan Industries)*

FIGURE 4–7. Mobile concrete dispensers allow the production of custom concrete mixes on-site. *(Courtesy Cement Tech, Inc.)*

these mixers are generally in cubic feet, and they are used for small concrete placement or when time of placement must be controlled.

The biggest problem with concrete produced with construction mixers on-site is quality control. Most contractors do not bother to weigh batch ingredients, but instead use volumetric batching. Though the total cubic yards placed by this method today is rather small, the same care should be exercised by those contractors who still use it as is used in a modern concrete producer's plant.

Paving mixers are concrete mixers mounted on crawler treads. The materials are fed into the mixer from dry-batch trucks, and the machine travels along the finish grade and deposits fresh concrete behind itself to be screeded and finished by the rest of the paving train.

Shotcrete is a nonproprietary term used to describe mortar or concrete that is placed by high-velocity compressed air and adheres to the surface on which it is projected. In the dry-mix process, the dry materials are thoroughly mixed with enough water to prevent dusting. The dry mixture is forced through the delivery hose by compressed air, and the water is added at the mixing nozzle. The wet mix utilizes wet mortar or concrete forced through the delivery hose to the nozzle, where compressed air is introduced to increase the velocity of the material. Because of the velocity at impact, a certain amount of material bounces off the surface of the structure; this material is called *rebound*. With dry-mix shotcrete, rebound may average 30 percent on overhanging surfaces or squaring corners, about 25 percent for vertical surfaces, and on nearly level surfaces about 20 percent. The rebound material must be cleared away from the application area, as it is mostly sand and coarse particles rather than cement that make up rebound. In fact, the cement content of the shotcrete in place is higher than that of the rebound.

Either sand or coarse aggregate shotcrete can be applied to the surfaces of various materials. The primary uses of shotcrete are in repairing and strengthening existing structures, as protective coatings for structural steel and masonry, in tunnel and canal construction, and in the building of free-form swimming pools.

Shotcrete's strength is usually evaluated by cores removed from sample panels or from the application itself, if the shotcrete is thick enough.

The most familiar concrete production system is ready mix. Ready mix is concrete delivered to the jobsite ready for placement. Ready-mix concrete is produced by one of three methods: central mix, transit mix, and shrink mix. Central-mix concrete is mixed in a stationary mixer at the producer's yard and delivered to the jobsite in a transit mixer operating at agitating speeds, in an agitator truck, or in dump trucks (Figure 4–8). Transit-mixed concrete is completely mixed in the truck. The ingredients are batched, water is added, and the concrete is mixed in the drum mounted on a truck (Figure 4–9). Shrink-mixed concrete is a combination of central mix and transit mix, with mixing requirements split between the central plant and the transit truck.

Concrete mixing equipment loads should not exceed rated capacities. When capacities are exceeded to increase production, the quality of the concrete produced is lowered. The interior of the mixing drum should be periodically checked for worn blades and hardened concrete buildup, as both factors decrease mixing efficiency.

The final quality of the concrete produced by a ready-mix supplier is determined by the contractor's operations, which include forming, placing, and curing. This joint involvement in material production can cause problems. If the contractor's placing and curing operations are faulty, concrete which should have met all requirements may not meet job service requirements.

The division of responsibilities generally requires a third party, the certified materials testing laboratory, to become involved in concrete construction. The materials laboratory will monitor the concrete production, placement, and curing operations by performing ASTM or other specified tests during these operations. Most ready-mix firms have quality control programs to monitor their products'

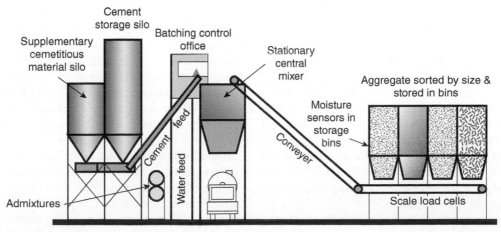

FIGURE 4–8. Central mix plant. *(Courtesy of Wayne Deyo)*

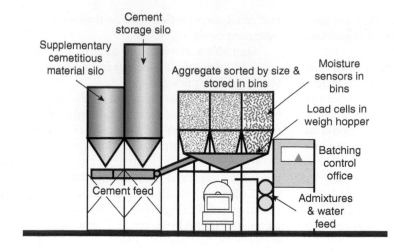

FIGURE 4–9. Transit mix plant. *(Courtesy of Wayne Deyo)*

performance. When concrete is produced by any method—jobsite or ready mixed—several basic tests are performed to measure quality. It is advised that the technician read the specifications carefully and note the who, what, when, and where of the tests. Compliance tests require that the technician be certified by ACI as a Concrete Field Testing Technician, Grade 1 or equivalent.

TESTING CONCRETE

It is important that the samples of concrete to be tested be representative samples. If they are not, then the results obtained by testing will not represent the concrete placed. ASTM makes provision for sampling fresh concrete in C172. Samples of concrete to be tested should be obtained from two or more regularly spaced intervals from the middle of the batch within 15 minutes and in the case of revolving drum mixers or agitating trucks, the samples should be obtained after all of the water and admixtures have been added. The minimum composite sample size, when cylinders are made, should be 1 cu ft (28 l); however, smaller samples may be used for routine, slump, temperature, and air content tests. After obtaining the composite sample, transport it to the test area and remix; during this procedure, protect the sample from contamination and rapid evaporation. Within 5 minutes of obtaining the composite sample, start tests for slump, temperature, and air content and within 15 minutes of obtaining the same, begin cylinder casting. When the concrete contains coarse aggregate that is larger than appropriate for the test method, the oversize aggregates must be removed by wet-sieving.

A major requirement of fresh concrete is *workability*, a composite term used to denote the ease with which concrete can be mixed, transported, placed, and finished without segregation. Workability is a relative term because concrete which satisfies its requirements under one set of conditions may not satisfy its requirements under different conditions. A concrete used in a nonreinforced footing may not be usable in a reinforced wall or column.

A major requirement of fresh concrete is *consistency*, denoted by the fluidity of the concrete as measured by the

Slump test. If the slump of a concrete mix is controlled, the consistency and workability necessary for proper placement and indirectly the water–cement ratio can be controlled. Changes in water content have a pronounced effect on slump. A 3 percent change in water content will increase the slump about 1 in. If the surface moisture on the sand changes by about 1 percent, the slump may increase 1 to $1\frac{1}{2}$ in. The slump may also be increased 10 mm by adding about 2 kg of water per cubic meter (1 in. with the addition of 8–10 pounds of water per cubic yard).

In conjunction with slump, the term *workability* is often used to denote the ease with which concrete can be mixed, transported, placed, and finished without segregation. The workability of a concrete mix can be estimated by the Slump test and by observations of the concrete for stickiness and harshness. Wet concretes are usually more workable than dry concretes, but concretes of the same slump may vary in workability depending on the paste and aggregates involved.

The ASTM C143 test for slump of portland cement concrete details the procedure for performing Slump tests on fresh concrete (Figure 4–10). A slump cone is filled in three layers of equal volume so that the first layer is $2\frac{5}{8}$ in. (76 mm) high, and the second layer is $6\frac{1}{8}$ in. (155 mm) high. Each layer is rodded 25 times with a tamping rod 24 in. (600 mm) long and $\frac{5}{8}$ in. (16 mm) in diameter, with a hemispherical tip of $\frac{5}{8}$ in. diameter. The rodding is uniformly distributed and is full depth for the first layer and just penetrating previous layers for the second and third layers. If the level of concrete falls below the top of the cone during the last rodding, add more concrete as required to keep an excess above the top of the mold. Strike off the surface of concrete by a screeding motion and rolling the rod across the top of the cone. In 5 ± 2 seconds, raise the cone straight up. Set the inverted slump cone next to the concrete, and measure the difference in height between the slump cone and the original center of the specimen. With the rod set on the cone, the slump measurement can be read to the nearest $\frac{1}{4}$ in. (6 mm). The test from filling of the slump cone to measuring the slump should take no longer than $2\frac{1}{2}$ minutes. If two

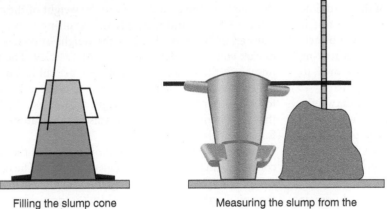

Filling the slump cone

Measuring the slump from the
bottom of the rod to the displaced
center of the concrete

FIGURE 4–10. Filling Slump Cone.
(Courtesy of Wayne Deyo)

consecutive tests on a sample show a falling away of a portion of the sample, the concrete probably lacks the cohesiveness for the Slump test to be applicable.

The *air content* of fresh concrete is a very important value. If a mix design for a certain air content is to meet exposure requirements, that air content must be maintained. As shown in Figure 4–11, if air content decreases, durability decreases; if it increases, strength decreases about 3 to 5 percent for each 1 percent of air. There are three field procedures in use today: the ASTM C173 test for air content of freshly mixed concrete by the volumetric system, the ASTM C231 test for air content of freshly mixed concrete by the pressure method, and the ASTM C138 unit weight, yield, and air content (gravimetric) of concrete.

The volumetric method can be used with all types of aggregates and is recommended for lightweight or porous aggregates. A sample of concrete is placed in the bowl in two layers, each layer is rodded 25 times, and the side of the bowl is tapped with a mallet 10 to 15 times after each rodding. After filling and rodding, strike off the excess concrete, clean the bowl flange, and clamp the top section onto the bowl. Using the dispersion funnel, add one pint of water followed

by the appropriate amount of isopropyl alcohol, add additional water to fill the sight tube, and adjust the water level so that the bottom of the meniscus is zeroed. Place the cap on the meter and invert and shake for 5 seconds, return meter to upright position, repeat this process for a minimum of 45 seconds, then set meter on edge at about 45 degrees and roll $\frac{1}{4}$ to $\frac{1}{2}$ turn vigorously; during the rolling process, turn base edge about $\frac{1}{3}$ turn listening for the aggregate moving in the meter. Stand meter up and loosen cap if there is more than 2 percent foam in the meter or if it is leaking, disregard test and start over using more alcohol. When the liquid level falls below the range of the meter, add calibrated cups of water to bring liquid up to a readable level and record the number of cups used. When the liquid level is stable, that is, it has not changed by more than 0.25 percent, record the level as the initial reading. Tighten cap and repeat 1-minute rolling procedure; if the second reading has not changed by more than 0.25 percent, record this reading as the final one, disassemble the meter, and dump the contents out—there should be no concrete left in the bowl for a valid test. If the second reading differs by more than 0.25 percent, repeat rolling and record this procedure of rolling burping a minimum of two times

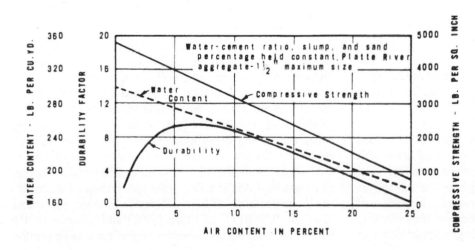

FIGURE 4–11. Effects of air content on durability, compressive strength, and required water content of concrete. Durability increases rapidly to a maximum and then decreases as the air content is increased. Compressive strength and water content decrease as the air content is increased. 288-D-1520 *(Courtesy U.S. Department of the Interior, Water and Power Resources Service)*

and a maximum of three times; if at the end of the third time the level changes by more than 0.25 percent the test is invalid. The recorded air content is determined by subtracting the alcohol correction factor from the final meter reading and adding the number of cups of water used, if required, to bring liquid level above the 9 percent range of the meter.

The pressure method cannot be used with lightweight or porous aggregates because the water will be forced into the aggregate's pore structure and erroneous results will be obtained. Correction factors are relatively constant for normal-weight aggregates and, even though small, should be determined and applied when air contents are determined by the pressure method. The pressure method of determining air content is based on Boyle's law, which relates pressure to volume. When the pressure is applied, the air in the concrete sample is compressed. ASTM C231 recognizes two types of apparatus, Meter Type A and Meter Type B.

The operation of the Type A meter involves the introduction of water in a glass tube to a predetermined height above a sample of concrete of known volume and the application of a predetermined air pressure over the water. By observing the reduction in the water level, the percentage of air in the concrete can be read from the glass tube.

The Type B meter involves the equalizing of a known volume of air at a known pressure in a sealed air chamber with the unknown volume of air in the concrete sample, the dial and the pressure gauge being calibrated in terms of percentage of air for the pressure at which equalization takes place.

A nonstandardized test used to measure air content of plastic concrete is the Chace air indicator. The utilization of this simple, inexpensive test is usually restricted to determining the presence of or lack of entrained air in plastic concrete. The testing device consists of a graduated glass tube and a rubber stopper that has a brass cup attached to it. The brass cup is filled with cement paste containing no aggregate particles larger than one-tenth of an inch and inserted into the bottom of the glass tube. The glass tube is filled with isopropyl alcohol up to the zero line and with the thumb held over the open end is shaken to remove the air from the mortar. The drop in alcohol level is recorded and multiplied by a factor based on the mortar content of the concrete mix to determine the air content. Because the test is not standardized no concrete should be rejected; however, the load in question should be tested by a standard method to determine its air content.

The unit weight of fresh concrete and yield determinations are covered by ASTM C138. The *unit weight* of fresh concrete is determined in pounds per cubic foot or kilograms per cubic meter, and yield is the number of cubic feet or cubic meters of concrete produced from a mixture of known quantities of materials.

The unit weight determination is made with a standard cylindrical measure which is filled in three equal layers with each layer rodded 25 times. The sides of the container must be rapped 10 to 15 times until no large air bubbles appear and the holes left from rodding have closed. The surface of the concrete must be struck off, the sides of the container cleaned, and the container full of concrete weighed. Subtract the weight of the container from the weight of the container full of concrete and multiply the resulting number by the calibration factor or dividing the weight of concrete by the bucket volume as determined by ASTM C29. The resulting number is the weight of the concrete per cubic foot or cubic meter. The unit weight of concrete is often used as a guide for air content. If the unit weight falls, it usually indicates that the air content is rising. It is also an indicator of strength for a given class of concrete.

Yield determinations are important in that the volume of concrete being produced for a given batch can be checked. If the total weight of all of the ingredients batched is divided by the unit weight of concrete, the result is the yield in cubic feet.

Example

$$\text{Batch Weights}$$

Water	265 lb
Cement	510
C/A	1917
F/A	1350
	4042 lb total

Unit weight = 149.2 lb/cu ft

$$\text{Yield} = \frac{4042 \text{ lb}}{149.2 \text{ lb/cu ft}} = 27.1 \text{ cu ft}$$

Batch mass = 14,611 kg

Density = 2290 kg/m^3

$$\text{Yield} = \frac{14,611 \text{ kg}}{2298 \text{ kg/m}^3} = 6.36 \text{ m}^3$$

Weight of materials batched 35,866 lb

Unit weight of concrete   147.6 lb/cu ft

$$\text{Yield in cu ft} \quad \frac{35,866 \text{ lb}}{147.6 \text{ lb/cu ft}} = 243 \text{ cu ft}$$

$$\text{Yield in cu yd} \quad \frac{243 \text{ cu ft}}{27 \text{ cu ft/cu yd}} = 9 \text{ cu yd}$$

The volume of concrete produced as determined by yield calculations enables a ready-mix producer to check his production, and the contractor can check that he is getting the material he is paying for. The yield calculation can be used to solve disputes concerning the cubic yards of concrete required to fill a set of forms. If the forms move during placement, the yardage required to fill them will increase and the contractor may claim he was shortchanged. If the yield determinations are available, then some problems can be easily resolved.

While the tests are being completed on a sample of concrete, the temperature of the concrete should also be taken and recorded, since placement temperature may govern setting times, curing, and strength development. In hot weather, the maximum placement temperature is often limited to 90°F; in cold weather, a minimum temperature of 55 or 60°F is sometimes specified. ASTM C1064 is the test method for determining the temperature of fresh concrete; a thermometer with a range of 30°F (0°C) to 120°F (50°C) and accurate to 1°F (0.5°C) is placed into the concrete so that the sensing portion will be encased by a minimum of 3 in. of concrete. The concrete around the thermometer penetration is sealed by pressing the concrete around the thermometer stem. The temperature is read while the device is in the concrete for a period of 2 to 5 minutes and recorded to the nearest 1°F (0.5°C). When the concrete contains 3 in. or larger aggregate, it may take up to 20 minutes to obtain an accurate temperature reading.

COMPRESSIVE STRENGTH TESTS

A very important property of hardened concrete is its compressive strength. *Compressive strength* is the measured maximum resistance to axial loading, expressed as force per unit of cross-sectional area in pounds per square inch (psi).

The designer of a concrete structure selects a desired compressive strength (f_c') for the concrete. The various parts of the structure are designed so that f_c', reduced by an appropriate safety factor, is not exceeded. Suitable safety factors for various parts of the structure (beams, columns, floors, etc.) and for various kinds of loads (static, moving traffic, etc.) are determined and published by organizations such as the American Concrete Institute (ACI). The concrete that is placed in the structure is tested at 28 days and evaluated based on ASTM or ACI criteria to determine that it has the required compressive strength (f_c').

ASTM C31 is the test procedure for making compressive test cylinders. The standard test cylinder is 6 in. in diameter by 12 in. high for aggregates up to 2 in. For larger aggregates, the diameter should be at least three times the aggregate size and the height at least twice the diameter. The molds used are generally waxed cardboard or plastic, and ASTM C470 is the governing specification.

The 6 in. test cylinders are filled in three equal layers and each layer is rodded 25 times. Concretes of 1- to 3-in. slump may be rodded or vibrated; concretes under 1-in. slump must be vibrated, and concretes over 3-in. slump must be rodded. Concrete test cylinders for acceptance testing, quality control, or verifying mix proportions for strength evaluation

must be cured in accordance with ASTM C31 curing requirements. Immediately after casting, specimens shall be stored at 60°F (16°C) to 80°F (27°C) for up to 48 hours. During this period, the cylinders must be in an environment that is capable of controlling temperature, as well as, preventing moisture loss. When specified strengths are 6000 psi (40 MPa) or greater, the temperature range is 68°F (20°C) to 78°F (26°C) for the initial curing period. Final curing must begin with 30 minutes of mold removal at a temperature of $73 \pm 3°F$ ($23 \pm 2°C$) in a moist curing room at 100 percent relative humidity or in water saturated with calcium hydroxide. Cylinder must be transported to a laboratory for final curing, must not be moved until at least 8 hours after final set, and must be protected from jarring, freezing, or moisture loss. The transportation time from the jobsite to the laboratory should not exceed 4 hours. The strengths of cylinders are generally taken at 7 and 28 days. Figures 4–12 to 4–14 indicate strength properties of concrete subjected to different curing conditions. Figure 4–15 represents a typical stress–strain diagram for hardened concrete.

The speed with which construction takes place has cast some doubt on the usefulness of the 7- and 28-day concrete strength tests. If concrete strengths are not adequate in the footings of a building, the engineer may not receive the information until the walls have already been cast on the questionable footing concrete.

Accelerated tests are being developed to allow earlier acquisition of strength information. Systems currently being tested or approved include 2-day breaks of cylinders that have been cured in boiling water or in autogenous curing boxes, 5-hour breaks of cylinders cured under heat and pressure, and a system that will chemically analyze plastic concrete before placement for its potential strength. These systems will require extensive correlation testing with 28-day cylinders before they gain wide acceptance. The compressive strength of the concrete is determined by loading the cylinders to failure.

For acceptance testing of concrete strengths below 1500 psi (10 MPa) or above 12,000 psi (85 MPa), the use of ASTM C1231 unbonded caps is not permitted. Unbonded cap is

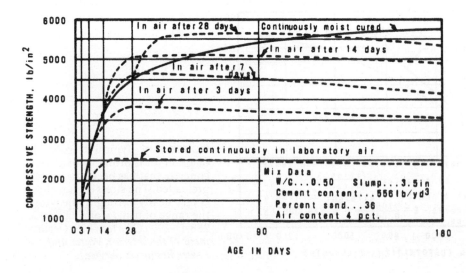

FIGURE 4–12. Compressive strength of concrete dried in laboratory air after preliminary moist curing. 288-D-2644 (*Courtesy U.S. Department of the Interior, Water and Power Resources Service*)

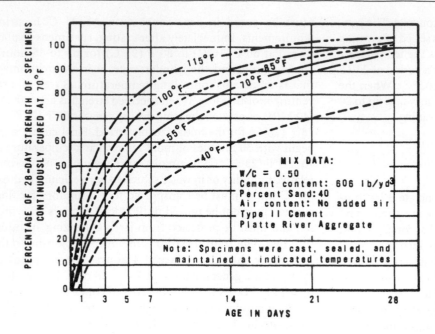

FIGURE 4–13. Effect of curing temperature on compressive strength of concrete. 288-D-2645 (*Courtesy U.S. Department of the Interior, Water and Power Resources Service*)

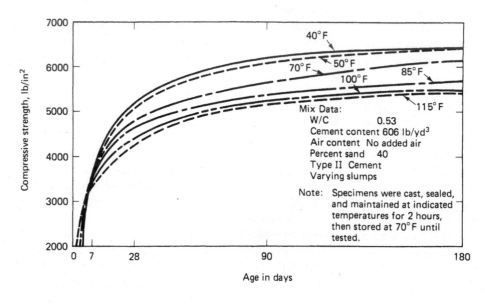

FIGURE 4–14. Effect of initial temperature on compressive strength of concrete. 288-D-2646 (*Courtesy U.S. Department of the Interior, Water and Power Resources Service*)

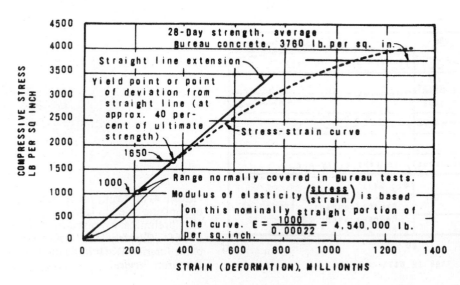

FIGURE 4–15. Typical stress–strain diagram for thoroughly hardened concrete that has been moderately preloaded. The stress–strain curve is very nearly a straight line within the range of usual working stresses. 288-D-799 (*Courtesy U.S. Department of the Interior, Water and Power Resources Service*)

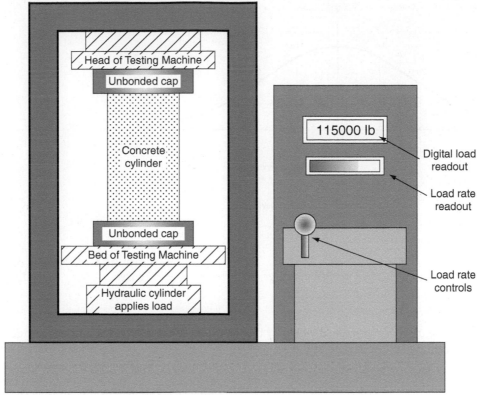

Safety doors not shown for clarity, doors must always be closed when running compression tests

FIGURE 4–16. Compression testing machine. *(Courtesy of Wayne Deyo)*

defined as a metal retaining ring with an elastomeric pad insert of a specified Shore A Durometer hardness. When concrete strengths exceed 7000 psi (50 MPa), the use of sulfur mortar, high-strength gypsum plaster and other materials are used with manufacturers' documentation that the materials comply with ASTM C617 statistical requirements(Figure 4–16).

Example

Cylinder diameter = 6 in.
   Load at failure = 115,000 lb

$$s = \frac{P}{A}$$

   s = compressive strength
   P = load in lb
   A = area in sq in.

$$s = \frac{115,000 \text{ lb}}{28.27 \text{ sq in.}}$$

   s = 4070 lb/sq in.

Cylinder diameter = 4 in.
   Load at failure = 54,900 lbs

$$s = \frac{54,900 \text{ lb}}{12.57 \text{ sq in.}} = 4368 \text{ psi}$$

   s = .00689476 × 4368 psi = 30.12 MPa

Note: 4 in. diameter cylinders usually yield higher strengths. Therefore, many specifiers require comparison testing between 4 in. and 6 in. diameter cylinders before approving the use of 4 in. diameter cylinders. To convert pounds per square inch (psi) to megapascals (MPa), multiply by .00689476.

The specimens made under controlled curing conditions are generally the specimens used to judge the quality of the concrete for the job and are called *record cylinders*.

The contractor may have field cylinders made and cured under jobsite conditions to determine when forms may be stripped or the structure may be put into service.

The evaluation of compression test results are based on ASTM C94 or ACI 318 criteria. Both ASTM and ACI define a strength test as the average strength of two 6 × 12 in. cylinders or three 4 × 8 in. cylinders. The average strength of any three consecutive tests shall be equal to or greater than the specified strength f_c'. When the required strength is 5000 psi (35 MPa) or less no single test can fall more than 500 psi (3.5 MPa) below the required strength. However, if the required strength is greater than 5000 psi (3.5 MPa) no single test can fall below 0.90 f_c'.

Statistical computations used to control the quality of concrete can be performed manually or by using simple calculators. The results of concrete strength tests will, if plotted, assume the "normal distribution"—that is, the familiar bell-shaped curve (Figure 4–17). The curve can be described

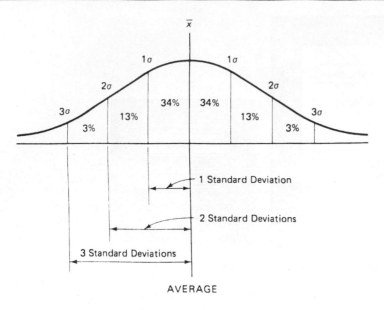

FIGURE 4–17. Normal distribution curve.

by two characteristics: the mean, or average, denoted by the letter $\bar{x}$, and the standard deviation denoted by the Greek letter sigma, σ.

$$\text{Mean } \bar{x} = \frac{\Sigma x}{n}$$

$$\text{Standard deviation } \sigma = \sqrt{\frac{\Sigma(x - \bar{x})^2}{n - 1}}$$

where $\bar{x}$ = mean

x = test strength

n = number of tests

S or σ = standard deviation

The smaller the standard deviation, the steeper the curve, indicating results grouped tightly around the mean; the higher the standard deviation, the shallower the normal distribution curve, indicating widespread test results.

Whatever the value of the standard deviation, vertical lines drawn at one, two, and three standard deviations on either side of the mean always include the same proportion of area under the curve. For example, if $\bar{x} = 3500$ psi and $\sigma = 250$ psi, approximately 68 percent of the test results would fall between 3750 and 3250 psi, 94 percent between 4000 and 3000 psi, and almost 100 percent (99.73 percent) between 4250 and 2750 psi.

A statistical measure of the uniformity of concrete production and testing is the coefficient of variation (V). The producer of the concrete must control factors such as aggregate variations, admixture doses, mixing, and delivery. The testing of the concrete must conform to the stipulations of standards, such as ASTM that govern sampling, handling, transportation, curing, and finally the actual testing of the representative samples of concrete. Therefore, it is possible that uniformly produced concrete will have a high coefficient

of variation due to the impact of poor testing. When the evidence suggests such a situation, it is customary to have a second testing laboratory shadow the original laboratory, performing the same tests to determine if procedural testing errors are contributing to the high coefficient of variation.

Estimated Ranges for Coefficients of Variation

V	Control Level
Under 5%	Associated with well-controlled laboratory tests
5 to 10%	Excellent
10 to 15%	Good
15 to 20%	Fair
Over 20%	Poor

Example

$$V = \frac{\sigma}{\bar{x}} \times 100$$

V = Coefficient of variation

σ = standard deviation

$\bar{x}$ = mean

A concrete producer has a mean of $\bar{x} = 3400$ psi and a standard deviation of $\sigma = 450$ psi on a concrete construction project. Calculate the coefficient of variation.

$$V = \frac{420 \text{ psi}}{3400 \text{ psi}} \times 100 = 12.4\%$$

The coefficient of variation calculated indicates good concrete production and control testing.

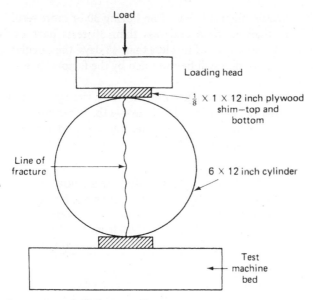

FIGURE 4–18. Splitting tensile test.

Direct tests for tensile strength of concrete are seldom made, but a convenient, reliable test to determine indirectly the tensile strength of concrete is in use today. Developed in Brazil and standardized by ASTM C496, the test gives a splitting tensile strength value which is about 15 percent higher than values obtained through direct tensile tests.

The method utilizes standard 6-by-12-in. cylinders which are loaded along the length of the cylinder (Figure 4–18). The splitting tensile strength is determined by using a formula based on the theory of elasticity.

$$f'_{sp} = \frac{2P}{\pi l d}$$

where f'_{sp} = splitting tensile strength, psi

P = maximum load, lb

l = length of cylinder, in.

d = diameter of cylinder, in.

Example

Calculate splitting tensile strength if a standard 6×12 in. concrete specimen fails at 42,800 lbs.

$$f'_{sp} = \frac{2(42,800 \text{ lb})}{\pi \times 6 \text{ in.} \times 12 \text{ in.}} = 378 \text{ psi}$$

When tests are not conducted on concrete specimens, reasonable approximations of tensile and splitting tensile strengths may be obtained using these empirical equations:

$$\text{Tensile strength } = f'_t = 4.5 f \sqrt{f'_c}$$

$$\text{Splitting tensile strength } = f'_{sp} = 6.5 \sqrt{f'_c}$$

Concrete used for pavement slabs is subjected to bending loads, and the flexure strength or modulus of rupture of the concrete is usually determined with 6-by-6 concrete beam specimens. The test procedure is ASTM C78 and utilizes simple beams with third-point loading (Figure 4–19). When the full ASTM procedure is not used, an adequate estimate of flexure strength using the compressive strength of the concrete can be determined.

$$f'_R = 7.5 \sqrt{f'_c}$$

ASTM C78 Standard Test Method for Flexure Strength of Concrete (Using Simple Beam with Third-Point Loading) is one test used to determine flexure strength and defines three failure modes. The beams are cast using C31 criteria and when tested are marked with four lines to create three equal 6-in. spaces on the tension face of the beam. The beams are set in the apparatus (Figure 4-19) and a constant

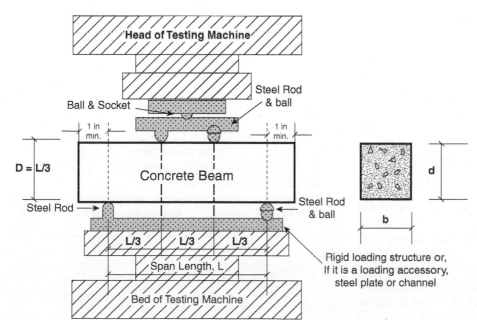

FIGURE 4–19. Apparatus and beam set for third point loading. (*Courtesy of Wayne Deyo*)

rate load is applied to failure. When the rupture occurs within the middle third of the beam, the modulus of rupture is calculated using this equation:

MODE 1: $R = \dfrac{PL}{bd^2}$

R = modulus of rupture, psi or MPa

P = load at failure in lbs or N

L = span length in in. or mm

b = average width of beam, in., or mm, at the fracture

d = average depth of beam, in., or mm at the fracture

a = average distance from fracture line to the nearest support

If the failure occurs outside the middle by no more than 5 percent of the span length, calculate the modulus of rupture as follows:

MODE 2: $R = \dfrac{3Pa}{bd^2}$

MODE 3: If the failure occurs beyond the 5 percent of the span zone, the test is disregarded.

Example

A flexure beam is set up with an 18-in. span, the beam is 6 in. deep and 6 in. wide and the failure occurs at a load of 6128 lb. Within the middle third:

$$R = \frac{PL}{bd^2}$$

$$R = \frac{6128\ \text{lb}\ (18\ \text{in.})}{6\ \text{in.}\ (6\ \text{in.})^2} = 511\ \text{psi}$$

A second beam is tested and the rupture is 6.3 in. from the nearest support at a failure load of 5685 lb.

$$R = \frac{3Pa}{bd^2}$$

$$R = \frac{3(5685\ \text{lb})6.3\ \text{in.}}{6\ \text{in.}\ (6\ \text{in.})^2} = 497\ \text{psi}$$

A third beam is tested and failure occurs outside the 5 percent L secondary failure zone.

Disregard this test.

REQUIRED COMPRESSIVE STRENGTH (f'_{CR})

Based on ACI 318 requirements, concrete mix designs for new projects may use current mix designs or designs which have been used in the past. Concrete mix design required average strengths f'_{cr} are based on standard deviation values calculated from strength test records not more than 12 months old. The tests values shall represent quality control procedures, similar conditions and materials, as well as strengths that are at the required strength f'_c or within 1000 psi of f'_c.

These requirements are based on having 30 or more results; if more than 15 tests and less than 30 tests have been recorded over a span of not less than 45 days, the calculated standard deviation will be modified by the factors shown in Table 4–11a.

The required average compressive f'_{cr} strength is calculated using the equations in Table 4–11b. When no usable concrete mix history is available, the required average is calculated using Table 4–11c.

Table 4–11a. Modification factor for standard deviation when less than 30 tests are available

Number of tests*	Modification factor for standard deviation**
Less than 15	
15	1.16
20	1.08
25	1.03
30 or more	1.00

*Interpolate for intermediate number of tests.

**Modified standard deviation to be used to determine required average strength, f'_{cr}.

Source: Adapted from ACI 318.

Table 4–11b. Required average compressive strength when data are available to establish a sample standard deviation

Specified compressive strength, psi	Required average compressive strength, psi	
$f'_c \le 5000$	Use the larger value computed from Eq. (5-1) and (5-2)	
	$f'_{cr} = f'_c + 1.34S_s$	(5-1)
	$f'_{cr} = f'_c + 2.33S_s - 500$	(5-2)
$f'_c > 5000$	Use the larger value computed from Eq. (5-1) and (5-3)	
	$f'_{cr} = f'_c + 1.34S_s - 500$	(5-1)
	$f'_{cr} = 0.90f'_c + 2.33S_s - 500$	(5-3)

Table 4–11c. Required average compressive strength when data are not available to establish a sample standard deviation

Specified compressive strength, psi	Required average compressive strength, psi
$f'_c < 3000$	$f'_{cr} = f'_c + 1000$
$3000 \le f'_c \le 5000$	$f'_{cr} = f'_c + 1200$
$f'_c > 5000$	$f'_{cr} = 1.10f'_c + 700$

Source: Courtesy of the American Concrete Institute.

Example

Calculate the f'_{cr} for a concrete mix that has a specified 4000 psi strength requirement if previous data indicate a standard deviation of 325 psi based on 20 tests for a similar concrete mix.

$$f'_{cr} = 4000 \text{ psi } + 1.34 \ (325 \text{ psi})(1.08) = 4470 \text{ psi}$$

$$f'_{cr} = 4000 \text{ psi } + 2.33 \ (325 \text{ psi})(1.08) - 500 \text{ psi} = 4318 \text{ psi}$$

The larger of the two calculated f'_{cr} values will be used to determine the concrete proportions for the new concrete mix. This procedure will limit the probability of strength test failures to 1 in 100 tests.

NONDESTRUCTIVE TEST METHODS

The need to determine the in-place strength of concrete occurs frequently in the construction industry. For example, when a building is going to be renovated to serve a new purpose, part of the design procedure may require strength testing of the structure's existing concrete. The other main use of in-place testing is during the construction process where the tests are utilized to monitor and evaluate concrete. In this situation, the testing may be used to determine when formwork and/or shoring may be removed safely or to assist in the evaluation of concrete that has not reached design strength requirements. The most widely used in-place test methods are as follows:

1. The rebound hammer
2. The penetration probe
3. Pullouts
4. Ultrasound

Each of the tests used for nondestructive testing has limitations and requires the development of empirical correlations to the standard compressive strength test (Table 4–12).

The Rebound Hammer

The impact rebound test ASTM C805 is often referred to as the Schmidt or Swiss Hammer test. The rebound hammer is

Table 4–12. Comparison among five nondestructive tests

Item to Compare	Test Device				
	Rebound Hammer	Ultrasonic Pulse Velocity	Penetration Probe	Pullout	Pin Penetration
Cost of purchase	Inexpensive	Expensive	Expensive	Expensive	Inexpensive
Cost of operation	Inexpensive	Inexpensive	Expensive	Expensive	Inexpensive
Time required to perform test	About 10 to 20 sec per shot	About 1 to 2 min per reading	About 3 to 4 min per shot	About 2 to 3 min per pull	About 1 min per shot
Appearance of surface of concrete after testing	Leaves an indentation on the surface	Grease used is difficult to remove and leaves stain on concrete	Leaves a small hole, and may cause some minor cracking	Leaves a large hole in the concrete— damage to the concrete must be repaired	Leaves tiny hole
Surface finish of concrete	Trowelled surface gives higher values than formed surface	Smooth surface is required	Not very important	Not important	Not important
Moisture content of the concrete	Dry concrete gives higher values than wet concrete	Pulse velocity increases with increased moisture content	Dry concrete gives higher value than wet concrete	Dry concrete gives higher value than wet concrete	Dry concrete gives higher value than wet concrete
Temperature	Frozen concrete will give very high values and must be thawed before testing. Temperature of the hammer will also affect the rebound number	Not sensitive to temperature in the range 5 to 30°C. At higher temperatures, the pulse velocity is decreased, and at temperature below freezing it is increased	Not investigated	Not investigated	Not investigated
Effect of carbonation of the surface	Can increase the ardness values by as much as 50 percent	Does not have great effect on the test	Relatively not important	Does not have any effect on the test	Not investigated

(Continued)

Table 4–12. (continued)

Item to Compare	Test Device				
	Rebound Hammer	Ultrasonic Pulse Velocity	Penetration Probe	Pullout	Pin Penetration
Effect of hard aggregate	Different size of aggregate affects the rebound number	Different calibration chart is required with change in source of aggregate	Tends to give higher compressive strength	Sometimes might increase the strength	Minor effect
Length and size of specimen tested	Small test pieces give consistently lower rebound number and higher scatter of results	Pulse velocity depends on the path length and size of structure is limited by the leads and transducer size	Not suitable for very small structures	Not suitable for very small structures	Minor effect
Existence of steel bars inside the structure	Does not have great effect	Tends to increase pulse velocity	Does not have great effect	Does not have great effect	Does not have great effect
General remarks	Useful in checking the uniformity of concrete and in comparing one concrete against another	Efficient for determining concrete quality	Device must be calibrated for the material tested; also different type of probes are required for different types of aggregate	Gives direct measure of the strength. It is suitable to evaluate existing structures where the concrete quality is suspect	Useful in checking the uniformity of concrete and in determining its quality

Source: The American Concrete Institute.

held perpendicular to the concrete surface and pushed toward the surface, stretching the internal spring to its limit. At the limit the spring releases and pulls the hammer toward the concrete and the plunger rebounds off the concrete. The plunger is locked in its rebound position, and the rebound number is read and recorded. Ten readings are taken and no two readings should be obtained closer together than 1 in. The 10 readings are averaged, and any reading that differs by more than 6 from the average is discarded; then average the remaining values. If more than 2 readings differ from the average by more than 6, discard all 10 readings and repeat the test. Most hammers have a graph printed on them that indicates compressive strengths of concrete based on the rebound numbers obtained by test. Even though the test results are not precise, they indicate relative strengths of in-place concrete; that is, for the same class of concrete, higher numbers represent stronger concretes than lower numbers. To obtain more usable information from the rebound test, concrete specimens of the class of concrete are restrained in a compression testing machine while rebound numbers are read from the specimen surface. The specimen is then loaded to failure and a correlation graph is plotted of compressive strength versus rebound numbers.

Though the test is simple to perform, the user should be aware of the factors that influence the rebound numbers. One of these factors is that the test is a surface test and may not be representative of the interior concrete conditions. Also, the concrete surface smoothness, surface carbonation, coarse aggregate type, moisture conditions, as well as the age, size, shape, and rigidity of the concrete being tested all have varying degrees of influence on the test results. Despite its limitations, the rebound hammer is still a very useful tool for determining relative strengths and uniformity of in-place concrete materials.

The Penetration Probe

The penetration probe method ASTM C803 is usually referred to by its commercial name, the Windsor Probe. The test uses a powder-activated gun to drive a hardened alloy probe into concrete. ASTM C803 requires that the exit muzzle velocity of the probes has a coefficient of variation no greater than 3 percent based on ten approved method ballistic tests.

The test procedure can be used on lightweight concretes of less than 125 pcf and normal-weight concretes using different probes. The gold probe has a 56 percent greater cross-sectional area than the silver probe and is used for lightweight concrete. The power setting of low is used for concretes that have an expected compressive strength below 3200 psi and the high power setting for concretes above 3200 psi. The probes are fired into concrete through a triangular template, which allows for the rejection of an anomalous reading, and by shifting the template setting of a new probe. The average penetration of the three probes is measured using a pair of triangular plates attached to the probes. At the completion of the test, the probes are removed from the concrete and the holes are patched. Charts designed to adjust compressive strength values based on Moh's hardness of the coarse aggregate, type of probe, and power setting are used with the average penetration value to determine the compressive strength.

Pullout

The pullout test ASTM C900 is often referred to as the Danish Pullout test. A steel rod with an enlarged head is cast in the concrete to be tested. When the concrete is ready for testing, a tension jack is placed over the protruding steel rod and a tensile force is applied to the rod until failure occurs in the concrete.

The feature of the pullout test is that it produces a well-defined failure and measures a static strength property of concrete that can be analyzed. The measured strength value determined by the pullout test can be correlated to the compressive strength of concrete provided that the test configuration and concrete materials are held constant.

Because the standard pullout test requires the inserts to be preplaced in the concrete, their use requires preplanning as to location. This insert requirement also means they cannot be used to test existing concrete; however, at this time various expanding-type anchors are being investigated for use in testing existing concrete.

Ultrasonic

The pulse velocity test ASTM C597 measures the velocity of a sound wave through concrete. The time required for the pulse to travel through the concrete from the transmitter to the receiver is divided into the straight-line distance the signal travels to determine the pulse velocity. The pulse velocity can be correlated to the elastic modulus and mass density of the concrete being tested. The compressive strength can then be empirically inferred from the test results.

The test results are affected by the age of the concrete, moisture content, cracks, voids, and reinforcing steel. The pulse velocity through steel is about 40 percent greater than through concrete. Therefore, if reinforcement bars run parallel to the pulse, the velocity readings will be much higher, indicating a higher compressive strength.

The test method for cores and sawed beam specimens is ASTM C42. The method covers the removal of the concrete specimens from the structure, followed by their preparation and testing to failure. The area of concrete to be tested should be aged at least 14 days to assure adequate bond development between the coarse aggregate and mortar.

The preparation of the specimens before test requires the removal of projections from the ends of the core, checking end diameter values, and correcting any deficiencies by sawing

or tooling. The specimens are then submerged in saturated lime water at 73.4 ± 3°F for at least 40 hours prior to test. This requirement may be altered by the specifying authority for the project so that specimens may also be tested at their normal moisture content. The core's length-to-diameter (L/D) ratio is important. If a capped specimen has an L/D ratio greater than 2.10, it must be trimmed so that the L/D ratio falls between 2.10 and 1.94. When the L/D ratio of a specimen falls below 1.94, a correction factor is applied to the compressive strength. The specification requires that a core having a maximum height of 95 percent of its diameter before capping or a height less than its diameter after capping shall not be tested. Chapter 4 of the Building Code ACI 318 specifies that the average strength of three cores equals at least 85 percent of the design strength f'_c and no single core strength falls below 75 percent of the design strength f'_c. When these criteria are met, the concrete represented by the core tests is considered structurally adequate. Locations represented by erratic core strength tests may be retested to verify testing accuracy.

Example
Three cores were tested in accordance with ASTM C42. Determine if the concrete meets strength requirements if the required strength f'_c was 4500 psi.

Core Strengths:
1. 4120 psi
2. 3995 psi
3. 4215 psi

$$x = \frac{12,330 \text{ psi}}{3} = 4110 \text{ psi}$$

$$4500 \text{ psi} \times .85 = 3825 \text{ psi}$$

$$4500 \text{ psi} \times .75 = 3375 \text{ psi}$$

4110 psi is greater than 3825 psi meets ACI 318 criteria. No single core below 3375 psi meets ACI 318 criteria.

If the conditions of ACI 318 Chapter 4 are not met, then the specifying authority may order load tests as described in ACI 318 Chapter 20 or take other appropriate actions (Figure 4–20).

If a load test is required, the test procedures should be performed under the direction of a qualified engineer with experience in structural investigations. The specified total test load shall be 85 percent of the dead and live load design strength required by the code.

The general criterion for acceptance of test load results is that the structure show no visible signs of failure. Visible signs of failure are evidenced by cracking, spalling, or excessive deflections. There are no simple rules developed that apply to all structures and situations, but if sufficient damage is evident no retest is permitted. If no visible signs of damage or failure occur, then the structure's adequacy is governed by deflection recovery. Except for obvious failures, the specifying authority may permit the use of a structure at a lower load rating if based on load test results; the structure is considered safe.

FIGURE 4–20. Concrete masonry units used here to apply a uniform load to a roof deck, for deflection testing. *(Courtesy American Concrete Institute)*

CONCRETE FORMWORK

The temporary structure used to contain fresh concrete is the form; its function is to contain the concrete until it develops adequate strength to be self-supporting. The total system of forms in contact with fresh concrete plus all supports, hardware, and bracing is referred to as the formwork. Formwork typically accounts for 35 to 65 percent of concrete construction costs and is designed to satisfy three major requirements: economy, quality, and safety.

Economical form design involves material selection and fabrication and construction techniques. The reuse and adaptability of a forming system to different uses will also have a large impact on forming costs. The designer of a structure may help lower forming costs, for example, by specifying consistent column sizes with a higher strength concrete and heavier reinforcement for lower level columns and reducing these requirements on upper levels, thus allowing the contractor to build standard-size column forms to be used throughout the structure.

Accurate formwork construction will govern size, shape, finish, and alignment of all structural elements. The formwork must be strong enough to resist excessive deflection, rigid enough to resist bulging and leakage, which may require expensive refinishing work, as well as prevent movement of individual elements that may affect structural integrity.

Another significant aspect of formwork is safety because the system must support the weight of the fresh concrete and formwork as well as the equipment and labor required to place and finish the concrete. Formwork design for structural loading requirements must be performed by design professionals, using specified codes, who will produce stamped formwork construction drawings that the contractor will use to build the formwork system.

PLACEMENT OF CONCRETE

During the testing of fresh concrete, the contractor's crews will be placing the concrete in the forms. Depending on the job conditions, various methods of concrete placement are utilized. Equipment will vary from simple chutes, wheelbarrows, and buggies to sophisticated conveyor or pump systems. Figures 4–21 and 4–22 illustrate correct and incorrect methods of concrete placement utilizing different types of equipment. Regardless of how the concrete is placed, extreme care must be taken to ensure that the concrete is not changed by transporting it and that it does not become segregated during placement. *Segregation* or the separation of coarse aggregate from the mortar or the water from the ingredients can be detrimental to the quality of the hardened concrete.

One versatile method of handling plastic concrete on many construction sites is the concrete pump. The concrete pump transports plastic concrete through a pipeline system from the ready-mix truck to the point of placement without changing the basic characteristics of the concrete mix.

The normal pumping distances will range from 300 to 1000 ft horizontally or 100 to 300 ft vertically. In some instances, concrete has been successfully pumped over 2000 ft horizontally and 900 ft vertically. Curves, vertical lifts, and harsh mixes tend to reduce maximum pumping distances. A 90° bend in the pipe is the equivalent of about 40 ft of straight horizontal line, and each 1 ft of vertical lift is the equivalent of about 8 ft of horizontal line.

The pipe used to carry the concrete is generally steel and/or rubber. Aluminum pipe, which originally was introduced as a labor-saving device because of its light weight, is no longer used because concrete passing through the pipe grinds aluminum particles from the pipe wall. These aluminum particles, reacting with lime from the concrete, create hydrogen gas, which increases the voids in the concrete and substantially reduces concrete strengths.

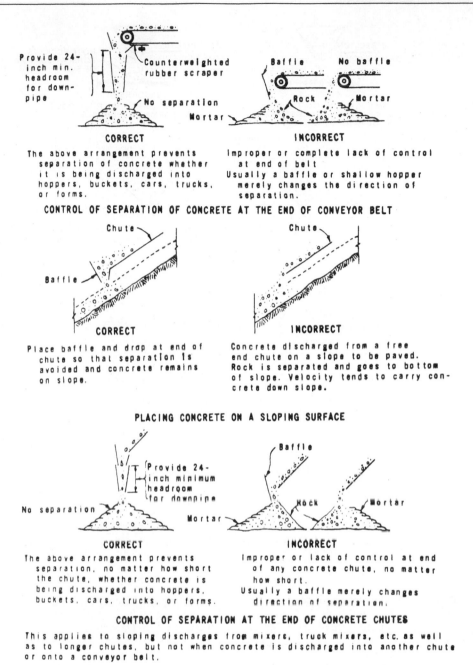

Provide 24-inch min. headroom for down-pipe

Counterweighted rubber scraper

No separation

Mortar

CORRECT

The above arrangement prevents separation of concrete whether it is being discharged into hoppers, buckets, cars, trucks, or forms.

Baffle   No baffle

Rock   Mortar

INCORRECT

Improper or complete lack of control at end of belt
Usually a baffle or shallow hopper merely changes the direction of separation.

CONTROL OF SEPARATION OF CONCRETE AT THE END OF CONVEYOR BELT

Chute

Baffle

CORRECT

Place baffle and drop at end of chute so that separation is avoided and concrete remains on slope.

Chute

INCORRECT

Concrete discharged from a free end chute on a slope to be paved. Rock is separated and goes to bottom of slope. Velocity tends to carry concrete down slope.

PLACING CONCRETE ON A SLOPING SURFACE

Provide 24-inch minimum headroom for downpipe

No separation

Mortar

CORRECT

The above arrangement prevents separation, no matter how short the chute, whether concrete is being discharged into hoppers, buckets, cars, trucks, or forms.

Baffle

Rock   Mortar

INCORRECT

Improper or lack of control at end of any concrete chute, no matter how short.
Usually a baffle merely changes direction of separation.

CONTROL OF SEPARATION AT THE END OF CONCRETE CHUTES

This applies to sloping discharges from mixers, truck mixers, etc. as well as to longer chutes, but not when concrete is discharged into another chute or onto a conveyor belt.

FIGURE 4–21. Correct and incorrect methods of concrete placement using conveyor belts and chutes. Proper procedures must be used if separation at the ends of conveyors and chutes is to be controlled. 288-D-854 *(Courtesy U.S. Department of the Interior, Water and Power Resources Service)*

Normal pump capacities range from 10 to 125 cu yd/hour, with special pumps having capacities in excess of 250 cu yd/hour. Aggregate sizes are important; generally, the maximum-size aggregate should not exceed 40 percent of the diameter of the pipe if the aggregate has a well-rounded shape. Because ideally shaped aggregates are not always available, further reductions in maximum aggregate size may be required for flat and elongated aggregates. Pumping lightweight aggregate concrete presents no problem, provided that it has been presoaked. Without the presoak, pump pressures tend to force water into the aggregate during pumping, and the concrete discharge becomes dry and unworkable. Slump ranges for lightweight concretes run between 2 and 5 in. Most pumps will handle concretes with slumps of 3 to 4 in. In this slump range the concrete discharge will exhibit no segregation. In fact, any concrete that does not segregate before pumping will not tend to segregate during pumping. If a concrete exhibits

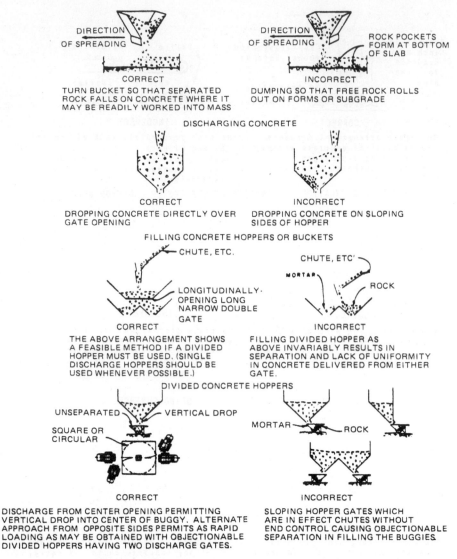

FIGURE 4–22. Correct and incorrect methods for loading and discharging concrete buckets, hoppers, and buggies. Use of proper procedures avoids separation of the coarse aggregate from the mortar. 288-D-3276 *(Courtesy U.S. Department of the Interior, Water and Power Resources Service)*

segregation before pumping, it usually is not a pumpable mix (Table 4–13).

Before actual pumping begins, the pump lines are lubricated with a concrete mix that contains no coarse aggregate. The amount of mortar used will depend on the length of the run. A cubic yard of mortar will lubricate approximately 1000 ft of pipe. If delays occur during the pumping process, the pump operator will have to move some concrete through the system at regular intervals to prevent plugs from forming in the system (Figure 4–23).

After completion of the concrete placement the pump and lines can be washed out with water. Some lines may require the use of a *go-devil*, a dumbbell-shaped insert placed in the pipe which will push out the concrete, leaving clean interior walls in the pipe system.

Conveyor belts have been used by the concrete industry to move plastic concrete for a number of years. The first successful use of a belt conveyor appears to be a 1929 concrete placement utilizing a 600-ft conveyor to transport concrete on a bridge job. Early belt conveyors had capacities of 30 to 40 cu yd/hour, whereas today equipment with capacities of up to 300 cu yd/hour is available for massive concrete placements. The volume of concrete transported by a conveyor is determined by the belt width, conveyor speed, angle of incline or decline, and the properties of the concrete mix itself, such as aggregate size and shape, mix proportions, and slump. Conveyors, charging hoppers, transfer devices, and belt wipers generally do not modify any of the important characteristics of the concrete being carried to the placement area. However, if placement is delayed excessively or weather

Table 4–13. Methods and equipment for transporting and handling concrete

Equipment	Type and Range of Work for Which Equipment Is Best Suited	Advantages	Points to Watch for
Belt conveyors	For conveying concrete horizontally or to a higher or lower level. Usually positioned between main discharge point and secondary discharge point.	Belt conveyors have adjustable reach, traveling diverter, and variable speed both forward and reverse. Can place large volumes of concrete quickly when access is limited.	End-discharge arrangements needed to prevent segregation and leave no mortar on return belt. In adverse weather (hot, windy) long reaches of belt need cover.
Belt conveyors mounted on truck mixers	For conveying concrete to a lower, horizontal, or higher level.	Conveying equipment arrives with concrete. Adjustable reach and variable speed.	End-discharge arrangements needed to prevent segregation and leave no mortar on return belt.
Buckets	Used with cranes, cableways, and helicopters for construction of buildings and dams. Convey concrete directly from central discharge point to formwork or to secondary discharge point.	Enable full versatility of cranes, cableways, and helicopters to be exploited. Clean discharge. Wide range of capacities.	Select bucket capacity to conform to size of the concrete batch and capacity of placing equipment. Discharge should be controllable.
Chutes on truck mixers	For conveying concrete to lower level, usually below ground level, on all types of concrete construction.	Low cost and easy to maneuver. No power required; gravity does most of the work.	Slopes should range between 1 to 2 and 1 to 3 and chutes must be adequately supported in all positions. End-discharge arrangements (downpipe) needed to prevent segregation.
Cranes and buckets	The right equipment for work above ground level.	Can handle concrete, reinforcing steel, formwork, and sundry items in bridges and concrete-framed buildings.	Has only one hook. Careful scheduling between trades and operations is needed to keep crane busy.
Dropchutes	Used for placing concrete in vertical forms of all kinds. Some chutes are one piece tubes made of flexible rubberized canvas or plastic, others are assembled from articulated metal cylinders (elephant trunks).	Dropchutes direct concrete into formwork and carry it to bottom of forms without segregation. Their use avoids spillage of grout and concrete on reinforcing steel and form sides, which is harmful when off-the-form surfaces are specified. They also will prevent segregation of coarse particles.	Dropchutes should have sufficiently large, splayed-top openings into which concrete can be discharged without spillage. The cross section of dropchute should be chosen to permit inserting into the formwork without interfering with reinforcing steel.
Mobile batcher mixers	Used for intermittent production of concrete at jobsite, or where only small quantities are required.	A combined materials transporter and mobile batching and mixing system for quick, precise proportioning of specified concrete. One-man operation.	Trouble-free operation requires good preventive maintenance program on equipment. Materials must be identical to those in original mix design.
Nonagitating trucks	Used to transport concrete on short hauls over smooth roadways.	Capital cost of nonagitating equipment is lower than that of truck agitators or mixers.	Concrete slump should be limited. Possibility of segregation. Height is needed for high lift of truck body upon discharge.
Pneumatic guns (shotcrete)	Used where concrete is to be placed in difficult locations and where thin sections and large areas are needed.	Ideal for placing concrete in freeform shapes, for repairing structures, for protective coating, thin linings, and building walls with one-sided forms.	Quality of work depends on skill of those using equipment. Only experienced nozzlemen should be employed.
Pumps	Used to convey concrete directly from central discharge point at jobsite to formwork or to secondary discharge point.	Pipelines take up little space and can be readily extended. Delivers concrete in continuous stream. Pumps can move concrete both vertically and horizontally. Truck-mounted pumps can be delivered when necessary to small or large projects. Tower-crane mounted pump booms provide continuous concrete for tall building construction.	Constant supply of freshly-mixed concrete is needed with average consistency and without any tendency to segregate. Care must be taken in operating pipeline to ensure an even flow and to clean out at conclusion of each operation. Pumping vertically, around bends, and through flexible hose will considerably reduce the maximum pumping distance.

(Continued)

Table 4–13. (continued)

Equipment	Type and Range of Work for Which Equipment Is Best Suited	Advantages	Points to Watch for
Screw spreader	Used for spreading concrete over large flat areas, such as in pavements and bridge decks.	With a screw spreader a batch of concrete discharged from bucket or truck can be quickly spread over a wide area to a uniform depth. The spread concrete has good uniformity of compaction before vibration is used for final compaction.	Screw spreaders are usually used as part of a paving train. They should be used for spreading before vibration is applied.
Tremies	For placing concrete underwater.	Can be used to funnel concrete down through the water into the foundation or other part of the structure being cast.	Precautions are needed to ensure that the tremie discharge end is always buried in fresh concrete, so that a seal is preserved between water and concrete mass. Diameter should be 250 to 300 mm (10 to 12 in.) unless pressure is available. Concrete mixture needs more cement, 390 kg/m^3 (658 lb/yd^3) and greater slump, 150 to 230 mm (6 to 9 in.), because concrete must flow and consolidate without any vibration.
Truck agitators	Used to transport concrete for all uses in pavements, structures, and buildings. Haul distances must allow discharge of concrete within $1\frac{1}{2}$ hours, but limit may be waived under certain circumstances.	Truck agitators usually operate from central mixing plants where quality concrete is produced under controlled conditions. Discharge from agitators is well controlled. There is uniformity and homogeneity of concrete on discharge.	Timing of deliveries should suit job organization. Concrete crew and equipment must be ready onsite to handle concrete.
Truck mixers	Used to transport concrete for uses in pavements, structures, and buildings. Haul distances must allow discharge of concrete within $1\frac{1}{2}$ hours, but limit may be waived under certain circumstances.	No central mixing plant needed, only a batching plant, since concrete is completely mixed in truck mixer. Discharge is same as for truck agitator.	Timing of deliveries should suit job organization. Concrete crew and equipment must be ready onsite to handle concrete. Control of concrete quality is not as good as with central mixing.
Wheelbarrows and buggies	For short flat hauls on all types of onsite concrete construction, especially where accessibility to work area is restricted.	Very versatile and therefore ideal inside and on jobsites where placing conditions are constantly changing.	Slow and labor intensive.

Source: Portland Cement Association.

FIGURE 4–23. Concrete pumps placing concrete for a water storage tank slab. *(Courtesy D. J. Rossetti, Inc.)*

FIGURE 4–24. Concrete placement using a conveyor system. *(Photo Courtesy of GOMACO Corporation, Ida Grove, Iowa, USA)*

conditions are not optimum, some provision may have to be made to cover the conveyor system.

The conveying system must be properly designed with enough power to start and stop with fully loaded belts during placement delays. The individual sections are designed for high mobility because the delivery of fresh concrete must be continuous over the placement area without excessive construction joints (Figure 4–24).

On construction projects utilizing pumps or conveyors, there tends to be some discussion as to the proper location for concrete testing. Should the tests be performed on the concrete as discharged from the truck into the transporting system, or after the concrete has traveled through the system to the placement area? The general recommendation is to make tests at both locations; if satisfactory correlation can be made, tests may be performed at the most accessible location as long as placement conditions do not change.

Like all materials used in construction, concrete expands and contracts under different conditions of moisture and temperature. To control random cracking, joints must be placed in the concrete to allow cracking to occur at the proper location. Figure 4–25 illustrates the types of joints.

A *contraction* or *control joint* is a cut made into the surface of the concrete. The slab is weakened at that point and cracks should develop in the joint rather than randomly. The joint may be sawed, made with a groover, or formed with divider strips. If the contraction joint is saw cut, conventional saw cuts are made within 4 to 12 hours, and with dry-cut early entry saws 1 to 4 hours after the concrete has been finished. The ACI recommendations for the spacing of joints for nonreinforced slabs is 24 to 36 times the thickness of the slab in each direction. Common guidelines for contraction joint spacing in feet is to place the joints 2-1/2 times the depth of the slab in inches. For example, a 6-in. slab would require joint spacing at 15 ft and a 10-in. slab would require 25 ft joint spacing. The joint spacing should create as near as possible square areas but in no case should the length to width ratio exceed 1-1/2 to 1. The depth of the saw-cut for fiber-reinforced slabs is usually increased to $\frac{1}{3}$ to $\frac{1}{2}$ the depth of the slab.

The *isolation joint* is designed to physically separate areas of concrete from one another or from columns, poles, and walls; this separation allows for differential settlement. The joint is usually formed with a premolded filler which is left in place just slightly below the concrete surface's tooled edges.

Construction joints are used when concrete cannot be placed continuously. They separate one concrete placement from the next concrete placement. Construction joints should be located in the concrete so that they may act as control joints. Some load-transfer device must be used to carry loads across the joint. Dowels or keyways can be used.

After the concrete has been properly placed and consolidated by rodding or vibration and the final screeding operations have been completed, the contractor must apply the proper surface finish and curing system (Figure 4–26).

During concrete placement the contractor must consolidate the plastic concrete. The consolidation must eliminate as far as practical the voids in the concrete. Well-consolidated concrete is free of rock pockets or honeycomb or bubbles of entrapped air and is in close contact with forms, reinforcement, and other embedded items such as anchor bolts and pipe sleeves.

Vibrators may be either immersion or external form-mounted systems powered by air or electricity. The contractor must determine a vibrator pattern and the amplitude and

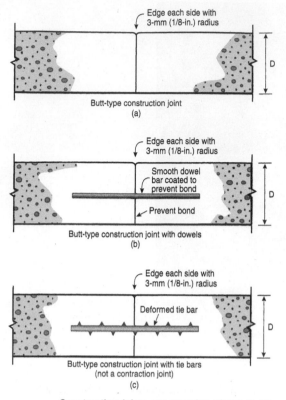

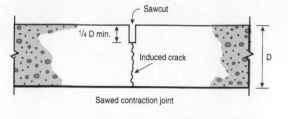

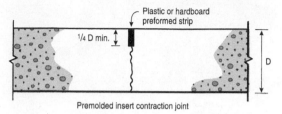

Contraction joints provide for horizontal movement in the plane of a slab or wall and induce controlled cracking caused by drying and thermal shrinkage.

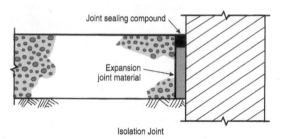

Isolation joints permit horizontal and vertical movements between abutting faces of a slab and fixed parts of a structure.

Slab thickness, in.	Maximum-size aggregate less than 3/4 in	Maximum-size aggregate 3/4 in. and larger
4	8	10
5	10	13
6	12	15
7	14	18**
8	16**	20**
9	18**	23**
10	20**	25**

Spacings are appropriate for slumps between 4 in. and 6 in. If concrete cools at an early age, shorter spacings may be needed to control random cracking. (A temperature difference of only 10°F may be critical.) For slumps less than 4 in., joint spacing can be increased by 20%

**When spacings exceed 15 ft, load transfer by aggregate interlock decreases markedly.

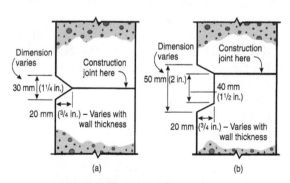

Horizontal construction joints in walls with V-shaped (a) and beveled (b) rustication strips.

FIGURE 4–25. Joints used in concrete placement. *(Courtesy Portland Cement Association)*

frequency of vibration, as well as the depth of the vibrator into the concrete, to ensure good consolidation. Care must be taken so as not to overvibrate fresh concrete, which will cause the coarse aggregate to settle and leave a wet mortar film at the surface of the concrete placement.

There is considerable evidence to indicate that revibration is beneficial to concrete, provided that the concrete is brought back to a plastic condition. The revibrated concrete

exhibits a higher strength and less settlement cracking, and the effects of internal bleeding are reduced.

Self-compacting concretes (SCC) are made with viscosity-modifying admixtures and/or new-generation superplasticizers and aggregate to sand ratios of about 1 to 1, plus additional fines. The concrete produced exhibits no segregation, flows freely, and is able to fill heavily reinforced formwork. Self-compacting concrete is widely used in the precast

FIGURE 4–26. Laser screed® maintains accurate elevation control while screeding concrete. *(Courtesy Somero Enterprises, Inc.)*

industry and has virtually eliminated the use of vibration equipment.

The finish of a concrete surface may vary from a wood float, broomed finish to a hard-troweled finish. Wood float and broom finishes are usually used on exterior flatwork, while the trowel finish is an interior finish. Contractors may choose to finish small areas by hand trowels and larger areas by power trowels. The power trowels are usually gasoline powered and have three or four steel blades which rotate on the surface of the concrete (Figure 4–27). Concrete pavements are placed and finished with slipform pavers (Figure 4–28).

CURING CONCRETE

Proper curing must begin after surfaces have been worked to proper finish for concrete to adequately gain its design strength, increase its resistance to freeze–thaw, and improve its watertightness and wear resistance. This requires that hydration of the cement be continued. During the curing period, drying shrinkage may occur if the concrete was placed at an excess water content. Figure 4–29 shows the relationship of water content per cubic yard of concrete to drying shrinkage. The continued hydration of the cement requires moisture and favorable temperatures to be maintained for an

FIGURE 4–27. Concrete slipform paver. *(Courtesy CMI Corporation)*

FIGURE 4–28. Ride-on power trowels being removed from finished slab. *(Courtesy D. J. Rossetti, Inc.)*

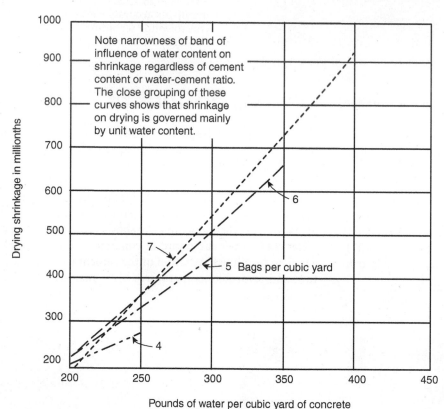

Note narrowness of band of influence of water content on shrinkage regardless of cement content or water-cement ratio. The close grouping of these curves shows that shrinkage on drying is governed mainly by unit water content.

6

7

5 Bags per cubic yard

4

Drying shrinkage in millionths

Pounds of water per cubic yard of concrete

FIGURE 4–29. The interrelation of shrinkage, cement content, and water content. The chart indicates that shrinkage is a direct function of the unit water content of fresh concrete. 288-D-2647 *(Courtesy U.S. Department of the Interior, Water and Power Resources Service)*

adequate time. The time required will depend on type of cement, mix proportions, design strength, size and shape of the concrete structure, and future exposure conditions.

The optimum concrete temperature at placement will vary with conditions, but generally 90°F (32.2°C) is set as the upper limit. To obtain the specified placement temperature in hot weather often requires the use of prechilled aggregates and possibly the substitution of shaved ice for mix water. During cold weather, the aggregates and mixing water may be heated to raise the concrete temperature.

During hot-weather concreting, the loss of moisture after placement is critical, and various methods can be used to prevent the moisture loss such as wind screens, fog misting systems or evaporation retarding chemicals or to add additional curing water to the concrete to determine if slabs have the potential for plastic shrinkage cracking (Figure 4–30).

Methods used to prevent moisture loss may include the use of waterproof papers, plastic film, and liquid curing compounds which form a membrane and the leaving of forms in place. During hot weather, dark coverings should not be used, as they will absorb the sun's rays.

The additional water methods are by ponding, sprinkling, and using wet coverings such as burlap, sand, and

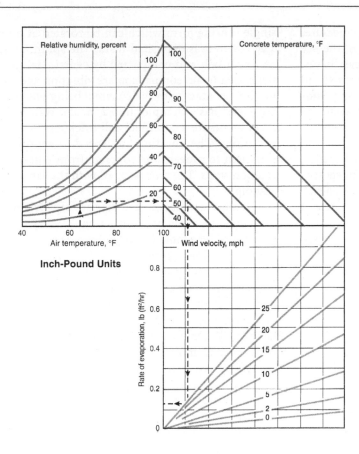

Inch-Pound Units

FIGURE 4–30. Effect of concrete and air temperatures, relative humidity, and wind velocity on rate of evaporation of surface moisture from concrete. There is no way to predict with certainty when plastic shrinkage cracking will occur. However, when the rate of evaporation exceeds 0.2 lb/sq ft/hour, precautionary measures are almost mandatory. Cracking is possible if the rate of evaporation exceeds 0.1 lb/sq ft/hour. (*Courtesy of Portland Cement Association*)

straw. The methods utilized will vary, but care should be taken so that the entire concrete surface is protected, especially corners and edges, and that the material used as a curing system will not stain the concrete.

Cold-weather concreting requires the maintenance of internal heat or the use of additional heat to provide the proper curing temperatures. To maintain internal heat, insulating blankets and straw may be used. The external heat may be supplied by salamanders, space heaters, or live steam. If fuel burning heaters are used, care must be taken to see that they are properly vented to prevent *carbonation*. The carbon dioxide produced by the combustion of fossil fuels reacts with the calcium hydroxide in the fresh concrete to form a calcium carbonate layer on the surface of the concrete. This surface weakness will cause the floor to dust when put into service. Some common materials that come into contact with concrete and their effect on hardened concrete are shown in Table 4–14.

No two concrete jobs are alike, and the specifications must be checked carefully to determine what will be required for hot- or cold-weather concrete placement and what methods of curing will be allowed.

PRECAST CONCRETE PRODUCTS

Precast concrete products are construction items usually manufactured off-site and delivered to the construction site ready for installation into the structure. During the

Table 4–14. Effects of various substances on hardened concrete

Substance	Effect on Unprotected Concrete
Petroleum oils, heavy, light, and volatile	None
Coal tar distillates	None, or very slight
Inorganic acids	Disintegration
Organic materials	
Acetic acid	Slow disintegration
Oxalic and dry carbonic acids	None
Carbonic acid in water	Slow attack
Lactic and tannic acids	Do
Vegetable oils	Slight or very slight attack
Inorganic salts	
Sulfates of calcium, sodium, magnesium, potassium, aluminum, iron	Active attack
Chlorides of sodium, potassium	None*
Chlorides of magnesium, calcium	Slight attack*
Miscellaneous	
Milk	Slow attack
Silage juices	Do
Molasses, corn syrup, and glucose	Slight attack*
Hot distilled water	Rapid disintegration

*Absence of moisture.

Source: Courtesy of U.S. Department of the Interior, Water and Power Resources Services.

manufacturing process, the same quality control measures applied to site-cast concrete are used to ensure the use of quality materials in the production of precast items. Precast concrete pipe, catch basins, septic systems, and structural elements such as beams, columns, and floor units are concrete items that can be precast in standard sizes and shapes and marketed as products ready for installation at the construction site (Figure 4–31).

Precast concrete pipe is classified by the production method utilized to manufacture the pipe. Cast concrete pipe is usually 48 in. in diameter or larger, with varying lengths. The split steel forms stand upright, and the concrete is placed between the inner and outer steel form. The pipe is reinforced with steel rebar and wire mesh. The concrete usually has a 3-in. or less slump and is consolidated by external form-mounted vibrators. Special care must be taken to ensure tight form connections; otherwise, as the concrete is vibrated, objectionable mortar leaks will occur at the joints. The concrete pipe will be removed from the mold and cured, steam or moist curing being used.

The centrifugally spun system is used to produce reinforced pipe 42 in. in diameter or less. The system utilizes a

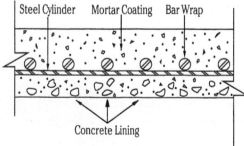

Steel Cylinder   Mortar Coating   Bar Wrap

Concrete Lining

FIGURE 4–31. Centrifugal lining: A measured amount of concrete or cement mortar is fed into the rotating cylinder. After placement, the rotation of the cylinder is increased to the packing velocity and maintained at that velocity until the excess water is removed, resulting in a densely compacted lining with exceptional long-term flow characteristics. *(Courtesy Ameron International® Concrete and Steel Pipe Group, Rancho Cucamonga, California)*

single outside form which can be rotated at high speeds. The concrete is deposited inside the spinning mold by a conveyor belt and compacted by centrifugal force. Variations used to aid compaction are vibration and steel rollers in direct contact with the concrete being placed. The concrete used has a slump of 0 to 2 in. and is deposited in the form to ensure a specified wall thickness with minimum variations. The duration and speed of spinning must be sufficient to prevent the concrete from sagging when the rotation stops. The pipe is cured with the same methods as cast pipe.

Tamped and packerhead pipe are usually nonreinforced concrete pipe made by compacting very dry concrete into steel molds. The mold is split and can be removed as soon as the pipe section is completed without damaging the pipe. The packerhead system utilizes a stationary mold, with the packerhead placing and shaping the interior surface. The tamped system uses a fixed interior cylinder with the outside mold rotating while the concrete is compacted by vertical tampers. When the pipe unit is completed, the form is removed and moist or steam curing is used to produce required concrete strengths.

Lined cylinder prestressed concrete pipe is used for high-pressure water distribution systems. This pipe is made in stages. The core of the pipe is produced by means of the centrifugal system. When the core has cured, it is wrapped with high-tensile-strength reinforcing steel under tension. Some systems utilize longitudinal steel in tension also. The wrapped core is then coated with mortar and cured to complete the pipe assembly.

Precast structural elements such as beams, columns, and highway median barriers are produced in casting beds of varying shapes and lengths (Figures 4–32 to 4–34). The forms or casting beds are usually set with the top of the form at grade; external vibrators are mounted on them. The forms are adjusted to the required dimensions of the finished beam; all required reinforcement is set, and the concrete is deposited in the form. Generally, precast operations attempt a 24-hour turnover or cycle for form use. To aid the initial set of the concrete, external heat is applied to the mold with electric heaters, a pipe system containing heated oil, or live steam systems.

Using high-early-strength cements, water reducers, and heated curing, precast plants attain concrete strengths of 3000 psi or more within a 24-hour cycle. The concrete beams cast are removed from the molds and stored on the plant site for delivery when needed, and the casting cycle is repeated.

Whether site cast or precast, the concrete structure or product must be constructed with careful attention and quality control to ensure a long service life with a minimum of maintenance and repairs. Therefore, while an understanding of concrete technology and the use of standardized techniques will usually result in satisfactory concrete, problems do sometimes arise. Jobsite problems may be relative to fresh concrete characteristics or type of construction, for example, wall or slab construction, and

FIGURE 4–32. Placing concrete by bucket into a segmental bridge form. *(Courtesy Unistress Corp.)*

FIGURE 4–33. Consolidation of concrete in form using hand-held vibrator. *(Courtesy Unistress Corp.)*

concrete testing. Therefore, using the checklist presented in Table 4–15 is recommended as a basic reference and starting point to preventing and solving concrete-related problems.

PERVIOUS CONCRETE

Pervious concrete mixes contain coarse aggregate, small quantities of fine aggregate, water and cementations materials in very carefully controlled amounts. The mortar paste produced coats and binds the aggregate particles together while leaving interconnected voids in the order of 15 to 25 percent of the hardened concrete. This void system allows for the quick passage of water through the pervious concrete into the underlying granular subbase, where it may be collected by a piping system for further treatment or allowed to percolate into the in situ subgrade soil. Flow rates for pervious concretes generally range between 3 gal/ft^2/minute to 5 gal/ft^2/minute depending on the void content of the placed pervious concrete.

The low mortar paste and the high void content reduce strength when compared to conventional mixes; however, with proper designs suitable strengths can be achieved for

FIGURE 4–34. Two of the match-cast segments being prepared for shipment. *(Courtesy Unistress Corp.)*

Table 4–15. Checklist of common field problems, their cause and prevention

Problem	Cause	Prevention or Correction
Fresh Concrete		
Excessive bleeding	Insufficient fines in mix	Increase percent of fines—cement, fly ash, sand content. Introduce or increase air entrainment.
	Excess mix water	Reduce water content
	Vapor retarder directly under slab	Compact 3 in. of sand or crusher-run gravel on the vapor retarder (also applicable to plastic shrinkage cracks and curling).
Segregation	High slump (excess water)	Reduce water content. Use superplasticizer for desired slump.
	Overvibration	Don't vibrate concrete that is already flowable (superplasticized expected).
	Inadequate vibration	Insert vibrator at closer intervals. Vibrate until concrete is flowable.
	Excessive drop in placing	Reduce free drop (use drop chutes).
	Lack of homogeneity in mix	Introduce air entrainment. Reduce coarse aggregate proportion in mix.
Sticky finish	High air content and/or oversanded mix	Promote bleeding—reduce air, percent sand.
	Rapid drying of the surface	Dampen subgrade and forms. Apply fog spray.
Rapid set (hot weather)	High concrete temperature High ambient temperature	Cool water, add ice or liquid nitrogen. Cool aggregate piles by sprinkling. Use maximum retarder dosage. Consider fly-ash mixes.
	High cement content	Introduce a water-reducing retarder and/or fly ash into mix.
	Trucks waiting in sun	Schedule trucks for shortest wait, in shaded areas if possible. Sprinkle outside of mixer drum.
Slow set (cool weather)	Lean mix—especially with fly ash or slag	Increase cement. Use accelerator. Heat aggregates and water.
	Cold or wet subgrade	Place plastic type (polyethylene) sheet on subgrade. Protect subgrade (cover with straw or mats).
Plastic shrinkage cracking	Rapid evaporation of water from the surface primarily from wind and low humidity	Fog spray on surface at time of finishing. Induce water gain on surface (not to excess) by reducing sand and/or air entrainment. Avoid use of vapor retarder, if possible, or cover with 3 in. of compacted sand or crusher-run gravel. Reduce mixing water. Provide windbreaks. Reduce concrete temperature.
Low yield	Incorrect batch	Confirm accuracy of scales. Confirm specific gravities of aggregates. Confirm mix design for correct batch weights. Confirm yield with field unit weight test.
	Low air content	Increase air-entrainment admixture.
	Inaccurate jobsite measure	Avoid rutting (displacing of subgrade), deflection of deck, bulge in forms.
	Waste concrete	Account for spillage.

Table 4–15. (continued)

Problem	Cause	Correction
Flat Slabs		
Shrinkage cracks	Inadequate spacing or depth of contraction joints	Space joints in feet at about $2\frac{1}{2}$ times the slab thickness in inches. Cut joints $\frac{1}{4}$ of the slab depth. Place joints at sharp change in direction of change in thickness or width of the slab.
	Late joint sawing	Saw as soon as possible (may be slight raveling). Avoid sawing during drop in concrete temperature.
	Bond not broken between slab and walls, columns, or other structure	Break bond (expansion joint material) where slab abuts walls or columns.
	Excessive shrinkage of concrete	Reduce water in the concrete mix. Cure immediately after finishing (paper, wet burlap, polyethylene, curing membrane).
Hairline surface cracking	High-slump concrete	Reduce water content.
	Excessive and/or early troweling	Don't trowel or overwork a wet surface.
	Rapid drying of a wet surface	Provide immediate cure after finishing.
Dusting floor	Same reasons as hair line cracking, with special emphasis on the problem created by bringing excess water to surface during placing and finishing. Correct minor to moderate dusting with the early application of a chemical surface hardener. The exposure of fresh concrete to carbon dioxide from unvented heaters is also a major cause, for which hardeners are not a cure.	
Blistering	Surface closed too soon Rising air and water trapped beneath surface	Delay start of floating, troweling. Use wood float to open surface. Press or pound blisters down.
Slab curling	Uneven drying (top dries and shrinks; lower part of slab retains moisture)	Use stiffer mix. Avoid use of vapor retarder, if possible. Cure without delay; extend curing as long as possible. Reduce space between contraction joints.
Discoloration (dark areas)	Calcium chloride not uniformly mixed	Chloride should be added in solution.
	Hard troweling, prolonged machine finishing, especially with calcium chloride in the mix.	When using calcium chloride, do not overtrowel. When possible, finish with broom or burlap drag.
	Uneven drying or uneven compaction of concrete	Maintain uniform curing; avoid uneven coverage with curing membrane or partial surface contact with polyethylene sheets.
Pavement scaling	Freeze–thaw exposure with application of de-icer salts (if not air entrained, concrete will scale regardless of other qualities)	Maintain air content at 5 to 8 percent, higher with less than $\frac{3}{4}$ in. maximum size aggregate. Test frequency for air content.
	Excessive finishing or overwet surface	Protect against rain; minimize surface finishing—use broom or burlap finish. Use minimum slump, about 4 in. maximum.
	Inadequate curing and drying before deicers are applied	Pavement should have 30 days of drying after normal curing before exposure to deicers.
	Quality of concrete mix in addition to air entrainment	Use the minimum slump necessary for placing. Strength of concrete should be about 3500 to 4000 psi before exposure to de-icers.

(Continued)

Table 4–15. (continued)

Problem	Cause	Correction
Wall Surfaces		
Honeycomb	Inadequate vibration for the workability of the mix	Insert vibrator more frequently. Vibrate near the form surface. For heavily reinforced or hard-to-reach areas, consider superplasticizer. Check coarse aggregate gradation and adequacy of the sand content. Reduce lift height. Increase fines, air entrainment in mix.
Sand streaking	Excessive bleeding Water loss at form opening	Check forms for watertightness.
Surface voids (bugholes)	Air or water trapped against form surface and not worked out during placement	Reduce lift height. Reduce percent of sand in mix. Reduce air content. Check form treatment. Insert vibrator more frequently. Vibrate upward as near to form face as possible.
Cylinder Tests		
Reported low cylinder strengths	Nonstandard testing procedures for molding and handling test cylinders on the job	Check test methods: Protect the cylinders against temperature extremes or disturbance while stored on the job. Take test sample from mid-part of the load (approximate). Deposit test sample in wheelbarrow and remix. Follow procedures in ASTM C 31.
	Low-strength mix	Check the mix proportions and check for water added on the job.
	Incorrect test evaluation for compliance with the specification	Confirm method of evaluation with ACI 301 or ACI 318. The average of any three successive tests must meet the specified strength. No single test may be more than 500 psi below the specified strength (a test is an average of two cylinders). Follow-up testing to confirm doubtful results may be used. The rebound hammer, penetration test, and coring are covered in ACI and ASTM documents.

Source: Courtesy American Concrete Institute.

applications, such as pavements, parking areas, driveways, and side walks. The benefits of using pervious concrete are numerous, including the reduction of the heat island effect, which is the reduction of stored heat in pavements, the breakdown of the hydrocarbons contained in runoff by bacteria and fungi living in the pore structure of the pavement and reduced need for retention ponds, thereby increasing buildable area.

The Environmental Protection Agency (EPA) recommends the use of pervious concrete as a best management practice (BPM) for managing storm water runoff. Because of the benefits of pervious concrete as a storm water and pollution control system, its use may help earn a credit point in the U. S. Green Building Councils Leadership in Energy & Environmental Design (LEED) Green Building Rating System.

Concrete as a construction material is a complex subject. In order to deal with it, many organizations have been formed which share technical information with the users of concrete. The serious student of concrete should study the literature of organizations such as the American Concrete Institute, the Portland Cement Association, the National Ready Mix Concrete Association, the Precast/Prestressed Concrete Institute, and others to keep abreast of new developments and to help understand the behavior of concrete as a construction material.

Review Questions

1. List the basic types of portland cement and describe their characteristics and uses.

2. What is the heat of hydration? What factors affect the rate of heat generation?

3. What factors influence the air content of concrete in the plastic state?

4. What are the four basic chemical compounds that make up portland cement, and what effect do they have on portland cement concrete?

5. If water from an untested source is to be used for concrete manufacturing, what test must be made for strength?

6. What are admixtures and what are they used for?

7. What modification is made to concrete to increase its resistance to freeze–thaw damage?

8. Why is concrete such a widely used construction material?

9. What factors are important during hot- and cold-weather concrete placement?

10. What is the maximum aggregate size that can be used in concrete for the following conditions:

    a. Nonreinforced walls

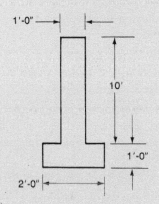
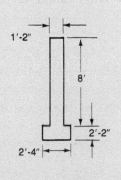

    b. Slabs on grade

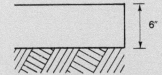

    Portland Cement Concrete

    c. Reinforced beams

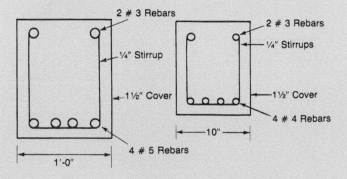

11. What is the difference between the types of air meters used to measure air content of fresh concrete, and why is the difference important?

12. Do the results of the listed strength tests satisfy ASTM C94 requirements? The concrete strength specified was 3000 psi.

3100 psi	3210 psi
2850 psi	2985 psi
3200 psi	3170 psi
2950 psi	2890 psi
3050 psi	2920 psi

13. Calculate the percentage of free moisture on a fine aggregate based on the given data.

    $$\text{Percent absorption} = 1.7$$
    $$\text{Wet weight} = 503.7 \text{ g}$$
    $$\text{Oven-dry weight} = 480.2 \text{ g}$$

14. Calculate the mean, standard deviation, and coefficient of variation for the following test data:

3700 psi	2920 psi
4310 psi	3680 psi
3890 psi	4010 psi
4100 psi	3980 psi

15. Determine the yield of the following concrete batches:

    a. $\gamma_{conc} = 147.6$ pcf

Water	265 lb
Cement	530 lb
Coarse aggregate	1900 lb
Fine aggregate	1290 lb

    b. $\gamma_{conc} = 149.2$ pcf

Water	325 lb
Cement	500 lb
Coarse aggregate	1875 lb
Fine aggregate	1330 lb

16. Determine the compressive strengths of the following:

    a. A 6-in.-diameter concrete cylinder that failed at a test load of 135,000 lb

    b. A 3-in.-diameter specimen that failed at a test load of 30,000 lb

17. Design a concrete mix to satisfy the following requirements: $f_c' = 2000$ psi

    Non-air-entrained   Interior slab on grade
    Nonreinforced   Maximum aggregate size = 1 in.

Aggregate Data					
			Bulk Unit Weight	Percent Moisture	
	Sp. Gr.	F. M.	(pcf)	ABS	Free
CA	2.68	6.00	95	0.5	1
FA	2.59	2.70	105	1	3

18. Design a concrete mix to satisfy the following requirements: $f'_c = 4000$ psi

Air-entrained   Exterior slab on grade
                     Moderate exposure
Nonreinforced   Maximum aggregate size $= 1\frac{1}{2}$ in.

Aggregate Data					
			Bulk Unit Weight	Percent Moisture	
	Sp. Gr.	F. M.	(pcf)	ABS	Free
CA	2.71	6.10	98	0.4	0.7
FA	2.65	2.80	112	0.9	2.1

19. Design a concrete mix to satisfy the following requirements: $f'_c = 5000$ psi

Air-entrained   Exterior column
                     Severe exposure
Reinforced   Maximum aggregate size $= \frac{3}{4}$ in.
Slump 5 in.

Aggregate Data					
			Bulk Unit Weight	Percent Moisture	
	Sp. Gr.	F. M.	(pcf)	ABS	Free
CA	2.62	6.70	96.5	0.3	0.9
FA	2.58	2.90	111.6	1.2	6.2

20. Design a concrete mix to satisfy the following requirements: $f'_c = 3000$ psi

Non-air-entrained   Interior slab on grade
Nonreinforced   Maximum aggregate size $= 1$ in.

Aggregate Data					
			Bulk Unit Weight	Percent Moisture	
	Sp. Gr.	F. M.	(pcf)	ABS	Free
CA	2.69	6.10	95	0.4	107
FA	2.59	2.70	110	0.9	3.0

21. Design a concrete mix to satisfy the following requirements: $f'_c = 30$ Mpa

Non-air-entrained   Maximum aggregate size $= 25$ mm
Slump 75 mm

Aggregate Data					
			Dry-Rodded Unit Weight	Percent Moisture	
	Sp. Gr.	F. M.	(kg/m^3)	ABS	Free
CA	2.67	6.70	1545	0.3	0.9
FA	2.63	2.90	1762	1.2	3.1

22. a. Calculate the required compressive strength for a concrete mix design if $f'_c = 5000$ psi and data indicate a standard deviation of 600 psi based on 15 tests for a similar concrete mix.

    b. Calculate the required compressive strength for a 6000 psi mix design with no data from previous designs available.

23. Calculate the modulus of rupture for a by-in. square concrete flexure beam with a span length of 18 in. if failure occurs:

    a. within the middle third at a load of 5150 lb.
    b. 0.5 in. outside of the middle third.
    c. 1.2 in. outside of the middle third.

24. Determine the approximate modulus of rupture of a concrete with a compressive strength of 4100 psi.

25. A splitting tensile test was performed on a standard by-12-in. cylinder; the cylinder fractured at 49,800 lb. Calculate the splitting tensile strength and the approximate direct tensile strength of the concrete.

26. Three concrete cores were tested in accordance with ASTM C42. Determine the structural adequacy of the concrete if the required $f'_c = 3500$ psi.

Core	Strength (psi)
1	3150
2	2975
3	3220

27. Calculate the compressive strength of a 4 in. concrete test cylinder if failure occurs at 76,500 lb.

28. Calculate the quantity of concrete required to place a 40-by-40-ft 6-in.-thick slab in cubic yards and cubic meters.

CHAPTER FIVE

IRON AND STEEL

Iron in its various forms, including steel, is by far the most important of the metals used in the construction industry. The term *ferrous metals* includes all forms of iron and steel. They are manufactured to meet a wide variety of specifications for various uses. Chemical composition and internal structure of ferrous metals are closely controlled during manufacturing. Therefore, strength and other mechanical properties can be determined with a high degree of reliability.

Ferrous products are fabricated in shops to desired size and shape. The finished products are ordinarily delivered to a construction site ready to be installed, with inspection and testing completed. Ferrous metals are seldom damaged during transportation because of their strength and hardness. Therefore, people in the construction field have little opportunity to control the quality of iron or steel. Compared to aggregates, asphalt concrete, or portland cement concrete, all of which are partially "manufactured" during installation at the construction site, there is little that can be done to improve or harm a ferrous metal product once it leaves the fabrication shop.

STRUCTURE AND COMPOSITION

Iron and steel appear to be smooth and uniform, yet they consist of particles called *grains* or *crystals* that can be distinguished under a microscope. The grains are formed as the metal passes from the liquid to the solid state. This internal crystalline structure called the *constitution* determines to a great extent what mechanical properties the metal will have. Each grain consists of a symmetrical pattern of atoms which is the same in all types of iron and steel. The grains are not all similar because they press on each other as they form, causing variations in size, shape, and arrangement, and this accounts for many of the differences in the behavior of various irons and steels.

Some types of iron and steel also contain a different kind of grain interspersed among the typical grains. These have an influence on the material's behavior. The internal structure is determined by the way the metal is cooled and by the way the metal is given its final shape. Grain size, shape, and arrangement are generally the same throughout a finished piece of metal, but special procedures can be used to make them different in different areas of the same piece.

The strength of the metal depends on the cohesion of the atoms in each crystal and the cohesion between adjacent crystals. In this respect, the structure is somewhat like that of aggregate surrounded by adhesive to make concrete. Instead of adhesion holding the crystals or grains together, an atomic bond which is much stronger holds them. Iron and steel therefore have a higher tensile strength than any aggregate–adhesive combination.

Strain of any kind consists of movement of the atoms, closer together in compression or farther apart in tension. Atoms arranged close together allow more stretching or, in other words, more ductility than less-concentrated atomic arrangements. As long as the atoms retain their spatial relationships, even in a distorted way, they return to their original positions when stress is removed. The extent to which the atoms can move and still return to their original positions is the limit of elastic deformation. Beyond this extent, the pattern cannot be distorted without slippage along a plane or parallel planes through the grains. Any distortion in this range is plastic or permanent.

The final temperature and rate of heating do not affect the internal structure at the time materials are melted to make pig iron or when pig iron is melted to make iron or steel. However, the rate of cooling is important. Rapid cooling causes large crystals. Metal with large crystals is more brittle and does not have the strength, ductility, or shock resistance of metal with the smaller crystals caused by slower cooling. However, large crystals produce better machinability. Any elongation and alignment of grains in one direction will increase the strength of the metal to resist stresses in

Table 5–1. Ferrous metal properties

Element	Common Content	Effects
Carbon	Up to 0.90%	Increases hardness, tensile strength, and responsiveness to heat treatment with corresponding increases in strength and hardness.
	Over 0.90%	Increases hardness and brittleness; over 1.2%, causes loss of malleability.
Manganese	0.50% to 2.0%	Imparts strength and responsiveness to heat treatment; promotes hardness, uniformity of internal grain structure.
Silicon	Up to 2.50%	Same general effects as manganese.
Sulfur	Up to 0.050%	Maintained below this content to retain malleability at high temperatures, which is reduced with increased content.
	0.05% to 3.0%	Improves machinability.
Phosphorus	Up to 0.05%	Increases strength and corrosion resistance, but is maintained below this content to retain malleability and weldability at room temperature.

Source: "Construction Lending Guide," courtesy U.S. League of Savings Associations.

that direction. The means of producing ferrous metals, both the refining with heat and the working into final shape, affect the mechanical properties of the material.

No ferrous metal is pure iron. All include the elements shown in Table 5–1, which have great effect on the properties of the metal, even if present in very small amounts. The chemical content of the metal is determined by the composition of the iron ore, the way in which the metal is heated, and the elements added to it at different stages. Iron ore contains varying percentages of manganese, silicon, and sulfur, and may or may not contain some phosphorus. Carbon comes from the burning coke, and additional carbon may be added to the molten metal. Excess sulfur may be removed by the addition of manganese.

Generally speaking, longer or hotter treatment in a furnace decreases the percentages of carbon, manganese, phosphorus, silicon, and sulfur. Increases are made by adding the desired element to the liquid metal. Other elements also may be added to the liquid metal (refer to Table 5–3).

PRODUCTION OF FERROUS METALS

The first step in the manufacture of iron or steel is to produce a low grade of iron in a continuously operating furnace called a *blast furnace*. These furnaces are about 200 ft high and about 50 ft in diameter (Figure 5–1). Iron ore, coke, and limestone are loaded continuously at the top. Iron ore is an oxide of iron found in nature mixed with rock or soil called *gangue*. Coke is produced by heating coal to drive the impurities out. It then burns with greater heat than coal. Limestone is a type of rock that occurs in nature. Burning the coke and supporting the combustion with a strong blast of hot air melt the iron ore and limestone at a temperature of about 1500°F (815°C). The heat melts the iron, frees it of oxygen, and forms carbon monoxide gas, which imparts carbon to the liquid iron.

Melting permits separation of iron from the gangue, which combines with the molten limestone to form slag. Iron is much heavier than slag, so there is a natural separation of the two as they melt. Iron flows to the bottom of the furnace

and molten slag floats on the iron. Iron is removed from a tap near the bottom and slag from a tap slightly higher. These are removed a half dozen times per 24 hours of operation. Use of the slag as an aggregate is discussed in Chapter 2. The iron flows into molds and is allowed to solidify into shapes called *pigs,* or it is taken in a ladle while still liquid to be refined into steel or a better grade of iron. In either case, the product of the blast furnace is called *pig iron.*

The makeup of the iron resulting from this process is not accurately controlled. It contains about 4 percent carbon, 2 percent silicon, 1 percent manganese, and 0.05 percent sulfur. It may contain up to 2 percent phosphorus depending on the type of ore used.

Pig iron is not useful for construction because it is weak and brittle, although it is very hard. The general term *iron* refers to a ferrous metal that is of a higher quality than pig iron. To produce useful iron or steel, a second melting is needed for further purification.

Emerging technologies are promising to reduce the cost and environmental impact of iron and steel production. Using graphite electrodes, electric arc furnaces (EAFs) induce an electrical current into iron or scrap steel to produce molten steel. The advantages of EAFs over blast furnaces is that they do not require continuous operation and can produce smaller amounts of finished material at mini-mills rather than the large-scale production required to maintain the economic viability of larger blast furnaces. Because they often use recycled (scrap) steel, EAFs have historically been used for specialty steel making, including stainless steel. Improved recycling methods to remove impurities from scrap steel allow increased production of construction-grade steels at EAF mills.

Iron

It is possible to refine pig iron until it is nearly pure iron containing little more than traces of impurities. In this form, iron is suitable for construction. It is highly resistant to corrosion, highly ductile, and readily machined. It is drawn into wires and rolled into sheets for roofing, siding, and corrugated pipe. Vitreous enamel coatings adhere well to this type of iron.

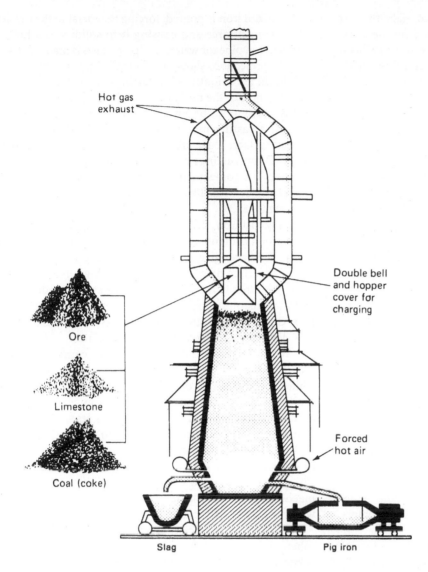

Hot gas exhaust

Ore

Limestone

Coal (coke)

Double bell and hopper cover for charging

Forced hot air

Slag

Pig iron

FIGURE 5–1. A typical blast furnace.

Despite these qualities, the high cost of refining prevents this type of iron from being one of the major construction materials.

The types of iron more common to the construction industry are gray and white cast iron, malleable cast iron, and wrought iron. *Cast iron* is a general term denoting ferrous metals composed primarily of iron, carbon, and silicon, and shaped by being cast in a mold. They are too brittle to be shaped any other way. The brittleness is caused by the large amount of carbon present, which also increases strength.

Wrought iron is highly refined iron with slag deliberately incorporated but not in chemical union with the iron. The slag forms one-directional fibers uniformly distributed throughout the metal. Chemical compositions of various types of iron are shown with cast steel in Table 5–2.

Table 5–2. Typical composition of ferrous metals

Metal	Typical Composition (Percent)				
	C	Si	Mn	P	S
Cast steel	0.5–0.9	0.2–0.7	0.5–1.0	0.05	0.05
Gray cast iron	2.5–3.8	1.1–2.8	0.4–1.0	0.15	0.10
White cast iron	1.8–3.6	0.5–2.0	0.2–0.8	0.18	0.10
Malleable cast iron	2.0–3.0	0.6–1.3	0.2–0.6	0.15	0.10
Wrought iron	<0.035	0.075–0.15	<0.06	0.10–0.15	0.006–0.015
Pure iron	0.015	Trace	0.025	0.005	0.025
Pig iron	3–5	1–4	0.2–1.5	0.1–2.0	0.04–0.10

Pig iron is remelted in small furnaces to make the cast metals. Chemical composition is controlled by the addition of scrap iron or steel of various kinds and of silicon and manganese as needed. The molten metal flows from the furnace to a ladle from which it is poured into molds to be formed into useful shapes. This operation is called *casting*. The materials of which molds are made are listed below:

molding sand: a cohesive mixture of sand and clay.

loam: a cohesive mixture of sand, silt, and clay.

shell mold: a mold consisting of a mixture of sand and resin that hardens when heated prior to the casting.

metal dies: molds machined to the proper mold shape.

The first three types of mold are used once and broken to remove the casting. The dies may be used thousands of times. The first two types are formed around a *pattern*, which is usually made of wood. For a shell mold the pattern is made of metal which is heated to solidify the mold material. The size of the pattern in all cases must allow for cooling shrinkage of the casting. Patterns may be reused, whether wood or metal.

The mold material is packed around the pattern, which has been heated in the case of a shell mold. Removal of the pattern leaves a mold of the desired shape. The molds, except a few very simple ones, are made in two parts and placed together for the casting. Sometimes more than two parts are needed and intermediate sections are placed between the upper and lower molds. Cores are inserted to supplement the mold when necessary. A typical mold is shown in Figure 5–2. Manhole frames and covers, storm water inlet grates, pump casing, fire hydrants, sinks, and bathtubs are made of iron castings.

Iron is also cast in centrifugal molds, which are of cylindrical shape with metal or sand linings. They are spun rapidly as the molten iron is poured, forcing the metal to the outside by centrifugal force and causing it to solidify as a hollow cylinder. Iron pipe for water, sewage, and gas is made this way.

The projections shown in Figure 5–2 must be broken off and machined smooth. After casting, the metal surface has the roughness of the mold material. Even if die cast, the surface is not smooth because of cooling shrinkage. Certain areas of a casting may be required to fit tightly against another surface. These areas must be machined to a smooth finish. Often castings are made in two parts and bolted together. The contact surfaces are machined for a tight fit, and bolt holes are drilled through the flanges of the two castings (Figure 5–3).

Gray cast iron, the most widely used type of iron, has a high carbon content and contains large numbers of graphite flakes. The flakes give a gray appearance to a fractured surface. Properties of gray iron include low viscosity when molten (so that fairly intricate castings can be made), excellent machinability, high resistance to abrasion, and rather poor ductility and toughness. ASTM A48, Gray Iron Castings, contains specifications for gray cast iron.

White cast iron contains its carbon completely combined with the iron. A fractured surface appears bright white. The advantages of white iron over gray iron are that it is harder and more resistant to wear from abrasion. However, it is more difficult to machine, less resistant to corrosion, more brittle, and more difficult to cast. By controlling chemical composition and cooling rate, castings with cores of gray iron and surfaces of white iron can be made. These are called *chilled iron* castings. White iron is used in machinery such as crushers, grinders, chutes, and mixers where resistance to abrasion is critical.

Cast iron with the carbon reformed from flakes into tiny spheroids by the addition of magnesium to the molten iron is known as *ductile iron*. The basic nature of the iron is not

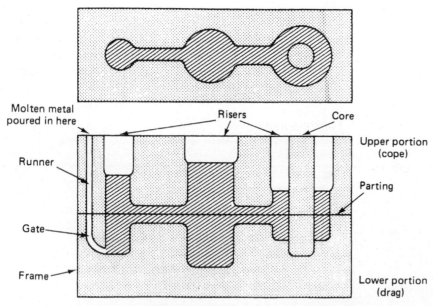

Molten metal in the risers fills the mold as cooling contraction takes place.

FIGURE 5–2. Mold and casting.

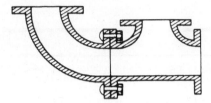

FIGURE 5–3. Cast iron pipe fittings bolted together.

FIGURE 5–4. Ornamental wrought iron grille.

changed, except that tensile strength, ductility, and the ability to withstand shock loads are greatly increased. Ductile iron pipe is used for water, sewage, and gas.

Malleable cast iron consists of white iron made tough and ductile by *annealing,* which consists of heating to about 1600°F (870°C), holding that temperature for a time, and cooling very slowly to about 1275°F (710°C). This process requires several days. During the entire process, carbon is precipitated from the solution as small lumps in the metal until there is no combined carbon. Some carbon may be allowed to remain combined to increase hardness, strength, and resistance to abrasion. There is then a loss of ductility and toughness. Brittleness is eliminated by removal of carbon from solution, and machinability is improved by the carbon lumps. Malleable iron is used for pipe fittings, guardrail fittings, and other items which require machining and which are subject to shock loads.

Wrought iron is made by refining pig iron in a furnace in a manner similar to the refining of steel. The iron silicate slag is melted, and the relatively pure molten iron is poured into the slag. A pasty mixture of the two is formed, with the slag evenly distributed as individual particles. The mixture forms a semisolid ball, which is dumped out and pressed into a rectangular block, squeezing the excess slag out in the process. The block is rolled to the desired shape, aligning the slag as strings or ribbons in the direction of rolling. It is then readily shaped further by drawing, bending, or forging, and can be made into thin, intricate shapes. It is easily welded and machined.

The fibrous structure of wrought iron results in a material with different mechanical properties parallel and perpendicular to the fibers or the axis of the grains. Tensile strength is 10 to 15 percent lower across the fibers than parallel to them, and ductility is only about 20 percent as much. Shearing strength across the fibers is much greater than that along the fibers. A rod, twisted until it fails in torsion, comes apart along the axis, separating between fibers.

Wrought iron has excellent corrosion resistance which is greater on faces that have been rolled than on sheared or machined faces. Wrought iron pipe is used extensively where corrosion resistance is needed. This type of pipe has threaded joints because wrought iron is easily machined. Cast and ductile iron pipe joints must be of another kind. Wrought iron is also used extensively for ornamental ironwork, as well as for miscellaneous ironwork where corrosion resistance, machinability, or ductility and malleability are needed (Figure 5–4).

Steel

Pig iron is further oxidized in another furnace at about 3000°F (1650°C) to produce steel. Most steel is made by the basic oxygen process, electric-arc process, open-hearth process, or vacuum process. Each has unique features, but in all cases, pig iron, scrap steel, and sometimes iron ore are melted together with a flux of limestone or lime. The process is a repetition of the blast-furnace operation with variations. The purpose is the same—to remove impurities. Impurities are removed as gases and in the slag.

Phosphorus and sulfur are reduced to less than 0.05 percent of the steel. Manganese content is reduced to an amount from 0.2 to 2.0 percent; silicon from 0.01 to 0.35 percent. The final amounts depend on the specifications for the steel. Carbon is the key element in controlling the properties of ordinary steel called *carbon steel.* Strength and hardness increase with an increase in carbon up to about 1.2 percent. Brittleness increases and ductility decreases as carbon increases. Usually an amount less than 1.2 percent is specified in order to obtain a product satisfactory in all respects.

Carbon in an amount up to 2 percent is completely soluble in molten iron, and when cooled, the mixture forms a solid chemical solution. Carbon in greater amounts forms separate grains of graphite or iron carbide throughout the metal.

Steel is defined as a chemical union of iron and carbon (carbon is, therefore, less than 2 percent by weight) plus other elements. This definition does not exclude every kind of iron. However, almost all kinds of iron contain carbon in excess of 2 percent and steel usually contains less than 1.2 percent carbon. Therefore, it is usually obvious whether a ferrous metal is iron or steel. Customary terminology should be used for borderline cases. The term *iron* is used to refer to cast iron, malleable cast iron, ductile iron, wrought iron, or pure iron. As can be seen in Table 5–2, some of these metals may have less than 2 percent carbon.

Carbon, manganese, silicon, phosphorus, and sulfur are considered impurities because generally they must all be

Table 5–3. Effects of alloying elements

Element	Amount	Effect
Aluminum	Variable	Promotes small grain size and uniformity of internal grain structure in the as-cast metal or during heat treatment.
Copper	Up to 0.25%	Increases strength and corrosion resistance.
Lead	0.15% to 0.35%	Improves machinability without detrimental effect on mechanical properties.
Chromium	0.50% to 1.50%	In alloy steels, increases responsiveness to heat treatment and hardenability.
	4.0% to 12%	In heat-resisting steels, causes retention of mechanical properties at high temperatures.
	Over 12%	Increases corrosion resistance and hardness.
Nickel	1.0% to 4.0%	In alloy steels, increases strength, toughness, and impact resistance.
	Up to 27.0%	In stainless steels, improves performance at elevated temperatures and prevents work hardening.
Molybdenum	0.10% to 0.40%	In alloy steels, increases toughness and hardenability.
	Up to 4.0%	In stainless steels, increases corrosion resistance and strength retention at high temperatures.
Tungsten	17% to 20%	In tool steels, promotes hardness at high cutting temperatures; in stainless steels, smaller amounts assure strength retention at high temperatures.
Vanadium	0.15% to 0.20%	Promotes small grain size and uniformity of internal grain structure in the as-cast metal or during heat treatment; improves resistance to thermal fatigue and shock.
Tellurium	Up to 0.05%	Improves machinability when added to leaded steels.
Titanium	Variable	Prevents loss of effective chromium through carbide precipitation in "18-8" stainless steels.
Cobalt	17.0% to 36.0%	Increases magnetic properties of alloy steels. In smaller amounts, promotes strength at high temperatures in heat resisting steels.

Source: "Construction Lending Guide," courtesy U.S. League of Savings Associations.

reduced below the amount found in iron ore. However, when present in the correct amount, each of these elements improves the final product. In some cases, due to the characteristics of the iron ore, there is a deficiency of one or more of these elements, and they must be added to the molten steel.

Any added element is considered an *alloying element*, but when only the aforementioned five elements are involved, the steel is not considered an alloy steel. Other elements may be added to impart certain properties to steel. These are also called alloying elements, and the steel that results is called *alloy steel*. Table 5–3 shows the effects of alloying elements.

Steels are identified according to a classification system of the Society of Automotive Engineers (SAE). Each type of steel is designated by a group of numbers. The first digit indicates the class of steel. For example, carbon steel is designated by No. 1 and nickel steel by No. 2. The next one or two digits indicate the approximate percentage of the major alloying element for alloy steels. The last two or three digits indicate the carbon content in hundredths of a percent. The classification system is outlined in Table 5–4.

Besides the key information shown directly by the classification number, percentage ranges for all impurities and alloying elements are also designated indirectly when the system is used. The system provides a simplified way to specify steel. For example, a 1018 steel is a carbon steel containing 0.15 to 0.20 percent carbon, 0.60 to 0.90 percent manganese, 0.040 maximum percent phosphorus, and 0.050 maximum percent sulfur; and 4320 steel is a molybdenum steel containing 0.17 to 0.22 percent carbon, 0.45 to 0.65 percent manganese, 0.040 percent phosphorus, 0.040 percent sulfur, 0.20 to 0.35 percent silicon, 1.65 to 2 percent nickel, 0.40 to 0.60 percent chromium, and 0.20 to 0.30 percent molybdenum.

The American Iron and Steel Institute has adopted the SAE system with some variations and has added letter prefixes to designate the steel-making process used and other letters to designate special conditions. These designations are also shown in Table 5–4.

Table 5–5 shows the American Institute of Steel Construction, Inc. (AISC) system of designating structural steel. Each steel is known by the number of the ASTM standard that describes it. However, manufacturers may call them by trade names. The figure shows which rolled plates, bars, and shapes are made of each type of steel. For example, it can be seen that A529 Gr. 50 steel is carbon steel with a yield stress of 50 Ksi or more and ultimate tensile stress of 65–100 Ksi, and that only shapes listed in group 1, table A of ASTM A6 and plates and bars up to $\frac{1}{2}$ in. thickness are manufactured. ASTM A529 specifies manufacturing methods, chemical content, and mechanical properties required of steel to be supplied when A529 steel is specified.

The chemical composition of steel is determined by the composition of the materials used, the temperature,

Table 5–4. Classification of steels (SAE and AISI)

SAE Classification System	
Carbon steels	1xxx
Plain carbon	10xx
Free-cutting (screw stock)	11xx
Free-cutting, manganese	X13xx*
High-manganese	T13xx**
Nickel steels	2xxx
0.50% nickel	20xx
1.50% nickel	21xx
3.50% nickel	23xx
5.00% nickel	25xx
Nickel-chromium steels	3xxx
1.25% nickel, 0.60% chromium	31xx
1.75% nickel, 1.00% chromium	32xx
3.50% nickel, 1.50% chromium	33xx
3.00% nickel, 0.80% chromium	34xx
Corrosion- and heat-resisting steels	30xxx
Molybdenum steels	4xxx
Chromium	41xx
Chromium-nickel	43xx
Nickel	46xx and 48xx
Chromium steels	5xxx
Low-chromium	51xx
Medium-chromium	52xxx
Chromium-vanadium steels	6xxx
Tungsten steels	7xxx and 7xxxx
Triple-alloy steels	8xxx
Silicon-manganese steels	9xxx

*X indicates manganese or sulfur content has been varied from the standard for that number.

**T indicates manganese content has been varied in 1300 range steels.

Additional Symbols Used in AISI System	
10xx	Basic open-hearth and acid Bessemer carbon steel grades, nonsulfurized and nonphosphorized.
11xx	Basic open-hearth and acid Bessemer carbon steel grades, sulfurized but not phosphorized.
12xx	Basic open-hearth carbon steel grades, phosphorized. Prefix
B	Acid Bessemer carbon steel.
C	Basic open-hearth carbon steel.
CB	Either acid Bessemer or basic open-hearth carbon steel at the option of the manufacturer.
D	Acid open-hearth carbon steel.
E	Electric furnace alloy steel.

the length of time in the furnace, the medium surrounding the steel (whether air, oxygen, or vacuum), and whether open flame or heat. The surrounding medium depends on the process used, and the other variables can be controlled for each process. The steel is tested at intervals during the process, and adjustments are made. Alloying elements are added just before the melt is tapped to flow from the furnace to a ladle. The steel may be poured directly from the ladle into molds to make castings. Steel castings are made the same way and used for the same purposes as iron castings. Steel is stronger and tougher, but more expensive.

Table 5–5. Applicable ASTM Specifications for Various Structural Shapes (AISC)

Steel Type	ASTM Designation		F_y Min. Yield Stress (ksi)	F_u Tensile Stress[a] (ksi)	Applicable Shape Series							HSS		Pipe
					W	M	S	HP	C	MC	L	Rect.	Round	
Carbon	A36		36	58–80[b]										
	A53 Gr. B		35	60										
	A500	Gr. B	42	58										
			46	58										
		Gr. C	46	62										
			50	62										
	A501		36	58										
	A529[c]	Gr. 50	50	65–100										
		Gr. 55	55	70–100										
High-Strength Low-Alloy	A572	Gr. 42	42	60										
		Gr. 50	50	65[d]										
		Gr. 55	55	70										
		Gr. 60[e]	60	75										
		Gr. 65[e]	65	80										
	A618[f]	Gr. I & II	50[g]	70[g]										
		Gr. III	50	65										
	A913	50	50[h]	60[h]										
		60	60	75										
		65	65	80										
		70	70	90										
	A992		50–65[i]	65[i]										
Corrosion Resistant High-Strength Low-Alloy	A242		42[j]	63[j]										
			46[k]	67[k]										
			50[l]	70[l]										
	A588		50	70										
	A847		50	70										

■ = Preferred material specification.

■ = Other applicable material specification, the availability of which should be confirmed prior to specification.

■ = Material specification does not apply.

[a] Minimum unless a range is shown.

[b] For shapes over 426 lb/ft, only the minimum of 58 ksi applies.

[c] For shapes with a flange thickness less than or equal to $1\frac{1}{2}$ in. only. To improve weldability a maximum carbon equivalent can be specified (per ASTM Supplementary Requirement S78). If desired, maximum tensile stress of 90 ksi can be specified (per ASTM Supplementary Requirement S79).

[d] If desired, maximum tensile stress of 70 ksi can be specified (per ASTM Supplementary Requirement S91).

[e] For shapes with a flange thickness less than or equal to 2 in. only.

[f] ASTM A618 can also be specified as corrosion-resistant; see ASTM A618.

[g] Minimum applies for walls nominally $\frac{3}{4}$- in. thick and under. For wall thicknesses over $\frac{3}{4}$ in., F_y = 46 ksi and F_u = 67 ksi.

[h] If desired, maximum yield stress of 65 ksi and maximum yield-to-tensile strength ratio of 0.85 can be specified (per ASTM Supplementary Requirement S75).

[i] A maximum yield-to-tensile strength ratio of 0.85 and carbon equivalent formula are included as mandatory in ASTM A992.

[j] For shapes with a flange thickness greater than 2 in. only.

[k] For shapes with a flange thickness greater than $1\frac{1}{2}$ in. and less than or equal to 2 in. only.

[l] For shapes with a flange thickness less than or equal to $1\frac{1}{2}$ in. only.

Source: AISC Steel Manual 13th Edition

Most of the steel is poured into ingot molds prior to further shaping. The ingots are of various sizes and shapes depending on their future use. Their weight ranges from hundreds of pounds to many tons. They are tall compared to their cross sections, which are square or rectangular. A common size is about 6 ft tall with a cross section 2 ft by 2 ft and a weight over 4 tons. An ingot is cooled to a uniform temperature throughout in a *soaking pit*, which is a furnace where the steel temperature is allowed to decrease to about 2300°F (1260°C). It is then taken to a mill to be given its final shape.

Steel properties are influenced to a great extent by the mechanical operations that change an ingot of steel into a useful shape. The operations are rolling, extruding, drawing, forging, and casting. All except casting may be performed while the steel is in a plastic condition at a temperature of about 2000°F (1090°C), or as low as room temperature. The operations are called *hot working* or *cold working*.

Hot working breaks up coarse grains and increases density by closing tiny air holes and forcing the grains closer together. Cold working elongates grains in the direction of the steel elongation, increases strength and hardness, and decreases ductility. Cold working results in more accurately finished products, because there is no cooling shrinkage to be estimated. The surfaces are smoother, because oxide scale does not form as it does during hot working. Overworking, whether hot or cold, causes brittleness.

For all but very large objects, working to final shape is done in two stages. The first stage consists of squeezing the ingot into a smaller cross section between two rollers, called *blooming rolls,* which exert a very high pressure. This operation is always performed while the steel is hot. In the second stage, the ingot is rolled into a much longer piece with a square or rectangular cross section closer to its final size. The desired cross section is obtained by turning the ingot 90° to be rolled on the sides as it is passed back and forth through the rolls. If it is approximately square in cross section, it is called a *bloom* if large (over 36 sq in.) and a *billet* if smaller. It is called a *slab* if the width is twice the thickness or more. The appropriate shape is used to manufacture beams, rails, plate, sheets, wire, pipe, bolts, or other items by one of the following methods.

Rolling consists of compressing and shaping an ingot into a useful shape by squeezing it through a succession of rollers, each succeeding set of rollers squeezing the material smaller in cross section and closer to the final shape. The piece being rolled becomes longer and wider as it is compressed. It may be made narrower by cutting or by rolling after turning 90° so that the rolling reduces the width. A wide variety of cross sections useful for construction of buildings and bridges can be rolled in long pieces by means of specially shaped rollers (Figure 5–5). Flat sheets can be rolled by rollers of a constant diameter. Corrugated sheets can be rolled from flat sheets by using corrugated rollers. Corrugated steel roof deck can be seen in Figure 5–15. Hot rolling usually precedes cold rolling until the steel is close to its final shape. Hot rolling is usually the first step in reducing the size of an ingot prior to extruding, drawing, or forging.

To vary the area and weight within a given nominal size, the flange width, the flange thickness, and the web thickness are changed.

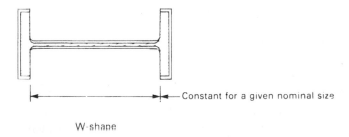

Constant for a given nominal size

W-shape

To vary the area and weight within a given nominal size, the web thickness and the flange width are changed by an equal amount.

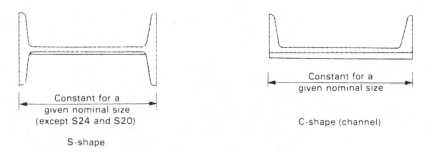

Constant for a given nominal size (except S24 and S20)

S-shape

Constant for a given nominal size

C-shape (channel)

To vary area and weight for a given leg length, the thickness of each leg is changed. Note that leg length is changed slightly by this method.

L-shape (angle)

FIGURE 5–5. Typical rolled sections (AISC).

Extrusion consists of forcing a billet of hot, plastic steel through a die of the desired shape to produce a continuous length of material of reduced cross section in the shape of the die. The resulting product has the shape of a rolled product; that is, it is long with a constant cross section. However, more intricate shapes can be formed by extrusion than by rolling, and the surface is of higher quality. An extrusion is made in one operation rather than repetitive operations as in the case of rolling. An extruded section can sometimes be used in place of a section that requires several operations if formed any other way (Figure 5–6). Extrusions can be made with cross sections having a maximum dimension of nearly 2 ft.

Drawing consists of pulling steel through a small die to form wire or a small rod of round, square, oval, or other cross section. Steel is hot rolled to form a rod of a size not much larger than the shape to be drawn. It is then finished by cold drawing. Seamless steel pipe may also be finished by cold drawing over a round, bullet-shaped mandrel to form a hollow center and through a die to form the outside. The advantages of cold drawing are a smoother finish, more accurate size, more strength, and better machinability.

Forging consists of deforming steel by pressure or blows into a desired shape. The forging may be made from an ingot or from a rolled shape. The steel is usually heated to a semi-solid state at a temperature over 2000°F (1090°C). In some cases it is forged cold. It is forced to fill the shape between dies by pressure or blows of the upper die upon the lower one. The shape may be formed more accurately by successive forgings, each succeeding operation performed with smaller dies closer to the desired final shape. Instead, the final shape may be achieved by machining. Many shapes can be either cast or forged. Economics often determine which method is used. However, forging is preferred if strength of the part is important. Forging improves the mechanical properties of the metal, as does other hot working or cold working, and

produces a stronger, more ductile, more uniform product with smaller grain size than is produced by casting.

After steel cools and is given its final shape, further heating and cooling processes can change the internal structure and thereby impart certain properties. Heat treatment consists of heating, holding the metal at the high temperature, and cooling. Even here the rate of heating is not important, except for high-carbon alloyed steels. The metal is held at the upper temperature so that it can be heated to a uniform temperature throughout. The rate of cooling is very important.

Normalizing consists of heating the steel to a temperature of about 1500°F (815°C) or higher, depending on the type of metal, and cooling several hundred degrees slowly in air. This process increases uniformity of structure.

Annealing consists of heating the steel to a temperature slightly lower than for normalizing and cooling it several hundred degrees very slowly, usually in a furnace. Methods vary somewhat depending on the purpose, which may be to soften the metal, produce a special structure, facilitate machining and cold shaping, or reduce stresses.

Quenching consists of cooling steel very rapidly in oil, water, or brine from a temperature of about 1500°F (815°C). Quenching increases hardness and strength, but reduces ductility and toughness. Residual stresses are introduced by quenching and should be relieved by tempering.

Tempering consists of reheating the quenched steel to a temperature of 300 to 1200°F (150 to 650°C) and cooling in air to reduce the residual stresses and increase ductility. Heating to the lower temperature range produces greater hardness, strength, and wear resistance, whereas higher heat produces greater toughness.

STEEL TENSILE TEST

Mechanical tests for steel include tension, bending, hardness, and impact. For structural steel the tension, or tensile, test is the most important (Figure 5–7). The purchaser may specify that the test be performed on steel from the furnace, after rolling, or after fabrication. Specimens for testing are poured separately as an ingot is being made or are cut from the waste material of a rolled member. Specimens may be of various sizes.

The typical tensile test specimen is a 0.500-in.-diameter cylinder machined to a smooth, accurate circular cross section. The specimen is clamped at each end or threaded into a testing machine that applies an axial pull at a uniformly increasing rate until the specimen breaks. As the pulling proceeds, the force is constantly indicated in digits or by a dial on the machine. Tensile stress is calculated by dividing the force by the original cross-sectional area.

Before the force is applied, two marks are made on the specimen 2 in. apart in the direction of the applied force. The two marks are drawn farther apart as the specimen deforms under the tension. Strain is calculated by dividing the increase in distance between the marks by the original 2-in. distance.

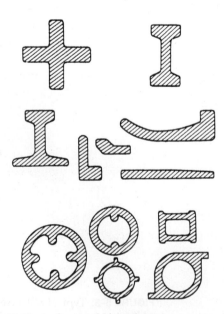

FIGURE 5–6. Typical extruded sections.

FIGURE 5–7. The application of tensile loads to a steel specimen; note the stress-strain recorder. (*Courtesy Al Gaudreau*)

Stress and strain are determined at regular intervals from readings of force and the measured increase in distance between the marks. A curve of stress versus strain is plotted to determine whatever information is desired. The yield stress and rupture stress are often specified as lower limits for acceptance of steel. (Refer to Figure 1-3a for a typical stress–strain diagram for steel.) In many cases, a complete stress–strain curve is plotted by an automatic recorder as the test proceeds.

Automatic devices are available to determine yield strength by noting the strength at the correct plastic (permanent) deformation as the sample is being tested without plotting a stress–strain curve.

STEEL PROTECTION

Rusting is oxidation or combining of the iron with oxygen, which occurs in the presence of moisture. It proceeds more rapidly where there is noticeable dampness, but it occurs in any air with a relative humidity higher than 70 percent. It progresses more rapidly in salt air and in an industrial atmosphere. The thickness of metal lost by rusting is $\frac{1}{2}$ mil (a mil is 0.001 in.) or less per year in average conditions. It may be much higher in the presence of industrial air pollution. The rust is formed from the solid metal, reducing its size so that the member becomes weaker and loses any decorative finish it might have. The rust penetrates deeper as time goes on. Painting the steel prevents rust. However, painting is expensive and paint must be replaced periodically at additional cost. Carbon steel may be made more rust resistant by the addition of copper as an alloying element.

Ordinary steel cannot be used where it is exposed to high temperatures. At a very high temperature it melts to the liquid state, and it begins the process of liquifying at moderately high temperatures. Steel weakens at 800 to 900°F (430 to 480°C) and cannot support any load at 1200°F (650°C). Steel may be used economically for structural support of industrial furnaces, incinerators, and other heat-producing devices, but it must be insulated from the heat.

Structural steel that is ordinarily subjected to normal temperatures may be subjected to great heat during an accidental fire. The steel will then melt or at least become soft from the heat unless protected by rated insulation. Most steel-framed buildings constructed before 1950 encased structural steel in concrete. While this was an effective means of fireproofing, it added greatly to the structural dead load. Fireproofing materials containing asbestos were also used, but these materials proved to be a significant health hazard to occupants and have been removed from most buildings in the United States. Today, lightweight, cementitious based sprayed on fireproofing is most often used for structural steel applications. These materials typically contain a mixture of gypsum, treated cellulose, and an agent that facilitates drying once the material is applied. Lightweight fiberboard fireproofing systems are also available for spaces where steel components are articulated as an element of the building design. Steel members may be exposed in highway and railroad bridges, and in buildings where the occupancy does not require specific fire ratings of structural elements.

The protection of ordinary steel in a moderately corrosive atmosphere requires thorough cleansing of the steel followed by the application of three coats of various types of oil-based paint with a total thickness of 4 mils (0.004 in.). The prevention of rust can be made easier by careful design to avoid pockets or crevices that hold water, spots that are inaccessible for repainting, and sharp edges that are difficult to coat with the required thickness of paint.

Another type of protection commonly used is galvanizing, or coating with zinc or with zinc-pigmented paints. Zinc adheres readily to iron or steel to form a tight seal against the atmosphere. It prolongs the life of iron and steel because it corrodes much more slowly than they do. Zinc continues to protect the iron or steel even after it has been eaten through in spots, because corrosion will take place in the zinc in preference to the ferrous metal. The zinc coating has a shiny, silvery appearance which is not suitable for all uses.

Bituminous coatings are used to protect iron and steel from the effects of atmosphere, water, or soil. The usefulness of bitumens as protective coatings is discussed in Chapter 3. The appearance and odor of bitumens make them unsuitable for many applications.

Stainless steels, which are known as high-alloy steels, have chromium and nickel as their chief alloying elements. They contain 16 to 28 percent chromium and may contain up to 22 percent nickel. Chromium may be used alone or in combination with nickel. Adding manganese increases the ductility and acid resistance of the steel. Stainless steels have high resistance to corrosion, and their wide variety of finishes, from dull to mirrorlike, last indefinitely. They are used where appearance or sanitation is important, such as in kitchens, laboratories, and exterior building trim. They are also used for mechanisms in wet or corrosive atmospheres, such as rockers and rocker plates for bridges, water valves and gates, and smokestack controls. Corrosion resistance of stainless steels is due to the forming of a thin, transparent coating of chromium oxide over the surface. Stainless steel may be made harder at the cost of some of the corrosion resistance.

Stainless steels are available with various characteristics, such as good corrosion resistance at high temperatures, no magnetic property, good weldability, and a low coefficient of thermal expansion. For some kinds of stainless steel, the tendency toward galvanic corrosion when in contact with other metals and a higher-than-normal coefficient of thermal expansion can lead to trouble unless precautions are taken in design.

Heat-resisting steel contains chromium as the primary alloying element. This steel does not lose a significant amount of strength at temperatures up to 1100°F (590°C). It can be hardened by heat treatment.

STRUCTURAL STEEL

Structural steel includes rolled shapes and plates used in structural frames, connectors, plates, or bracing needed to hold the frame in place, and most other steel attached directly to the frame. Exceptions are grating and metal deck, open-web steel joists, ornamental metal, stacks, tanks, and steel required for assembly or erection of materials supplied by trades other than structural steel fabricators or erectors.

Figure 5–8 shows a typical structural steel frame building under construction. Steel erection consists of connecting precut pieces of rolled sections (see Figure 5–5) together as shown on construction drawings to form a structure.

Yield stress (see Figure 1-3a for explanation) is used as a basis for the design of all steel structures. It is the yield point for all structural steels that produce a yield point on their stress-strain curves. This includes most structural steels. It is the yield strength determined at a plastic (permanent) strain of 0.002 (0.2 percent) for steels having no yield point. Of the

FIGURE 5–8. Steel frame building. (Note various steel shapes and connections.) *(Courtesy J. Coffey)*

steels listed in Table 5–5, only A514 is of this type. Yield stresses range from 24 to 100 Ksi for structural steels.

Maximum allowable stresses (described in Chapter 1) in compression, tension, and shear for various types of structural members and connections and for various types of loads are published by the AISC, American Association of State Highway and Transportation Officials (AASHTO), and other organizations. These published stresses are all computed by reducing the yield stress for each type of steel by a safety factor. (Safety factors are discussed in Chapter 1.) The designer selects the type of steel and the size and arrangement for each member and connection so that the expected loads are carried economically, without exceeding the published allowable stresses.

Structural steels designated as carbon steel in Table 5–5 are mild-carbon steels containing 0.29 maximum percent carbon, 1.20 maximum percent manganese, 0.04 maximum percent phosphorus, and 0.05 maximum percent sulfur. Silicon is required only for plates at 0.15 to 0.40 percent. These steels have yield stresses from 32 to 42 Ksi. Strength of these steels is closely related to carbon content. Their corrosion resistance may be doubled by adding copper. A36 is an all-purpose steel, widely used for buildings and bridges.

Alloying elements added to steel can impart properties impossible to impart by heating and cooling or by working (see Table 5–3). An important reason for adding alloying elements is to improve mechanical properties. High-strength, low-alloy steels contain alloying elements that improve mechanical properties and resistance to rust. The total amount of alloying elements added is 2 or 3 percent. Specifications for these steels require that they meet certain performance standards rather than certain chemical formulas. However, each manufacturer has his own method of producing a product that meets the standards.

Those designated as high-strength, low-alloy steel in Table 5–5 have the carbon content of mild-carbon steel with columbium, vanadium, nitrogen, or copper, or combinations of them added as alloys in small amounts. Maximum percentages of the alloying elements are as follows: carbon, 0.26; manganese, 1.65; phosphorus, 0.04; sulfur, 0.05; and silicon, 0.40. These steels have yield stresses from 40 to 65 Ksi. Strength is increased beyond that of ordinary carbon steel by means of a finer structure that occurs during cooling with no additional heat treatment.

Those designated as corrosion-resistant, high-strength, low-alloy steels in Table 5–5 have four times the corrosion resistance of carbon structural steel or two times that of carbon structural steel with copper and are also known as *weathering steels.*

Exposure to the atmosphere forms a thin, rust-colored, protective coating of iron oxide, which eventually becomes blue-gray and which prevents any further corrosion. Two or three years are required for the coating to form completely. The coating does not protect against corrosion from concentrated, corrosive industrial fumes, from wetting with salt water, or from being submerged in water or buried in the ground. During the early stages of weathering, rainwater dripping from the steel carries corroded materials which stain concrete, brick, and other light-colored, porous materials.

Weathering steels have the carbon content of mild-carbon steel or low-carbon steel, with various small percentages of chromium, nickel, silicon, vanadium, titanium, zirconium, molybdenum, copper, and columbium added as alloys. Yield stresses range from 42 to 50 Ksi. These steels also attain high strength through finer structure that occurs during cooling.

Those designated as quenched and tempered alloy steel in Table 5–5 are available only as plates. They also have the carbon content of mild-carbon steel or low-carbon steel and alloys of silicon, nickel, chromium, molybdenum, vanadium, titanium, zirconium, copper, and boron. These steels have yield stresses of 90 or 100 Ksi. These steels are intended for use in welded structures, and special precautions must be taken to weld them. The steel is fully killed and fine grained. It must be heated to at least 1650°F, then quenched and tempered by reheating to not less than 1150°F, and allowed to cool slowly. Quenching produces martensite causing great strength and hardness along with brittleness. Tempering greatly improves ductility and toughness, but at the cost of some reduction in strength and hardness.

Judging by the high strengths of the alloy steels in Table 5–5, one might wonder why carbon steel is still used. Carbon steel is so much cheaper by the pound that a member made of it may be cheaper than a smaller member of the same strength made of stronger, more expensive steel. In addition to this, the size of many members is dictated by considerations other than strength, and they cannot be smaller, no matter how strong the material is. In these cases, the least expensive material is used unless corrosion resistance or heat resistance is of great importance.

Steel is produced in mills to standard shapes; plates and bars of standard thicknesses; pipe, tubing, plain rods, deformed rods, and wire of standard diameters; and various rolled shapes of standard dimensions (see Table 5–6). Designers ordinarily select standard shapes that best suit their purposes and adapt their designs to conform to the shapes that are available. In some cases they modify standard shapes to conform to their design requirements. Some examples of modified standard shapes are T shapes cut from W or S shapes, L's cut from C shapes, and tapered members fabricated from various shapes.

Steel produced in a mill is not sent directly to a construction project; it goes next to a fabricating shop. Detail drawings that show exactly how separate parts fit together to form a structure are used to determine the finished lengths and any fabrication needed to facilitate the assembly of the parts at the jobsite. Fabrication takes place in the shop as much as possible to make fieldwork simpler.

The operations performed in the fabricating shop include cutting or shearing to the correct length. Steel is cut with an oxygen torch machine or sheared with a mechanical cutting device with a large blade that drops onto the steel to shear it to the correct length. When smoother ends are

Table 5–6. Designations of rolled steel shapes

Designation	Type of Shape
W 24 × 76	W shape
W 14 × 26	
S 24 × 100	S shape
M 8 × 18.5	M shape
M 10 × 9	
M 8 × 34.3	
C 12 × 20.7	American standard channel
MC 12 × 45	Miscellaneous channel
MC 12 × 10.6	
HP 14 × 73	HP shape
L6 × 6 × $\frac{3}{4}$	Equal leg angle
L6 × 4 × $\frac{5}{8}$	Unequal leg angle
WT 12 × 38	Structural Tee cut from W shape
WT 7 × 13	
ST 12 × 50	Structural Tee cut from S shape
MT 4 × 9.25	Structural Tee cut from M shape
MT 5 × 4.5	
MT 4 × 17.15	
PL $\frac{1}{2}$ × 18	Plate
Bar 1 ▢	Square bar
Bar 1$\frac{1}{4}$ φ	Round bar
Bar 2$\frac{1}{2}$ × $\frac{1}{2}$	Flat bar
Pipe 4 Std.	Pipe
Pipe 4 X—Strong	
Pipe 4 XX—Strong	
TS 4 × 4 × 0.375	Structural tubing: square
TS 5 × 3 × 0.375	Structural tubing: rectangular
TS × OD × 0.250	Structural tubing: circular

needed, the cut or sheared surfaces are finished by planing or grinding to a smooth surface. Holes for rivets or bolts are drilled through thick sections and punched through thinner ones. Holes may also be punched and reamed smooth. If connections are to be made by welding, the work is partially done in the fabricating shop. As a general rule, the less riveting, bolting, or welding left for the field, the more efficient the entire operation is. Examples of fabrication are shown in Figure 5–9.

Open-web steel joists are fabricated of angles and rods and are stocked in standard lengths, grouped according to the roof load or floor load that they are able to support. They are Warren trusses with a top chord fabricated to support a roof deck or floor. They are designed to be spaced close together and are available in lengths from 4 to 96 ft. Open-web steel joists can be seen in Figure 5–15.

Steel, especially the more complicated shapes, may cool unevenly while being rolled. The final cooling then causes a variation in the amount of shrinkage of various areas, which results in curved or twisted members. These are straightened in the mill after rolling. Shapes that are to

be used as beams are then deliberately curved upward so that when they deflect downward in use they will be approximately straight. The curve or *camber* is a smooth curve, roughly part of a circle, and may be applied over almost the entire length of the beam or between any two points specified, depending on how the beam is expected to deflect in use. The designer specifies the curvature by designating the ordinate (also called camber) at the middle of the curve.

Usually sufficient camber is induced to balance the expected dead load deflection so that in use the beam will be straight except when deflected by live load. Camber is induced by bending the beam while the steel is cold. Approximately a quarter of the mill-induced camber is lost before the beam reaches the construction site; the remainder is permanent.

Camber for complicated loading arrangements may be built into beams and also trusses at the fabricating shop. It is done by rapid heating with a blowtorch of a short section of the bottom of the beam, which attempts to expand but is restrained by the cold steel on each side, resulting in sufficient compression to cause permanent plastic shortening in the heated area when it cools. As a result of the bottom flange shortening, the beam or truss bows upward. This method is also used in the shop and field to straighten members that are accidentally bent.

Steel is painted in the fabricating shop with a "shop coat" of paint having the required dry film thickness in mils (one thousandth of an inch). The shop coat is the prime coat of the protective paint system and will not protect against corrosive atmospheres or against prolonged exposure to normal atmosphere. It is meant to protect the steel until it is erected and painting can be completed.

The steel must be cleaned by wire brushing or comparable means before painting. The paint may be applied with a brush or roller or by spraying, dipping, or flow coating. Required dry thickness of the paint film may be 1 mil or more. Touch-up painting required to replace paint removed by handling after the shop coat is applied is the responsibility of the builder.

Paint thickness may be measured with a dial comparator, dial indicator, or micrometer by measuring the thickness of painted metal, removing a small portion of paint, and measuring the thickness of bare metal; the difference is the paint thickness.

Another method depends on the reduction in magnetic force caused by a nonmagnetic coat of paint between a permanent magnet and the magnetic metal covered by the paint. The thicker the coating, the greater the reduction in magnetic force. A small magnet is placed in contact with the paint, and its force field, which passes through the paint in order to reach the steel, is measured. The value obtained is converted to paint thickness by referring to a calibration curve for that apparatus. A variation of this method is to measure the impedance to electric eddy currents set up by magnetism. Impedance to the flow of electricity through the paint is proportional to its thickness.

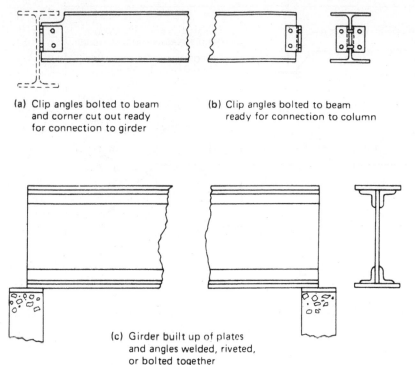

(a) Clip angles bolted to beam and corner cut out ready for connection to girder

(b) Clip angles bolted to beam ready for connection to column

(c) Girder built up of plates and angles welded, riveted, or bolted together

FIGURE 5–9. Examples of fabricated steel.

STRUCTURAL CONNECTIONS

Steel members must be connected to form a structure. They may be connected by riveting, welding, or bolting. A connection is not designed on a purely rational basis as the structural members are. Stresses at connections are so complex that the connections are designed according to empirical methods based on successful experience. A connection includes a group of rivets or bolts or a predetermined length of weld.

Riveting

Riveting was once the most common method of making connections. It is now used only for shop connections, and that too only rarely. Holes are punched or drilled through the members to be connected, and a steel rivet (Figure 5–10) slightly smaller than the holes is heated to a cherry red color

(1000 to 1950°F) and inserted through the holes. The head is braced and the shank end is hammered until it flattens to a head, compressing the members between the two rivet heads, as shown in the figure. Cooling of the rivet causes it to shorten, compressing the members still further.

Welding

A welded connection is neat in appearance, and the metal of a weld is stronger than the metal being connected. Weld metal is manufactured to more demanding specifications than structural steel and is protected from the atmosphere while cooling. The weld metal also benefits somewhat by combining with constituents of the welding rod coating. The result is a steel with better crystalline structure and higher mechanical properties.

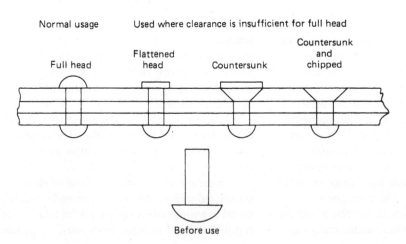

Normal usage   Used where clearance is insufficient for full head

Full head

Flattened head

Countersunk

Countersunk and chipped

Before use

FIGURE 5–10. Rivets.

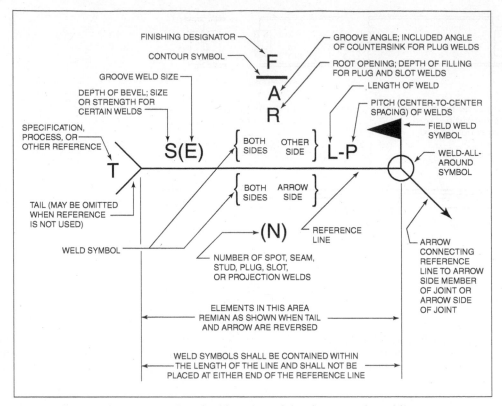

FIGURE 5–11. AWSA2.4: 2007, Standard location of the elements of a welding symbol, Figure 3.

(Reproduced with permission of the American Welding Society (AWS), Miami, Florida.

Figure 5–11 shows typical welding arrow symbols used to describe required weld properties. Chemical and mechanical properties of weld metal must be matched to the metal being welded. Therefore, a wide variety of welding electrodes is available. The American Welding Society and ASTM have established a numbering system for electrodes. All designations begin with the letter E, which is followed by a four- or five-digit number. The first two or three digits indicate the minimum tensile strength in Kips per square inch. The next digit indicates the recommended welding positions. Digit 3 is for flat only; 2 is for flat and horizontal; and 1 is for all positions including vertical and overhead. The next digit indicates current supply and recommended welding techniques. The four welding positions are shown in Figure 5–12.

Welding consists of heating the two pieces to be joined until they melt enough to fuse. The heat comes from an electric arc that is formed between a welding rod and the two pieces to be welded. A portable electric generator is connected to the structural steel and to the welding rod, and either an alternating or a direct current is passed through the rod and the structural members when the rod touches or nearly touches the members. The tip of the rod and some depth of the base metal, called *penetration,* are melted. The two metals combine and harden upon cooling. The liquid metal rapidly absorbs oxygen and nitrogen, which causes it to be brittle and lose its resistance to corrosion unless it is protected from the atmosphere.

Four welding methods are allowed by the AISC for structural work. In the *shielded metal-arc method,* the metal

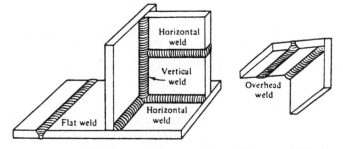

FIGURE 5–12. The four welding positions. *(Courtesy U.S. Department of the Interior, Bureau of Reclamation)*

welding rod is coated with a *flux,* which melts as the weld metal melts and covers the molten metal, shielding it from the atmosphere. The flux is partially converted to gas, which surrounds the working area, helping to protect the weld from oxygen and nitrogen. Shielded metal-arc welding is a manual method suited for field use. The three other methods discussed in the following section are suitable for semiautomatic or automatic use.

In the *submerged-arc method,* powdered flux is automatically spread ahead of the electrode and completely covers the welding arc and also protects the new weld metal.

In *gas metal-arc welding,* a coil of electrode wire is constantly fed to a holder as the electrode melts. The new weld metal is protected from the air by CO_2 or other gas constantly fed to the location as the welding proceeds.

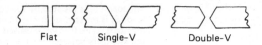

Flat Single-V Double-V

Butt or groove weld preparations

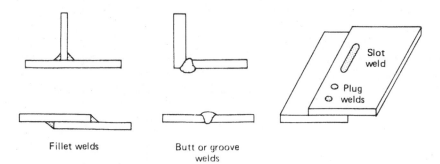

Slot weld

Plug welds

Fillet welds

Butt or groove welds

FIGURE 5–13. Weld types.

In *flux-cored arc welding,* the welding rod consists of a core of flux surrounded by weld metal. This is used to facilitate continuous feeding of the electrode as welding takes place.

The fillet weld is the most frequently used type. Other types commonly used are the butt or groove weld and the plug or slot weld, as shown in Figure 5–13. The *fillet weld* is triangular in cross section and is placed at a right-angle joint formed by the pieces to be connected.

For a *butt* or *groove weld,* the ends to be connected are butted together and welded. The abutting edges may be smooth, flat surfaces, or they may be shaped to form a groove. Grooves of several shapes are used. Flat edges and the two most common grooves are shown in the figure. Most butt welding is done to join plates edge to edge. Butt welding requires that the pieces to be welded be cut precisely to size or they will not meet properly. This expensive, precise cutting is avoided if the pieces are lapped and welded with fillet welds. This economy accounts for the greater popularity of fillet welds, even though butt welds are stronger.

A *plug* or *slot weld* consists of filling with weld metal a circular or oblong hole in one piece, which is positioned on top of the piece to which it is to be connected. This type of weld is useful for joining pieces that must act together over an area too large to be satisfactorily connected at the edges, such as elements of the flange of a plate girder. They are also used on plate joints that are overlapped to avoid overhead welding in the field. Fillet welds may be used inside the plugged or slotted holes.

Welds, and other ferrous materials, are tested for cracks or any other discontinuity by magnetic particle examination; methods are described in ASTM E709. This type of test is useful for testing welds or steel members for acceptance. It is useful for checking ferrous products from pig iron to finished castings or forgings. It is also useful for preventive maintenance examinations of structures already in use. The method is accurate enough to test whether or not a specimen meets measurable standards.

The test method depends on the principle that magnetic lines of force are distorted by a discontinuity in the material through which the lines of force pass. Magnetic particles spread loosely on a ferrous metal surface will collect at points of discontinuity at or near the surface when magnetic lines of force are present in the ferrous material. An entire object or any part of it may be tested with proper size equipment and proper placing.

Three steps are required for magnetic particle examination: the part to be tested must be magnetized, the proper type of magnetic particles must be applied, and particle accumulations must be observed and evaluated.

The part may be magnetized by passing an electric current through it or by placing it within a magnetic field from a separate source. A narrow crack parallel to the lines of force may not be detectable. The test must often be performed twice or more in different directions to catch all the cracks.

Finely divided ferromagnetic particles, colored and even fluorescent for visibility, are used dry or suspended in a liquid. To give a clear picture, they must have high attraction to a discontinuity in the material and low attraction to each other. For field inspection, dry powder sprayed or dusted onto the object is commonly used. Particles suspended in water or oil are used normally indoors with a reservoir and recirculation of the particle-bearing liquid. The liquid may also be sprayed for use only once. Wet particles are more sensitive to very small discontinuities because of their smaller size, but dry particles are less affected by heat or cold and more easily carried and applied on the jobsite. Fluorescent particles require a black light for viewing so are not often used on a construction site. Fluorescent particles are the easiest to see against any background, so they are often used in liquid indoors.

Heavy oil containing flake-like particles may be brushed into overhead or steeply inclined surfaces before magnetizing. Because of its high viscosity, the suspension remains in place long enough to be tested. It may even be used under water.

Particle-bearing liquid polymers that solidify in place forming a thin permanent record are useful for investigating hard-to-see areas. The thin coating containing the particles gathered at discontinuities is removed for examination.

When dry particles are used, the part is magnetized before the particles are applied because they do not move readily once in contact with the metal to be tested. With wet particles it is more convenient to flood the part with the suspended particles and then magnetize the part briefly.

The magnetic field must be strong enough to cause all discontinuities larger than allowable to be discovered and must be reasonably consistent for the same reason. Strength depends on the source of magnetism (electric current or magnetic field) and on the size, shape, and type of steel being tested.

Determining and holding the proper strength can be difficult. It may be necessary to experiment with identical pieces of material having discontinuities of known size in order to set the proper strength.

Bolts

A *bolt* is manufactured with a head at one end and threads at the other end, to which a nut can be threaded. Washers may be used. They fit loosely on the shank of the bolt at either or both ends and increase the area that bears on the member when the nut is tightened. When the bolt is in position to hold two members together, a tightening of the nut pulls the bolt with a tensile force and presses inward on the two members, causing friction between them to resist movement (Figure 5–14).

A bolted joint is subject to two types of loading from the structure. Tension tends to pull the plates apart in a line parallel with the bolt axis, and shear tends to make the plates slip against the friction between their surfaces in a direction perpendicular to the bolt axis. The friction available to resist shear is proportional to the bolt tension. Movement perpendicular to the bolt axis cannot occur until the friction between the members is overcome. If the friction is overcome, there is slight movement, and both members bear on the bolt. A bearing connection is satisfactory, provided slight movement is permissible and the load is static. If the joint is subject to load changes, stress reversal, impact, or vibration, a bearing connection is unsatisfactory, because it may loosen. The joint must then be designed so that sufficient friction is developed between the members to resist any load perpendicular to the bolt axis and so that the load parallel with the bolt is not sufficient to stretch it and thereby reduce that friction.

Like rivets, common bolts have an uncertain tension, and therefore the friction caused between members cannot be accurately determined. Nevertheless, common bolts and rivets have been and still are used successfully. In many connections where little strength is required and there are no vibrations, impact loads, or stress cycles, common bolts should be used because they are less expensive and easier to install than high-strength bolts. However, common bolts are much weaker than high-strength bolts; they loosen under vibration, impact loads, or cyclic loads, especially if there is stress reversal.

High-Strength Bolts

High-strength steel bolts can be installed to produce a predetermined tension greater than that of a common bolt. Because of the great friction developed between members, there is very little movement between them when loads are applied to the members. High-strength bolts are several

The load (*T*) is resisted by the bolt in tension. The load (*S*) is resisted by friction between the two members. The friction is caused by compression of the members between the bolt head and the nut.

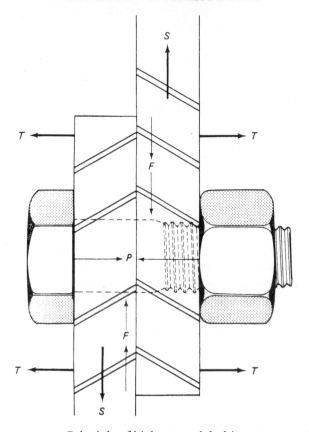

FIGURE 5–14. Principle of high-strength bolting.

times as strong in tension as common bolts. They are the most recently developed fastening devices, but have become the most popular by far, especially for field connections.

Tables of minimum tension loads for various size high-strength bolts have been prepared by AISC. In a properly constructed joint, nuts are tightened until each bolt has at least the minimum tension. Greater tension is permitted because the tabulated minimum tension loads are only 70 percent of the specified minimum tensile strength of the bolts. The designer must select the number and size of bolts so that the friction required at that joint is provided when all bolts are under the minimum allowable tension.

Bolts are tightened with wrenches powered by electricity or compressed air. These power wrenches operate by impacting against the object to be turned and are called *impact wrenches*. The nut is usually turned, but the bolt may be turned by its head if more convenient. The opposite end is held with a hand wrench while tightening takes place.

Each nut in a group of high-strength bolts must be tightened until bolt tension is equal to or greater than that specified for the bolt size by the AISC. Three methods of obtaining the required bolt tension are approved. The turn-of-nut method depends on the fact that tightening of the nut elongates the bolt, inducing tension in the bolt and therefore friction between the members in proportion to the bolt elongation. All nuts are first tightened as tight as they can be by one person using an ordinary spud wrench. This *snug-tight* position is the starting position for all methods of final tightening and may also be obtained by a few impacts of an impact wrench. Any turning of the nut beyond this original tightening draws the end of the bolt into the nut, thereby stretching the bolt. All nuts are then rotated the additional amount prescribed by AISC (1/3 turn to one full turn, depending on bolt size) to obtain correct elongation and tension.

Tightening may be done with a calibrated wrench. To calibrate a wrench, at least three bolts of each size to be used on that job are tightened in a calibrating device that indicates the tension in each bolt. Torque delivered by the wrench becomes greater as the nut is turned. The wrench is set to stall at the average torque load that produces the correct tension. Each bolt on the job is then turned until the wrench stalls.

Tightening may also be controlled by the use of a direct tension indicator on each bolt as it is tightened. One type of indicator is a load indicator washer manufactured by Bethlehem Steel Corporation. The washer has small projections on one of its flat faces that bear against the bolt head (or nut if the bolt head is to be turned).

As the nut is turned, the washer projections are compressed, causing the washer to move closer to the bolt head. The width of the gap can be related to bolt tension. The gap will be a predetermined width at proper bolt tension. The gap is measured with a feeler gauge to determine when proper tension is reached.

Inspection of high-strength bolt installation starts with witnessing the calibration of the wrench and proper tightening of each bolt. After completion of the bolting, the inspector also checks to see that no bolt is skipped. Nut surfaces are examined to determine that none was missed. The impact wrench almost always leaves its mark on the nut or bolt head.

A *torque wrench* is used to inspect a percentage of bolts chosen at random. A torque wrench is a long, hand-operated wrench that includes an indicator dial to indicate torque while the wrench is used. The wrench is calibrated to determine the torque needed to produce the required bolt tension. This wrench is applied to each bolt to be tested until the bolt moves slightly in the tightening direction. The torque required to do this must be as large as the torque determined by calibration.

Bolts installed using load-indicating washers are tested by the insertion of a metal feeler gauge. If the gauge fits into the opening, the washer projections have not been compressed enough and nuts must be tightened until the gap is too narrow for the gauge.

High-strength bolts are of two types: A325, described in ASTM A325, and A490, described in ASTM A490. The A325 bolts are of carbon steel and the A490 bolts are of alloy steel. There are four types of A325 bolts: medium-carbon steel, low-carbon martensite, weathering steel, and hot-dip galvanized. There are two types of A490 bolts: alloy steel and weathering steel. An A325 bolt has the strength of one and one-half rivets of the same size, and an A490 bolt has the strength of one and one-half A325 bolts of the same size. Figure 5–15 shows a typical connection made with high-strength bolts. Clip angles

FIGURE 5–15. Steel framing with bolted column-to-beam connection.

were shop-welded to the beams, and the angles were bolted to the column in the field. Also shown are open-web steel joists and corrugated sheet steel roof decking.

REINFORCING STEEL

Steel is used as the reinforcing in combination with portland cement concrete for reinforced concrete structural members. Members are constructed so that steel resists all tension, and compression is resisted by concrete or by concrete and steel together. Reinforcing steel is shown being tied in place for a concrete wall in Figure 5–16.

FIGURE 5–16. Completion of reinforcing steel work before outside form is placed. *(Courtesy MLB Industries, Inc.)*

Reinforcing bars are made either plain or deformed. Figure 5–17 shows standard bar markings. Deformed bars create a better bond between concrete and steel. They are designated by the number of eighths of an inch in their diameter and are available in sizes from No. 2, or $\frac{1}{4}$ in. diameter, to No. 18, or $2\frac{1}{4}$ in. diameter. A table of rebar sizes and weights is given in Chapter 4.

Rebars are made from Bessemer or open-hearth carbon steel with 0.40 to 0.70 percent carbon; from scrap carbon steel axles of railroad cars; and from standard section T rails. The bars are hot-rolled and furnished in the grades shown in Table 5–7. Standard reinforcing bars are referred to as *black bars* and are used in most reinforced concrete. Reinforcing steel is also available with epoxy or galvanized coatings, which are used where corrosion of the rebars may cause structural and other problems with the concrete. Figure 5–18 shows steel reinforcing bars being bent.

FERROUS METAL PIPE

Both gray cast iron and ductile iron have been used extensively for pipe installed underground. Ductile iron has all but replaced cast iron for this purpose because of its greater ability to withstand the stresses of handling and underground installation. It is used to carry liquids or gases under pressure and also to carry nonpressurized flows of liquid. Water, sewage, and natural gas are frequently transported through iron pipe.

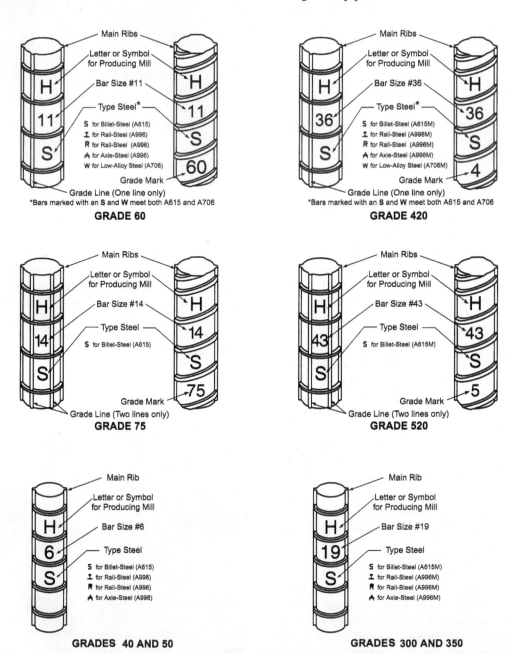

FIGURE 5–17. Identification marks*—ASTM standard bars. *(Courtesy Concrete Reinforcing Steel Institute)*

Table 5–7. Mechanical requirements for standard ASTM deformed reinforcing bars

ASTM SPECIFICATIONS—BAR SIZES, GRADES, AND TENSILE AND BENDING REQUIREMENTS

Type of Steel and ASTM Specification	Bar Sizes	Grade	Minimum Yield Strength, psi [MPa]	Minimum Tensile Strength, psi [MPa]	Minimum Percentage Elongation in 8 in. [203.2 mm]	Bend Test Pin Diameter (d = nominal diameter of specimen)
Billet-Steel A615/A615M	#3 to #6 [#10 to #19]	40 [300]	40,000 [300]	70,000 [500]	#3 [#10] . . . . 11 #4, #5, #6 [#13, #16, #19] 12	#3, #4, #5 [#10, #13, #16] . . . . 3½d #6 [#19] 5d
	#3 to #18 [#10 to #57]	60 [420]	60,000 [420]	90,000 [620]	#3, #4, #5, #6 [#10, #13, #16, #19] . . . . 9 #7, #8 [#22, #25] . . . . 8 #9, #10, #11, #14, #18 [#29, #32, #36, #43, #57] 7	#3, #4, #5 [#10, #13, #16] . . . . 3½d #6, #7, #8 [#19, #22, #25] . . . . . . 5d #9, #10, #11 [#29, #32, #36] . . . . . . 7d #14, #18 (90°) [#43, #57 (90°)] 9d
	#6 to #18 [#19 to #57]	75 [520]	75,000 [520]	100,000 [690]	#6, #7, #8 [#19, #22, #25] . . . . 7 #9, #10, #11, #14, #18 [#29, #32, #36, #43, #57] 6	#6, #7, #8 [#19, #22, #25] . . . . . . 5d #9, #10, #11 [#29, #32, #36] . . . . . . 7d #14, #18 (90°) [#43, #57 (90°)] 9d
Low-Alloy Steel A706/A706M	#3 to #18 [#10 to #57]	60 [420]	60,000 [420]	80,000 [550]	#3, #4, #5, #6 [#10, #13, #16, #19] . . . . 14 #7, #8, #9, #10, #11 [#22, #25, #29, #32, #36] . . . . . . 12 #14, #18 [#43, #57] 10	#3, #4, #5 [#10, #13, #16] . . . . 3d #6, #7, #8 [#19, #22, #25] . . . . . . 4d #9, #10, #11 [#29, #32, #36] . . . . . . 6d #14, #18 [#43, #57] 8d

For low-alloy steel reinforcing bars, the ASTM A706/A706M specification prescribes a maximum yield strength of 78,000 psi [540 MPa] and tensile strength must be 1.25 times the actual yield strength.

Bend tests are 180° except ASTM A615/A615M permits 90° for bar sizes #14 and #18 [#43 and #57].

(Courtesy Concrete Reinforcing Steel Institute)

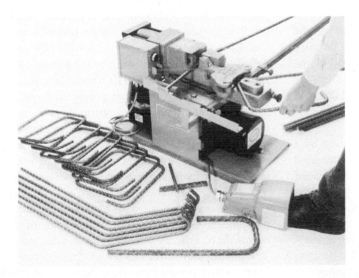

FIGURE 5–18. Steel reinforcing bars can be cut and bent on portable machines or shop fabricated. (*Courtesy Fascut Industries, Inc.*)

Pipe is cast in a cylindrical, water-cooled steel mold, with or without resin lining, and with no core. The mold is nearly horizontal but is inclined slightly so that liquid metal flows toward one end. At the other end, molten metal is poured into a trough that is suspended inside the mold and extends to the opposite, lower end. As the molten iron flows out the lower end of the trough into the mold, the mold spins rapidly, forming a uniform layer on its inside surface. The mold is withdrawn horizontally from the trough at a rate matched to the rate of flow so that the desired pipe wall thickness is obtained.

Unequal settling and shifting of the earth may subject buried pipes to tensile stresses under beam or arch loading conditions. Handling pipe between the manufacturing plant and the jobsite often causes impact or shock loads due to bumping and jarring of the pipe. The ability of ductile iron to withstand these stresses is very high compared to its cost.

Corrosion resistance is important for any material to be buried underground. Ductile iron and gray cast iron are much more resistant to corrosion than steel, though not as resistant as clay and plastics that are also used for pipes. The two are about equal in corrosion resistance, with ductile iron possibly a little more resistant. It is, nevertheless, standard manufacturing procedure to coat the outside of both types of pipe with a 1-mil (one-thousandth of an inch)-thick coat of bituminous material.

A loose tube of polyethylene has proved to be an effective deterrent to all types of corrosion in gray cast iron or ductile iron pipe. The tube is inserted over the pipe at the time of installation. The covering need not be watertight, although contact between soil and pipe must be prevented. Groundwater that seeps between the polyethylene sheet and the pipe surface does not cause a measurable amount of corrosion. Sheets of polyethylene may be wrapped around the pipe and secured with tape. However, use of the tubes is usually more economical. Standards for protection with loose polyethylene encasement are contained in American Water Works Association Standard C105.

Steel pipe is also used as an underground conduit for liquids and gases under pressure. Narrow, rolled sheets of steel are wound spirally and are butt welded at the spiral joint to form a tube. Steel pipe is also fabricated with one straight, longitudinal joint butt welded. Steel has a greater tensile strength, toughness, and ductility than ductile iron, but is not as resistant to corrosion as gray cast iron or ductile iron. However, all these characteristics in a pipeline depend on wall thickness, bedding and backfill conditions, and corrosion protection. All must be properly designed to suit the installation conditions. That is, the pipe wall must be thick enough not to be overstressed by the external load or the internal pressure of the liquid or gas; the soil must be arranged under, around, and over the pipe so that it does not concentrate forces on the pipe; and the material must be protected from external and internal corrosion.

Steel pipe is protected from external corrosion by a coating of coal tar enamel. The coal tar enamel is protected from scratches or abrasion by tough paper or fabric wrapped tightly around the enamel.

Coal tar is also used as a lining for iron and steel pipe. However, cement mortar made of portland cement and sand is much more popular for coating pipe interiors. A lining is necessary to protect the ferrous metal from being corroded by some of the many substances that are transported through pipes.

Many natural waters contain dissolved iron on which iron bacteria feed and multiply, forming solid deposits on iron or steel pipe walls. These deposits, called *tubercles,* look somewhat like rust and grow nearly large enough to fill the pipe so that flow is inadequate. *Tuberculation* does not take place if the metal is protected from contact with the water. Most water and sewer pipe is cement-mortar lined to prevent tuberculation and corrosion. A tube full of mortar is inserted into a rapidly spinning horizontal length of pipe, and the mortar is dumped uniformly along the pipe length. The pipe continues to spin, compacting the mortar to a uniform thickness by centrifugal force. The finished thickness is 3/16 in. or thinner, depending on the pipe size.

The mortar may instead be discharged from the end of the tube and spread with a trowel attached to the tube as the pipe spins; the tube is withdrawn at a rate matched to the rate of discharge of mortar so that the desired thickness of mortar is applied. Centrifugal force and vibration combine to form a dense lining of uniform thickness with the entrapped air and excess water driven out.

A wide variety of materials can be used to line iron and steel pipe when special protection is needed; polypropylene, polyethylene, and glass linings are among those available.

Steel, usually a copper or other rust-resisting alloy, is also used for corrugated pipe up to 10 ft in diameter for storm drains, sanitary sewers, and other nonpressurized flows. The pipe is made of thin metal sheets (0.05 to 0.17 in. thick) which have been corrugated between rollers. They are made either of rectangles curved to circular shape and riveted together or of long, narrow strips wound spirally and joined with one long, spiral lock seam.

The strength of corrugated steel pipe comes partly from the deep section given to it by the corrugations. The corrugations cause the pipe to be much stronger than a noncorrugated pipe of the same wall thickness. However, most of its strength is due to the fact that it is flexible. The thin metal deflects extensively under load before it breaks. To achieve its potential strength, it must be confined all around by soil. A heavy weight on flexible, corrugated pipe flattens the top and bottom and bulges the sides outward unless they are restrained. A properly constructed pipeline is surrounded by solidly tamped soil so that the sides cannot bulge and, therefore, the top cannot deflect. Nonflexible pipe cracks before it deflects noticeably, and therefore soil pressure from the sides causes very little increase in its resistance to loading.

Corrugated steel pipe is protected from corrosion by galvanizing as a standard procedure for ordinary usage. It is coated inside and out by dipping in hot asphalt in addition to the galvanizing for protection from corrosive soil or corrosive water. Asbestos fibers may be used to join the zinc and asphalt, thereby improving the bond between them.

The interior may be paved with asphalt thick enough to cover the corrugations with a smooth surface to decrease resistance to flow and to protect the inside of the pipe from corrosion or erosion by solid particles carried in the liquid. The paving may cover the entire circumference or just the lower part, depending on where flow improvement and protection are needed.

Elliptical cross sections and cross sections approximating a circle with the bottom flattened are popular for special uses and are easily fabricated of corrugated steel. Very large sizes of all the cross sections are shipped as curved plates completely fabricated for erection by bolting the plates together at the jobsite.

Review Questions

1. What two treatments determine the internal structure of ferrous metals?

2. Describe the raw materials and the product of a blast furnace.

3. What conditions of use would require a chilled iron casting?

4. What is the difference between steel and iron, chemically and in physical properties?

5. How is steel improved by cold working?

6. What are the advantages of extrusion compared to rolling?

7. How does forging improve steel compared to casting?

8. What improvements can be made in steel by heat treatment?

9. How does steel fail in a fire?

10. In what environments does weathering steel need protection from corrosion?

11. What is the chief characteristic of stainless steel?

12. Sketch the three most common types of weld.

13. Describe the two methods of tightening high-strength bolts.

14. Determine the number of pounds and kilograms of rebar required for a construction project with the following bar list:
    700 No. 6—20 ft long
    300 No. 5—8 ft long
    400 No. 4—4 ft long

15. What methods can be used to inhibit corrosion in buried steel pipe?

16. Calculate the stress in a No. 7 rebar subjected to a tensile load of 10,600 lb.

17. Calculate the stress in a No. 7 rebar subjected to a tensile load of 9700 lb in psi and MPa.

CHAPTER SIX

WOOD

Wood has universal appeal. It is generally pleasant in appearance, no matter how it is cut or finished. Certain types can be extremely beautiful if properly cut and finished.

Most wood has a pleasant odor for a long time after it is removed from the forest. Because of its poor heat-conducting properties, wood does not remove heat from the hand and so feels warm to the touch. Metals, plastics, stone, and concrete conduct heat rapidly from the hand whenever they are lower in temperature than the human body. Because normal room temperature is lower than body temperature, these materials feel unpleasantly cold to the touch when compared to wood.

Wood is an excellent material for doors, door and window frames, flooring, trim, and other items where appearance is of primary importance. Wood is used extensively in dwellings where feelings of comfort and well-being are especially desired. Some uses of wood are shown in Figure 6–1.

GROWTH OF TREES

Trees are either *deciduous*, having broad leaves and usually shedding them in the fall, or *coniferous*, having needles and cones containing seeds. In the wood industry the deciduous trees are called *hardwoods* and the conifers are called *softwoods*. Hardwoods are generally harder than softwoods, but there are many exceptions. For example, Douglas fir and southern pine are harder than some hardwoods, and basswood and poplar are softer than many softwoods.

The way a tree grows explains much about the characteristics of lumber (Figure 6–2). Wood consists of long, narrow, hollow cells called *fibers*. New fibers grow at the outside of the tree, increasing the diameter layer by layer. The thin, growing layer is called the *cambium layer*. The *sapwood* is within the cambium layer, and its fibers are active in the life processes of the tree, but they do not grow. The *heartwood* at the center of the tree consists of dead fibers.

Heartwood and sapwood do not differ in mechanical properties. However, heartwood is a darker shade because of the resins, gums, and minerals it contains. At the center is a thin vein of soft tissue called *pith* extending the length of the tree and having no strength. Outside the cambium layer is the *bark*, which protects the growing cells from insects but is useless as a structural material.

The fibers of all wood species consist of a structurally sound material called *cellulose* (approximately 70 percent of the volume) cemented with *lignin* (approximately 25 percent), plus miscellaneous substances that are of less importance. Paper is made of the cellulose from wood.

The forming of new fibers and their growth proceed rapidly in the spring and early summer, forming large, thin-walled fibers with a small percentage of wall material in their cross section. In the summer and autumn, growth is slower with a greater percentage of fiber wall material. In cold weather, there is no growth. Springwood and summerwood form alternating concentric bands around the tree. This circular pattern is easily seen when the tree is sawed transversely. Summerwood is heavier, darker, harder, and stronger than springwood, because of the thicker fiber walls.

The age of the tree can be determined by the number of rings. Something about the weather during the tree's lifetime can be told by the width of the individual rings of springwood and summerwood. They indicate whether or not conditions were favorable for growth. The strength of a piece of wood can be judged by the percentage of summerwood compared to springwood showing in cross section. The strength can also be judged by the unit weight or specific gravity. A higher percentage of summerwood results in greater strength and weight.

Some fibers grow transversely, forming radial lines from the center. These lines are called *rays* and are useful in the life processes of the tree. They do not affect strength, but they do have an influence on the appearance of finished lumber.

(a)

(b)

FIGURE 6–1. (a) Church interior with 82-ft glu-lam timber trusses. *(Courtesy American Institute of Timber Construction)* (b) Two-span pedestrian bridge with a total length of 240 ft, the deck width of 12 ft and the arch radius of 85 ft. *(Courtesy American Institute of Timber Construction)*

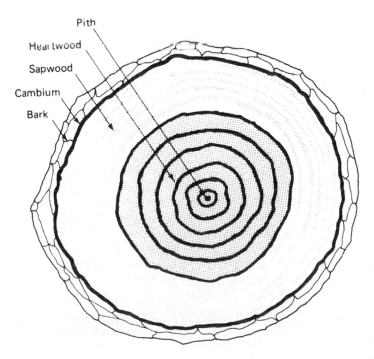

FIGURE 6–2. Cross section of a tree.

The hollow fiber structure of wood makes it an excellent insulator against the passage of heat and sound. Generally, wood is not used primarily as an insulator, but the insulation is an added benefit when wood is selected for the siding or roofing of a building. However, fiber boards intended primarily as insulation are manufactured of pressed wood chips and sawdust.

The solid matter in wood is heavier than water. Wood floats because it consists of hollow fibers containing air that cause its overall unit weight to be less than that of water. Wood with the fibers saturated with water is heavier than water and will sink in it. A pack of drinking straws held together with rubber bands approximates the fibrous structure of wood.

The hollow fibers are easily crushed or pulled apart in a transverse direction. They have a great deal more strength longitudinally because fiber walls make up a greater percentage of the cross section in this direction than in a transverse direction. Figure 6–3 shows how there is more cell wall material to resist stresses along the axis of a tree or the axis of a piece of wood sawed from a tree. Notice how lateral forces can collapse the long, narrow fibers without crushing any fiber walls. This is not true in the longitudinal direction, where the resisting walls are too long and close together. Refer to Table 6–3 for a listing of the allowable stresses parallel and perpendicular to the grain of various species of wood.

If stress is applied in any direction between longitudinal and transverse, the resisting strength varies from maximum in the longitudinal direction to minimum in the transverse direction.

The coefficient of thermal expansion of wood varies from 0.000001 to 0.000003 per °F (0.0000005 to 0.000002 per °C) parallel to the grain, and from 0.000015 to 0.000035 per °F (0.000008 to 0.000019 per °C) perpendicular to the grain. It varies with the species and is seldom important because expansion and contraction due to moisture changes are greater unless the humidity is nearly constant and the temperature varies greatly.

The hollow cells of a tree contain water both within the hollow space and within the cell walls. The total weight of water often exceeds the weight of solid material. Moisture evaporates from the wood after the tree is dead until an equilibrium moisture is reached that depends on air temperature and humidity. The *free water* in the hollow space evaporates first with no change in the volume of the wood. When the free water is gone, *absorbed water* evaporates from the cell walls, causing shrinkage of the wood. This shrinkage is the cause of cracks in the wood called *checks* (see Figure 6–13).

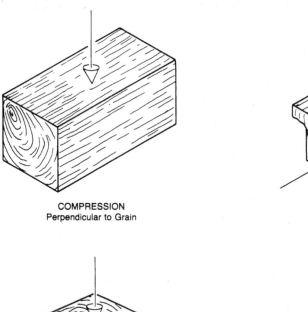

COMPRESSION
Perpendicular to Grain

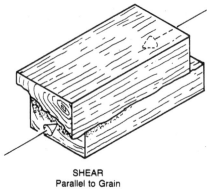

SHEAR
Parallel to Grain

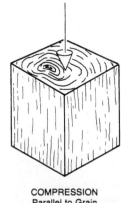

COMPRESSION
Parallel to Grain

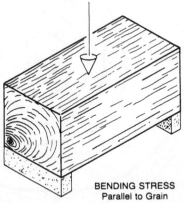

BENDING STRESS
Parallel to Grain

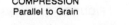

FIGURE 6–3. Wood stress. *(Courtesy Timothy Dennis)*

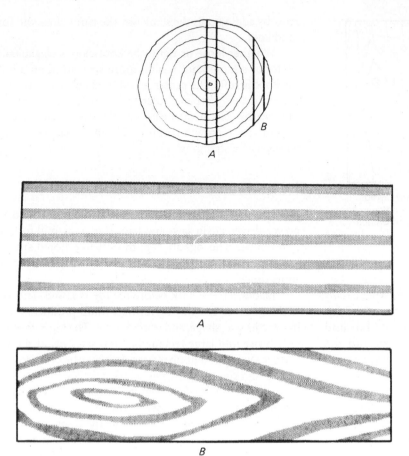

FIGURE 6–4. Pattern of grain depends on how wood is cut from log.

When wood is sawed parallel to the length of the tree, the light and dark annual rings appear as stripes called *grain*. The width of the grain is determined by the angle between the rings and the sawed surface. The stripes are the same width as the annual rings if the cut is perpendicular to the rings and become wider as the cut varies more from the perpendicular (Figure 6–4).

The grain (and therefore the fibers) may be inclined to the axis of the piece of wood because of a branch nearby or because of a bend in the tree. This deviation in direction of grain is called *slope of grain*. Figure 6–5 shows how a branch causes slope of grain. A bend in a tree trunk or a fork in a tree obviously causes the grain to slope in a piece of wood sawed straight. Because nearly all major

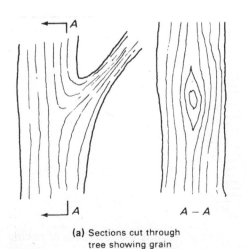

(a) Sections cut through tree showing grain

(b) Slope of grain— 1 in 6 and 1 in 2

FIGURE 6–5. Slope of grain caused by a branch.

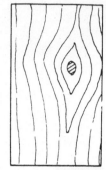

(a) Board sawed on axis of branch showing spike knot

(b) Board sawed perpendicular to axis of branch

FIGURE 6–6. Knots in lumber.

stresses on a piece of wood are parallel to the long axis, a slope in grain away from the axis weakens the wood for normal usage. The greater the angle between axis and grain, the weaker the wood. (See the discussion of the Hankinson formula and the Scholten nomographs later in the chapter.)

The outward growth of new cells partially encloses branches that have started to grow and forms a discontinuity in the annual ring pattern. The discontinuity appears as a knot in a piece of lumber (Figure 6–6). The branch may die and the branch stub become completely enclosed within the trunk, forming a loose knot. The branch may continue to grow by adding cells the same way the trunk does and form a solid knot.

Whether loose or tight, the knot causes weakness. The center of a loose knot has no more strength than a hole. If the knot is solid, the center is still weak, because its cells are approximately perpendicular to the long axis of the wood. The nearby cells are at various angles caused by the spreading of the new growth around the branch, and they therefore have varying strengths.

LUMBER PRODUCTION

Trees are felled and cut and trimmed into logs with portable, power-driven chain saws and hauled to a sawmill for sawing into lumber. The logs are kept moist while stored at the mill to prevent shrinkage cracks, unless they will be stored only a short time.

Blades remove bark before the log is sawed. The sawyer decides how to saw the log for maximum production on the basis of its size, shape, and irregularities. The log is then sawed lengthwise into large rectangular and semiround shapes for further sawing into lumber sizes, or it may be cut directly into lumber thickness by gang saws cutting the log into many slices at once. Saws called *edgers* trim the rough-edged slabs longitudinally to the desired lumber width and cut off the rounded edges. Saws called *trimmers* then saw the lumber transversely into desired lengths and trim away defective portions.

Figure 6–7 shows the two basic methods of sawing a log into lumber and also a combination method. The

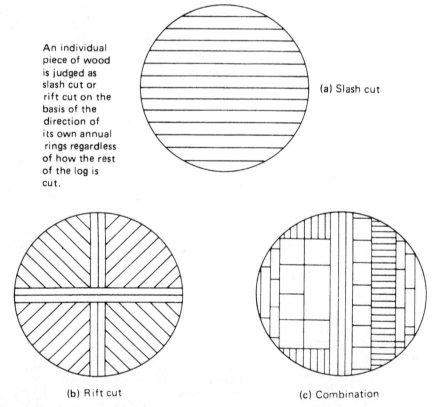

An individual piece of wood is judged as slash cut or rift cut on the basis of the direction of its own annual rings regardless of how the rest of the log is cut.

(a) Slash cut

(b) Rift cut

(c) Combination

FIGURE 6–7. Methods of sawing logs into boards.

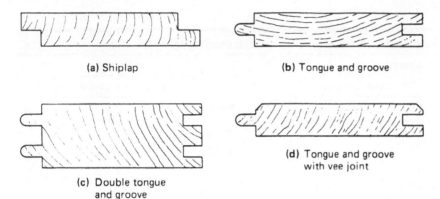

(a) Shiplap

(b) Tongue and groove

(c) Double tongue and groove

(d) Tongue and groove with vee joint

FIGURE 6–8. Typical cross sections of worked lumber.

slash-cut lumber is called *plain sawed* in the terminology of the hardwood industry and *flat grain* in softwood terminology. The rift-cut lumber is called *quarter sawed* when referring to hardwoods and *edge grain* when referring to softwoods.

Lumber is finished in one of several ways. *Rough lumber* remains as sawed on all four sides with no further finishing. *Dressed lumber* or *surfaced lumber* is planed or surfaced on at least one face. It is designated as S1S if surfaced on one side, S1E if surfaced on one edge, and S1S1E if surfaced on one side and one edge. The abbreviations S2S, S2E, S1S2E, S2S1E, and S4S are used for other combinations of sides and edges that are surfaced. *Worked lumber* is dressed and also worked to provide tongue-and-groove or shiplap joints and/or to change the cross section in some other way. Figure 6–8 shows typical cross sections of worked lumber.

Lumber to be finished may be planed while green, or it may be seasoned first and planed later. Seasoning is the process of reducing the moisture until a suitable moisture level is reached. The cross section of the wood becomes smaller during seasoning. If lumber is planed to the proper size after seasoning, it will remain the proper size. If it is planed first, it must be left larger than the proper size to allow for shrinkage to the proper size.

Lumber is sawed to nominal sizes (usually to the whole inch), but the width of the saw blade reduces the size somewhat. Planing reduces the size further to the correct net or finished size at which the lumber is sold. Nominal sizes and net sizes before and after seasoning are shown in Table 6–1.

Quantities of lumber are measured and sold by the *foot board measure* (*fbm*). One *board foot* is a quantity 1 sq ft by 1 in. thick. When fbm of finished lumber is calculated, nominal dimensions are used so width and thickness are always in full inches. Finished lumber less than an inch thick is considered 1 in. thick when computing board feet. Lumber length is taken to the next lower foot if its length is a number of whole feet plus a fraction. For example, a 12 ft 3 in. length of 2×4 lumber contains $12 \times 2 \times 4\frac{1}{4} = 8$ board ft. Multiply the length in feet by the thickness in inches by the width in feet or fraction of a foot.

Table 6–1. Standard sizes of yard lumber and timbers

	NOMINAL AND MINIMUM-DRESSED SIZES OF BOARDS, DIMENSION, AND TIMBERS						
	Thickness				Face Widths		
		Minimum Dressed				Minimum Dressed	
Item	Nominal	Dry** (in.)	Green** (in.)	Nominal	Dry* (in.)	Green* (in.)
Boards	1	$\frac{3}{4}$	$\frac{25}{32}$	2	$1\frac{1}{2}$	$1\frac{9}{16}$
	$1\frac{1}{4}$	1	$1\frac{1}{32}$	3	$2\frac{1}{2}$	$2\frac{9}{16}$
	$1\frac{1}{2}$	$1\frac{1}{4}$	$1\frac{9}{32}$	4	$3\frac{1}{2}$	$3\frac{9}{16}$
				5	$4\frac{1}{2}$	$4\frac{5}{8}$
				6	$5\frac{1}{2}$	$5\frac{5}{8}$
				7	$6\frac{1}{2}$	$6\frac{5}{8}$
				8	$7\frac{1}{4}$	$7\frac{1}{2}$
				9	$8\frac{1}{4}$	$8\frac{1}{2}$
				10	$9\frac{1}{4}$	$9\frac{1}{2}$
				11	$10\frac{1}{4}$	$10\frac{1}{2}$
				12	$11\frac{1}{4}$	$11\frac{1}{2}$

(continued)

Table 6–1. (continued)

NOMINAL AND MINIMUM-DRESSED SIZES OF BOARDS, DIMENSION, AND TIMBERS

Item	Thickness			Face Widths		
	Nominal	Minimum Dressed		Nominal	Minimum Dressed	
		Dry** (in.)	Green** (in.)		Dry* (in.)	Green* (in.)
				14	$13\frac{1}{4}$	$13\frac{1}{2}$
				16	$15\frac{1}{4}$	$15\frac{1}{2}$
Dimension	2	$1\frac{1}{2}$	$1\frac{9}{16}$	2	$1\frac{1}{2}$	$1\frac{9}{16}$
	$2\frac{1}{2}$	2	$2\frac{1}{16}$	3	$2\frac{1}{2}$	$2\frac{9}{16}$
	3	$2\frac{1}{2}$	$2\frac{9}{16}$	4	$3\frac{1}{2}$	$3\frac{9}{16}$
	$3\frac{1}{2}$	3	$3\frac{1}{16}$	5	$4\frac{1}{2}$	$4\frac{5}{8}$
				6	$5\frac{1}{2}$	$5\frac{5}{8}$
				8	$7\frac{1}{4}$	$7\frac{1}{2}$
				10	$9\frac{1}{4}$	$9\frac{1}{2}$
				12	$11\frac{1}{4}$	$11\frac{1}{2}$
				14	$13\frac{1}{4}$	$13\frac{1}{2}$
				16	$15\frac{1}{4}$	$15\frac{1}{2}$
Dimension	4	$3\frac{1}{2}$	$3\frac{9}{16}$	2	$1\frac{1}{2}$	$1\frac{9}{16}$
	$4\frac{1}{2}$	4	$4\frac{1}{16}$	3	$2\frac{1}{2}$	$2\frac{9}{16}$
				4	$3\frac{1}{2}$	$3\frac{9}{16}$
				5	$4\frac{1}{2}$	$4\frac{5}{8}$
				6	$5\frac{1}{2}$	$5\frac{5}{8}$
				8	$7\frac{1}{4}$	$7\frac{1}{2}$
				10	91	$9\frac{1}{2}$
				12	$11\frac{1}{4}$	$11\frac{1}{2}$
				14	—	$13\frac{1}{2}$
				16	—	$15\frac{1}{2}$
Timbers	5 and thicker	—	$\frac{1}{2}$ off	5 and wider	—	$\frac{1}{2}$ off

*The thicknesses apply to all widths and all widths to all thicknesses.

**Dry lumber is defined as lumber which has been seasoned to a moisture content of 19 percent or less.

Green lumber is defined as lumber having a moisture content in excess of 19 percent.

Source: National Forest Products Association.

Example

Calculate the number of board feet in 36 2 × 4s which are 8 ft long.

$$\text{Board feet} = \frac{\text{number of pieces} \times \text{thickness in inches} \times \text{width in inches} \times \text{length in feet}}{12}$$

$$\text{Board feet} = \frac{36 \times 2 \times 4 \times 8}{12} = 192 \text{bf.}$$

SEASONING

Although the moisture of green wood is very high, it will eventually reach equilibrium with the surrounding atmosphere by losing moisture to the atmosphere (refer to Table 6–2, Figures 6–9 and 6–10).

Moisture in wood is expressed as *moisture content (m.c.)*, which is the weight of water in the wood expressed as a percentage of the weight of the oven-dry wood. Moisture content can be determined by weighing a piece of wood including the water, and then driving all water out by drying in an oven and weighing the same piece of wood in the oven-dry condition. The difference between weight with water and weight without water is the weight of the water. Dividing this weight by the oven-dry weight gives a ratio of weight of water to oven-dry weight of wood. This ratio, expressed as a percentage, is the moisture content.

Reaching equilibrium takes months or years for green wood. Wood may be seasoned in the air, or much time may be saved by seasoning it in a kiln. Wood being dried in a kiln is subjected to warm, moist air or steam and loses sufficient moisture in a period of several days to several months. This is far less than the 4 years sometimes needed for air-drying some hardwoods.

Seasoning can take place through the use of hygroscopic chemicals applied to the wood. This method reduces shrinkage cracks (checks) because the chemical keeps the surface of the wood moist and partially replaces moisture lost from the cell walls with the chemical. After the chemical is applied, the wood is dried in a solution of the chemical, in a kiln, or in the air.

Table 6–2. Average moisture content of green heartwood and sapwood of 20 species of American trees

Species	Average Moisture Content (Percent of Dry Weight)	
	Heartwood	Sapwood
Hardwoods		
Ash, white	38	40
Beech	53	78
Birch, yellow	68	71
Elm, American	95	92
Gum, black	50	61
Maple, silver	60	88
Maple, sugar	58	67
Softwoods		
Douglas fir	36	117
Fir, lowland white	91	136
Hemlock, eastern	58	119
Hemlock, western	42	170
Pine, loblolly	34	94
Pine, lodgepole	36	113
Pine, longleaf	34	99
Pine, Norway	31	135
Pine, ponderosa	40	148
Pine, shortleaf	34	108
Redwood	100	210
Spruce, Engelmann	54	167
Spruce, Sitka	33	146

Source: U.S. Forest Products Laboratory.

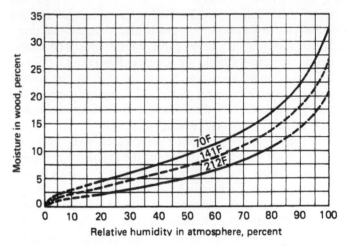

FIGURE 6–10. Relation of the equilibrium moisture content of wood to the relative humidity of the surrounding atmosphere at three temperatures. (*Courtesy U.S. Forest Products Laboratory*)

be used and on whether it is to be used indoors or outdoors (Figure 6–9). The m.c. of wood does not exceed 19 percent in normal outdoor usage.

In use, wood never ceases to change its moisture content with the seasons or other influencing factors. It takes moisture from the atmosphere or releases its moisture to the atmosphere continuously in response to changing atmospheric conditions. However, it never again approaches the high moisture content it had while living. Figure 6–10 shows how wood m.c. varies with temperature and humidity of the surrounding air.

The strength of a given piece of wood increases as the moisture content decreases. This fact is acknowledged by allowing higher stresses in seasoned wood (see Table 6–5). This is sufficient reason for drying wood before using it, and also sufficient reason for specifying maximum allowable moisture contents when ordering wood to be used where strength is important.

Another reason for drying wood before use is that it shrinks in size when dried, but it does not shrink uniformly. Longitudinal size change is negligible. Transverse shrinkage is greatest in the curved direction of the annual rings and

Seasoning is continued until a moisture content is reached that approximates the average equilibrium m.c. for the conditions in which the lumber will be used. This varies, depending on the geographical area in which the lumber will

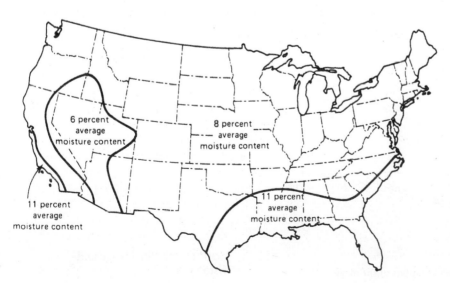

FIGURE 6–9. Recommended moisture content averages for interior finish woodwork in various parts of the United States. (*Courtesy U.S. Forest Products Laboratory*)

varies to a minimum of one-half to one-third this amount radially or in a direction perpendicular to the annual rings. For this reason, a log develops large radial cracks (*checks*) if dried too rapidly. Checks are shown in Figure 6–11. Because the direction of maximum shrinkage is along the annual growth rings, shrinkage is greater as the circumference of the ring gets larger. Thus, the cracks are open wider at the outside, becoming narrower toward the center.

The splitting or checking is increased by the fact that drying takes place at the outside of the log before the inside dries. The outer portion is restrained from shrinking inward by the unyielding center and the cracks are larger as a result. It is therefore advantageous to saw wood into smaller sizes before seasoning to minimize checking. After seasoning, when there will be very little additional change in size, the wood is finished to its final cross section size.

Differential shrinkage results in distortion as the wood dries. In many cases the distortion is insignificant, but it is sometimes sufficient to cause construction difficulties. If the wood is too distorted, it cannot be used for construction. However, if green wood is fastened in place, it pulls fastenings

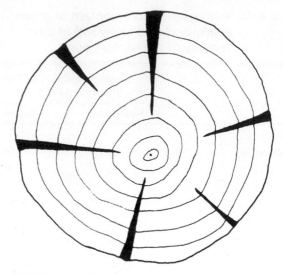

FIGURE 6–11. Checks in a log.

loose and distorts the structure as it dries. The various distortions caused by differential shrinkage are shown in Figure 6–12.

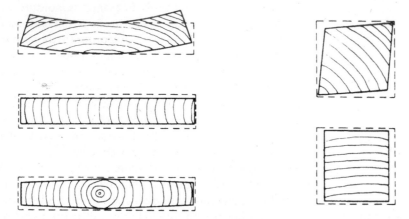

(a) Typical unequal shrinkage determined by direction of annual rings

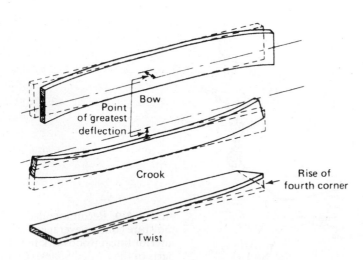

(b) Warping caused by various combinations of shrinkage

FIGURE 6–12. Shrinkage and distortion.

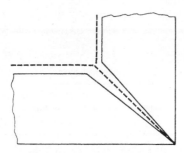

FIGURE 6–13. Open joint due to shrinkage.

Seasoned wood also changes shape with changes in moisture content. Doors that swell and stick in damp weather and shrink, leaving open cracks, when the weather is dry are common evidence of volume change with change in moisture content. The shrinkage is most noticeable indoors when wintertime heating causes the air to be very dry.

Joints in wooden door and window frames open during a period when the air is dry and close again when moisture is normal. Figure 6–13 shows how a joint opens. The shrinkage is a certain percentage of the original width for a given moisture loss, and therefore the total shrinkage is greater where the wood is wider. Longitudinal shrinkage is negligible, so length of wood has no noticeable influence. This explains why the joint opens as shown in the figure.

If the wood is cut so that the wide dimension is radial to the annual rings (rift cut), there is less change in width than if the wood is cut tangential to the annual rings (slash cut) (see Figure 6–12 for illustration).

If the wood gains moisture greatly in excess of the amount it contained when it was installed, the resulting expansion can cause damage. For example, if the wood at the joint shown in Figure 6–13 expanded instead of contracting, the corner of the frame would be pushed apart, cracking the wood or loosening the nails. Shrinkage may be unsightly, but damage from it is less likely. For this reason, the m.c. at the time of construction should approximate the upper level the wood is expected to have during ordinary usage.

STRENGTH

The important stresses and corresponding strengths of wood are divided into the following six types:

modulus of elasticity (E): This is a measure of the stiffness or resistance to deflection. It is not usually considered a measure of strength, but is a measure of the ability to resist failure due to excessive deformation. It is used to predict movement under load and avoid failure due to excessive movement.

extreme fiber stress in bending (F_b): This is the stress (compression at the top, tension at the bottom) that must be resisted in a beam undergoing bending.

tension parallel to grain (F_t): This is the stress induced by pulling apart in a longitudinal direction. Resistance to tension perpendicular to grain is so weak that it is usually considered negligible.

compression parallel to grain (F_c): This is the stress induced by pressing together longitudinally.

compression perpendicular to grain ($F_{c\perp}$): This is the stress induced by pressing together in a transverse direction. There is no appreciable difference in strength to resist this compression perpendicular to the annual rings or parallel to them.

horizontal shear (F_v): This is the stress induced by the tendency for upper fibers to slide over lower fibers as a beam bends.

The first step in obtaining useful allowable stresses and an accurate modulus of elasticity is to test small, perfect samples of wood to determine the stresses at failure and the modulus of elasticity for each species. This testing is done in accordance with ASTM D2555, Methods for Establishing Clear Wood Strength Values. The samples tested have no defects to reduce their strength or stiffness.

Allowable stresses for lumber are determined by reducing the stresses in the samples at failure to provide a safety factor of approximately 2.5. By this means, basic allowable stresses are established for each species of tree. These, along with moduli of elasticity, are published in ASTM D245, Methods for Establishing Structural Grades for Visually Graded Lumber, as Basic Stresses for Clear Lumber Under Long-Time Service at Full Design Load.

Because individual pieces of lumber contain defects that reduce strength and stiffness, studies and tests are conducted to determine the reduction in strength and stiffness caused by the various kinds of defects. The allowable stresses and correct modulus of elasticity are determined for wood with defects by reducing the values for clear wood using the procedures of ASTM D245.

These procedures require determination of the location, magnitude, and condition of defects that reduce strength and stiffness in order to determine the total amount of that reduction. Each defect reduces the strength somewhat, and the total effect of these reductions is expressed as a ratio that represents the strength of the piece of lumber being graded compared to the strength of clear wood. This ratio is used to establish usable allowable stresses by multiplying the allowable stresses for clear wood and the modulus of elasticity for clear wood by the ratio.

The National Design Specification for Stress-Grade Lumber and Its Fastenings includes grades established for each species of wood with usable allowable stresses for each grade. A sample of allowable stresses from the National Design Specification (NDS) is shown in Table 6–3. These

Table 6–3. Allowable unit stresses—visual grading

Species or Group	Grade	Extreme Fiber Stress in Bending "F_b" Single	Tension Parallel to Grain "F_t"	Horizontal Sheer "F_v"	Compression Perpendicular "$F_{c\perp}$"	Parallel to Grain "$F_{c\,/\!/}$"	Modulus of Elasticity "E"
Douglas Fir-Larch	Select Structural	1450	1000	95	625	1700	1,900,000
	No. 1 & Btr.	1150	775	95	625	1500	1,800,000
	No. 1	1000	675	95	625	1450	1,700,000
Douglas Fir	No. 2	875	575	95	625	1300	1,600,000
Western Larch	No. 3	500	325	95	625	750	1,400,000
	Construction	1000	650	95	625	1600	1,500,000
	Standard	550	375	95	625	1350	1,400,000
	Utility	275	175	95	625	875	1,300,000
	Stud	675	450	95	625	825	1,400,000
Douglas Fir-South	Select Structural	1300	875	90	520	1550	1,400,000
	No. 1	900	600	90	520	1400	1,300,000
	No. 2	825	525	90	520	1300	1,200,000
Douglas Fir South	No. 3	475	300	90	520	750	1,100,000
	Construction	925	600	90	520	1550	1,200,000
	Standard	525	350	90	520	1300	1,100,000
	Utility	250	150	90	520	875	1,000,000
	Stud	650	425	90	520	825	1,100,000
Hem-Fir	Select Structural	1400	900	75	405	1500	1,600,000
	No. 1 & Btr.	1050	700	75	405	1350	1,500,000
Western Hemlock	No. 1	950	600	75	405	1300	1,500,000
Noble Fir	No. 2	850	500	75	405	1250	1,300,000
California Red Fir	No. 3	500	300	75	405	725	1,200,000
Grand Fir	Construction	975	575	75	405	1500	1,300,000
Pacific Silver Fir	Standard	550	325	75	405	1300	1,200,000
White Fir	Utility	250	150	75	405	850	1,100,000
	Stud	675	400	75	405	800	1,200,000
Spruce-Pine-Fir (South)	Select Structural	1300	575	70	335	1200	1,300,000
	No. 1	850	400	70	335	1050	1,200,000
Western Species	No. 2	750	325	70	335	975	1,100,000
Engelmann Spruce	No. 3	425	200	70	335	550	1,000,000
Sitka Spruce	Construction	850	375	70	335	1200	1,000,000
Lodgepole Pine	Standard	475	225	70	335	1000	900,000
	Utility	225	100	70	335	650	900,000
	Stud	575	250	70	335	600	1,000,000
Western Cedars	Select Structural	1000	600	75	425	1000	1,100,000
	No. 1	725	425	75	425	825	1,000,000
Western Red Cedar	No. 2	700	425	75	425	650	1,000,000
Incense Cedar	No. 3	400	250	75	425	375	900,000
Port Orford Cedar	Construction	800	475	75	425	850	900,000
Alaska Cedar	Standard	450	275	75	425	650	800,000

(continued)

Table 6–3. (continued)

Species or Group	Grade	Extreme Fiber Stress in Bending "F_b" Single	Tension Parallel to Grain "F_t"	Horizontal Sheer "F_v"	Compression Perpendicular "$F_{c\perp}$"	Parallel to Grain "$F_{c//}$"	Modulus of Elasticity "E"
	Utility	225	125	75	425	425	800,000
	Stud	550	325	75	425	400	900,000
Western Woods	Select Structural	875	400	70	335	1050	1,200,000
	No. 1	650	300	70	335	925	1,100,000
Any of the species in the first four species groups above plus any or all of the following:	No. 2	650	275	70	335	875	1,000,000
	No. 3	375	175	70	335	500	900,000
	Construction	725	325	70	335	1050	1,000,000
	Standard	400	175	70	335	900	900,000
Idaho White Pine	Utility	200	75	70	335	600	800,000
Ponderosa Pine	Stud	500	225	70	335	550	900,000
Sugar Pine							
Alpine Fir							
Mountain Hemlock							

Note: These represent base values for western dimension lumber (2 to 4 in. thick by 2 in. and wider). They should be used in conjunction with appropriate adjustments given in Table 6–4. The design values are given in pounds per square inch.

Source: National Forest Products Association.

stresses are derived by multiplying the basic allowable clear wood stresses by adjustments (Table 6–4).

Each manufacturers' association then establishes grading rules to categorize its lumber products according to the various grades. The grading rules take into account the strength and modulus of elasticity reductions due to defects and describe the location, magnitude, and condition of defects permitted for each grade. Each piece of lumber is assigned to the correct grade according to the association's rules and may be used with the specified allowable stresses for that grade.

Lumber may also be stress-graded by machine with a supplementary visual inspection (see Table 6–5). Lumber so graded is called *machine stress-rated* (MSR). Each piece is subjected to beam loading, and the modulus of elasticity is determined from the resulting deflection without damaging the piece. The piece is then examined for characteristics and manufacturing imperfections that fall below standards. If it is acceptable, it is stamped with the value of modulus of elasticity and allowable fiber stress in bending.

The fiber stress and the tension and compression parallel to grain are inferred from the modulus of elasticity and

verified by visual inspection. The E values recognized by the Western Wood Products Association are shown in Table 6–3. Note that the allowable compression perpendicular to grain and horizontal shear are constant for each species, regardless of the modulus of elasticity.

Allowable compression is much greater parallel to grain than perpendicular to it (see Table 6–3) and is intermediate in value for any other direction. The Hankinson formula is used to compute the allowable compression or bearing at other angles:

$$F_T = \frac{F_c \times F_c'}{F_c \sin^2 \theta + F_c' \cos^2 \theta}$$

where F_c = allowable stress in compression parallel to grain

F'_c = allowable stress in compression perpendicular to grain

T = angle between direction of grain and direction of load

F_T = allowable compressive stress at angle θ with the direction of grain

Table 6–4. Adjustment factors for base values

BASE VALUE EQUATIONS

The basic difference between using BASE VALUES and the design values that were published for dimension lumber prior to the results of the In-Grade Testing Program, is that BASE VALUES must be adjusted for SIZE before conditions of use. The table below shows how the adjustments are applied to BASE VALUES.

BASE VALUE EQUATIONS

Apply to Dimension Lumber Values in Table 1

Base Value	x	Size Adjustment Factor	x	Routine Adjustment Factors	x	Special Use Factors	=	Design Value
F_b	x	C_F	x	C_D x C_r x	C_M x C_R x C_t x C_{fu}		= F'_b
F_t	x	C_F	x	C_D	C_M x C_R x C_t		= F'_t
F_v			x	C_D x C_H x	C_M x C_R x C_t		= F'_v
$F_{c\perp}$*					C_M x C_R x C_t		= $F'_{c\perp}$
$F_{c/}$	x	C_F	x	C_D x	C_M x C_R x C_t		= $F'_{c/}$
E					C_M x C_R x C_t		= E'

* For $F_{c\perp}$ value of 0.02" deformation basis, see Table F.

Note:
- C_F = Size Factor
- C_r = Repetitive Member Factor
- C_H = Horizontal Shear
- C_D = Duration of Load
- C_{fu} = Flat Use Factor
- C_M = Wet Use Factor
- C_R = Fire Retardant Factor, refer to the National Design Specification
- C_t = Temperature Factor, refer to the National Design Specification

SIZE FACTORS (C_F)   Table A

Apply to Dimension Lumber Base Values

Grades	Nominal Width (depth)	F_b 2" & 3" thick nominal	F_b 4" thick nominal	F_t	$F_{c/}$	Other Properties
Select Structural, No. 1 & Btr., No. 1, No. 2 & No. 3	2", 3" & 4"	1.5	1.5	1.5	1.15	1.0
	5"	1.4	1.4	1.4	1.1	1.0
	6"	1.3	1.3	1.3	1.1	1.0
	8"	1.2	1.3	1.2	1.05	1.0
	10"	1.1	1.2	1.1	1.0	1.0
	12"	1.0	1.1	1.0	1.0	1.0
	14" & wider	0.9	1.0	0.9	0.9	1.0
Construction & Standard	2", 3" & 4"	1.0	1.0	1.0	1.0	1.0
Utility	2" & 3"	0.4	—	0.4	0.6	1.0
	4"	1.0	1.0	1.0	1.0	1.0
Stud	2", 3" & 4"	1.1	1.1	1.1	1.05	1.0
	5" & 6"	1.0	1.0	1.0	1.0	1.0

REPETITIVE MEMBER FACTOR (C_r)   Table B

Apply to Size-adjusted F_b

Where 2" to 4" thick lumber is used repetitively, such as for joists, studs, rafters and decking, the pieces side by side share the load and the strength of the entire assembly is enhanced. Therefore, where three or more members are adjacent or are not more than 24" apart and are joined by floor, roof or other load distributing elements, the F_b value can be increased 1.15 for repetitive member use.

REPETITIVE MEMBER USE
F_b x 1.15

DURATION OF LOAD ADJUSTMENT (C_D)   Table C

Apply to Size-adjusted Values

Wood has the property of carrying substantially greater maximum loads for short durations than for long durations of loading. Tabulated design values apply to normal load duration. (Factors do not apply to MOE or $F_{c\perp}$).

LOAD DURATION	FACTOR
Permanent	0.9
Ten Years (Normal Load)	1.0
Two Months (Snow Load)	1.15
Seven Day	1.25
One Day	1.33
Ten Minutes (Wind and Earthquake Loads)	1.6
Impact	2.0

Confirm load requirements with local codes. Refer to Model Building Codes or the National Design Specification for high-temperature or fire-retardant treated adjustment factors.

HORIZONTAL SHEAR ADJUSTMENT (C_H)   Table D

Apply to F_v Values

Horizontal shear values published in Tables 1-6 are based upon the maximum degree of shake, check or split that might develop in a piece. When the actual size of these characteristics is known, the following adjustments may be taken.

2" THICK LUMBER	3" AND THICKER LUMBER
For convenience, the table below may be used to determine horizontal shear values for any grade of 2" thick lumber in any species when the length of split or check is known and any increase in them is not anticipated.	Horizontal shear values for 3" and thicker lumber also are established as if a piece were split full length. When specific lengths of splits are known and any increase in them is not anticipated, the following adjustments may be applied.

When length of split on wide face is:	Multiply Tabulated Fv value by:	When length of split on wide face is:	Multiply Tabulated Fv value by:
No split	2.00	No split	2.00
1/2 of wide face	1.67	1/2 of narrow face	1.67
3/4 of wide face	1.50	1 of narrow face	1.33
1 of wide face	1.33	1½ of narrow or more	1.00
1½ of wide face or more	1.00		

Table 6–4. (continued)

FLAT USE FACTORS (C_{fu}) — Table E
Apply to Size-adjusted F_b

NOMINAL WIDTH	NOMINAL THICKNESS 2" & 3"	4"
2" & 3"	1.00	—
4"	1.10	1.00
5"	1.10	1.05
6"	1.15	1.05
8"	1.15	1.05
10" & wider	1.20	1.10

ADJUSTMENTS FOR COMPRESSION PERPENDICULAR TO GRAIN ($C_{c\perp}$) — Table F

For Deformation Basis of 0.02"
Apply to $F_{c\perp}$ Values

Design values for compression perpendicular to grain ($Fc\perp$) are established in accordance with the procedures set forth in ASTM Standards D 2555 and D 245. ASTM procedures consider deformation under bearing loads as a serviceability limit state comparable to bending deflection because bearing loads rarely cause structural failures. Therefore, ASTM procedures for determining compression perpendicular to grain values are based on a deformation of 0.04" and are considered adequate for most classes of structures. Where more stringent measures need to be taken in design, the following formula permits the designer to adjust design values to a more conservative deformation basis of 0.02".

$$Y_{02} = 0.73 \, Y_{04} + 5.60$$

EXAMPLE: Douglas Fir-Larch: Y_{04} = 625 psi
Y_{02} = 0.73 (625) + 5.60 = 462 psi

WET USE FACTORS (C_M) — Table G

Apply to Size-adjusted Values

The design values shown in the accompanying tables are for routine construction applications where the moisture content of the wood does not exceed 19%. When use conditions are such that the moisture content of dimension lumber will exceed 19%, the Wet Use Adjustment Factors below are recommended.

	PROPERTY	ADJUSTMENT FACTOR
F_b	Extreme Fiber Stress in Bending	0.85*
F_t	Tension Parallel to Grain	1.0
F_c	Compression Parallel to Grain	0.8**
F_v	Horizontal Shear	0.97
$F_{c\perp}$	Compression Perpendicular to Grain	0.67
E	Modulus of Elasticity	0.9

*Fiber Stress in Bending Wet Use Factor 1.0 for size adjusted F_b not exceeding 1150 psi
** Compression Parallel to Grain in Wet Use Factor 1.0 for size adjusted F_c not exceeding 750 psi

Source: Western Wood Products Association.

Table 6–5. Allowable unit stresses—machine stress-rated lumber

MSR LUMBER DESIGN VALUES* — Table 3
2" and less in thickness, 2" and wider
Grades described in Section 52.00 of *Lumber Grading Rules.*

Grade Designation[1]	Extreme Fiber Stress in Bending "F_b"[2] Single	Modulus of Elasticity "E"	Tension Parallel to Grain "F_t"	Compression Parallel to Grain "$F_c //$"
2850 Fb-2.3E	2850	2,300,000	2300	2150
2700 Fb-2.2E	2700	2,200,000	2150	2100
2550 Fb-2.1E	2550	2,100,000	2050	2025
2400 Fb-2.0E	2400	2,000,000	1925	1975
2250 Fb-1.9E	2250	1,900,000	1750	1925
2100 Fb-1.8E	2100	1,800,000	1575	1875
1950 Fb-1.7E	1950	1,700,000	1375	1800
1800 Fb-1.6E	1800	1,600,000	1175	1750
1650 Fb-1.5E	1650	1,500,000	1020	1700
1500 Fb-1.4E	1500	1,400,000	900	1650
1450 Fb-1.3E	1450	1,300,000	800	1625
1350 Fb-1.3E	1350	1,300,000	750	1600
1200 Fb-1.2E	1200	1,200,000	600	1400
900 Fb-1.0E	900	1,000,000	350	1050

*Design values in pounds per square inch. Design values for compression perpendicular to grain, $F_{c\perp}$, and horizontal shear, F_v, are the same as assigned visually graded lumber of the appropriate species.
[1] For any given value of F_b, the average modulus of elasticity, E, may vary depending upon species, timber source and other variables. The E value included in the F_b-E grade designations in the table are those usually associated with each F_b level. Grade stamps may show higher or lower E values (in increments of 100,000 psi), if machine rating indicates the assignment is appropriate. When an E value varies from the designated F_b level in the table, the tabulated F_b, F_t, and F_c values associated with the designated F_b value are applicable.
[2] The tabulated F_b values are applicable to lumber loaded on edge. When loaded flatwise, refer to Table E.

ADJUSTMENTS FOR MSR LUMBER — Checklist 3

- ☐ Repetitive Member Use Factor (C_r) — Table B,
- ☐ Duration of Load (C_D) — Table C,
- ☐ Horizontal Shear (C_H) — Table D,
- ☐ Flat Use Factor (C_{fu}) — Table E,
- ☐ Compression Perpendicular ($C_{c\perp}$) — Table F,
- ☐ Wet Use Factor (C_M) — Table G
  (only when appropriate)

Source: Western Wood Products Association.

Example

Determine the allowable F_b value for the following condition: Douglas fir 2 × 4 stud grade, repetitive framing with permanent loading, using Tables 6–3 and 6–4.

Allowable F_b = Base value × size adjustment factor (C_F) × repetitive framing factor (C_R) × load duration factor (C_D)

Allowable F_b = 675 psi × 1.1 × 1.15 × .9 = 768 psi

The Scholten nomographs in Figure 6–14 are used to solve the equation. The equation or nomograph is used with allowable stresses F_c and F'_c to find the allowable stress at any angle desired.

The design of light frame wood construction is simplified by the use of allowable span tables, for floor and ceiling joists. The tables are based on the wood species or group, grade, size and loading conditions. With this information, allowable spans are read from the design tables (Table 6–6). If the span is known, the tables will give the designer the needed information regarding framing lumber requirements.

Bearing strength of wood at angles to the grain (Hankinson formula)
The compressive strength of wood depends on the direction of the grain with respect to the direction of the applied load. It is highest parallel to the grain, and lowest perpendicular to the grain. The variation in strength, at angles between parallel and perpendicular, is determined by the Hankinson formula. The Scholten nomographs, shown here, are a graphical solution of this formula which is —

$$F_n = \frac{F_c F_{c\perp}}{F_c \sin^2 \theta + F_{c\perp} \cos^2 \theta}$$

F_c Unit stress in compression parallel to the grain.

$F_{c\perp}$ Unit stress in compression perpendicular to the grain.

θ Angle between the direction of grain and direction of load normal to the face considered.

F_n Unit compressive stress at inclination θ with the direction of grain.

The difference between the two charts is in scale, the one on the right to units of 1000 pounds, and the one on the left to units of 100 pounds. These units may be applied to allowable lumber stresses in pounds per square inch, or to total loads in the case of bolts, timber connectors, or lag screws.

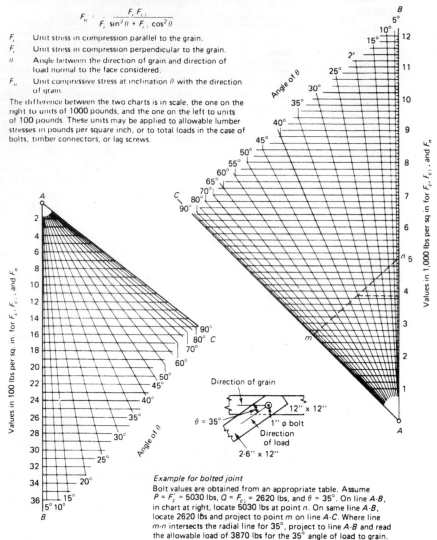

Example for bolted joint
Bolt values are obtained from an appropriate table. Assume $P = F_c' = 5030$ lbs, $Q = F_{c\perp} = 2620$ lbs, and $\theta = 35°$. On line *A-B*, in chart at right, locate 5030 lbs at point *n*. On same line *A-B*, locate 2620 lbs and project to point *m* on line *A-C*. Where line *m-n* intersects the radial line for 35°, project to line *A-B* and read the allowable load of 3870 lbs for the 35° angle of load to grain.

FIGURE 6–14. Scholten nomographs. *(Courtesy National Forest Products Association)*

LUMBER CLASSIFICATION

Standards have been established for the classification of lumber according to appearance, strength, shape, and use. Each piece of lumber is assigned to a grade according to certain rules. Softwoods come mainly from the southeastern area of the United States, the western part of the United States, or Canada, where they are graded to meet widely recognized standards and are used throughout the world. Hardwoods are produced throughout a greater area, although total production is much less and distribution from any one mill is usually not so extensive.

Organizations

Grading standards for softwoods are published by the U.S. Department of Commerce in Product Standard PS 20 (American Softwood Lumber Standard), which is recognized by the associations that issue grading rules throughout the United States and Canada. Grading rules for each region are established by organizations whose rules conform to PS 20 with additions for special conditions of each region.

The Western Wood Products Association, consisting of lumber producers within the states of Arizona, California, Colorado, Idaho, Montana, Nevada, New Mexico, Oregon, South Dakota, Utah, Washington, and Wyoming, provides grading rules for the following: Douglas fir, Engelmann spruce, mountain hemlock, western hemlock, Idaho white pine, incense cedar, larch, lodgepole pine, ponderosa pine, sugar pine, the true firs, and western red cedar.

The California Redwood Association is similar, but covers only 12 northern California counties and provides rules for redwood and some other softwood species.

Table 6–6. Example of allowable floor joist and ceiling joist spans

FLOOR JOIST SPANS

40# Live Load
10# Dead Load

L/360

Design Criteria: *Strength* - 10 lbs. per sq. ft. dead load plus 40 lbs. per sq. ft. live load.
Deflection - Limited in span in inches divided by 360 for live load only.

Species or Group	Grade*	2 x 6			2 x 8			2 x 10			2 x 12		
		12" oc	16" oc	24" oc	12" oc	16" oc	24" oc	12" oc	16" oc	24" oc	12" oc	16" oc	24" oc
Douglas Fir-Larch	1 & Btr	11-2	10-2	8-10	14-8	13-4	11-8	18-9	17-0	14-5	22-10	20-5	16-8
	1	10-11	9-11	8-8	14-5	13-1	11-0	18-5	16-5	13-5	22-0	19-1	15-7
	2	10-9	9-9	8-1	14-2	12-7	10-3	17-9	15-5	12-7	20-7	17-10	14-7
	3	8-8	7-6	6-2	11-0	9-6	7-9	13-5	11-8	9-6	15-7	13-6	11-0
Douglas Fir-South	1	10-0	9-1	7-11	13-2	12-0	10-5	16-10	15-3	12-9	20-6	18-1	14-9
	2	9-9	8-10	7-9	12-10	11-8	10-0	16-5	14-11	12-2	19-11	17-4	14-2
	3	8-6	7-4	6-0	10-9	9-3	7-7	13-1	11-4	9-3	15-2	13-2	10-9
Hem-Fir	1 & Btr	10-6	9-6	8-3	13-10	12-7	11-0	17-8	16-0	13-9	21-6	19-6	16-0
	1	10-6	9-6	8-3	13-10	12-7	10-9	17-8	16-0	13-1	21-6	18-7	15-2
	2	10-0	9-1	7-11	13-2	12-0	10-2	16-10	15-2	12-5	20-4	17-7	14-4
	3	8-8	7-6	6-2	11-0	9-6	7-9	13-5	11-8	9-6	15-7	13-6	11-0
Spruce-Pine-Fir (South)	1	9-9	8-10	7-8	12-10	11-8	10-2	16-5	14-11	12-5	19-11	17-7	14-4
	2	9-6	8-7	7-6	12-6	11-4	9-6	15-11	14-3	11-8	19-1	16-6	13-6
	3	8-0	6-11	5-8	10-2	8-9	7-2	12-5	10-9	8-9	14-4	12-5	10-2
Western Woods	1	9-6	8-7	8-0	12-6	10-10	8-10	15-4	13-3	10-10	17-9	15-5	12-7
	2	9-2	8-4	7-0	12-1	10-10	8-10	15-4	13-3	10-10	17-9	15-5	12-7
	3	7-6	6-6	5-4	9-6	8-3	6-9	11-8	10-1	8-3	13-6	11-8	9-6

*Spans were computed for commonly marketed grades and species. Spans for other grades and Western Cedars can be computed using the WWPA *Span Computer.*

CEILING JOIST SPANS

20# Live Load (Limited attic storage)
10# Dead Load

L/240

Design Criteria: *Strength* - 10 lbs. per sq. ft. dead load plus 20 lbs. per sq. ft. limited storage.
Deflection - Limited in span in inches divided by 240 for live load only.

Species or Group	Grade*	2 x 6			2 x 8			2 x 10			2 x 12		
		12" oc	16" oc	24" oc	12" oc	16" oc	24" oc	12" oc	16" oc	24" oc	12" oc	16" oc	24" oc
Douglas Fir-Larch	1 & Btr	16-1	14-7	12-0	21-2	18-8	15-3	26-4	22-9	18-7	30-6	26-5	21-7
	1	15-9	13-9	11-2	20-1	17-5	14-2	24-6	21-3	17-4	28-5	24-8	20-1
	2	14-10	12-10	10-6	18-9	16-3	13-3	22-11	19-10	16-3	26-7	23-0	18-10
	3	11-2	9-8	7-11	14-2	12-4	10-0	17-4	15-0	12-3	20-1	17-5	14-3
Douglas Fir-South	1	14-5	13-0	10-8	19-0	16-6	13-6	23-3	20-2	16-5	27-0	23-4	19-1
	2	14-1	12-6	10-2	18-3	15-9	12-11	22-3	19-3	15-9	25-10	22-4	18-3
	3	10-11	9-6	7-9	13-10	12-0	9-9	16-11	14-8	11-11	19-7	17-0	13-10
Hem-Fir	1 & Btr	15-2	13-9	11-6	19-11	17-10	14-7	25-2	21-9	17-9	29-2	25-3	20-7
	1	15-2	13-5	10-11	19-7	16-11	13-10	23-11	20-8	16-11	27-9	24-0	19-7
	2	14-5	12-8	10-4	18-6	16-0	13-1	22-7	19-7	16-0	26-3	22-8	18-6
	3	11-2	9-8	7-11	14-2	12-4	10-0	17-4	15-0	21-3	20-1	17-5	14-3
Spruce-Pine-Fir (South)	1	14-1	12-8	10-4	18-6	16-0	13-1	22-7	19-7	16-0	26-3	22-8	18-6
	2	13-8	11-11	9-8	17-5	15-1	12-4	21-3	18-5	15-0	24-8	21-4	17-5
	3	10-4	8-11	7-4	13-1	11-4	9-3	16-0	13-10	11-4	18-6	16-1	13-1
Western Woods	1	12-9	11-1	9-0	16-2	14-0	11-5	19-9	17-1	14-0	22-11	19-10	16-3
	2	12-9	11-1	9-0	16-2	14-0	11-5	19-9	17-1	14-0	22-11	19-10	16-3
	3	9-8	8-5	6-10	12-4	10-8	8-8	15-0	13-0	10-7	17-5	15-1	12-4

*Spans were computed for commonly marketed grades and species. Spans for other grades and Western Cedars can be computed using the WWPA *Span Computer.*

Source: Western Wood Products Association

The Southern Forest Products Association, consisting of lumber producers within the states of Alabama, Arkansas, Florida, Georgia, Louisiana, Maryland, Mississippi, Missouri, North Carolina, Oklahoma, South Carolina, Texas, and Virginia, provides grading rules for shortleaf, longleaf, loblolly, slash, Virginia, and ponderosa pine.

The National Lumber Grades Authority, a Canadian government agency, provides grading rules effective throughout Canada for Sitka spruce, Douglas fir, western larch, eastern hemlock, tamarack, eastern white pine, western hemlock, amabilis fir, ponderosa pine, red pine, white spruce, red spruce, black spruce, Engelmann spruce, lodgepole pine, jack pine, alpine fir, balsam fir, western red cedar, Pacific coast yellow cedar, western white pine, and several hardwoods.

The Northeastern Lumber Manufacturers' Association, consisting of lumber producers within the states of Connecticut, Maine, Massachusetts, New Hampshire, New York, Pennsylvania, Rhode Island, and Vermont, provides grading rules for red spruce, white spruce, black spruce, balsam fir, white pine, jack pine, Norway pine, pitch pine, hemlock, tamarack, white cedar, and several hardwoods.

The National Hardwood Lumber Association, consisting of lumber producers throughout the United States, provides grading rules for hardwoods. The lumber produced to meet these rules is used to manufacture products such as furniture, paneling, tool handles, flooring, and a wide variety of other products. Hardwood used directly in construction is usually purchased from local mills according to specifications agreed on between producer and user.

Grading

Lumber is a term that includes all finished or semifinished wood shaped with parallel longitudinal surfaces. Those pieces $1\frac{1}{2}$ in. or less in thickness and 2 in. or more in width are *boards*. Pieces at least 2 in. thick and less than 5 in. thick and 2 in. or more wide are *dimension* lumber. Pieces 5 in. or more in thickness and width are *timbers*. These are nominal dimensions and finished sizes are smaller.

Each piece of lumber is assigned on the basis of expected use to the category of *factory and shop lumber* or *yard lumber*. Factory and shop lumber includes pieces to be used in making sash, doors, jambs, sills, and other millwork items. This lumber is graded on the basis of how much is usable and how much must be wasted because of defects. Yard lumber is the one which is used structurally and includes most of that used in construction. Figure 6–15 illustrates the categories into which it is divided.

Yard lumber is divided according to size and shape into boards, dimension lumber, and timbers as shown in the figure. Boards are used for roofs, floors, siding, paneling, and trim. Dimension lumber is used for joists, studs, and rafters. Timbers are used infrequently as part of a structure. They are used more often for shoring earthwork or mine tunnels, or as bracing for concrete forms.

Dimensions and timbers are almost always *stress-graded*, which means each piece is assigned to a grade depending on its strength. Allowable stresses are specified for each grade and the design stresses cannot exceed these.

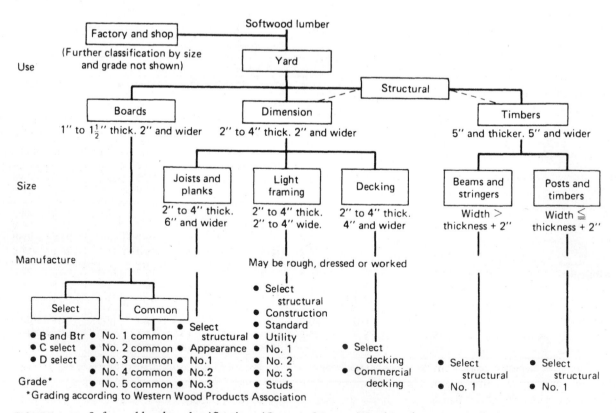

FIGURE 6–15. Softwood lumber classification. (*Courtesy Western Wood Products Association*)

FIGURE 6–16. Wood frame building under construction. *(Courtesy APA—The Engineered Wood Association)*

Note: Glu-lam garage door header.

A structure is designed by selecting members of cross section sizes and stress grades in the most appropriate combinations to perform the required functions. Each piece may have design stresses up to the allowable stresses for its grade. (See Table 6–3 for an example of commercial grades with their allowable stresses.) The lumber for the structure must be of the correct grade. A lower grade is too weak and a higher grade is unnecessarily expensive.

Wood frame construction consists of studs, joists, and rafters which act in unison to support loads on the structure (Figure 6–16). If one member is weaker and sags or moves out of line more than adjacent members, those adjacent members receive a greater load and the sagging one is relieved of part of its load. If the added load on the adjacent members causes them to move excessively, they are helped out by the next members. Therefore, these members, called *repetitive members*, are not required to support the design load independently. If most of them can bear a slightly greater load, the occasional weaker one is assisted by them. Therefore, a smaller safety factor is justified.

Refer to Table 6–3, which shows that the NDS's allowable stresses provide a lower safety factor for repetitive members by specifying a higher allowable stress for the extreme fiber in bending. This means that lumber of any particular commercial grade can be designed with a higher stress when used for repetitive members subjected only to beam loading.

The grade for each piece of lumber depends on its appearance and strength. The two are related. Often a flaw that is considered to mar the appearance is also a weakness. Standards are established for each grade. The standards consist of the maximum size or degree of each characteristic and manufacturing imperfection allowed for each grade. *Characteristics* are natural marks resulting from the tree's growth or from seasoning, and manufacturing imperfections are imperfections resulting from sawing, planing, or other manufacturing operations. Some of the characteristics affect appearance, others affect strength, and still others affect both appearance and strength. Manufacturing imperfections affect only appearance.

Each piece of lumber is visually inspected to compare its natural characteristics and manufacturing imperfections with the standards. It is then assigned to the correct grade, which is stamped onto the piece. Some grade stamps are shown in Figure 6–17.

If a graded piece is cut into parts, each part must be regraded, because the parts do not necessarily rate the same grade as the original.

Boards.   Boards are ordinarily graded strictly on the basis of appearance. Factors considered in grading boards are listed here:

stains: Discolorations that affect appearance but not strength. They are allowed to some extent in all grades.

checks: Lengthwise separations of the wood normally occurring across the annual growth rings and caused by seasoning or by the flattening of a piece of cupped lumber between rollers. They affect appearance and strength. Some are allowed in all grades (Figure 6–18).

shakes: Lengthwise separations caused by slippage occurring between the annual growth rings and sometimes outward from the pith through the rings. Shakes are caused while the tree is growing. They affect appearance and strength, and are permitted only in the lower grades (Figure 6–18).

cup and crook: Distortions caused by unequal shrinkage during seasoning. They affect appearance and are permitted to some extent in all grades (see Figures 6–12 and 6–18).

wane: Bark or missing wood on an edge or corner of a piece of lumber. It affects appearance and strength, and is permitted to some extent in all grades (see Figure 6–18).

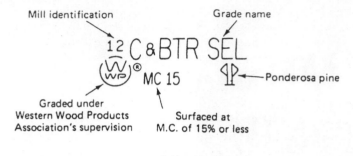

Mill identification

Grade name

12 C&BTR SEL

(W WP)® MC 15   ∂P ← Ponderosa pine

Graded under
Western Wood Products
Association's supervision

Surfaced at
M.C. of 15% or less

(a)

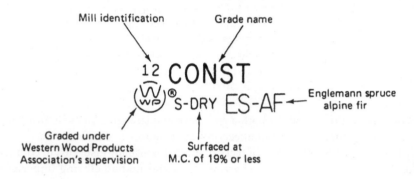

Mill identification

Grade name

12 CONST

(W WP)® S-DRY ES-AF ← Englemann spruce
alpine fir

Graded under
Western Wood Products
Association's supervision

Surfaced at
M.C. of 19% or less

(b)

FIGURE 6–17. Typical grade stamps. *(Courtesy Western Wood Products Association)*

splits: Separation of the wood due to the tearing apart of wood cells. They affect appearance and strength, and are permitted to some extent in all but the highest grades.

pitch: An accumulation of resinous material. It affects appearance and is permitted to some extent in all grades.

pockets: Well-defined openings between the annual growth rings which develop during the growth of the tree and usually contain pitch or bark. They affect appearance and strength, and are permitted to some extent in all grades.

pith: The small, soft core in the structural center of a log. It affects appearance and strength, and is permitted to some extent in most grades.

holes: These may occur and extend partly or completely through the wood. They affect appearance and strength, and are permitted to some extent in most grades.

knots: Portions of branches over which the tree has grown. They affect appearance and strength, and are permitted to some extent in all grades. They may be desirable for appearance. The sizes and conditions of knots are important. A knot may be decayed or sound, have a hole in it, be loose or tight, and be intergrown with the growth rings of the surrounding wood or "encased" with no intergrowth.

unsound wood: The result of disintegration of the wood substance due to action of wood-destroying fungi. It is also known as rotten or decayed wood.

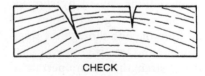

CHECK

SHAKE

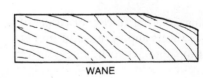

WANE

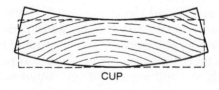

CUP

FIGURE 6–18. Defects in wood due to uneven shrinkage and other factors. *(Courtesy Timothy Dennis)*

There are several kinds of unsound wood caused by fungi that cease activity when the tree is felled. Additional decay is no more likely to take place in this kind of lumber than in lumber with no decay. Unsound wood affects appearance and strength, and is permitted only in lower grades.

torn grain: An irregularity in the surface of a piece where wood has been torn or broken out by surfacing. It affects appearance and is permitted to some extent in all grades. It could affect strength only if very severe, in which case it would not be acceptable from an appearance standpoint either.

raised grain: An unevenness between springwood and summerwood on the surface of dressed lumber. It affects appearance and is permitted to some extent in nearly all grades.

skips: Areas that did not surface cleanly. They affect appearance and are permitted to some extent in all grades.

These characteristics and manufacturing imperfections do not detract from a good appearance if they are on the side of the lumber that does not show. Most pieces have one side with a better appearance than the other, and because only one side shows in normal usage, the grading rules allow the grade to be determined by judging the better side for many of the grading factors.

Rules for each of the grades shown in Figure 6–15 describe the maximum extent to which each characteristic or imperfection is permitted. Some are not permitted at all in the higher grades. Requirements become more lenient for each succeeding grade from Select B and better to No. 5 common. Stress-rated boards are available for trusses, box beams, structural bracing, and other engineered construction.

Dimension Lumber.   Dimension lumber is graded for use as joists and planks, light framing, or decking according to the same characteristics and manufacturing imperfections as boards. Some additional characteristics are also considered. Because of the greater thickness of dimension lumber, bow and twist (see Figure 6–12) are subject to limits. They are not considered for boards because a board is flexible enough to be forced into a flat, untwisted position and nailed into place.

Two other characteristics are considered because strength is a major consideration. The direction of grain is not permitted to vary by too great an angle from the axis of the piece of lumber, and a minimum density of wood is required. *Slope* of grain is the deviation between the general grain direction and the axis of the lumber. It is determined by measuring a unit length of offset between grain and axis, and measuring the length along the axis in which the deviation takes place. It is the tangent of the angle between grain and axis expressed as a fraction with 1 as the numerator. For example, the slope of grain may be expressed as 1 in 8.

Determining the density is a way of determining the proportion of summerwood compared to the total amount of wood. It is determined by measuring the thickness of summerwood and springwood annual rings or by determining the specific gravity.

For dimension joists and planks and light framing, checks, shakes, and splits are limited only at the ends of the lumber pieces, and larger knots are allowed along the longitudinal centerline than near the edges. See the following discussion on timber stress grading for an explanation.

Timbers.   Timbers are graded as beams and stringers or as posts and timbers. *Beams* and *stringers* are pieces with a width more than 2 in. greater than the thickness, and *posts* and *timbers* have a width not more than 2 in. greater than the thickness. The grading rules for each take into account the proposed usage. A piece graded under beams and stringers rules can be used most efficiently by being stressed to its allowable stress under beam loading. If graded under posts and timbers rules, it can be used most efficiently by being stressed to its allowable stress under column loading. However, any member can sustain any of the allowable stresses listed for its grade, and therefore it may be used in any way as long as it is not stressed beyond the allowable stresses.

The chief difference in grading requirements between the two categories is in the allowed locations for characteristics that lower the strength of the member. The greatest tension and compression fiber stresses (due to bending) under beam loading are at the center of the beam's length at the top and bottom edges of the beam. The greatest tendency for the fibers of a beam to move relative to each other (fail in shear) under beam loading is at both ends along the center line between top and bottom. A column is stressed more uniformly throughout its entire length and cross section.

This difference is reflected in the grading requirements. Larger knots are allowed near the longitudinal centerline than near the top and bottom edges of beams and stringers where they have more effect in lessening the resistance to fiber stress. Checks, shakes, and splits are limited only near the ends of beams and stringers and, even there, not in the vicinity of the top and bottom edges where unit shearing stress is negligible.

For posts and timbers, the knot location is not considered. Checks, shakes, and splits are measured at the ends for posts and timbers. The shorter of the two projections of a shake onto the faces of the member is considered in grading posts and timbers; and for beams and stringers, the horizontal projection of the shake while the member is in use is considered.

Post and Beam Construction.   Timber framing, also known as post and beam or post and lintel construction, has been in use for many centuries. In the nineteenth century, most commercial and residential buildings were timber framed with exterior load bearing masonry walls. With the use of cast iron and, ultimately, steel construction, heavy timber framing saw diminished use for commercial buildings. Most residential buildings in the late nineteenth and twentieth centuries were framed using dimension lumber. Only agricultural buildings continued to see regular use of timber framing. In the past 30 years, timber framing has seen a revival, especially for custom and vacation homes and a variety of commercial structures including ski lodges, religious buildings, and specialty retail structures.

Post and beam framing is characterized by regularly spaced posts (vertical supports) and beams (horizontal members) using panelized walls which may be **structural insulated panels** (SIPs) or other panels to fill in between the openings. Timber framing is usually shop fabricated using mortise and tendon joinery to allow the transfer of loads between horizontal and vertical members. In the past, wood pegs were used to "join" these elements, but contemporary codes often require steel pins. Figure 6–19 shows a typical post and beam connection and photos in Figures 6–20 and 6–21 illustrate typical post and beam framing prior to the application of panelized walls. The ability to articulate the post and beam structure on the interior of the finished building is one of the qualities of post and beam construction that is most desirable.

Special-Use Wood

Wood meets the requirements for several special construction items so perfectly that entire industries are in operation to produce these items.

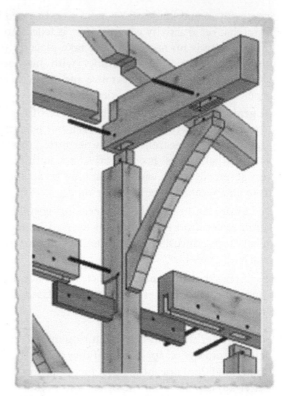

FIGURE 6–19. Typical post and beam connection. *(Illustration Courtesy of Timberpeg®)*

Wood *shingles* and *shakes* are used for roofing and siding. Most are made of western red cedar, a species with high resistance to decay and weathering and highly impermeable to water. This wood has a high strength-to-weight ratio. This is advantageous for any roofing material

FIGURE 6–20. Typical post and beam framing for a single family home. *(Photo Courtesy of Timberpeg® and Samyn-D'Elia Architects)*

FIGURE 6–21. Typical post and beam detail. These elements will remain exposed on the interior. *(Photo Courtesy of Timberpeg® and Samyn-D'Elia Architects)*

because it means that the structural frame has only a lightweight roof to support. The close, straight grain of the cedar allows the wood to be sawed into thin shingles parallel to the grain with very little cutting across the grain. This is an advantage over most other woods because each shingle has its flexure loading parallel to grain or nearly so.

The grain permits thin shakes to be split from a block of wood either by hand or with a machine. Shakes are thicker than shingles and have a hand-hewn look. Shingles and shakes can be treated with fire-resistant chemicals so that they do not support combustion.

Wood is frequently used for *flooring.* Both hardwoods and softwoods of many kinds are used, and both rift-cut and slash-cut lumber are used. Abrasion resistance is important for any floor, and close-grained species are preferred. The wood is installed in strips or blocks with side grain showing where appearance is important, or in blocks with end grain showing where appearance is not important. Wear resistance is greater on end grain.

The pieces of wood are usually of tongue-and-groove construction or interlocked in some way to prevent uneven warping. Wood is highly valued as flooring where appearance (in homes) or proper resilience (for basketball courts) is important. Wood is difficult to maintain under heavy usage if appearance is important, but easy to maintain even under heavy industrial usage if appearance is not important. It is valued for industrial floors in certain cases because of its resistance to acids, petroleum products, and salts. It is also nonsparking and is less tiring to the feet and legs of workers than harder floors.

DETERIORATION OF WOOD

Wood has four major enemies: insects, marine borers, fungi, and fire. Any of the four can destroy the usefulness of wood. However, wood can be protected to a great extent from any of them by the use of chemicals. Certain woods, such as cedar, cypress, locust, and redwood, are more resistant to fungi (decay) than other wood, and have long been used for fence posts and foundations of small buildings. The heartwood of all trees is more resistant to decay and insect attack than the sapwood because some of the substances that give the heartwood its color are poisonous to fungi and insects. However, sapwood absorbs preservative chemicals better because its cells are not occupied by these poisonous substances. The sapwood can be made more resistant with chemicals than the untreated heartwood.

Insects of various types damage wood by chewing it— termites are the best known. They are antlike creatures that consume cellulose as food, digesting the cellulose content of cardboard, paper, and cloth, as well as wood. They do not fly ordinarily, but do fly in swarms to form new colonies in the spring and lose their wings thereafter. They are most easily seen before a new colony is established and therefore are often seen with wings. Once they establish a colony, they do not venture out into daylight, and in fact cannot survive long in the sun's rays. Once they enter a piece of lumber, they are capable of consuming most of it without ever eating their way to the outside. A structural member that appears sound may be eaten hollow and may fail before the termites are discovered.

Subterranean termites, which are found throughout the United States, eat wood but live in the earth. They cannot live without the moisture they find there. Because they die quickly by drying out in sunlight, they construct passageways from their dwelling place in the moist soil to the food supply. These passageways may be tunnels underground directly to wood in contact with the ground. If the wood is not in contact with the ground, they construct a mud tube attached to a wall, pier, or whatever provides a route to the wood. The tube may even be built from ground to wood across a vertical gap. The termites are protected from the sun as long as they use the passageway for their travels. The tube passageway, which can easily be seen, gives away the presence of these termites.

Nonsubterranean termites are found in the southern part of the United States and are far fewer in number than the subterranean kind. They can live in wood, whether damp or dry, without returning to the ground for moisture. They therefore do not give themselves away by building passageways and are not as easily discovered as the subterranean variety. The termites' excrement of very tiny pellets resembling sawdust and their discarded wings are signs that indicate their presence inside wood structural members. Wings are found when the termites have just moved in, usually in late spring or early summer. These termites may be within lumber that is delivered to the construction site, and so lumber should be inspected for evidence of infestation before it is used.

Two methods of protecting wood structures from termites are to provide either a physical barrier or a chemical barrier to keep them from reaching the wood. The physical barrier consists of concrete or steel supports that provide a space between the ground and the entire wood. A projecting metal shield may be installed around the supports. A physical barrier is not guaranteed to keep termites out, but makes it possible to discover their passageways through regular inspections.

A chemical barrier consists of saturating the soil adjacent to the structure with poison. This method is used against subterranean termites when their passageways are found. The termites cannot pass the poison barrier. Therefore, those that are in the wood will dry out and die, and those in the soil will have to leave or starve to death.

The wood adjacent to the soil may be saturated with poison to get rid of either type of termite. It is better to treat the wood before construction and prevent the entrance of termites. A coat of paint prevents nonsubterranean termites from entering wood; but repainting is required and may be difficult to do in the areas under a building where termites are likely to enter.

Carpenter ants are a problem in some parts of the United States. They excavate hollows for shelter within wood, although they do not eat wood. They can be as destructive as termites, but do not attack the harder woods.

Marine borers, which include several kinds of mollusks and crustaceans, attack wood from its outer surface, eating it much more rapidly than termites would. They are found occasionally in fresh water, but are much more numerous in salt water and are particularly active in warm climates. Docks and other structures are often supported over water on wood piles, and their protection is a major problem. Heavy impregnation of the wood fibers with creosote oil or creosote–coal tar solution slows the attack of borers, but does not prevent it completely. The only complete protection available is complete encasement of the wood within concrete.

Fungi, which are microscopic, plantlike organisms, feed on wood fibers, leaving a greatly weakened residue of rotten wood. All forms of rot or decay are caused by some type of this microorganism, which is carried on air currents and deposited on the wood. Fungi require air, moisture, and a temperature above 40°F (4°C) to be active. They are most active at temperatures around 80°F (27°C). Wood with a moisture content of 19 percent or below, which is normal outdoors, is not subject to decay. Wood which is indoors is normally drier.

Wood kept below water does not decay because there is no air for the fungi. Logs that have been at the bottom of lakes or rivers since the earliest logging days have been recovered and found to be completely sound. Wood continuously below the groundwater table does not decay because it lacks air. Wood marine piles in water and foundation piles in soil remain sound below water. They do not decay above water either because of the lack of moisture.

The length of pile between high and low water elevations is alternately above water and below water. If in tidal water, it never dries completely when out of water, and when submerged, it is seldom, if ever, so saturated that there is no air. Even underground, with a fluctuating water table, moisture and air are both available much of the time. Fungi thrive in this zone, and the wood may be completely rotten there while the dry wood above and the saturated wood below are still sound. Wood is often used for the lower part of a pile and concrete or steel for the part above low water level. Various chemicals are used to protect wood from fungi when the wood must survive conditions favorable to the fungi.

FIRE

Fire is the obvious enemy of wood, yet buildings of wood construction can be as fire-safe as buildings of any other type of construction. Large timbers do not support combustion except where there are corners or narrow openings next to them, and when they do burn, the square edges burn first. This is demonstrated by the way logs burn in a fireplace. One log alone does not burn unless the fire is constantly fed by other fuel, but two or three logs close together support combustion in the spaces between them with no additional fuel. The sharp edges of a split log burn better than a rounded natural log.

The charred exterior caused by fire is a partially protective coating for the inner wood, cutting the rate of consumption by fire to less than half the rate for unburned wood. A large timber may have enough sound wood that remains effective after a protective covering is built up.

A construction method called *heavy timber construction* has been devised to take advantage of these observations. It consists of using large structural members with their exposed edges trimmed to remove the square shape and with corners blocked in to remove "pockets" of heat concentration. Siding, decking, and flooring are thick, with tight joints that permit no cracks to support combustion and allow no heat or flames to burst through. A building of this type does not burn unless the contents maintain a very hot fire for an extended period of time. In that case, no other kind of building can resist the fire either. Even though other building materials do not burn, they fail by melting, expanding excessively, or simply losing strength.

Protected construction consists of ordinary wood frame construction with a protective coating of plaster, gypsum board, or acoustical tile protecting the joists and with plywood used for walls, floors, and roof. With joists having some protection, structural failure is delayed so that there is time for people within the building to escape. The plywood is a barrier to the passage of flames and heat from the source of the fire to other parts of the building.

Wood can be impregnated with fire-retardant chemicals which decrease flame spread and smoke generation. Burning occurs when wood becomes heated sufficiently (300°F or 149°C) and escaping gases burst into flame. The absorbed chemicals prevent this from happening by changing these combustible gases to water plus noncombustible gases.

Fire-retardant paints which have an appearance satisfactory for a finish coat can also provide reduction in flame spread and smoke generation through their insulating properties. Impregnation is more effective than painting and is not detrimental to appearance or strength. However, if wood is to be impregnated with a chemical, it must be done before it is part of a structure.

Therefore, paint is usually used to increase the fire resistance of existing buildings, and impregnation is used for new construction. It is done by saturating the wood cells with a fire-retardant chemical dissolved in water. Wood swells when it absorbs the water and shrinks when it dries again. These size changes can cause problems if no allowance is made for them.

Because of the way a fire is fueled, the rapidity with which it causes damage, and the ways in which it causes damage, there is much more involved in fire protection than the selection of materials. Building contents in many cases include trash, cleaning supplies which often burn readily, flammable waxes and polishes applied to floors and furniture, drapes, spilled liquids which may be flammable, and many kinds of flammable stored materials.

People cannot survive in a room with a fire for more than a few minutes, even if the space is large. They are killed by heat, smoke, or lack of oxygen in less than 10 minutes and sometimes in a minute or two. The contents of a room are very quickly damaged by smoke and heat, even if not actually burned. Water damage may also be extensive. The material that burns and the damage done by the burning are largely beyond the control of the designer of the building.

Therefore, the strategy employed by designers and by firefighters is to delay the collapse of any part of a building so that all occupants can get out, to provide safe passageways for them to get out, and to prevent the spread of the fire to other spaces. It is not feasible to accomplish more than this unless the contents of the building do not burn. The fire safety of a building depends on the contents and how they are stored, good housekeeping, sprinkler and fire alarm systems, firefighting methods, and the type of construction more than on building materials.

PRESERVATION OF WOOD

Preserving wood involves treating it with a poison so that fungi and insects do not consume it. Strength is not affected by the poisonous preservative. The preservative may be brushed onto the surface of the wood. This provides a small amount of protection. More protection is obtained by dipping the wood into a solution containing the preservative. These two methods are used only for small projects or for wood requiring little protection.

The best protection is obtained by forcing the preservative into the fibers of the wood. A certain retention of preservative in pounds per cubic foot of wood is required, depending on the type of preservative, species of wood, and conditions under which the wood is to be used. Some wood must be punctured with sharp points (*incised*) before the operation to retain the specified amount of preservative. All wood must have the bark removed before being treated because the preservatives do not penetrate bark well. If at all possible, wood is cut to final size before treatment to avoid wasting preservative on excess wood and to avoid later cutting into parts of the wood not penetrated by the preservative. If later cutting exposes any untreated wood, it should be painted with a heavy coat of the preservative.

The water repellency of treated wood surfaces becomes less effective after exposure to weather, foot traffic, and the sun's ultraviolet rays (Figure 6–22). Therefore, it is advisable to recoat exposed wood surfaces with a commercially available sealer every 2 years.

The preservatives and manufacturing methods used to pressure-treat wood must meet all Environmental Protection Agency (EPA) requirements. The most common treatment used today is chromated copper arsenate (CCA). Effective December 31, 2003, by voluntary agreement with the EPA, manufacturers of CCA will cease the production of CCA-treated wood for consumer applications. Residential construction affected by the agreement includes play structures, decks, patios, picnic tables, fencing, landscape timbers, and walkways. To avoid the use of treated wood, playground equipment, decks, and other structures may be constructed with cedar or redwood lumber, which are naturally resistant to insects and microbial agents. However, CCA-treated wood will still be available for highway construction, saltwater (marine) use, utility poles, piles, and some engineered wood products.

FIGURE 6-22. Treated wood deck with 2 × 8 joists, double 2 × 10 beams with $\frac{1}{2}$ plywood filler, and 4 × 4 posts on 8-in.-diameter piers, 4 ft deep. *(Courtesy Tedmar Construction Corp.)*

Coating existing CCA-treated wood with oil-based, clear, or semitransparent stains on a regular basis may prevent the migration of preservative chemicals from treated wood. The use of nonpenetrating stains that form a film on the wood are not recommended because subsequent peeling and flaking may have an impact on exposure to the preservatives and durability of the treated wood.

Two common nonarsenic preservatives registered with the EPA are ammonium copper quaternary (ACQ) and copper boron azole (CBA). Wood treated with these preservatives is currently available in many retail lumber outlets.

The chemicals are inert within the wood and provide moisture, termite, and decay protection. All pressure-treated wood used in the United States is produced in accordance with American Wood Preservers' Association (AWPA) standards. The four EPA-approved wood preservatives in use for pressure-treated wood are waterborne, oil-borne, creosote,

and fire retardant solutions. The treatment process does not alter the basic characteristics of the wood; therefore, treated wood should not shrink, swell, check, or warp any more than untreated wood. The selection of a preservative is based on the wood's use with respect to required durability, color, amount of animal or human contact, and other factors.

Typical Treated Wood Products

1. *Dimension lumber*—Wood, 2 to 5 in. thick, sawn on all four sides, including the common 2 × 4, 2 × 6, 2 × 8, 2 × 10, 2 × 12, and 4 × 4. These are commonly used as sill plates, joists, beams, rafters, trusses, and decking.

2. *Heavy timbers*—Wood sawn on all four sides, including 4 × 6, 6 × 6, 8 × 8, and larger timbers. These are commonly used in heavy timber-frame construction, landscaping, and marine construction.

3. *Round stock*—Round posts and poles, including round fence posts, building poles, marine piling, and utility poles.

4. *Plywood*—Sheets of wood laminated to provide higher strength plywood, usually produced in 4 ft by 8 ft sheets in varying thicknesses. Plywood is used in many sheathing and utility applications.

5. *Specialty items*—Many wood items are now available to enhance backyard deck and gazebo construction. These include lattice, hand rails, spindles, radius edge decking, turned posts, and ball tops.

Treated wood is graded by certified graders and marked with an AWPA stamp. The stamp includes the AWPA standard, the plant name or number, the preservative used, the logo of the accrediting agency, and the retention, which is the amount of preservative retained in the wood, measured in pounds per cubic foot (pcf) (Table 6–7). Grades indicate strength and/or how attractive the wood is. Higher grades are used where strength and/or appearance is important. Lower grades are used for hidden framing, rustic appearance, and nonstructural or low-structural requirements.

Contractors, as should anyone working with treated woods, are advised to follow the EPA-approved consumer information sheet guidelines available from the project lumber supplier.

Preservatives that are suitable for indoor use are inoffensive in appearance and odor, and the wood can be painted after it is treated. There are some disadvantages, however. The wood expands and increases in weight when water is injected, and some time is required to dry the wood to an acceptable m.c. for use. Because these preservatives are soluble in water, the wood treated with them cannot be used in water or exposed to rain without the loss of some of the salts by leaching. Therefore, they are not suitable for outdoor use.

Poisonous organic materials such as pentachlorophenol or copper naphthenate dissolved in petroleum oil are used to protect wood from fungi and insects. Since these poisonous materials are not soluble in water, they are effective on wood used in wet places. The oil evaporates rapidly after treatment so that the wood is ready for use sooner than that treated with waterborne salts. The wood cannot be painted after treatment, but does not have an unpleasant appearance. Required retention is in the range of $\frac{1}{2}$ lb/cu ft.

Creosote, a liquid by-product of the refining of tar, is the most effective preservative against fungi and insects and the only one effective at all against marine borers. It is used alone, mixed with coal tar, or mixed with petroleum. Railroad ties and utility poles are treated with one of the three. Mixing with petroleum is an economy measure to increase the quantity of preservative for uses where less effective preservation is satisfactory. Creosote and creosote–coal tar mixtures are used for more demanding conditions and are the only types used in salt water.

Creosote is a black or dark brown, foul-smelling, sticky substance and is therefore not satisfactory for all uses. Wood treated with creosote is very difficult to paint. Creosote after application burns very readily until it cures by drying. Generally, it is not a fire hazard by the time the wood is put to use. Creosote must be used in larger quantities than the other preservatives. However, it is so much cheaper that it is more economical to use it wherever its unpleasant appearance and odor are not objectionable.

Typical retention requirements for creosote are 8 lb/cu ft for installation above ground, 10 lb/cu ft for wood in contact with the ground, and 25 lb/cu ft for piling in salt water.

Pentachlorophenol-, copper napthenate-, or creosote-treated wood should not be used for occupied spaces, where prolonged human or animal contact is possible. Therefore these treated woods would not be used for farm, commercial, educational or residential facilities. Caution must also be exercised when disposing of waste and scrap materials; they should never be burned because of the toxicity of the ash and smoke, but disposed of at landfill or construction waste disposal sites.

Pressure treatment of wood includes placing seasoned wood into a pressure vessel, flooding it with the preservative, and forcing the preservative into the cells of the wood with air pressure. The full-cell method results in the cell walls being completely coated and hollow cells being nearly filled with preservative. The empty-cell method results in the cells being nearly empty but with the cell walls completely coated. The weight of preservative retained is greater after the full-cell operation. It is more costly, but it is necessary to achieve the high retention needed for the most severe conditions.

Wood exposed to the weather becomes eroded, often turns gray, and develops checks. It may be protected from the weather by a surface coating of paint, enamel, varnish, or

Table 6–7. Standard retention (pcf)

Location of Use	Standard Retention (pcf)
Aboveground use (deck boards and fence boards)	0.20
Ground or fresh water contact (deck posts and fence posts)	0.40
Salt water splash (decking for docks and marinas)	0.60
Wood foundations and heavy structural use (foundations and poles)	0.60
Foundation piles (a key part of high-rise foundation construction)	0.80
Salt water immersion (docks and marinas)	2.50

sealer. All should be applied with the wood at the moisture content it will have in use so that there will be a minimum of shrinkage or swelling to loosen the coating. Satisfactory performance requires the finish to be hard enough for protection, be flexible enough not to crack when the wood shrinks and swells, and adhere strongly to the wood.

These same finishes may be used primarily for appearance rather than protection. They may be designed for interior use solely to improve appearance by providing a desired color, enhancing the natural wood appearance, hiding blemishes, making the surface smooth, or improving lighting with a more reflective surface. Paints are classified for specified use as exterior or interior and by the solvent used as water- or oil-based coatings.

paint: Finely divided solids suspended in a liquid vehicle to allow it to spread over a surface, where it dries to a solid, protective film covering the wood from view. It is normally applied in two coats: the first (prime coat) to bond to the wood, and the second (cover coat) to provide protection.

enamel: Similar to paint and provides a smoother, harder, and more brittle surface. Because of the brittleness, it is not suitable for exterior use in wet or cold climates because swelling and shrinking or expansion and contraction cause it to crack and chip.

varnish: Resins dissolved in a liquid vehicle. Varnish provides a clear coat through which the wood grain shows.

sealer: A water-repellant substance dissolved in a solvent and used to seal and moisture-proof the wood by penetration into the pores while allowing the grain to show.

semitransparent stain: A stain that alters the color of wood while allowing the texture and grain to show. The term also applies to interior wiping stains.

opaque stain: Exterior stain that hides the color and grain of wood, but allows the texture to show.

polyurethane varnish: A very durable modified alkyd resin coating used to clear-coat wood.

GLUED LAMINATED WOOD

Construction with sawn wood is limited by the size, shape, and characteristics of readily available trees. Lumber longer than 24 ft (7.2 m) or with a cross-sectional dimension greater than 12 in. × 12 in. is difficult to obtain in large quantities. When lumber of these sizes is sawed, each piece is more likely to have some characteristic that seriously reduces strength or lowers the quality of its appearance than is likely with smaller pieces. Even if there are no serious weaknesses at the time of sawing, it is difficult and time consuming to season large timbers without producing serious checks, and difficult to season long pieces without serious warping. Heavy timber construction and long spans are not feasible with sawn lumber. Sawn lumber cannot be bent into curves except in small cross sections and is ordinarily used straight.

However, structural members of any length and cross section and with just about any desired curve can be made by gluing smaller pieces together. The smaller pieces, which are of standard lumber cross section, are glued one over the other, wide face to wide face, as laminations. No one lamination need be as long as the member. They are glued end to end to reach the full length. A structural member made this way is called a *glu-lam member* (short for glued and laminated), and construction with members of this kind is called *glu-lam construction*. The supporting members for heavy timber construction are glu-lam timbers. Almost all glu-lam members are made of Douglas fir or southern pine (Figure 6–23).

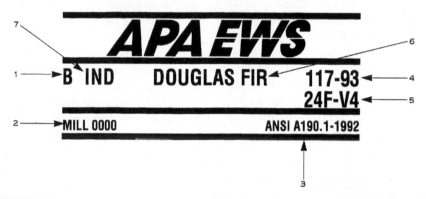

(1) Indicates structural use: B-Simple span bending member. C-Compression member. T-Tension member. CB-Continuous or cantilevered span bending member.

(2) Mill number.

(3) Identification of ANSI Standard A190.1, Structural Glued Laminated Timber. ANSI A190.1 is the American National Standard for glulam beams.

(4) Applicable laminating specification.

(5) Applicable combination number. In the example shown: 24F means the allowable bending stress in fiber is 2400 psi; V4 indicates that the laminations were visually graded and designates the grade of laminations used in the beam layup.

(6) Species of lumber used. Some common species used in APA EWS glulams include Douglas-fir, hem-fir, Alaska cedar, spruce-pine-fir (SPF) and southern pine.

(7) Designation of appearance grade. INDUSTRIAL, ARCHITECTURAL or PREMIUM.

FIGURE 6–23. Typical *APA EWS* trademark. (*Courtesy APA—The Engineered Wood Association*)

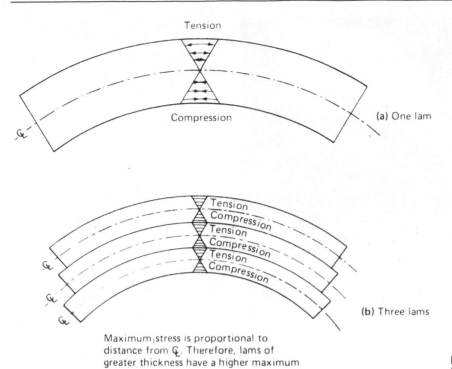

Tension

Compression

(a) One lam

Tension
Compression
Tension
Compression
Tension
Compression

(b) Three lams

Maximum stress is proportional to distance from $\mathcal{C}$. Therefore, lams of greater thickness have a higher maximum stress.

FIGURE 6–24. Stress due to bending laminations.

Lumber of 2 in. nominal thickness is ordinarily used for laminations. These laminations can be used straight in columns or beams and can be bent to form arches or for desired architectural effects. For sharper curves, thinner laminations are used. Bending lumber introduces deformation, as shown in Figure 6–24. Deformation causes stresses just as stresses cause deformation, both in accordance with the modulus of elasticity of the material. The figure demonstrates how the same curvature introduces greater stress in a thicker lamination. Sharper curves can be made using thinner lams as long as the allowable F_b is not exceeded. Horizontal shear is also induced by bending, and the allowable F_v must not be exceeded in the wood.

Preparation of wood to fabricate a glu-lam structural member is shown in Figure 6–25, and a finished product is shown in Figure 6–26. The individual laminations are placed so that:

1. Weak spots are separated from each other to avoid a concentration of weaknesses.
2. Disfiguring characteristics are hidden within the member.
3. End joints between lams are separated from each other to avoid a plane of weakness.
4. The strongest wood is placed where stresses are the highest.
5. The wood with the best appearance is placed where appearance is important.

Usually the lamination width is the full width of the member. If laminations must be placed side by side to equal the proper width, their side joints must be glued and not be located one over another. End joints and side joints are shown in Figure 6–27.

The use of glu-lam construction allows wood to be used in large projects for which only concrete or steel would have been considered previously. Glu-lam members can be made stronger than the strongest sawn lumber of the same size. Large members can be made entirely from small trees or from the small pieces left after the weak parts are cut out of large trees. Glu-lam has proven particularly popular for roof arches such as those for churches, auditoriums, and supermarkets.

FIGURE 6–25. Manufacturing of glu-lam arches. *(Courtesy American Institute of Timber Construction)*

FIGURE 6-26. Finished curved roof beam. *(Courtesy Weyerhaeuser Company)*

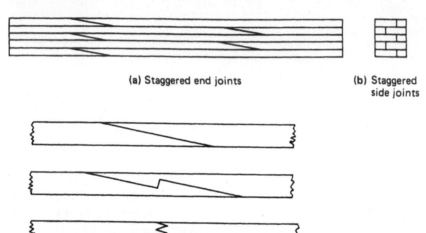

(a) Staggered end joints

(b) Staggered side joints

(c) Types of scarf joint

FIGURE 6-27. Joints in glu-lam members.

PLYWOOD

Plywood is another type of glued, laminated wood. The laminations are thin and arranged with the grain of each one running perpendicular to the grain of the laminations adjacent to it. The laminations are called *veneers*. The center veneer is called the *core*, the outside veneers are the *face* and *back*, and the inner veneers with grain perpendicular to the face and back are called *cross bands*. There is always an odd number of veneers, usually three or five.

Alternating the grain direction tends to equalize the strength in all directions, although the strength is less parallel to grain than it would be if the grain of all veneers were in the same direction. The advantages far outweigh this disadvantage, however.

The advantages of plywood over sawn lumber of comparable thickness are listed below:

1. Plywood's greater transverse strength stiffens or braces the entire structure to a greater degree than lumber sheathing when plywood is used over studs, joists, and rafters for wood frame construction.

2. Plywood's greater stiffness allows it to bridge wider spaces between beams without excessive deflection, thus reducing the number of beams required.

3. Plywood resists concentrated loads better because its mutually perpendicular grain structure spreads the loads over a larger area.

4. Plywood can be worked closer to the edges without splitting, allowing it to be bolted, screwed, or nailed with less excess material.

5. Desired appearance can be obtained by using thin veneers of high-quality wood only where they show. A matching repetitive pattern can be obtained by using one log for all the face veneer and wood with less desirable appearance for the other veneers.

6. Length change and warping due to moisture change are much less in plywood than in sawn lumber because shrinkage parallel to grain is negligible, and wood grain running both ways prevents excessive shrinkage either way. Using an odd number of veneers results in the grain on top and bottom running in the same direction so that shrinkage and expansion are equalized on the two faces and warping is prevented.

7. Plywood can be more easily bent to form curves for concrete forms or for curved wood construction.

8. Plywood is fabricated in large sheets that can be handled more efficiently at the construction site than sawn lumber. Because one sheet covers the same area as half a dozen boards, plywood requires less handling and fitting into place.

9. Plywood has demonstrated greater fire resistance than boards of the same thickness.

Plywood does not have equal strength in both directions. If there is an equal thickness of veneers in each direction, the tensile and compressive strengths are equal. However, bending strength is greater parallel to the grain of the face and back, unless there is a much greater thickness of cross bands with their grain in the other direction. If cross bands are thick enough to provide equal bending strength perpendicular to face grain, tensile and compressive strengths are much greater perpendicular to the face grain.

Plywood is usually made in 4-by-8-ft sheets with the face grain in the long direction. If the panel is subjected to bending across a 4-by-8-ft opening, the bending unit stress (F_b) is less in the 4-ft dimension than in the 8-ft dimension. The unit stresses resulting from a uniform load are therefore balanced. This means they reach roughly the same percentage of allowable unit stress in each direction, since the direction of greater allowable unit stress is also the direction of greater actual unit stress.

Veneers of plywood are manufactured by being cut from logs slightly longer than the 8-ft length of a finished plywood veneer. The cutting is done with a knife blade held in place while the log rotates against it. A thin, continuous layer is peeled from the ever-decreasing circumference and cut into sheet-size veneers.

The veneers cannot always be made 4 ft wide and must sometimes be made from more than one piece and neatly joined side by side. The desired combination of these veneers is dried in an oven, glued together, and pressed into plywood sheets. Sometimes veneers are cut in slices across the log parallel to the axis of the log in the same direction as slash-cut lumber to obtain a special appearance. The veneers are usually $\frac{1}{16}$ to $\frac{3}{16}$ thick.

Softwood Plywood

The types and grades of softwood plywood are described in the U.S. Department of Commerce's Product Standard PSI for Softwood Plywood—Construction and Industrial. Softwood plywood sheet thicknesses in common use are $\frac{1}{4}$, $\frac{5}{16}$, $\frac{3}{8}$, $\frac{1}{2}$, $\frac{5}{8}$ $\frac{3}{4}$, $\frac{7}{8}$, $1\frac{1}{8}$, and $1\frac{1}{4}$in. Veneers are classified into five *grades* based on finished appearance as judged by knots, pitch pockets, discoloration, and other characteristics similar to those used in grading lumber. The grades, starting with the best, are A, B, C, C plugged, and D. Only the face and back veneers are judged. These may be repaired by cutting out defects and replacing them with patches that will match so well in grain and color and fit so tightly in the hole that they are difficult to see.

Plywood is classified into five *species groups* according to strength and stiffness, with Group 1 the strongest and stiffest. The species of wood included in each group are shown in Table 6–8. The sheet is considered to be in a group if the face and back are of a species from that group, although the inner veneers may be of another group. This is logical because the stiffness and bending strength depend mostly on the outer veneers. Although more than 50 species are listed, most plywood is made from Douglas fir.

Two general *types* of plywood are manufactured: interior and exterior. *Interior type plywood* is made with glue that is adequate for service indoors, and *exterior type plywood* is made with hot, phenolic resin glue that is unaffected by water and resists weathering as well as wood does. Exterior type plywood does not include any veneers, outer or inner, below C grade. It is intended for permanent outdoor installation. Interior plywood may be made in any grade.

The interior type is available in three categories. Interior glue is used for plywood intended for use where it may be exposed briefly to weather. Intermediate glue is used for plywood intended for use in high humidity or for exposure to the weather for a short time. Exterior glue is used for plywood expected to be exposed to weather for an extended period during construction.

Plywood is graded for strength, although not in the same way as lumber. Based on the thickness and classification of species, strength is indicated by stamping two *identification index* numbers on the plywood sheet. The first number gives the maximum span if the sheet is used for a roof, and the second gives the maximum span if the sheet is used for a subfloor.

Plywood is specified and graded primarily either for appearance or for strength, depending on how it is to be used. Descriptions of the appearance veneer grades are listed in Table 6–9. If an appearance grade is desired, the grade, number of veneers, species group, type, and thickness are specified. If an engineered grade is desired for strength, the grade, identification index, number of veneers, and thickness are specified. These specified requirements are verified by a certification of quality stamped on the back or edge of

Table 6–8. Classification of species

Group 1	Group 2		Group 3	Group 4	Group 5
Apitong* **	Cedar, Port Orford	Maple, black	Alder, red	Aspen	Basswood
Beech, American	Cypress	Mengkulang*	Birch, paper	Bigtooth	Poplar, balsam
Birch	Douglas fir 2†	Meranti, red* ‡	Cedar, Alaska	Quaking	
Sweet	Fir	Mersawa*	Fir, subalpine	Cativo	
Yellow	Balsam	Pine	Hemlock, eastern	Cedar	
Douglas fir 1†	California red	Pond	Maple, bigleaf	Incense	
Kapur*	Grand	Red	Pine	Western red	
Keruing* **	Noble	Virginia	Jack	Cottonwood	
Larch, western	Pacific silver	Western white	Lodgepole	Eastern	
Maple, sugar	White	Spruce	Ponderosa	Black (western poplar)	
Pine	Hemlock, western	Black	Spruce	Pine	
Caribbean	Lauan	Red	Redwood	Eastern white	
Ocote	Almon	Sitka	Spruce	Sugar	
Pine, southern	Bagtikan	Sweetgum	Engelmann		
Loblolly	Mayapis	Tamarack	White		
Longleaf	Red lauan	Yellow poplar			
Shortleaf	Tangile				
Slashy	White lauan				
Tanoak					

* Each of these names represents a trade group of woods consisting of a number of closely related species.

** Species from the genus *Dipterocarpus* marketed collectively: Apitong if originating in the Philippines, Keruing if originating in Malaysia or Indonesia.

† Douglas fir from trees grown in the states of Washington, Oregon, California, Idaho, Montana, Wyoming, and the Canadian provinces of Alberta and British Columbia shall be classed as Douglas fir No. 1. Douglas fir from trees grown in the states of Nevada, Utah, Colorado, Arizona, and New Mexico shall be classed as Douglas fir No. 2.

‡ Red Meranti shall be limited to species having a specific gravity of 0.41 or more based on green volume and oven-dry weight.

Source: APA—The Engineered Wood Association.

each piece of plywood. Examples of such grade stamps are shown in Figure 6–28.

Structural members can be fabricated of plywood, or of plywood in combination with sawn lumber or glu-lam lumber. The plywood is fastened to the lumber by nailing or gluing. Glue may be held under pressure by specialized equipment while drying, or the pressure may be applied by nailing tightly immediately after gluing in a procedure called *nail gluing*.

Plywood is used for the gusset plates and splices of sawn wood trusses. It is also used for *box beams*, where it serves as the web with sawn or glu-lam lumber as the flanges. *Stressed-skin panels* are made with the materials reversed. The upper and

Table 6–9. Veneer grades used in plywood

A	Smooth, paintable. Not more than 18 neatly made repairs, boat, sled, or router type, and parallel to grain, permitted. Wood or synthetic repairs permitted. May be used for natural finish in less demanding applications.
B	Solid surface. Shims, sled or router repairs, and tight knots to 1 in. across grain permitted. Wood or synthetic repairs permitted. Some minor splits permitted.
C$_{plugged}$	Improved C veneer with splits limited to $\frac{1}{8}$-in. width and knotholes or other open defects limited to $\frac{1}{4} \times \frac{1}{2}$ in. Admits some broken grain. Wood or synthetic repairs permitted.
C	Tight knots to $1\frac{1}{2}$ in. Knotholes to 1 in. across grain and some to $1\frac{1}{2}$ in. if total width of knots and knotholes is within specified limits. Synthetic or wood repairs. Discoloration and sanding defects that do not impair strength permitted. Limited splits allowed. Stitching permitted.
D	Knots and knotholes to $2\frac{1}{2}$ in. width across grain and $\frac{1}{2}$ in. larger within specified limits. Limited splits allowed. Stitching permitted. Limited to Interior, Exposure 1, and Exposure 2 panels.

Source: APA—The Engineered Wood Association.

HOW TO READ THE BASIC TRADEMARKS OF *APA – THE ENGINEERED WOOD ASSOCIATION*

Product Standard PS 1-95 is intended to provide for clear understanding between buyer and seller. To identify plywood manufactured by association member mills under the requirements of Product Standard PS 1-95, four types of trademarks and one typical edge mark are illustrated. They include the plywood's exposure durability classification, grade and group, and class or Span Rating. Here's how they look, together with notations on what each element means.

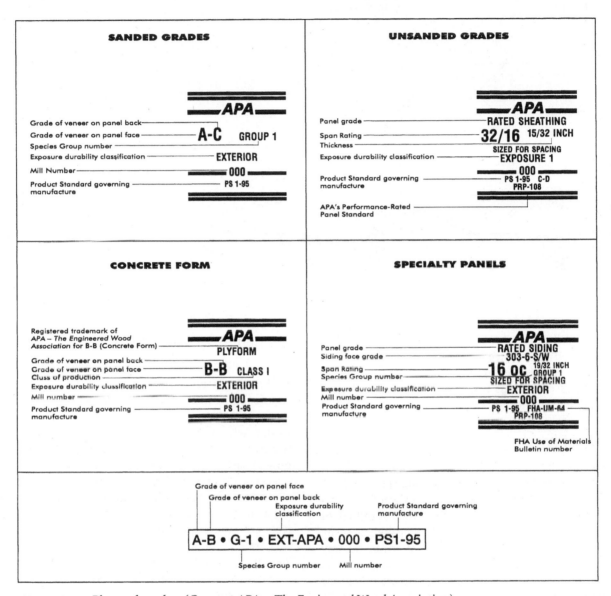

FIGURE 6–28. Plywood grades. (*Courtesy APA—The Engineered Wood Association*)

lower flanges (panel covers) are made of plywood with webs of sawn lumber. The assembly is very wide compared to its depth, but is stressed like a beam while having the shape of a panel.

Sandwich panels consist of plywood panel covers separated by a comparatively weak material with no need for sawn lumber except at the edges of the panels in some cases. The material between panel faces may be plastic foam or resin-impregnated paper in a honeycomb pattern. The material need only be strong enough to unite the load-bearing upper and lower panel faces, which act structurally like the upper and lower chords of a truss. Truss chords are held apart by a thin network of web members, and the panel faces are also held apart by a thin network of material. Figure 6–29 has illustrations of the various types of structural components.

Plywood siding is manufactured with face materials of various kinds to provide a particular effect while retaining the desirable properties of plywood. *Overlaid plywood* has a resin-treated surface, hot-bonded to the plywood sheet. The surface is smooth and intended for paint finish of the best quality. It is also used for concrete forms because its smooth surface makes a better concrete surface and allows forms to

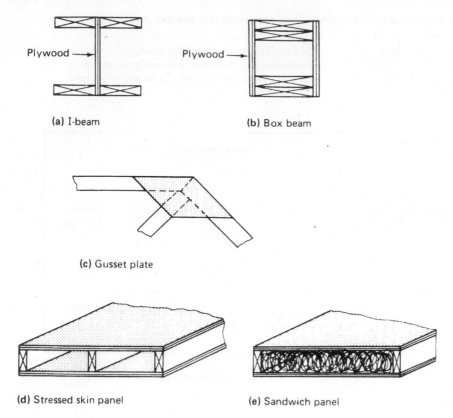

(a) I-beam   (b) Box beam

(c) Gusset plate

(d) Stressed skin panel   (e) Sandwich panel

FIGURE 6–29. Structural uses of plywood.

be removed easily. *Coated plywood* is surfaced with metal or plastic designed to produce a special effect. Coatings that simulate stucco or exposed aggregate are available as well as metal coatings of various textures and colors and clear plastic coating for protection only.

Hardwood Plywood

The types and grades of hardwood plywood are described in the U.S. Department of Commerce's Product Standard PS51 for Hardwood and Decorative Plywood. Hardwood plywood is usually selected for its appearance to be used as the face of cabinets, furniture, doors, flooring, and wall paneling. Many special effects can be obtained in wood pattern by the way the veneer is cut from the log. Knots, crooked grain, and other characteristics that indicate weakness are valued for their appearance.

Wall paneling is made with a face veneer selected for its appearance and then stained or treated for special effects. It may be sand blasted, antiqued, striated, or grooved to give the appearance of individual boards. It is made in sheets as thin as 1/8 in. with little strength and little care for the appearance of the back.

The core material behind hardwood veneers may be sawn lumber with a thin veneer on one or both faces for cabinets or furniture. It may be made from sawdust, shavings, and other scraps of wood mixed with glue and pressed into flat sheets called particle boards. It may also be made of noncombustible material for fire-resistant construction.

Oriented Strand Board (OSB)

Oriented strand board is a strong, stiff panel manufactured in a cross-ply pattern similar to plywood. The layers are formed from thin rectangular shaped wood strands oriented in layers at right angles to each other and bonded with waterproof resin adhesives. Because the strands are small, OSB can be manufactured from smaller, younger fast-growing trees. After debarking, the logs are processed into strands or wafers, dried, and screened to correct size. Glue is applied and the strands are oriented into mats. Face layer strands run along the panel, and core layers are random or run across the panel. The oriented mats are then bonded together under pressure and heat. The panels are then cut to size either 4 × 8 ft or 1.25 × 2.50 m. In some cases, panels are custom-sized for specialized applications.

OSB panels are manufactured to meet Voluntary Product Standard PS-2 or APA PRP-108 performance standards. The criteria for meeting the performance standard are structural adequacy, dimensional stability, and bond durability, and may include tests for uniform and concentrated loading, impact loading, racking, linear expansion, and fastener withdrawal and strength. OSB panels are rated for moisture exposure as exterior with fully waterproof bond for applications permanently exposed to weather. However, most panels are classified as Exposure 1 with a waterproof bond and used where construction delays will occur before providing panel protection.

OSB-rated sheathing is used for wall and roof sheathing as well as subflooring. OSB APA-rated Sturd-I-Floor, often with tongue and groove edges, is intended for single-layer floor systems under pad and carpet.

GLUE

Glues commonly used in the manufacturing of plywood and in other wood fabrication are listed here.

Animal glues have traditionally been used for gluing wood. They are made from hides, hoofs, and other animal parts. They are heated before application and pressed cold, set quickly, show little stain, develop high strength, do not dull tools excessively, are not moisture resistant, and are low priced. They are used mostly for furniture, millwork, and cabinet work.

Casein glues are made from milk. They are used cold, set quickly, stain some woods excessively, develop high strength, cause tools to become dull, and perform well in damp conditions. They are used for structural joining and glu-lam and plywood manufacturing.

Melamine resin, phenol resin, resorcinol resin, and *resorcinol resin–phenol resin* combinations are used when waterproof glues are needed. All except resorcinol require high temperatures for curing, although some resorcinol-phenol combinations may use temperatures as low as 100°F (38°C). All of them show little stain, develop very high strength, and are expensive. They are used for structural joining and glu-lam and plywood manufacturing.

Some *vegetable glues* are made from starch. They are used cold, set at various rates, stain some woods slightly, develop high strength, cause some dulling of tools, and have low resistance to moisture. They are used in manufacturing plywood.

MECHANICAL FASTENERS

Wood members are fastened together and to other materials with metal fasteners of various kinds. The traditional nails made of steel, and sometimes aluminum, copper, zinc, or brass, are widely used. Screws, nails, bolts, joist hangers, post caps, and bases as well as any other metal accessories used to construct treated-wood structures should be galvanized or coated, or be made of stainless steel. The standard lengths of nails are made in three diameters called box nails, common nails, and spikes, with box nails being the thinnest and spikes the thickest. Spikes are also made in longer sizes than the others. The tendency of wood to split when nailed is proportional to its specific gravity, and it is necessary to use several thinner nails rather than one thick nail in heavy wood. The three diameters allow flexibility in the design of nailed joints.

Unseasoned wood that is nailed causes the nails to loosen as it shrinks during seasoning. Wood must be nailed dry if it is to be dry in use. However, wet wood may be nailed if it is going to stay wet.

A nail may be subjected to an axial force tending to pull it out of the wood. The nail is said to be subjected to *withdrawal loading.* The resistance to withdrawal depends on the nail diameter and the length driven into the wood and on the specific gravity of the wood, which is much greater for heavier wood. Wood is much better able to resist withdrawal from side grain than from end grain. Nails subjected to withdrawal from end grain have half the resistance of those in side grain and are not permitted by the NDS. Nails are sometimes coated with cement or rosin to make them more resistant to withdrawal.

The most common load on the wood of a nailed joint is a lateral load in side grain. The load that can be resisted depends on the specific gravity of the wood and on the length and diameter of the nail embedded in the wood. It does not depend on the direction of load. The nails connecting flanges and webs of box beams and the nails connecting truss joints are loaded laterally. Resistance to lateral loads in end grain is two-thirds of the side grain resistance. Wood offers more resistance to lateral loads than to withdrawal, and nails should be loaded laterally in preference to withdrawal whenever possible. Figure 6–30 illustrates nailed joints.

Wood screws are used for the same purpose as nails and transmit withdrawal or lateral loads to the wood in much the same way. Strength depends on the length and width of the screw and on the specific gravity of the wood. It is independent of direction for lateral loads. Screws can resist stronger forces than nails of comparable size because of the grip of the threads in the wood. Resistance to lateral loads is greater,

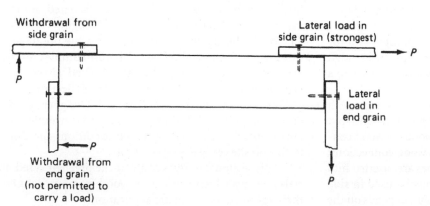

FIGURE 6–30. Loads on nails.

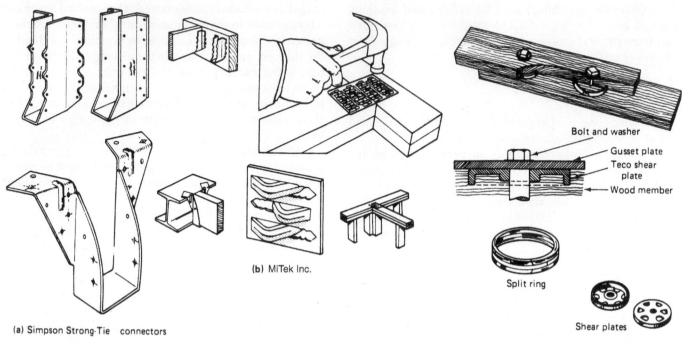

(b) MiTek Inc.

Bolt and washer

Gusset plate
Teco shear plate
Wood member

Split ring

Shear plates

(a) Simpson Strong-Tie connectors

(c) Timber Engineering Company

FIGURE 6–31. Typical connectors used with wood: (a) Simpson Strong-Tie® connectors *(Courtesy Simpson Co.)*; (b) top plate tie with truss clip *(Courtesy MiTek Inc.)*; (c) split rings and shear plates. *(Courtesy Timber Engineering Co.)*

and the design should require them rather than withdrawal loads if at all possible. Resistance to withdrawal from end grain is low, and screws are not permitted by the NDS to be loaded this way. Lateral resistance in end grain is two-thirds of the resistance in side grain.

Bolts are used to transmit lateral forces from one wood member to another and are not loaded in withdrawal. A bolt is inserted into a predrilled hole and held in place between the bolt head and a nut. The load transmitted by a laterally loaded bolt is different from that transmitted by a laterally loaded nail or screw. The resistance is a function of the compressive strength of the wood and the area (length × width of bolt) bearing against the wood. Just as wood is strongest in compression parallel to grain and weakest in compression perpendicular to grain, so the bolted joint is strongest when the load is parallel to grain and weakest when the load is perpendicular to grain, and has an intermediate strength at any other angle. The allowable load at other angles is determined by using the Hankinson formula or Scholten nomographs and the allowable loads parallel and perpendicular to grain.

Lag screws, which are threaded at the pointed end for one-half to three-quarters of the length and unthreaded toward the head, provide good withdrawal resistance like a screw and strong lateral resistance like a bolt.

Two types of connectors are used in combination with bolts to increase the area of wood in compression. One split ring or a pair of shear plates is used for wood-to-wood connections and one shear plate for metal-to-wood connections; several may be used in one joint. Both types are inserted into tight-fitting prefabricated holes. They may be used in side grain or end grain at any angle. Their strength depends on the

area of wood in contact with the connector and the compressive strength of the wood at the angle as determined by the Hankinson formula or Scholten nomographs. Split rings transfer the load from wood to wood through the ring, and the bolt takes no load, but only holds the joint together. Shear connectors fit tightly over the bolt, and the load is transferred from one member to another through the bolt. Many other specialized fasteners are designed for connecting wood rapidly and securely. Figure 6–31 illustrates some of the types in use.

PREENGINEERED WOOD PRODUCTS

Before the advent of preengineered wood products, conventional wood frame construction often utilized dimension lumber to fashion headers, joists, beams, and other structural members by doubling or tripling individual pieces. Such built-up members were mechanically fastened, sometimes using adhesives to provide additional strength.

While this practice continues today, the use of conventional wood framing for larger and more complex residential and light commercial buildings required alternatives to achieve greater span lengths with lower deflection rates and greater economy in the use of lumber. Designers desire the enhanced predictability of preengineered products which are manufactured under controlled conditions and delivered to the job site ready for installation.

This section describes some of the preengineered materials now available to construct wood frame structures and their advantages for specific applications.

Structural Composite Lumber (SCL)

Structural composite lumber (SCL), as described by APA-The Engineered Wood Association, consists of laminated strand lumber (LSL), parallel strand lumber (PSL), and laminated veneer lumber (LVL). All have similar characteristics in that they are manufactured utilizing wood strands, flakes, or veneers and, with bonding adhesives, are pressed into large billets which can then be cut to desired dimensions for specific uses.

LSL is manufactured from flaked strands and can be used for headers for window and door openings, wall studs, and posts and columns. Nominal dimensions for LSL headers are 3-$\frac{1}{2}$ in. in width to 11 in. in depth. Columns and posts vary from 3-$\frac{1}{2}$ in. × 3-$\frac{1}{2}$ in. to 3-$\frac{1}{2}$ in. × 8 in. PSL is composed of parallel veneer strands with uses including headers, joists, beams, and posts and columns. PSL headers and beams are manufactured in widths ranging from 2 in. to 7 in. and depths from 9-$\frac{1}{4}$ in. to 18 in. Columns and posts are available at 3-$\frac{1}{2}$ in. × 3-$\frac{1}{2}$ in. to 7 in. × 7 in. LVL bonds thin sheet veneer layers and is very often used for structural purposes, including as headers, beams, and rafters. Widths are typically 1-$\frac{3}{4}$ in., with depths ranging from 5 in. to 20 in.

When compared to dimension lumber, SCL has greater bending strength, more dimensional stability, and is free from the defects typically found in dimension lumber. Lengths of individual SCL members can be greatly increased over dimension lumber pieces. The manufacture of SCL provides opportunities to utilize an expanded variety of tree species while producing significantly less waste than conventional lumber milling. The primary disadvantage of SCL is its higher material cost. This can, however, be offset by lowered job-site labor costs needed to fabricate specialty structural components and the overall reduction in lumber use.

Structural Insulated Panels (SIP)

Structural insulated panels (SIP) are produced by combining two layers of $\frac{1}{2}$ in. OSB sandwiching a layer of foam board insulation (usually expanded polystyrene) (Figure 6–32). The resultant structural and insulating properties of SIPs allow their use for exterior wall and roof panels. Insulation cores can vary from 4 to 8 in. in thickness with R values averaging 4 per inch of thickness. Typical panel dimensions are 48 in. × 96 in. with specialty manufacture of larger sizes possible. These larger panels must be placed by a crane.

Principal advantages of SIPs are increased speed of erection compared to conventional framing methods and improved thermal properties because the insulation cores in SIPs are continuous and not interrupted by studs or rafters.

I-Joists

I-joists are manufactured by combining OSB web members with LVL flanges to produce an I-shaped structural composite which can be substituted for conventional floor joist and rafter framing (Figure 6–33). The primary advantages of I-joists are their significantly increased span lengths, lower deflection rates, and overall reduction in cross-sectional area, thereby reducing lumber use and structural dead loads. Table 6–10 illustrates typical span lengths for various I-joists ranging in depth from 9-$\frac{1}{2}$ in. to 16 in.

FIGURE 6–33. Joist LVL flange and OSB web. *(Courtesy of J. Coffey)*

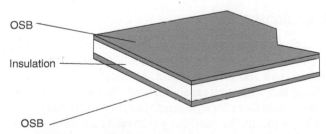

OSB

Insulation

OSB

FIGURE 6–32. Compound of a structural insulated panel (SIP). *(Courtesy of J. Coffey)*

Table 6–10. L/360 Live load deflection chart for TJI joist series iLevel Trus Joist® Joist Specifiers Guide TJ.4000 February 2009.

Depth	TJI®	40 PSF Live Load/10 PSF Dead Load				40 PSF Live Load/20 PSF Dead Load			
		12" o.c.	16" o.c.	19.2" o.c.	24" o.c.	12" o.c.	16" o.c.	19.2" o.c.	24" o.c.
9½"	110	18'-9"	17'-2"	15'-8"	14'-0"	18'-1"	15'-8"	14'-3"	12'-9"
	210	19'-8"	18'-0"	17'-0"	15'-4"	19'-8"	17'-2"	15'-8"	14'-0"
	230	20'-3"	18'-6"	17'-5"	16'-2"	*20'-3"*	18'-1"	16'-6"	14'-9"
11⅞"	110	22'-3"	19'-4"	17'-8"	15'-9"[(1)]	20'-5"	17'-8"	16'-1"[(1)]	14'-4"[(1)]
	210	23'-4"	21'-2"	19'-4"	17'-3"[(1)]	*22'-4"*	19'-4"	17'-8"	15'-9"[(1)]
	230	24'-0"	21'-11"	20'-5"	18'-3"	*23'-7"*	20'-5"	18'-7"	16'-7"[(1)]
	360	25'-4"	23'-2"	21'-10"	20'-4"[(1)]	*25'-4"*	*23'-2"*	*21'-10"[(1)]*	17'-10"[(1)]
	560	28'-10"	26'-3"	24'-9"	23'-0"	*28'-10"*	*26'-3"*	*24'-9"*	20'-11"[(1)]
14"	110	24'-4"	21'-0"	19'-2"	17'-2"[(1)]	22'-2"	19'-2"	17'-6"[(1)]	15'-0"[(1)]
	210	26'-6"	23'-1"	21'-1"	18'-10"[(1)]	24'-4"	21'-1"	19'-2"[(1)]	16'-7"[(1)]
	230	27'-3"	24'-4"	22'-2"	19'-10"[(1)]	*25'-8"*	22'-2"	20'-3"[(1)]	17'-6"[(1)]
	360	28'-9"	26'-3"	24'-9"[(1)]	21'-5"[(1)]	*28'-9"*	*26'-3"[(1)]*	22'-4"[(1)]	17'-10"[(1)]
	560	32'-8"	29'-9"	28'-0"	25'-2"[(1)]	*32'-8"*	*29'-9"*	*26'-3"[(1)]*	20'-11"[(1)]
16"	210	28'-6"	24'-8"	22'-6"[(1)]	19'-11"[(1)]	26'-0"	22'-6"[(1)]	20'-7"[(1)]	16'-7"[(1)]
	230	30'-1"	26'-0"	23'-9"	21'-1"[(1)]	*27'-5"*	23'-9"	21'-8"[(1)]	17'-6"[(1)]
	360	31'-10"	29'-0"	26'-10"[(1)]	21'-5"[(1)]	*31'-10"*	*26'-10"[(1)]*	22'-4"[(1)]	17'-10"[(1)]
	560	36'-1"	32'-11"	31'-0"[(1)]	25'-2"[(1)]	*36'-1"*	*31'-6"[(1)]*	26'-3"[(1)]	20'-11"[(1)]

[(1)] Web stiffeners are required at intermediate supports of continuous-span joists when the intermediate bearing length is *less* than 5¼" and the span on either side of the intermediate bearing is greater than the following spans:

(Courtesy iLevel by Weyerhauser)

Several manufacturers of I-joists have patented noise reducing floor systems with lowered deflection rates. By limiting the potential movement of the floor system, sounds produced by the resultant friction of flooring components can be greatly reduced or eliminated. Additionally, finished flooring materials such as ceramic tile are less subject to cracks in joints with such floor systems.

Shear Walls

Shear walls or brace walls systems are designed to resist lateral forces applied to buildings. Seismic (earthquake) and wind loading are typical of these forces. Shear walls or braces can be field fabricated using conventional framing methods including the use of plywood sheathing panels, let-in corner braces, and specialized connectors to foundations. However, meeting the requirements of the International Building Code (IBC) and International Residential Code (IRC) for lateral force resistance can be labor intensive and require multiple inspections by the design engineer to insure compliance. Premanufactured systems using panelized shear walls and SCL elements can simplify this process because the shear wall system is preengineered and manufactured to meet specific seismic and wind loading requirements. Shear

wall systems are also advantageous in buildings with large or repetitive wall openings such as garage doors in a multi-family dwelling.

Shear wall systems have several common elements: First, a positive connection to the foundation using bolted connections allowing transfer of lateral forces to the ground and preventing uplift (Figure 6–34 illustrates a typical foundation connection detail); second, the ability of the wall assembly to flex and absorb energy under extreme loading conditions; and finally, shear wall systems most have positive connection to headers and other structural elements that transfer loads over openings.

Wood Composites

Wood fibers can be combined with other materials, including plastics, to produce composites which can be used for outdoor deck and railing systems. Although these materials typically lack structural capabilities (i.e., they cannot be used for posts, joists, or stair stringers), they can be exposed to the weather without the need for preservative or finish treatments required of natural wood. The advantages of these systems are resistance to moisture and insects, lower maintenance, and lack of splintering.

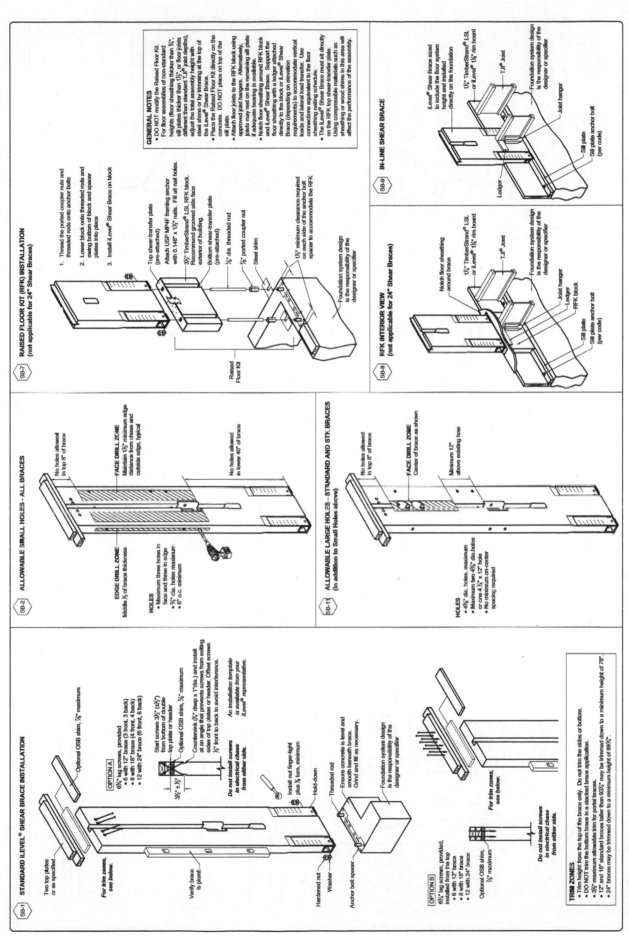

FIGURE 6-34. Shear brace installation detail. (*Courtesy I Level Weyerhauser.*)

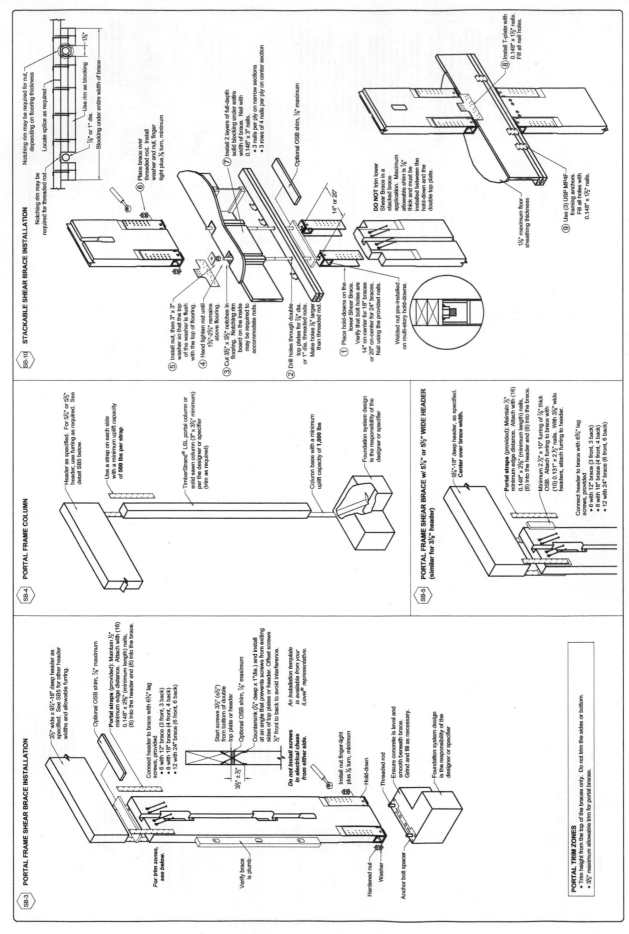

FIGURE 6-34. (*continued*)

Review Questions

1. At what time of year was the tree as shown in Figure 6–2 cut down? Explain your answer.

2. Indicate in a sketch the pattern of grain in a straight board sawed from a tree with a bend in it. Show boards sawed parallel to the plane of the bend and those sawed perpendicular to it.

3. Give the fbm of the following quantities of wood:
   a. 800 lineal ft of $\frac{3}{4} \times 8$ in.
   b. 15 pieces of 2×4 from 12 ft 0 in. to 12 ft 11 in. in length
   c. one post 10 ft long, 14×14 in. in cross section

4. Should a window frame be slash-cut or rift-cut to keep the joint opening to a minimum? Explain your answer.

5. In what ways do springwood and summerwood differ?

6. What are the moisture content values used to define greenwood and dry wood? Why is this an important distinction?

7. A sample of Douglas fir sapwood had a wet weight of 1315.8 g and an oven-dry weight of 612 g. Calculate the moisture content of the sapwood and determine if it is a reasonable value.

8. Define equilibrium moisture content.

9. Describe three methods of seasoning wood.

10. If the stress at failure in bending is 7200 psi for small, perfect samples of a certain species of wood, what is the allowable bending stress for a piece of lumber with defects that are judged to reduce the strength by 40 percent?

11. a. Determine the allowable fiber stress in bending for a Douglas fir south select structural grade 2 ¥ 4 for single and repetitive member use.
    b. What are the allowable spans for a Hem fir number 2, 2 ¥ 8 set at 12 in. o.c., 16 in. o.c., and 24 in. o.c.?

12. What methods are used to grade lumber? What agencies provide the grading rules used to perform lumber grading?

13. Explain the difference between single member and repetitive member allowable stresses in bending. Cite an example.

14. What are the differences among boards, dimension lumber, and timber?

15. What is the difference between a characteristic and a manufacturing imperfection?

16. If other factors are equal, which piece of lumber is weaker: one with a slope of grain of 1 in 5 or one with a slope of grain of 1 in 6?

17. How do preservatives protect wood from rotting, insects, and marine borers?

18. Discuss the advantages and disadvantages of creosote as a preservative.

19. Why can glu-lam timbers be stronger than sawn timbers?

20. Briefly describe the process used to manufacture plywood.

21. What characteristic of plywood is indicated by the veneer grade and by the species group?

22. What does the designation 24/16 mean on a plywood grade trademark?

CHAPTER SEVEN

MASONRY

asonry is an important aspect of the construction industry that utilizes manufactured products, field-produced materials, and skilled workmanship to produce walls and floor surfaces. Masonry construction must have adequate strength to carry loads, be highly durable, and, when used in exterior applications, be watertight.

Masonry construction is also a very aesthetic material, providing the designer with a number of variables to work with, such as size of units, color, texture, and type of bond. Masonry construction may also be used as an acoustical material and for the construction of fire-rated wall systems. The materials associated with masonry construction include concrete block, clay brick, clay tile, stone, mortar, and metal reinforcing systems.

CLAY MASONRY

One of the oldest manufactured building materials still in use today is clay brick. The remains of structures found in the Tigris-Euphrates basin indicate the use of sun-baked brick as early as 6000 B.C. By 600 to 500 B.C., bricks were being hard-burned in kilns to produce a more durable construction material.

The brickworks were usually owned by royal families, and they would have the family's name or insignia molded into the brick, much the same as some brick manufacturers still do today. Though the process of brick making has been improved through technological advances, the basic theory of hard-burned brick manufacturing has remained virtually unchanged.

Brick

The term *brick* is used to denote solid clay masonry units. Cored units are considered solid as long as the cores do not exceed 25 percent of the total cross-sectional area of the unit.

The cores, which vary in size and number, reduce the weight of the brick, increase bond strength, and allow a more even-drying of the units during the burning phase of brick manufacturing. Most brick produced today is of standard nominal modular sizes and shapes, as illustrated in Table 7–1 and Figure 7–1. The nominal dimensions of a masonry unit include the thickness of the mortar joint.

Tile

Hollow clay masonry units are called *clay tile* and are produced with core areas in excess of 25 percent of the gross cross-sectional area of the unit. The main classifications of hollow masonry are *structural clay tile* (Figure 7–2) and *structural facing tile,* the latter having a surface that has been worked with a ceramic glaze, color, or surface texture treatment. When the tile units are designed to be used with the cores horizontal, they are called *side construction tile;* when cores are set vertical, the units are called *end construction tile.*

Raw Materials for Clay Masonry

Chemically, the clays used to manufacture brick and tile are compounds of alumina and silica with differing amounts of metallic oxides and other impurities. The metallic oxides act as fluxes during burning and also influence the color of clay masonry units.

Manufacturers of brick blend clays from different locations as well as vary the manufacturing processes to reduce variations in the finished product. However, slight differences in the properties of clay masonry units are normal.

The clays used to produce brick and tile must have enough plasticity to be shaped and molded when wet and adequate tensile strength to retain the molded shape until the units are fired in the kilns. The clay particles must also fuse together when subjected to the elevated kiln temperatures.

Table 7–1. Common brick sizes

Unit Name		Actual Size (inches)	Actual Size (mm)	Modular Metric Size (mm)	Vertical Coursing
Modular	width	3–1/2	89	90	3:200 mm
	height	2–1/4	57	57	
	length	7–1/2	190	190	
Engineer Modular	width	3–1/2	89	90	5:400 mm
	height	2–3/4	70	70	
	length	7–1/2	190	190	
Closure Modular	width	3–1/2	89	90	2:200 mm
	height	3–1/2	89	90	
	length	7–1/2	190	190	
Roman	width	3–1/2	89	90	4:200 mm
	height	1–5/8	41	40	
	length	11–1/2	292	290	
Norman	width	3–1/2	89	90	3:200 mm
	height	2–1/4	57	57	
	length	11–1/2	292	290	
Engineer Norman	width	3–1/2	89	90	5:400 mm
	height	2–3/4	70	70	
	length	11–1/2	292	290	
Utility	width	3–1/2	89	90	2:200 mm
	height	3–1/2	89	90	
	length	11–1/2	292	290	
Standard*	width	3–5/8	92		3:200 mm
	height	2–1/4	57	57	
	length	8	203		
Engineer Standard*	width	3–5/8	92		5:400 mm
	height	2–13/16	71	70	
	length	8	203		
King*	width	3	76		5:400 mm
	height	2–3/4	70	70	
	length	9–5/8	245		

*Not a modular unit in either the inch-pound or the metric system.

Source: Construction Metrication Council of the National Institute of Building Sciences, Washington, D.C.

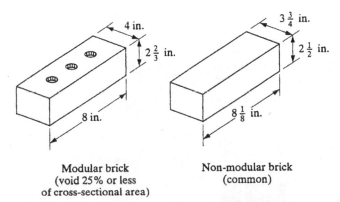

Modular brick
(void 25% or less
of cross-sectional area)

Non-modular brick
(common)

FIGURE 7–1. Typical clay brick. *(Somayaji, Shan, Civil Engineering Materials, 2nd ed., © 2001. Reprinted by permission of Pearson Education, Inc., Upper Saddle River, NJ)*

The raw materials used to manufacture brick and tile are as follows:

1. *Surface clays*—clays found at or close to the surface of the earth.

2. *Shales*—clays that have been subjected to high pressures and have hardened into a rock formation.

3. *Fire clays*—deep-mined clays having more uniform chemical and physical properties and fewer impurities, and are used to produce brick with refractory qualities (fire brick).

Manufacturing Clay Masonry

The manufacturer of brick and tile will use individual clays or blends of clays, depending on the specific end product required.

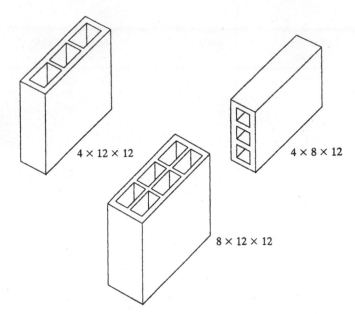

4 × 12 × 12

4 × 8 × 12

8 × 12 × 12

FIGURE 7–2. Structural clay tile. *(Somayaji, Shan, Civil Engineering Materials, 2nd ed., © 2001. Reprinted by permission of Pearson Education, Inc., Upper Saddle River, NJ)*

The steps in the manufacturing process of clay masonry units are illustrated in Figure 7–3.

Winning or mining is the process of obtaining the raw clays from surface pits or underground mines. The clays are blended for color, to increase uniformity, and to allow more control of the raw material's suitability for a given product run.

The clays are crushed to break up large chunks, screened to remove stones, and then pulverized with 4- to 8-ton grinding wheels. The clays are then screened to control particle sizes going to the pugmills. *Pugmills* are large mixing chambers where the clays are blended with water. The operation is called *tempering* and produces a plastic, relatively homogeneous mass ready for molding.

Three types of forming methods are used to produce clay masonry units: stiff-mud, soft-mud, and dry-press. The most common in use is the stiff-mud method. It accounts for all structural tile production and most brick production.

The *stiff-mud method* utilizes a clay blended with approximately 10 to 15 percent water by weight to produce a plastic mass which is then de-aired in a vacuum to reduce the air content of the wet clay. The clay is then extruded through a die. As it leaves the die, any required surface texture may be applied to the clay ribbon. Cutter wires, spaced to compensate for normal drying and burning shrinkage, cut the brick to size. The bricks are then sorted off of a continuous-belt conveyor, with acceptable bricks being placed on dryer carts and the rejects being returned to the pugmill.

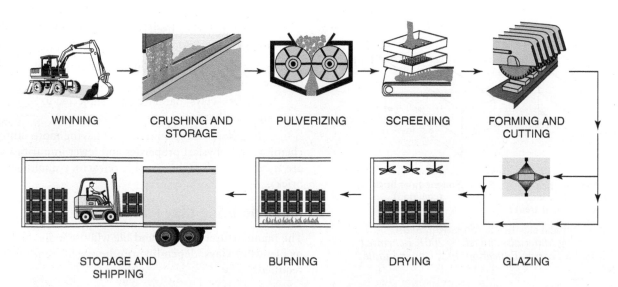

WINNING   CRUSHING AND STORAGE   PULVERIZING   SCREENING   FORMING AND CUTTING

STORAGE AND SHIPPING   BURNING   DRYING   GLAZING

FIGURE 7–3. Diagrammatic representation of the manufacturing process for clay masonry.

The *soft-mud process* is used for clays which contain 20 to 30 percent water in their natural state. The clays are mixed and then molded in forms. Either sand or water can be used as a release agent to prevent the clay from sticking to the mold. Brick produced this way is called water-struck or sand-struck, depending on the material used as a release agent.

The *dry-press process* is used for low-plasticity clays. These clays are blended with less than 10 percent water and formed in molds under pressures ranging from 500 to 1500 psi.

The wet units coming from the cutting or molding machines normally have moisture contents ranging from 7 to 30 percent. This water is removed in dryers at temperatures ranging from 100 to 300°F over a period of 1 to 2 days.

When brick or tile units require a glazed finish, two methods are utilized. High-fired glazes are applied to the units before or after drying and the units are then kiln-burned at normal temperatures. Low-fired glazes are applied after the clay unit has been kiln-burned and allowed to cool. The clay units are then sprayed with the glazing compound and refired at low temperatures to set the glaze.

Brick and tile are fired in continuous tunnel kilns or periodic kilns, with the process requiring 2 to 5 days. The *tunnel kiln* allows the brick to move through various temperature zones on special carts. The *periodic kiln* requires the brick to be stacked inside, with the temperature of the interior space fluctuating as required by the burning and cooling operations.

The burning of the brick is a very critical stage in the production process, and kiln temperatures are monitored constantly because variations during the burn will affect finish color, strength, absorption, and size of the units.

The first stage in the burning process is called *water smoking* and occurs at temperatures up to 400°F; the second stage is *dehydration* and occurs in the temperature range of 300 to 1800°F; the third stage is *oxidation* and occurs between 1000 and 1800°F; and the final stage, *vitrification*, occurs between 1600 and 2400°F. Near the end of the burning, the units may be flashed to produce different colors and color shading. *Flashing* is the creation in the kiln of a reducing atmosphere (oxygen reduction to reduce combustion).

The cooling down of the clay units normally requires 2 to 3 days in a periodic kiln and no more than 2 days in a continuous kiln. The rate of cooling will affect color, cracking, and checking of the clay unit and is therefore carefully controlled.

After cooling, the units are removed from the kilns, sorted, graded, and prepared for direct shipping or storage.

Strength of Clay Masonry

The compressive strength of brick and clay tile varies based on the clay source, method of manufacturing, and the degree of burning. Plastic clays used in the stiff-mud process generally yield higher compressive strengths than soft-mud or dry-press methods. When the same clay and methods of manufacturing are utilized, higher degrees of burning will yield higher compressive strengths. Compressive strengths of brick usually range from 1500 to 20,000 psi.

Absorption of Clay Masonry

The absorption of water by brick depends on the clay, the process of manufacturing, and the amount of burn to which the brick has been subjected. Plastic clays and higher degrees of burning generally produce brick units having low absorptions.

Brick which will be exposed to weathering, especially alternate freezing and thawing, should have low absorption capacities. High compressive strengths or low absorption values usually indicate brick or tile that will exhibit adequate durability when exposed to alternate freezing and thawing conditions.

Suction is the initial rate of absorption of clay masonry units and has a great influence on bond strength. When a brick is laid in a bed of mortar, water is drawn up into the brick's surface. If the brick is highly absorptive, this process will leave a dry bed of mortar which will not develop adequate bond between the brick units. If the unit has very low absorption, the water will allow the brick to float and when the mortar dries, inadequate bond strengths will result.

Maximum bond strength will generally occur when the brick suction rate does not exceed 0.7 oz. (20 g) per minute. Brick having suction rates in excess of this limit should be sprayed with water prior to use; however, the surfaces should be allowed to dry before the brick is used.

To determine in the field whether brick should be prewetted or not is relatively easy. Simply sprinkle a few drops of water on the flat side of a brick. If the drops are absorbed in less than a minute, prewetting is required. A more accurate field test requires a quarter, a wax marker, and an eyedropper. Draw a circle on the brick using the quarter and wax marker and place 20 drops of water within the circle. If the water is absorbed in less than $1\frac{1}{2}$ minutes, then the brick should be presoaked.

Clay Masonry Colors

Colors of brick cover a wide range and include tones of pearl gray, cream, red, purple, and black. The chemical composition of the clay, method of burning, and degree of burn all affect the color of brick. The most important oxide present in clay with respect to color is iron. Iron oxides in clays will produce red brick when exposed to an oxidizing fire, and purple when burned in a reducing atmosphere. Generally, lighter colors are the result of underburning. Overburning produces clinker brick, which is dark red, black, or dark brown, depending on the original clays. Underburned brick is softer and more absorptive and has lower compressive strengths than brick produced at higher temperatures.

Masonry Standards

Standard specifications for the numerous types and grades of brick, tile, and concrete masonry units have been developed by the American Society for Testing and Materials (ASTM) and are widely accepted. It is recommended that the appropriate ASTM specification be included by reference for all masonry construction. The Brick Institute of America (BIA) and the National Concrete Masonry Association (NCMA) also produce excellent references and guidelines for masonry construction.

CONCRETE MASONRY

The first concrete masonry units used in construction were large, cumbersome blocks of solid concrete that had been cast in wood forms. Late in the nineteenth century, builders began experimenting with the production of hollow concrete masonry units. These blocks were much lighter than the solid type, yet still retained adequate load-carrying capacity.

The first patent for a hollow concrete masonry unit mold and manufacturing process was issued to Harold S. Palmer in 1900. The molds were filled with a relatively dry concrete and hand-tamped. Production rates varied, but 80 units per day was the average. By the 1920s, automatic machines were producing about 3000 concrete masonry units per day.

It is not unusual today to see manufacturing equipment in a block plant producing over 20,000 block units per day, depending on size, shape, and materials being used to produce the block.

Typical shapes of concrete block are illustrated in Figure 7–4. The sizes of concrete blocks are stated as nominal dimensions. For example, an 8-in., 8 × 18 concrete block is actually $7\frac{5}{8}$ in. wide and $7\frac{5}{8}$ in. high by $17\frac{5}{8}$ in. long. The $\frac{3}{8}$-in. difference from nominal to actual represents the head and bed mortar joint when the block is installed. Therefore, when an 8 × 18-in. block is placed in a wall, the occupied wall surface area is 1 sq ft; when an 8 × 16 block is installed in a wall, it occupies 0.89 sq ft of wall surface.

The term *concrete masonry unit (CMU)* refers to molded concrete units used in construction to build load-bearing and nonload-bearing walls. The units are cored or solid and manufactured off-site. However, like most masonry construction, the units are integrated into the structure on-site, utilizing field materials (mortars) and skilled workmanship.

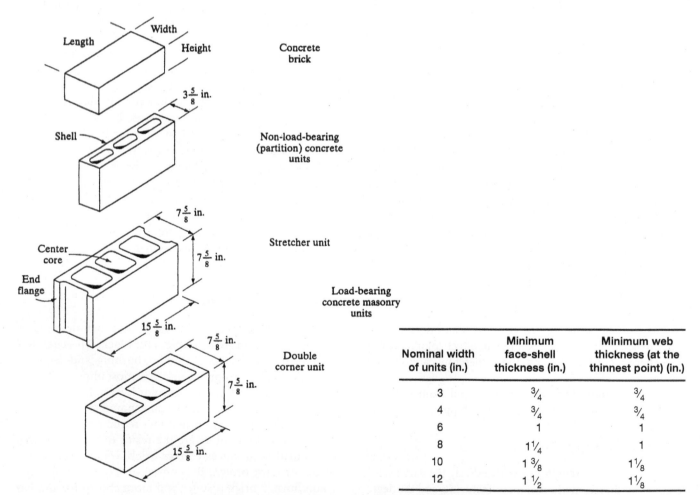

Nominal width of units (in.)	Minimum face-shell thickness (in.)	Minimum web thickness (at the thinnest point) (in.)
3	¾	¾
4	¾	¾
6	1	1
8	1¼	1
10	1⅜	1⅛
12	1½	1⅛

FIGURE 7–4. Typical shapes of concrete block. *(Somayaji, Shan, Civil Engineering Materials, 2nd ed., © 2001. Reprinted by permission of Pearson Education, Inc., Upper Saddle River, NJ)*

Raw Materials for Concrete Masonry Units

Concrete masonry units are manufactured from portland cement, water, aggregates, and in some cases admixtures, which may include coloring agents, air-entraining materials, water repellents, and other additives.

Most of the cementitious material used to manufacture block is Type I portland cement; however, Type III high-early-strength cements are also used to increase early strengths and reduce breakage during handling and delivery. Most block plants utilize liquid and powder admixture systems to alter the concrete properties rather than special cements. These systems allow greater control of the concrete mix properties going into the block molds.

Portland blast-furnace slag cements, fly ash, silica flour, and other pozzolanic materials may also be substituted for some of the Type I cement. When the pozzolanic materials are used, the strength of the concrete masonry unit takes longer to mature, depending on the curing method and type of aggregate used.

The aggregates used to manufacture concrete masonry units are very important because they make up approximately 90 percent of the unit by weight. The aggregate characteristics generally control the concrete masonry units' physical properties as well as their production costs. Generally, local availability governs aggregate use; however, when designers specify certain manufactured aggregates, they will be transported to the block manufacturing plant.

Desirable aggregate properties generally are the same as those required to produce quality concrete. These required properties include the following:

1. Toughness, hardness, and strength to resist impact, abrasion, and loading.

2. Durability to resist freezing and thawing and the expansion and contraction resulting from moisture and/or temperature changes.

3. Uniform gradation of fine and coarse aggregate sizes to produce an economical, moldable mixture and uniform appearance (aggregates used should not exceed one-third of the smallest block shell).

4. The aggregate should be free of any deleterious material which would affect strength or cause surface imperfections.

Although admixtures play a great role in the production of concrete, their use in concrete masonry unit production is limited. Air-entraining admixtures increase plasticity and workability of block mix concretes and the air-void system increases the concrete masonry units' ability to resist weathering. The air-entraining admixtures also allow greater compaction, producing denser units with more uniform surfaces and less breakage of freshly molded units. Metallic stearates are somewhat effective in reducing absorption rates and capillary action in concrete masonry units; however, their use is limited.

Calcium chloride and other accelerating admixtures are used by some manufacturers to allow faster production rates during cold weather.

Manufacturing Concrete Masonry Units

The production of concrete masonry units varies to some degree from plant to plant. However, the basic sequencing is fairly common to all facilities (Figure 7–5).

Raw materials are delivered to the production site by truck, railroad, or barge, where they are usually stored in

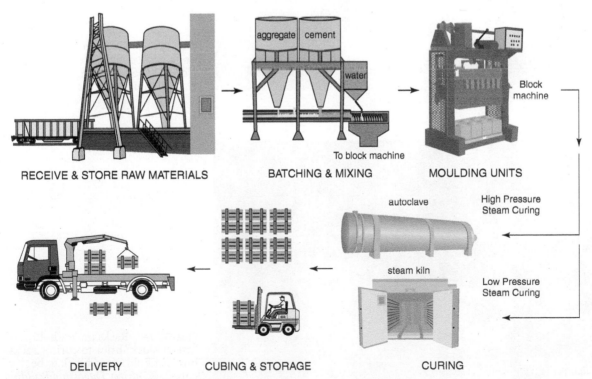

RECEIVE & STORE RAW MATERIALS   BATCHING & MIXING   MOULDING UNITS

aggregate   cement

water

To block machine

Block machine

High Pressure Steam Curing

autoclave

Low Pressure Steam Curing

steam kiln

DELIVERY   CUBING & STORAGE   CURING

FIGURE 7–5. Manufacturing flow chart for concrete masonry units (CMU).

open stockpiles. The cement is delivered by bulk tankers and blown into storage silos. The aggregates are fed by conveyor from stockpiles to bins located above the weigh hoppers. Aggregate gradations generally run toward the fine side, with maximum sizes rarely exceeding $\frac{3}{8}$ in.

The concrete block mix is proportioned by weight into a weigh batcher, with the ingredients carefully controlled to maintain a uniform mix from batch to batch. Electronic sensors determine the moisture content of the aggregates and vary accordingly the amount of mixing water required. The materials are dropped into the mixer, the required water is added, and the materials are mixed for 6 to 8 minutes. Admixtures are normally added with the mixing water. Because the block mix is relatively dry when compared to concrete, longer mixing or blending times are required. Concrete masonry units are produced with lower water–cement ratios and cement factors than structural concretes. Cement factors run from 250 to 375 lb/cu yd, and the mix is considered to be zero slump.

The completed batch is then deposited in a holding bin located over the block machine. The mix is fed into the block molds in measured quantities and then consolidated with pressure and vibration. The mold is then lifted from the molded block units (Figure 7–6) and repositioned for a repeat cycle. The green block (uncured block) sits on a steel tray pallet, which is then transferred to a curing rack (Figure 7–7). When a curing rack is filled, it is moved to the curing area (Figure 7–8).

Low-pressure steam curing accounts for approximately 80 percent of all block production, with the balance of production utilizing high-pressure steam curing and, when climatic conditions are favorable, moist curing at normal temperatures of 70 to 100°F.

Low-pressure steam curing is performed in tunnel kilns at atmospheric pressures, with steam temperatures ranging from 150 to 185°F. The curing sequence starts once the kiln is loaded with green units. The green units are allowed to obtain some initial hardening for a period of 1 to 3 hours

FIGURE 7–6. Concrete masonry units immediately after molding. Note: Shape retention. *(Courtesy Dagostino Building Blocks, Inc.)*

FIGURE 7–7. Racks on left hold "green block" prior to curing; racks on right hold cured block to be cubed. *(Courtesy Dagostino Building Blocks, Inc.)*

FIGURE 7–8. Curing area. *(Courtesy Dagostino Building Blocks, Inc.)*

prior to steam exposure. This time is customarily referred to as the *holding period.*

Following the holding period, the heating-up period or steaming-up period begins. According to a predetermined time-temperature program, saturated steam or moist air is fed into the kiln. The temperature of the units is raised at a rate not exceeding 60°F per hour, to a maximum of 150 to 165°F for normal aggregates.

When the units have reached the desired temperature in the kiln, the steam is turned off and the block undergoes a soaking period of 12 to 18 hours. The kiln operates during this timeframe on residual steam and heat, and the units are said to be *soaking.* After soaking, and if required, an artificial drying condition may be induced by elevating the kiln temperature for approximately 4 hours. The principal benefit of steam curing is economy, accelerating the strength gain of the units so they may be placed into inventory quicker, with 2- to 4-day strengths of 90 percent or more of their ultimate strength. The strength of steam-cured block is more than double that of moist-cured block, which at the 2- to 4-day age would normally exhibit 40 percent of its ultimate strength.

High-pressure steam curing accelerates the block setting time, using saturated steam at pressures ranging from 125 to 150 psi. The curing is performed in a pressure vessel called an *autoclave.* The green masonry units are allowed to gain initial hardening for a 2- to 3-hour holding period before being placed in the autoclave. Once the units are placed in the autoclave, the temperature is raised slowly over a period of 3 hours so that the green units are not subjected to full steam pressure too soon. When a 350°F temperature is reached, the block is allowed to soak for 5 to 10 hours, depending on unit sizes at a constant temperature and pressure.

The pressure release is called the *blowdown* and usually takes $\frac{1}{2}$ hour or less to perform. The rapid blowdown allows the masonry units to lose moisture quickly without building up shrinkage stresses in the individual units. Enough moisture is removed during the blowdown phase that the concrete masonry units are very close to a relatively stable air-dry condition.

The strengths of high-pressure-cured concrete masonry units at 1-day age are equal to the 28-day strengths of moist-cured block. The high-pressure steam process produces dimensionally stable units that exhibit less volume change than moist-cured block when subjected to environmental changes related to moisture conditions. The shrinkage of high-pressure steam-cured units, moving from a saturated condition to a relatively dry condition in a heated space, is about 50 percent less than for moist-cured concrete masonry units.

When the curing procedures have been completed, the concrete masonry units are moved to cubing stations. Concrete masonry units are handled in cubes which consist of six layers of block, each layer containing 15 to 18 blocks, depending on the individual unit's size. Normally a cube of 8 × 8 × 16 concrete block will contain both stretcher and corner blocks; however, larger units are usually cubed separately.

Concrete masonry units are normally stored outside two to three cubes high and require no special protection from the weather (Figure 7–9). However, architectural concrete masonry units are covered with protection. The inventory time usually runs from a few days to several weeks. If concrete masonry units are shipped to a construction project too soon, the breakage from handling may be quite high; therefore, most producers allow their inventories to age, thereby reducing contractor complaints about "green" block.

In general, much of the technical knowledge relating to concrete is applicable to the production of concrete masonry units. Like concrete, the physical properties of the concrete masonry units are determined by the physical properties of the hardened cement paste and the aggregate. Mix composition and consistency, consolidation methods, textural requirements, and curing also affect the physical properties of concrete masonry units.

FIGURE 7–9. Loading block delivery truck. Note: Extra capacity using trailer. *(Courtesy Dagostino Building Blocks, Inc.)*

Strength of Concrete Masonry Units

Compressive strengths of concrete masonry units are difficult to predict because the useful water–cement ratio concept is not valid for harsh block mixes. The production of concrete masonry units requires careful control of water quantities in the block mix. Wetter mixes are easier to mold and generally yield higher compressive strengths; however, breakage of green units increases during handling operations. The use of very dry mixes produces consolidation problems during molding operations and ultimately lowers compressive strengths. Therefore, each block-manufacturing facility, through experimentation, will develop mix designs that will produce concrete masonry units of adequate strength without sacrificing other required properties.

The factors which affect compressive strength values for concrete block include the type and gradation of aggregate, the type and amount of cementitious material, the degree of consolidation attained during molding, the curing method, the size and shape of the concrete masonry unit, and the conditions of the block with regard to moisture and temperature at the time of test.

Tensile strength, flexural strength, and modulus of elasticity values vary with the compressive strength values of a concrete masonry unit. Tensile strength normally ranges from 7 to 10 percent of the compressive strength, flexural strength from 15 to 20 percent of compressive strength, and modulus of elasticity from 300 to 1200 times the compressive strength. Strength and absorption requirements for concrete masonry units are given in ASTM C90.

Absorption of Concrete Masonry Units

Absorption tests provide a measure of the density of the concrete in concrete masonry units. The absorption value is calculated in pounds of water per cubic foot of concrete and varies over a wide range, depending on the aggregates used in the unit. Values for water absorption may vary from 4 lb/cu ft for dense sands and stones to as much as 20 lb/cu ft for lightweight aggregates.

The porosity of the concrete will also influence other properties, such as permeability, thermal conductivity, weight reduction, and acoustical properties. When these properties are required, the absorption values will rise. However, because these properties are usually required of interior or protected masonry, the higher absorption values are not detrimental to the durability of concrete masonry units. A high initial rate of absorption or suction indicates concrete masonry units of high permeability and low durability, because the concrete contains a large number of interconnected pores and voids. However, unconnected air-filled voids present in lightweight aggregates and air-entrained cement paste impart some of the desired porosity properties to the concrete masonry unit while limiting the permeability of the unit to water.

Even though concrete masonry units may have reasonably high suction rates, unlike clay masonry, they are never presoaked and in some cases may require a covering system to prevent moisture content changes in the units because of weather conditions.

Dimensional Changes in Concrete Masonry Units

Concrete masonry units normally undergo dimensional changes due to changes in temperature, moisture content, and a chemical reaction called *carbonation*.

Temperature changes cause units to expand and contract when heated and cooled. These volume changes are reversible through the same temperature ranges. The thermal expansion and contraction of concrete masonry units are governed primarily by the type of aggregate in the unit, since aggregates normally comprise 80 percent of the concrete volume. These volume changes, while relatively small in a single unit, can cause serious problems in the construction of long walls. Therefore, the designers of masonry structures will place control joints in long walls as relief areas for the compounded volume changes occurring in individual masonry units. The actual control joint locations are usually determined by the structural engineer and architect so as not to affect the building's structural system. Joints are typically spaced no more than 25 ft apart in walls without openings, 20 ft apart in walls with openings, and 10 to 15 ft from corners. Control joints should also be located in areas of high stress concentrations or potential wall weakness, such as changes in wall height or thickness, above floor or foundation joints, and below roof and floor joints that bear on the wall as well as one or both sides of door and window openings.

Moisture content changes cause concrete masonry units to expand when wet and shrink when dried. During the first few cycles of wetting and drying, the concrete masonry units may not return to their original sizes, because the concrete has a tendency toward a permanent contraction state. However, during subsequent wetting and drying cycles, the volume changes are reversible.

Original drying shrinkage is an important factor in crack development in concrete masonry walls. If concrete masonry units are placed in a wall before they have been allowed to shrink to a dimensionally stable volume, wherever the wall is restrained, tensile stresses will develop and cracks will occur. Drying shrinkage is greatly reduced by properly curing and drying units so that when they are placed in the structure, the moisture content of the unit is in equilibrium with the surrounding air.

Carbonation causes irreversible shrinkage in concrete masonry units when carbon dioxide is absorbed into the hardened concrete paste of masonry units. The changes in volume are approximately the same as those caused by moisture condition fluctuations. One method of reducing carbonation on the jobsite during cold weather masonry construction is to require all heat sources to be properly vented to the exterior of the work area.

Mortar

Mortar is a combination of one or more cementitious materials, a clean, well-graded sand, and enough water to produce a plastic mix. Mortar serves as the binding medium in masonry construction and its functions are to bond individual units together while sealing the spaces between units, compensate for size variation in units, cause metal ties and reinforcing and masonry units to act together in a structural system, and provide aesthetic qualities to the structure through the use of color and type of joint.

Mortar is used while plastic and then hardens; thus, both plastic and hardened properties determine a mortar's suitability for a specific construction project. The workability of a mortar is determined by its uniformity, cohesiveness, and consistency. A mortar is considered workable when the mix does not segregate easily, is easily spread, supports the weight of units, makes alignment easy, clings to vertical faces of masonry units, and is easily forced from mortar joints without excessive smearing of the wall. There are no tests in use that will measure the workability of a mortar mix; however, the mason in the field will be able to tell very quickly if a particular mix is workable or not.

Water retention in a mortar prevents rapid loss of water from mortar in contact with a highly absorptive masonry unit. If the mortar has a low water retention value, the plasticity of the mortar would be greatly reduced. When low-absorption units are placed on mortar, a high water retentivity is required to prevent bleeding, which is the formation of a thin layer of water between the unit and the mortar. The water causes the unit to float, drastically reducing bond strength.

Mortar flow is determined by a laboratory test using a truncated cone and flow table. A cone of mortar is formed on the table with an original base diameter of 4 in., then the table is raised and dropped 25 times in 15 seconds and the diameter of the mortar mass is measured. If the original 4-in. diameter measures 8 in. after the test, the mortar would have a flow of 100 percent. Flow values that range from 130 to 150 percent are required for construction projects.

Flow after suction is a test used to measure water retentivity and is stated as the ratio of flow after suction to initial flow, expressed as a percentage value. The initial flow test is repeated after the mortar sample being tested has been subjected to a vacuum for 1 minute, thereby removing some water from the mortar mix.

Bond strength is the most important property of hardened mortar and is affected by mortar properties, type and condition of masonry unit, workmanship, and curing.

When the air content of a mortar mix is increased, there usually will be a decrease in bond strength; however, water retention properties improve and the durability of the hardened paste increases.

The relationship of flow to tensile bond strength is direct for all mortars, with an increase in flow causing an increase in bond strength.

The compressive strength of mortar depends largely on the quantity of portland cement in the mix. Compressive strength increases with an increase in cement content and decreases with an increase in water content. However, because there have been very few reports of structural failure or distress associated with mortar, most masonry construction work is performed utilizing moderate-strength mortars. Tests have shown that concrete masonry wall compressive strengths increase only about 10 percent when mortar cube compressive strengths are increased 130 percent. Composite wall strengths increase 25 percent when mortar cube compressive strengths increase 160 percent. Generally, bond strength, workability, and water retentivity are considered more important than compressive strength and are usually given more consideration in specifications.

The strong concern placed on bond strength is evident when specifications allow retempering of mortar mixes. *Retempering* is the addition of water to mortar mixes that have lost water while sitting in mortar pans. The practice of retempering will reduce compressive strengths, but the loss is more than compensated for by the increase of the bond strength of the retempered mortar mix. The usual time limit for mortar use is around $2\frac{1}{2}$ hours from the time of initial mixing, after which time the mortar should be discarded. Many specifications limit the number of retemperings permitted as well as set a specified time limit for the use of a mortar mix.

Straight lime mortars—lime, sand, and water hardened at a slow, variable rate—have low compressive strengths and poor durability, but do have good workability and high water retentivity. The lime hardens when exposed to air and the hardening process occurs over long periods of time. Lime mortars have the ability to heal or recement small cracks and reduce water infiltration.

Portland cement, sand, and water combine to form portland cement mortars, which harden quickly and attain high compressive strengths with good durability; however, workability is poor and water retention is low.

Portland cement, lime, sand, and water are combined to produce mortars which have good durability, high compressive strengths, and consistent hardening rates. The lime component increases workability, elasticity, and water retentivity. Both cementitious materials contribute to good bond strength.

Masonry cement, sand, and water mortars are used for convenience. The proprietary masonry cement is pre-blended by the manufacturer and will normally include lime, an air-entraining agent, and other ingredients which produce desired properties in mortars utilizing masonry cements.

Individual mortar materials are required to conform to ASTM specifications, which are the result of extensive laboratory testing and field-use experiences over a long period of time.

Portland cement used in mortar mixes is governed by ASTM C150, and Types I, II, and III are permitted. Air-entraining portland cements may also be used in mortar mixes, with ASTM C175 as the governing standard. Types IA, IIA, and IIIA are available for use; however, experience has shown that wide variations in actual measured air contents may occur with those cements, so their use requires extreme caution.

ASTM C91 controls masonry cement properties, and though Type I and Type II masonry cements are manufactured, Type II is the recommended masonry cement for use in mortars.

The lime component of mortar mixes may be either quicklime or hydrated lime, with the latter being the preferred material. Quicklime is calcium oxide, which must be carefully mixed with water (slaked) and stored for as long as 2 weeks before use. When used on a jobsite, quicklime is prepared in barrels and added to the mortar at a soft, putty consistency. ASTM C5 covers the properties of quicklime. Hydrated lime is quicklime that has been slaked into a calcium hydroxide before packaging. Hydrated lime can be used without the delay of the slaking process and therefore is more convenient to use on construction projects. ASTM C207 covers Type S and Type N hydrated limes; however, only Type S hydrated lime is specified for use in mortars.

The sand used in mortars may be manufactured or natural sand and should meet ASTM C144 requirements. The sand used should be clean, sound, and well graded, with a top size of $\frac{1}{4}$ in. Both workability and durability of mortars are affected by the quality of sand used. Sands containing less than 5 to 15 percent fines produce unworkable, harsh mortars that will require additional cement or lime to make the mix usable, whereas sands deficient in top sizes or large particles tend to produce weak mortars.

Although there is no standard specification for the water used in mortar mixes, the general guidelines for water in concrete mixes are also applicable to mortars. Water suitable to produce mortars should be clean and free of deleterious acids, alkalis, or organic materials.

The selection of type of mortar depends on a number of variables. However, none of the recognized mortar types will produce a mortar that will rate highest in all properties required for a specific job requirement. Therefore, the properties of the various types of mortars are usually evaluated and a mortar type chosen which will reasonably satisfy end-use requirements (Figure 7–10).

FIGURE 7–10. Multi-wythe wall construction with wire joint reinforcement set in mortar. *(Courtesy DUR-O-WAL, Inc.)*

Table 7–2a. Recommended mortar types for various construction applications

Construction Applicaton	Recommended Minimum ASTM Mortar Types	Order of Relative Importance of Principal Properties		
		Plasticity*	Compressive Strength	Weather Resistance
Foundations, basements, walls, isolated piers**	M, S	3	2	1
Exterior walls	S, N	2	3	1
Solid masonry unit veneer over wood frame	N	2	3	1
Interior walls—load-bearing	S, N	1	2	3
Interior partitions—nonload-bearing	N, O	1	—	—
Reinforced masonry (columns, pilasters, walls, beams)	M, S†	3	1	2

*Adequate workability and a minimum water retention (flow after suction of 70 percent) assumed for all mortars.

**Also any masonry wall subject to unusual lateral loads for earthquakes, hurricanes, etc.

†Only portland cement-lime Type S and M mortars.

ASTM C270 currently recognizes four types of mortars for plain masonry: M, S, N, and O. The recommended construction applications are shown in Tables 7–2a and 7–2b.

Type M mortar is a high-strength mortar that has a greater durability than other mortar types. It is generally recommended for use below grade in foundation walls, retaining walls, walks, sewers, and manholes. It is also specified where high compressive strengths are required.

Type S mortar is a medium-high-strength mortar which is used where Type M is recommended, but where bond and lateral strength are more important than compressive strength. Tensile bond strength between brick and Type S

Table 7–2b. Proportion specification requirments

Mortar	Type	Proportions by Volume (Cementitious Materials)								Aggregate Ratio (Measured in Damp, Loose Conditions)
		Portland Cement or Blended Cement	Mortar Cement			Masonry Cement			Hydrated Lime or Lime Putty	
			M	S	N	M	S	N		
Cement-Lime	M	1	...	...	...	...	...	...	$1/4$	
	S	1	...	...	...	...	...	...	over $1/4$ to $1/2$	
	N	1	...	...	...	...	...	...	over $1/2$ to $1\,1/4$	
	O	1	...	...	...	...	...	...	over $1\,1/4$ to $2\,1/2$	
Mortar Cement	M	1	...	...	1	...	...	...	...	Not less than $2\,1/4$ and not more than 3 times the sum of the separate volumes of cementitious materials
	M	...	1	...	...	...	...	...	...	
	S	$1/2$	...	...	1	...	...	...	...	
	S	...	...	1	...	...	...	...	...	
	N	...	...	...	1	...	...	...	...	
	O	...	...	...	1	...	...	...	...	
Masonry Cement	M	1	...	...	...	...	...	1	...	
	M	...	...	...	...	1	...	...	...	
	S	$1/2$	...	...	...	...	...	1	...	
	S	...	...	...	...	...	1	...	...	
	N	...	...	...	...	...	...	1	...	
	O	...	...	...	...	...	...	1	...	

Note: Two air-entraining materials shall not be combined in mortar.

mortars approaches the maximum obtainable with cement-lime mortars. Type S mortar is used in reinforced and nonreinforced masonry where maximum flexural strengths are required.

Type N mortar is a medium-strength mortar recommended for use above grade in severe exposure conditions. Typical areas of use include chimneys, parapet walls, and exterior building walls.

Type O mortars are medium-low-strength mortars for general interior use in nonload-bearing walls. However, if compressive stresses are expected to stay below 100 lb/sq in., exposures are not severe, there are no high winds or other significant lateral loads, and the walls are solid masonry, they may be used in load-bearing wall systems. Type O mortars should never be used where they will be exposed to freeze–thaw cycling.

Two types of mortars, PM and PL, are specified for structural reinforced masonry, and are governed by ASTM C476, Standard Specification for Mortar and Grout for Reinforced Masonry. This type of construction utilizes standard deformed reinforcing rods, vertically and horizontally spaced in the wall system to produce high-strength masonry systems.

Mortar Production.   Mortars are usually mixed at the jobsite, and the production systems may range from a hoe and wheelbarrow or mortar box on a small project, to mechanical paddle mixes capable of producing 7 cu ft per batch on large projects. The materials required to produce mortars should be carefully batched by volume or weight to ensure uniformity. Typically mortar mixes are run for 3 to 5 minutes, which will yield a mortar with good workability. Decreases in mixing times reduce workability, uniformity, and water retention and result in lower-than-required air contents. Longer mixing times may adversely reduce air content, as well as lower the strength of the mortars. Therefore, when producing mortars, a skilled mason laborer should be placed in charge of the mix operations to ensure that the ingredients are consistently batched.

To overcome field mixing problems, contractors may elect to use preblended mortars meeting ASTM C270 or ASTM C1142 requirements. In some areas contractors can use trowel-ready mortar systems. The mortar is delivered to the jobsite in tubs, or by bulk delivery in a ready mix truck. The mortars contain set controllers, which affect setting time from $2\frac{1}{2}$ hours up to 72 hours, depending on the admixture. Trowel-ready mortars must meet ASTM C1142 and are designated RM, RS, RN, and RO, with 28-day compressive strengths of 2500, 1800, 750, and 350 psi, respectively, with a minimum water retention value of 75 percent and a maximum air content of 18 percent, except for structurally reinforced masonry. Where structural reinforcing is used in the mortar, the air content is limited to 12 percent, unless bond tests can justify higher values.

The dry, preblended mortars must meet ASTM C270. The systems use a jobsite storage silo with a capacity of seven 3000-lb bags of preblended dry mortars to larger 10-bay

FIGURE 7–11. A high lift recharging a storage silo with dry-blend mortar mix. *(Courtesy JCB, Inc.)*

silos, which may be designed to store two products at the same time, such as grout and mortar (Figure 7–11).

The silo is set over a paddle mixer. The bags holding all of the dry mortar ingredients, including sand, which is at 0 percent moisture content, are lifted up and emptied into the silo. The dry bulk system may also include preblended coloring agents for projects that require colored mortars. The mason laborer releases the proper amount of preblended materials, adds water as well as any other required liquids, and completes the mixing operation. Some silo systems incorporate a screw mixer at the bottom of the silo, thus eliminating the need for a paddle mixer.

Efflorescence

Efflorescence is a white, powdery deposit found on masonry walls, often appearing shortly after the walls are completed. Although it is mostly an aesthetic problem, if the deposits build up in the pores of the masonry materials, surface damage may occur.

The deposits are composed of soluble salts in the masonry materials brought to the surface by moisture. Prevention of efflorescence can be accomplished by pretesting materials for soluble salts, using clean equipment, and preventing moisture infiltration into and out of masonry systems.

The stains can usually by removed by dry brushing, washing with clean water and scrub brushes, or light sand blasting. When the stains are severe, a dilute solution of muriatic acid (1 to 10 or 12 percent) may be applied to damp walls, scrubbed with stiff brushes, and washed with clean water. When using muriatic acid, all of the manufacturer's safety precautions must be followed because the acid can affect eyes, skin, and, breathing.

Grout

ASTM C476 also governs construction grouts, which are an essential element of reinforced concrete masonry. Mortar is not grout and the two are not interchangeable; they have different characteristics and are used differently.

In reinforced load-bearing walls, grout is usually placed in the wall spaces or cores that contain steel reinforcement. The grout bonds the deformed steel reinforcement to the masonry units so that the two act together to resist imposed loads. Grout may also be used in cores that do not contain steel to further increase the load-carrying capabilities of a wall system. The strength of nonreinforced load-bearing walls may also be increased with the use of a grout mix by filling a portion or all of the cores in the wall assembly.

The size of the space as well as the height of the lift to be grouted will generally determine the specific grout mix selected. However, because building codes and standards differ on specific values of maximum aggregate size versus clear opening, the governing documents for the work should be consulted. The maximum size of the aggregate and mix consistency should be determined with regard to specific job conditions to ensure satisfactory grout placement and adequate embedment of the steel reinforcement.

The compressive strengths of grouts vary from 600 to 2500 psi. Grout strengths are typically affected by water content, sampling methods, and testing methods.

Grouts in place generally have strength values in excess of 2500 psi, because the surrounding masonry units will absorb water from the grout rapidly, thereby reducing the in-place water–cement ratio of the grout and causing an increase in compressive strength. Grout strengths are also aided by the moisture trapped in the surrounding masonry unit, which produces a moist condition essential to cement hydration and strength gain.

Grouts should be produced with a fluid consistency adequate for pouring or pumping without segregation. The grout should flow around the steel and into all masonry voids without bridging or honeycombing. ASTM C143 slump measurements can be used as guides for grout consistency for high- and low-absorption units, with 8-in. slumps recommended for low-absorption units and 10-in. slumps for high-absorption units.

Batching, mixing, and delivery of grout mixes should, whenever possible, follow ASTM C94 standard specifications for ready-mixed concrete.

Grout specimens for compressive strength testing are cast in molds with concrete masonry units having the same absorption characteristics and moisture content as the units being used in the construction. This molding system simulates the conditions existing in the wall system where the masonry units will absorb water from the grout, thereby reducing the grout's water–cement ratio and increasing its strength.

Refer to ASTM C1019 for complete information about casting grout specimens.

Fire Ratings of Concrete Masonry

Two concerns of fire safety codes deal with structural integrity during a fire and containment of the fire. Therefore, the materials used in construction must have measurable values for resistance to heat transmission and flame spread, depending on occupancy classifications, fire zone classifications, and the nature of combustible materials which may be in a structure. The structural integrity requirement includes resistance to thermal shock, impact, and overturning forces during the most severe fire possible for the building area's occupancy.

Fire-resistance or fire-rating values are expressed in terms of time that a material or type of construction will withstand the standard fire test and still perform its design purpose. A fire rating of 4 or more hours can be readily achieved utilizing concrete masonry walls.

The standard test for determining the fire ratings of construction materials is ASTM E119, Methods of Fire Tests of Building Construction and Materials. The extent and severity of the fire in the test apparatus must conform to ASTM E119 criteria.

Three criteria are used to develop a fire rating for a wall or partition assembly. Any one of the criteria will be decisive, should it occur first.

1. Structural failure of the system while carrying design loads when subjected to the standard fire test.

2. Heat transmission through the wall that will cause an average rise in temperature of the side not exposed to direct flame (nine locations are averaged) or a rise of 325°F at one location. This criterion governs the spread of a fire by ignition of combustible materials placed against or near the wall surface away from the fire.

3. Passage of flame or heated gases which will ignite cotton waste, or passage of water from a fire hose through the wall assembly. After cooling, but within 72 hours, the wall must carry twice the safe superimposed design load.

The generally accepted practice in building codes is to state the fire rating of concrete masonry walls in terms of equivalent solid thickness. Equivalent solid thickness is the solid thickness that would be obtained if the same amount of concrete contained in a hollow concrete masonry unit were recast without core holes. The percentage of solids used in the calculations can be determined from net area or net volume values obtained by using ASTM C140, Methods of Testing Concrete Masonry Units. When walls are plastered or otherwise faced with fire-resistant materials, the thickness of the materials is included in the calculations of the equivalent thickness effective for fire resistance.

Calculation of equivalent thickness is illustrated in Table 7–3. Estimated fire resistance ratings are shown in

Table 7–3. Calculating equivalent thickness of concrete masonry units

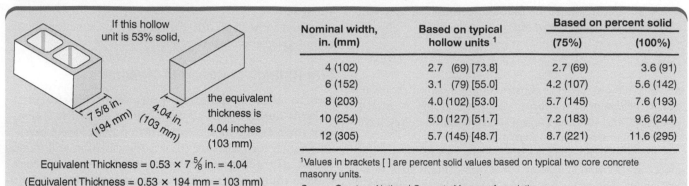

Nominal width, in. (mm)	Based on typical hollow units [1]	Based on percent solid	
		(75%)	(100%)
4 (102)	2.7 (69) [73.8]	2.7 (69)	3.6 (91)
6 (152)	3.1 (79) [55.0]	4.2 (107)	5.6 (142)
8 (203)	4.0 (102) [53.0]	5.7 (145)	7.6 (193)
10 (254)	5.0 (127) [51.7]	7.2 (183)	9.6 (244)
12 (305)	5.7 (145) [48.7]	8.7 (221)	11.6 (295)

If this hollow unit is 53% solid,

7 5/8 in. (194 mm)

4.04 in. (103 mm)

the equivalent thickness is 4.04 inches (103 mm)

Equivalent Thickness = 0.53 × 7 5/8 in. = 4.04
(Equivalent Thickness = 0.53 × 194 mm = 103 mm)

[1]Values in brackets [ ] are percent solid values based on typical two core concrete masonry units.

Source: Courtesy National Concrete Masonry Association.

Table 7–4. Fire resistance rating period of concrete masonry assemblies

Aggregate type in the concrete masonry unit[2]	Minimum required equivalent thickness for fire resistance rating, in. (mm)[1]						
	4 hours	3 hours	2 hours	1.5 hours	1 hour	0.75 hours	0.5 hours
Calcareous or siliceous gravel	6.2 (157)	5.3 (135)	4.2 (107)	3.6 (91)	2.8 (71)	2.4 (61)	2.0 (51)
Limestone, cinders or slag	5.9 (150)	5.0 (127)	4.0 (102)	3.4 (86)	2.7 (69)	2.3 (58)	1.9 (48)
Expanded clay, shale or slate	5.1 (130)	4.4 (112)	3.6 (91)	3.3 (84)	2.6 (66)	2.2 (56)	1.8 (46)
Expanded slag or pumice	4.7 (119)	4.0 (102)	3.2 (81)	2.7 (69)	2.1 (53)	1.9 (48)	1.5 (38)

[1]Fire resistance rating between the hourly fire resistance rating periods listed may be determined by linear interpolation based on the equivalent thickness value of the concrete masonry assembly.

[2]Minimum required equivalent thickness corresponding to the hourly fire resistance rating for units made with a combination of aggregates shall be determined by linear interpolation based on the percent by volume of each aggregate used in the manufacture.

Source: Courtesy National Concrete Masonry Association.

Table 7–4 and are for fully protected construction in which all structural members are of noncombustible materials. Where combustible members are formed into walls, equivalent solid thickness protecting such members should be not less than 93 percent of the thicknesses shown.

The fire resistance of masonry-constructed walls can be increased by filling core spaces with various fire-resistant materials. Standard fire tests have shown that by filling the cores in hollow masonry units or the air space in cavity walls with dry granular material, substantial reductions in heat transfer and increases in fire endurance are obtained.

Thermal Properties of Concrete Masonry

Concrete masonry walls offer insulation qualities combined with architectural appeal. Though the design of walls and insulation are important, the heat flow through wall systems is a small percentage of the total heat loss in building construction. One square foot of single-pane glass has a heat flow six or seven times as great as a square foot of lightweight concrete block wall with filled cores.

Resistance values of single-wythe 8 in. concrete masonry walls with cores empty or filled with various insulating materials as well as 8 in. walls with insulating materials applied to the wall face are given in Table 7–5. Resistance values for 6 in., 10 in., and 12 in. single-wythe walls can be found in NCMA TEK 6–2A. Table 7–6 data were used to develop Table 7–5.

Heat-transfer values increase as moisture content increases. When concrete masonry walls become saturated, heat transfer increases based on the masonry unit's density. Therefore, exterior masonry walls are usually protected from moisture.

Masonry construction is considered heavy-wall as opposed to light-wall construction of wood and metal stud. Heavy construction does not respond to temperature changes as rapidly as light construction, even though the two walls' systems may have the same U values (Figure 7–12). U values are the coefficients of total heat flow rate. They express the total amount of heat in British thermal units (Btu) that 1 sq ft of wall, ceiling, or floor will transmit per hour for each degree Fahrenheit of temperature difference between the warm and cool sides of the material or assembly.

Table 7–5. Resistance values ($R = 1/C$) of single-wythe concrete masonry walls with cores empty or filled with bulk insulation as well as surface insulation

Construction	Density of concrete, Pcf	Cores empty		Cores filled with[b]:							
				Loose-fill insulation				Polyurethane foamed insulation		Solid grouted	
				Perlite		Vermiculite					
		range	mid	range	mid	range	mid	range	mid	range	mid
Exposed block, both sides	85	2.4–2.7	2.5	6.3–8.2	7.1	5.9–7.5	6.6	6.9–9.4	8.0	1.9–2.1	2.0
	95	2.3–2.6	2.4	5.3–7.2	6.1	5.0–6.7	5.7	5.8–8.1	6.7	1.7–2.0	1.8
	105	2.1–2.4	2.2	4.5–6.3	5.2	4.3–5.9	4.9	4.8–7.0	5.6	1.6–1.9	1.7
	115	2.0–2.3	2.1	3.8–5.5	4.4	3.7–5.2	4.3	4.0–6.0	4.7	1.5–1.8	1.6
	125	1.9–2.2	2.0	3.2–4.8	3.8	3.1–4.6	3.7	3.3–5.1	4.0	1.5–1.7	1.5
	135	1.7–2.1	1.9	2.7–4.2	3.3	2.7–4.0	3.2	2.8–4.4	3.4	1.4–1.6	1.5
1/2 in. (13 mm) gypsum board on furring	85	3.8–4.1	3.9	7.7–9.6	8.5	7.3–8.9	8.0	8.3–10.8	9.4	3.3–3.5	3.4
	95	3.7–4.0	3.8	6.7–8.6	7.5	6.4–8.1	7.1	7.2–9.5	8.1	3.1–3.4	3.2
	105	3.5–3.8	3.6	5.9–7.7	6.6	5.7–7.3	6.3	6.2–8.4	7.0	3.0–3.3	3.1
	115	3.4–3.7	3.5	5.2–6.9	5.8	5.1–6.6	5.7	5.4–7.4	6.1	2.9–3.2	3.0
	125	3.3–3.6	3.4	4.6–6.2	5.2	4.5–6.0	5.1	4.7–6.5	5.4	2.9–3.1	2.9
	135	3.1–3.5	3.3	4.1–5.6	4.7	4.1–5.4	4.6	4.2–5.8	4.8	2.8–3.0	2.9
1 in. (25 mm) expanded polystyrene[c]	85	7.8–8.1	7.9	11.7–13.6	12.5	11.3–12.9	12.0	12.3–14.8	13.4	7.3–7.5	7.4
	95	7.7–8.0	7.8	10.7–12.6	11.5	10.4–12.1	11.1	11.2–13.5	12.1	7.1–7.4	7.2
	105	7.5–7.8	7.6	9.9–11.7	10.6	9.7–11.3	10.3	10.2–12.4	11.0	7.0–7.3	7.1
	115	7.4–7.7	7.5	9.2–10.9	9.8	9.1–10.6	9.7	9.4–11.4	10.1	6.9–7.2	7.0
	125	7.3–7.6	7.4	8.6–10.2	9.2	8.5–10.0	9.1	8.7–10.5	9.4	6.9–7.1	6.9
	135	7.1–7.5	7.3	8.1–9.6	8.7	8.1–9.4	8.6	8.2–9.8	8.8	6.8–7.0	6.9
1 in. (25 mm) extruded polystyrene[c]	85	8.8–9.1	8.9	12.7–14.6	13.5	12.3–13.9	13.0	13.4–15.8	14.4	8.3–8.5	8.4
	95	8.7–9.0	8.8	11.7–13.6	12.5	11.4–13.1	12.1	12.2–14.5	13.1	8.1–8.4	8.2
	105	8.5–8.8	8.6	10.9–12.7	11.6	10.7–12.3	11.3	11.2–13.4	12.0	8.0–8.3	8.1
	115	8.4–8.7	8.5	10.2–11.9	10.8	10.1–11.6	10.7	10.4–12.4	11.1	7.9–8.2	8.0
	125	8.3–8.6	8.4	9.6–11.2	10.2	9.5–11.0	10.1	9.7–11.5	10.4	7.9–8.1	7.9
	135	8.1–8.5	8.3	9.1–10.6	9.7	9.1–10.4	9.6	9.2–10.8	9.8	7.8–8.0	7.9
1 in. (25 mm) polyiso-cyanurate[d]	85	12.3–12.6	12.4	16.2–18.1	17.0	15.7–17.3	16.4	16.8–19.3	17.8	11.7–12.0	11.8
	95	12.1–12.4	12.3	15.2–17.1	16.0	14.9–16.5	15.6	15.6–18.0	16.6	11.6–11.9	11.7
	105	12.0–12.3	12.1	14.4–16.2	15.1	14.2–15.8	14.8	14.6–16.8	15.5	11.5–11.7	11.6
	115	11.9–12.2	12.0	13.7–15.4	14.3	13.5–15.1	14.1	13.8–15.8	14.6	11.4–11.6	11.5
	125	11.7–12.0	11.9	13.1–14.7	13.7	13.0–14.4	13.5	13.2–15.0	13.9	11.3–11.5	11.4
	135	11.6–11.9	11.7	12.6–14.0	13.1	12.5–13.9	13.0	12.7–14.3	13.2	11.3–11.5	11.4
2 × 4 furring with R13 batt & 1/2 in. (13 mm) gypsum board on furring	85	13.2–13.5	13.3	17.1–19.0	17.9	16.7–18.3	17.4	17.7–20.2	18.8	12.7–12.9	12.8
	95	13.1–13.4	13.2	16.1–18.0	16.9	15.8–17.5	16.5	16.6–18.9	17.5	12.5–12.8	12.6
	105	12.9–13.2	13.0	15.3–17.1	16.0	15.1–16.7	15.7	15.6–17.8	16.4	12.4–12.7	12.5
	115	12.8–13.1	12.9	14.6–16.3	15.2	14.5–16.0	15.1	14.8–16.8	15.5	12.3–12.6	12.4
	125	12.7–13.0	12.8	14.0–15.6	14.6	13.9–15.4	14.5	14.1–15.9	14.8	12.3–12.5	12.3
	135	12.5–12.9	12.7	13.5–15.0	14.1	13.5–14.8	14.0	13.6–15.2	14.2	12.2–12.4	12.3

[a] Notes: (hr ft^{2}° F/Btu) (0.176) 5 m^{2}K/W. Mortar joints are $^3/_8$ in. (10 mm) thick, with face shell mortar bedding assumed. Unit dimensions based on Standard Specification for Loadbearing Concrete Masonry Units, ASTM C 90 (ref. 2). Surface air films are included.

[b] Values apply when all masonry cores are filled completely. Grout density is 140 pcf (2243 kg/m^3). Lightweight grouts, which will provide higher R-values, are also available in some areas.

[c] Installed over wood furring. Includes 1/2 in. (13 mm) gypsum board and nonreflective air space.

[d] Installed over wood furring. Includes 1/2 in. (13 mm) gypsum board and reflective air space.

Source: National Concrete Masonry Association.

Table 7–6. Thermal data used to develop tables

Material:	Thermal resistivity (R-value per inch), hr ft²°F/Btu in (mK/W)
Vermiculite	2.27 (15.7)
Perlite	3.13 (21.7)
Expanded polystyrene	4.00 (27.7)
Extruded polystyrene	5.00 (34.7)
Cellular polyisocyanurate, gas-impermeable facer	7.04 (48.8)
Polyurethane foamed-in-place insulation	5.91 (41.0)
Wood	1.00 (6.9)
Concrete:	
85 pcf	0.23–0.34 (1.6–2.4)
95 pcf	0.18–0.28 (1.2–1.9)
105 pcf	0.14–0.23 (0.97–1.6)
115 pcf	0.11–0.19 (0.76–1.3)
125 pcf	0.08–0.15 (0.55–1.0)
135 pcf	0.07–0.12 (0.49–0.83)
140 pcf	0.06–0.11 (0.40–0.78)
Mortar	0.20 (1.4)

Material	R-value, hr ft²°F/Btu (m²K/W)
½ in. (13 mm) gypsum wallboard	0.45 (0.08)
Inside surface air film	0.68 (0.12)
Outside surface air film	0.17 (0.03)
Nonreflective air space	0.97 (0.17)
Reflective air space	2.38 (0.42)

Source: National Concrete Masonry Association.

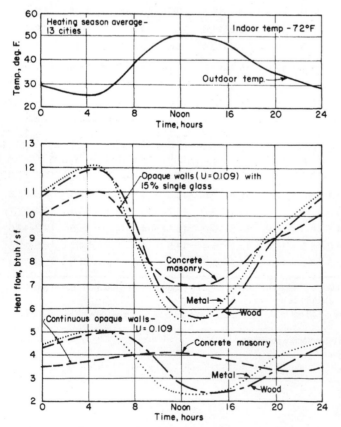

FIGURE 7–12. Heat loss through masonry and nonmasonry walls, U values being equal. *(Courtesy Portland Cement Association)*

Acoustical Properties of Concrete Masonry

Concrete masonry is an excellent sound barrier because of its density and, in some situations, its absorption qualities. Sound is composed of two elements: tone (frequency) and loudness (decibels). Tone is measured in vibrations per second, or frequency, and the unit of measure is the hertz (Hz). Loudness is measured in decibels (db). A 20 db increase in loudness indicates a tenfold increase in pressure. The human ear can adjust its sensitivity to noise somewhat, thereby reducing the pressure. The decibel as a unit of measure rates thunder at 120 db, average office noise 60 db, and normal breathing 10 db.

Building codes generally regulate the amount of noise stopped by floors, walls, and ceilings at 40 to 55 db of sound loss for airborne and impact sounds.

Concrete masonry units' ability to reduce sound transmission will vary depending on wall construction, type of block, painted or unpainted surfaces, and other characteristics. Therefore, the reader is encouraged to consult the National Concrete Masonry Association's TEK literature relative to sound-loss characteristics.

Estimating Masonry Materials

The quantities of masonry materials, such as concrete block and mortar, required to construct a wall are relatively easy to determine. The number of units required may be determined based on the face surface area of the concrete masonry unit or an estimating guide such as the one illustrated in Table 7–7 Usually corners are only counted once, any openings over 10 sq ft are deducted from the wall area, and variable waste factors are added to complete the determination of material quantities. Quantity determinations are part of a process that must take into account the costs of labor, material handling, equipment, and site conditions, all of which are covered in detail in estimating texts (Figure 7–13).

Example

Determine the number of 8 × 16-in. concrete masonry units required to build a wall 8 ft high and 44 ft long with four 3 × 4-ft openings.

$$\text{Total wall area} \quad 8 \times 44 \text{ ft} = 352 \text{ sq ft}$$
$$\text{Subtract the openings} \quad 4(3 \times 4 \text{ ft}) = -48 \text{ sq ft}$$
$$\text{Net wall area} \quad 304 \text{ sq ft}$$

$$\text{Number of CMUs} = \frac{304 \text{ sq ft}}{0.89 \text{ sq ft}} = 341.6 = 342$$

Using Table 7–7:

$$\text{Net wall area in squares} (100 \text{ sq ft}) = \frac{304 \text{ sq ft}}{100 \text{ sq ft}}$$
$$= 3.04 \text{ square}$$

$$3.04 \text{ square} \times 112.5 \text{ units/square} = 342 \text{ units}$$

The mortar required to place 342 concrete blocks from Table 7–7 is 7.5 cu ft per 100 units:

$$3.42 \times 7.5 \text{ cu ft} = 25.65 \text{ cu ft}$$

Table 7–7. Wall weights and material quantities for single-wythe concrete masonry construction*

Nominal Wall Thickness (in.)	Nominal Size (width × height × length) of Concrete Masonry Units (in.)	Average Weight of 100-sq-ft Wall Area (lb)**		Material Quantities for 100-sq-ft Wall Area		
		Units Made with Sand-Gravel Aggregate[†]	Units Made with Lightweight Aggregate[†]	Number of Units	Mortar (cu ft)	Mortar for 100 Units (cu ft)[‡]
4	4 × 4 × 16	4550	3550	225	13.5	6.0
6	6 × 4 × 16	5100	3900	225	13.5	6.0
8	8 × 4 × 16	6000	4450	225	13.5	6.0
4	4 × 8 × 16	4050	3000	112.5	8.5	7.5
6	6 × 8 × 16	4600	3350	112.5	8.5	7.5
8	8 × 8 × 16	5550	3950	112.5	8.5	7.5
12	12 × 8 × 16	7550	5200	112.5	8.5	7.5

* Based on 3/8-in. mortar joints.

** Actual weight of 100 sq ft of wall can be computed by the formula $WN + 145M$ where W is the actual weight of a single unit; N, the number of units for 100 sq ft of wall; and M, the mortar (cu ft) for 100 sq ft of wall.

† Using a concrete density of 138 pcf for units made with sand-gravel aggregate and 87 pcf for lightweight-aggregate units, and the average weight of unit for two- and three-core block.

‡ With face-shell mortar bedding. Mortar quantities include 10 percent allowance for waste.

Source: Portland Cement Association.

FIGURE 7-13. Masonry construction projects depend on high lift equipment to move materials on site. *(Courtesy JCB, Inc.)*

QUALITY OF MASONRY CONSTRUCTION

The material presented in Chapter 1 relative to the inspection and testing of construction materials and processes is a very important aspect of masonry construction.

The question of construction quality may arise during the design phase, construction phase, or after construction has been completed. Construction quality is governed by several factors, including, but not limited to, the knowledge and attitudes relative to required quality levels, the assignment of responsibility for quality construction, and familiar criteria to measure and evaluate levels of quality obtained.

Construction and protection requirements to ensure the quality of masonry construction are given in Tables 7-8a, 7-8b, 7-9a, and 7-9b, respectively.

Quality control is the term used to describe a contractor or manufacturer's effort to achieve a specified end result. It involves measuring, testing, and evaluating a product or process at regular intervals to maintain acceptable levels of quality. *Quality assurance* is the measuring, testing, and evaluation done by the purchaser to measure quality obtained in a product or structure.

In the construction industry, the roles of quality control and quality assurance inspectors entail much more than performing or witnessing materials tests and inspecting finished

Table 7-8a. Cold weather masonry construction requirements

Ambient temperature	Construction requirements
32 to 40°F (0 to 4.4°C)	Do not lay masonry units having a temperature below 20°F (−6.7°C). Remove visible snow and ice on masonry units before the unit is laid in the masonry. Remove snow and ice from foundation. Heat existing foundation and masonry surfaces to receive new masonry above freezing. Heat mixing water or sand to produce mortar temperatures between 40 and 120°F (4.4 and 48.9°C). Grout materials to be 32°F (0°C) minimum. Do not heat water or aggregates above 140°F (60°C).
25 to 32°F (−3.9 to 0°C)	Same as above for mortar. Maintain mortar temperatures above freezing until used in masonry. Heat grout aggregates and mixing water to produce grout temperatures between 70 and 120°F (21.1 and 48.9°C). Maintain grout temperature above 70°F (21.1°C) at time of grout placement.
20 to 25°F (−6.7 to −3.9°C)	Same as above, plus use heat masonry surfaces under construction to 40°F (4.4°C) and install wind breaks or enclosures when wind velocity exceeds 15 mph (24 km/hr). Heat masonry to a minimum of 40°F (4.4°C) prior to grouting.
20°F (−6.7°C) and below	Same as above, plus provide an enclosure for the masonry under construction and use heat sources to maintain temperatures above 32°F (0°C) within the enclosure.

Source: National Concrete Masonry Association.

Table 7–8b. Hot weather masonry preparation and construction requirements

Ambient temperature	Preparation and construction requirements
Above 100°F (37.8°C) or above 90°F (32.2°C) with a wind speed greater than 8 mph (12.9 km/hr)	Maintain sand piles in a damp, loose condition. Maintain temperature of mortar and grout below 120°F (48.9°C). Flush mixer, mortar transport container, and mortar boards with cool water before they come into contact with mortar ingredients or mortar. Maintain mortar consistency by retempering with cool water. Use mortar within 2 hours of initial mixing.
Above 115°F (46.1°C) or above 105°F (40.6°C) with a wind speed greater than 8 mph (12.9 km/hr)	Same as above, plus materials and mixing equipment are to be shaded from direct sunlight. Use cool mixing water for mortar and grout. Ice is permitted in the mixing water as long as it is melted when added to the other mortar or grout materials.

Table 7–9a. Cold weather masonry protection requirements

Mean daily temperature for ungrouted masonry Minimum daily temperature for grouted masonry	Protection requirements
25 to 40°F (−3.9 to 4.4°C)	Protect completed masonry from rain or snow by covering with a weather-resistive membrane for 24 hours after construction.
20 to 25°F (−6.7 to −3.9°C)	Completely cover the completed masonry with a weather-resistive insulating blanket or equal for 4 hours after construction (48 hr for grouted masonry unless only Type III portland cement used in grout).
20°F (−6.7°C) and below	Maintain masonry temperature above 32°F (0°C) for 24 hours after construction by enclosure with supplementary heat, by electric heating blankets, by infrared heat lamps, or by other acceptable methods. Extend time to 48 hours for grouted masonry unless the only cement in the grout is Type III portland cement.

Source: National Concrete Masonry Association.

Table 7–9b. Hot weather masonry protection requirements

Mean daily Temperature	Protection requirements
Above 100°F (37.8°C) or above 90°F (32.2°C) with a wind speed greater than 8 mph (12.9 km/hr)	Fog spray all newly construction masonry until damp, at least three times a day until the masonry is three days old.

Source: National Concrete Masonry Association.

work. Inspectors must have a very good working knowledge of the job drawings and specifications as well as a familiarity with reference specifications. On construction projects where quality is important, the architect and/or engineer or their authorized project representatives should be available on site during all phases of construction. ACI Committee 531, Concrete Masonry Structures—Design and Construction, provides for reductions in allowable stresses when on-site inspection does not exist for structural masonry construction.

Occasionally in masonry construction work the designer will require the building of one or more sample wall panels. These panels are usually 4 × 8 and should include all of the materials specified or selected for actual wall construction. Typically, the panels include the masonry units and reinforcement accessories, and show workmanship, coursing, bonding, wall thickness, color and texture range of units, and mortar joint color and tooling. After the designer or his representative approves the panel as fabricated, it is then used as a standard to measure actual construction work quality.

The appendix to this text includes selected ASTM specifications that are relevant to various aspects of masonry construction. The readers are encouraged to familiarize themselves with these specifications.

Review Questions

1. What is the difference between solid- and hollow-core masonry units?

2. What three clays are used to manufacture clay masonry?

3. Describe the manufacturing system most commonly used to manufacture clay masonry.

4. What factors influence the compressive strength of clay masonry?

5. Describe the procedures for field-testing clay brick suction rates.

6. What materials are used in the manufacture of concrete masonry units?

7. Describe the manufacturing process used to produce concrete masonry units.

8. What is the main advantage of autoclaved concrete masonry units?

9. What factors influence the compressive strength of concrete masonry units?

10. What functions does mortar serve in masonry construction?

11. Describe grout and how it is used in concrete masonry construction.

12. A 12-in. hollow masonry wall is constructed of expanded shale reported to be 55 percent solid. What is the estimated fire resistance of the wall?

13. What criteria are used to develop a fire rating for a wall or partition assembly?

14. What is the difference between heavy- and light-wall construction and why is this difference important in building construction?

15. What is the difference between quality control and quality assurance?

16. Why might an architect specify a sample wall panel for a masonry construction project? Explain.

17. Based on the data given, complete the required ASTM C140 calculations and compare them to the ASTM C90 requirements.

Concrete masonry unit 8 × 8 × 16 in.

Sampled weight	29.41 lb
Submerged weight	14.23 lb
Saturated surface-dry weight	31.74 lb
Oven-dry weight	28.21 lb

18. Three concrete masonry units of the type tested in the previous problem were loaded to failure with the listed results. Do these units meet ASTM C90 strength requirements?

MU	Load (lb)
1	221,490
2	212,610
3	219,000

19. Determine the cubic feet of mortar and number of 12 × 8 × 16-in. concrete blocks required to construct a 35-ft-wide, 55-ft-long, 8-ft-high foundation. Assume a 5 percent waste factor for the block and a 50 percent waste factor for the mortar.

20. What procedure can be used to determine if a cored brick is considered solid or hollow?

21. How many grades and types of brick are covered by ASTM C216? Summarize the major differences between the grades and types of brick.

22. What are the normally specified finish and appearance requirements for concrete block?

APPENDIX

Safety in Conducting Experiments

These experiments are intended for educational purposes. It is the responsibility of the persons directing and participating in these experiments to take appropriate measures to protect any one from harm. Certain testing equipment operate at high pressures and temperatures and all manufacturers' safety standards must be understood and followed. Safety and health practices as well as any regulatory limitations should be addressed prior to performing experiments.

WARNING: Contact with wet (unhardened) concrete, mortar, cement, or cement mixtures can cause SKIN IRRITATION, SEVERE CHEMICAL BURNS (THIRD-DEGREE), or SERIOUS EYE DAMAGE. Frequent exposure may be associated with irritant and/or allergic contact dermatitis. Wear waterproof gloves, a long-sleeved shirt, full-length trousers, and proper eye protection when working with these materials. If you have to stand in wet concrete, use waterproof boots that are high enough to keep concrete from flowing into them. Wash wet concrete, mortar, cement, or cement mixtures from your skin immediately. Flush eyes with clean water immediately after contact. Indirect contact through clothing can be as serious as direct contact, so promptly rinse out wet concrete, mortar, cement, or cement mixtures from clothing. Seek immediate medical attention if you have persistent or severe discomfort.

FIGURE 1   A typical warning label located on bagged mortar, cement or cement mixtures as well as ready mix concrete delivery tickets. (*Courtesy of the American Concrete Institute.*)

METRIC CONVERSION FACTORS

The following list provides the conversion relationship between U.S. customary units and SI (International System) units. The proper conversion procedure is to multiply the specified value on the left (primarily U.S. customary values) by the conversion factor exactly as given below and then round to the appropriate number of significant digits desired. For example, to convert 11.4 ft to meters: 11.4 × 0.3048 = 3.47472, which rounds to 3.47 meters. Do not round either value before performing the multiplication, as accuracy would be reduced. A complete guide to the SI system and its use can be found in ASTM E 380, Metric Practice.

To convert from	to	multiply by
Length		
inch (in.)	micron (μ)	25,400 E*
inch (in.)	centimeter (cm)	2.54 E
inch (in.)	meter (m)	0.0254 E
foot (ft)	meter (m)	0.3048 E
yard (yd)	meter (m)	0.9144
Area		
square foot (sq ft)	square meter (sq m)	0.09290304 E
square inch (sq in.)	square centimeter (sq cm)	6.452 E
square inch (sq in.)	square meter (sq m)	0.00064516 E
square yard (sq yd)	square meter (sq m)	0.8361274
Volume		
cubic inch (cu in.)	cubic centimeter (cu cm)	16.387064
cubic inch (cu in.)	cubic meter (cu m)	0.00001639
cubic foot (cu ft)	cubic meter (cu m)	0.02831685
cubic yard (cu yd)	cubic meter (cu m)	0.7645549
gallon (gal) Can. liquid	liter	4.546
gallon (gal) Can. liquid	cubic meter (cu m)	0.004546
gallon (gal) U.S. liquid**	liter	3.7854118
gallon (gal) U.S. liquid	cubic meter (cu m)	0.00378541
fluid ounce (fl oz)	milliliters (ml)	29.57353
fluid ounce (fl oz)	cubic meter (cu m)	0.00002957
Force		
kip (1000 lb)	kilogram (kg)	453.6
kip (1000 lb)	newton (N)	4,448.222
pound (lb) avoirdupois	kilogram (kg)	0.4535924
pound (lb)	newton (N)	4.448222
Pressure or stress		
kip per square inch (ksi)	megapascal (MPa)	6.894757
kip per square inch (ksi)	kilogram per square centimeter (kg/sq cm)	70.31
pound per square foot (psf)	kilogram per square meter (kg/sq m)	4.8824
pound per square foot (psf)	pascal (Pa)†	47.88
pound per square inch (psi)	kilogram per square centimeter (kg/sq cm)	0.07031
pound per square inch (psi)	pascal (Pa)†	6,894.757
pound per square inch (psi)	megapascal (MPa)	0.00689476
Mass (weight)		
pound (lb) avoirdupois	kilogram (kg)	0.4535924
ton, 2000 lb	kilogram (kg)	907.1848
grain	kilogram (kg)	0.0000648

To convert from	to	multiply by
Mass (weight) per length		
kip per linear foot (klf)	kilogram per meter (kg/m)	0.001488
pound per linear foot (plf)	kilogram per meter (kg/m)	1.488
Mass per volume (density)		
pound per cubic foot (pcf)	kilogram per cubic meter (kg/cu m)	16.01846
pound per cubic yard (lb/cu yd)	kilogram per cubic meter (kg/cu m)	0.5933
Temperature		
degree Fahrenheit (°F)	degree Celsius (°C)	$t_C = (t_F - 32)/1.8$
degree Fahrenheit (°F)	degree Kelvin (°K)	$t_K = (t_F + 459.7)/1.8$
degree Kelvin (°K)	degree Celsius (C°)	$t_C = t_K - 273.15$
Energy and heat		
British thermal unit (Btu)	joule (J)	1055.056
calorie (cal)	joule (J)	4.1868 E
Btu/°F · hr · ft²	W/m² · °K	5.678263
kilowatt-hour (kwh)	joule (J)	3,600,000. E
British thermal unit per pound (Btu/lb)	calories per gram (cal/g)	0.55556
British thermal unit per hour (Btu/hr)	watt (W)	0.2930711
Power		
horsepower (hp) (550 ft-lb/sec)	watt (W)	745.6999 E
Velocity		
mile per hour (mph)	kilometer per hour (km/hr)	1.60934
mile per hour (mph)	meter per second (m/s)	0.44704
Permeability		
darcy	centimeter per second (cm/sec)	0.000968
feet per day (ft/day)	centimeter per second (cm/sec)	0.000352

*E indicates that the factor given is exact.
**One U.S. gallon equals 0.8327 Canadian gallon.
†A pascal equals 1.000 newton per square meter.

Note:
One U.S. gallon of water weighs 8.34 pounds (U.S.) at 60°F.
One cubic foot of water weighs 62.4 pounds (U.S.).
One milliliter of water has a mass of 1 gram and has a volume of one cubic centimeter.
One U.S. bag of cement weighs 94 lb.

The prefixes and symbols listed below are commonly used to form names and symbols of the decimal multiples and submultiples of the SI units.

Multiplication Factor	Prefix	Symbol
$1,000,000,000 = 10^9$	giga	G
$1,000,000 = 10^6$	mega	M
$1,000 = 10^3$	kilo	k
$1 = 1$	—	—
$0.01 = 10^{-2}$	centi	c
$0.001 = 10^{-3}$	milli	m
$0.000001 = 10^{-6}$	micro	μ
$0.000000001 = 10^{-9}$	nano	n

THE CONSTRUCTION TRADES

Here are the metric units that will be used by the construction trades. The term *length* includes all linear measurements—length, width, height, thickness, diameter, and circumference.

	Quantity	Unit	Symbol
Surveying	length	kilometer, meter	km, m
	area	square kilometer	km²
		hectare (10 000 m²)	ha
		square meter	m²
	plane angle	degree (non-metric)	°
		minute (non-metric)	′
		second (non-metric)	″
Excavating	length	meter, millimeter	m, mm
	volume	cubic meter	m³
Trucking	distance	kilometer	km
	volume	cubic meter	m³
	mass	metric ton (1000 kg)	t
Paving	length	meter, millimeter	m, mm
	area	square meter	m²
Concrete	length	meter, millimeter	m, mm
	area	square meter	m²
	volume	cubic meter	m³
	temperature	degree Celsius	°C
	water capacity	liter (1000 cm³)	L
	mass (weight)	kilogram, gram	kg, g
	cross-sectional area	square millimeter	mm²

Source: The Construction Metrication Council of the National Institute of Building Sciences, Washington, D.C.

	Quantity	Unit	Symbol
Masonry	length	meter, millimeter	m, mm
	area	square meter	m^2
	mortar volume	cubic meter	m^3
Steel	length	meter, millimeter	m, mm
	mass	metric ton (1000 kg)	t
		kilogram, gram	kg, g
Carpentry Plastering	length	meter, millimeter	m, mm
	length	meter, millimeter	m, mm
	area	square meter	m^2
	water capacity	liter (1000 cm^3)	L
Glazing	length	meter, millimeter	m, mm
	area	square meter	m^2
Painting	length	meter, millimeter	m, mm
	area	square meter	m^2
	capacity	liter (1000 cm^3)	L
		milliliter (cm^3)	ml
Roofing	length	meter, millimeter	m, mm
	area	square meter	m^2
	slope	millimeter/meter	mm/m
Plumbing	length	meter, millimeter	m, mm
	mass	kilogram, gram	kg, g
	capacity	liter (1000 cm^3)	L
	pressure	kilopascal	kPa
Drainage	length	meter, millimeter	m, mm
	area	hectare (10 000 m^2)	ha
		square meter	m^2
	volume	cubic meter	m^3
	slope	millimeter/meter	mm/m

	Quantity	Unit	Symbol
HVAC	length	meter, millimeter	m, mm
	volume	cubic meter	m^3
	capacity	liter (1000 cm^3)	L
	airflow	meter/second	m/s
	volume flow	cubic meter/second	m^3/s
		liter/second	L/s
	temperature	degree Celsius	°C
	Force	newton, kilonewton	N, kN
	Pressure	kilopascal	kPa
	energy, work	kilojoule, megajoule	kJ, MJ
	rate of heat flow	watt, kilowatt	W, kW
Electrical	length	meter, millimeter	m, mm
	frequency	hertz	Hz
	power	watt, kilowatt	W, kW
	energy	megajoule	MJ
		kilowatt hour	kWh
	electric current	ampere	A
	electric potential	volt, kilovolt	V, kV
	resistance	Ohm	Ω

ROUNDING TABLE, 1/32 INCH TO 4 INCHES

Underline denotes exact conversion
Shaded figures are too exact for most uses

Inches	Nearest 0.1 mm (1/254″)	Nearest 1 mm (1/25″)	Nearest 5 mm (1/5″)
1/32″	0.8	1	
1/16″	1.6	2	
3/32″	2.4	2	
1/8″	3.2	3	
3/16″	4.8	5	
1/4″	6.4	6	
5/16″	7.9	8	
3/8″	9.5	10	
7/16″	11.1	11	
1/2″	<u>12.7</u>	13	
9/16″	14.3	14	
5/8″	15.9	16	
3/4″	19.0	19	
7/8″	22.2	22	
1″	<u>25.4</u>	25	25
1-1/4″	31.8	32	30
1-1/2″	<u>38.1</u>	38	40
1-3/4″	44.4	44	45
2″	<u>50.8</u>	51	50
2-1/4″	57.2	57	55
2-1/2″	<u>63.5</u>	64	65
2-3/4″	69.8	70	70
3″	<u>76.2</u>	76	75
3-1/4″	82.6	83	85
3-1/2″	<u>88.9</u>	89	90
3-3/4″	95.2	95	95
4″	<u>101.6</u>	102	100

Source: The Construction Metrication Council of the National Institute of Building Sciences, Washington, D.C.

ROUNDING TABLE, 4 INCHES TO 100 FEET

Underline denotes exact conversion
Shaded figures are too exact for most uses

Inches and Feet	Nearest 0.1 mm (1/254″)	Nearest 1 mm (1/25″)	Nearest 5 mm (1/5″)	Nearest 10 mm (2/5″)	Nearest 50 mm (2″)	Nearest 100 mm (2″)	1″ = 25 mm exactly
4″	101.6	102	100	100			100
5″	127	127	125	130			125
6″	152.4	152	150	150			150
7″	177.8	178	180	180			175
8″	203.2	203	205	200			200
9″	228.6	229	230	230			225
10″	254	254	255	250			250
11″	279.4	279	280	280			275
1-0″	304.8	305	305	300	300		300
2-0″	609.6	610	610	610	600		600
3″-0″	914.4	914	915	910	900		900
4″-0″	1219.2	1219	1220	1220	1200		1200
5″-0″		1524	1525	1520	1500		1500
6″-0″		1829	1830	1830	1850		1800
7″-0″		2134	2135	2130	2150		2100
8″-0″		2438	2440	2440	2450		2400
9″-0″		2743	2745	2740	2750		2700
10″-0″		3048	3050	3050	3050	3000	3000
15″-0″		4572	4570	4570	4550	4600	4500
20″-0″		6096	6095	6100	6100	6100	6000
25″-0″		7620	7620	7620	7600	7600	7500
30″-0″		9144	9145	9140	9150	9100	9000
40″-0″		12 192	12 190	12 190	12 200	12 200	12 000
50″-0″		15 240	15 240	15 240	15 250	15 200	15 000
75″-0″		22 860	22 860	22 860	22 850	22 900	22 500
100″-0″		30 480	30 480	30 480	30 500	30 500	30 000

FORMULAS FOR AREAS AND VOLUMES

CIRCLE

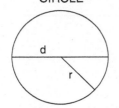

Circumference = πd

$$A = \pi r^2 = \pi \frac{d^2}{4}$$

π = 3.14159

RIGHT TRIANGLE

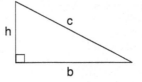

$$A = \frac{1}{2} bh$$

$$c^2 = h^2 + b^2$$

SQUARE

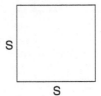

Perimeter = 4s

$$A = s^2$$

RECTANGLE

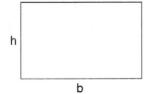

Perimeter = 2(b + h)

$$A = bh$$

TRAPEZOID

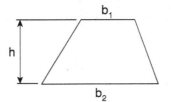

$$A = \frac{1}{2} (b_1 + b_2) h$$

ANY TRIANGLE

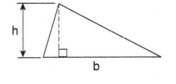

$$A = \frac{1}{2} bh$$

CONE

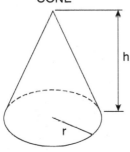

$$V = \frac{1}{3} \pi r^2 h$$

CYLINDER

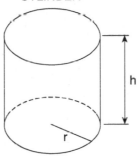

$$V = \pi r^2 h$$

RECTANGULAR SOLID

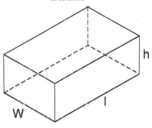

$$V = lwh$$

STUDENT-DIRECTED LABORATORY EXPERIMENTS

EXPERIMENT NO. 1   Verifying/Calibrating Balances to Be Used in Experiments

References

AASHTO M 231: Balances Used in the Testing of Materials

Objective

To verify the scales and/or balances to be used for future experiments are precise, and repeatable.

Background

Experiments to be conducted in class will utilize balances (scales). The results or output of those experiments are only as precise as the equipment and adherence to established procedures. In this experiment, balances will be verified for precision, or accuracy, as well as repeatability.

Equipment

>   Scales or balances
>   Standard analytical mass

Procedure

Step 1:   Choose standard analytical mass appropriate for the sensitivity and readability of the balance.

Step 2:   Weigh standard mass, and record.

Step 3:   Remove standard mass from scale.

Step 4:   Repeat steps 2 and 3 for a total of 3 measurements per scale/balance.

Calculations

Calculate the deviation of the known standard mass from the deviation from each repetition, and express as a percentage.

$$= \left(\frac{|S - R|}{S} \right) \times 100$$

where
> S = Known mass of standard analytical mass
> R = Recorded mass of standard analytical mass

Report

For each individual scale/balance evaluated, report equipment as readable if deviation within 0.1 percent. For each individual scale/balance, report equipment as repeatable if three measurements are within 0.1 percent. If measurements fall outside permissible tolerances, consult manufacturer's guide for calibration procedure.

EXPERIMENT NO. 2 Particle Size Analysis for Fine and Coarse Aggregates

References

ASTM C 136: Standard Test Method for Sieve Analysis of fine and Coarse Aggregates
ASTM D 75: Practice for Sampling Aggregates
AASHTO T 27: Sieve Analysis for Fine and Coarse Aggregates

Objective

To obtain the distribution of particle sizes within a given bulk sample of aggregate materials.

Equipment

Balance or scale accurate to 0.1 percent of the sample mass or sensitive to 0.1 g
Standard sieves meeting the requirements of ASTM C 136 Section 6.2
Mechanical sieve shaker
Suitable drying equipment—oven maintaining a uniform temperature of 110 ± 5°C (230 ± 9°F)
Containers and utensils

Background

Aggregate particle size determination is important in the design of many engineered systems such as filters and concrete and HMA mixtures. Aggregate particle size distribution is an important quality control tool. In the production of aggregates, should particle size change, surface area changes, and it could require a change in water–cement ratio in concrete, or asphalt content in HMA. In addition, aggregates are typically specified by their gradation or "grading" by size.

Individual particle size measurements would be time consuming and difficult for mix analysis purposes so we utilize ranges of particle sizes in the determination of aggregate gradation. Often sieve analysis data are used in determining compliance with engineering design specifications and construction quality control. A more in-depth discussion on aggregate size and gradation is presented in the section "Size and Gradation" in Chapter 2.

Use of the soil gradation in filtration design is extremely important to ensure proper flow of fluids through the soil medium. Care is required to ensure that the engineered systems do not undergo clogging. Clogging occurs when the pours in the aggregate fill with smaller size materials preventing fluid flow through the system. A more in-depth discussion of filtration can be found in the section "Permeability and Filters" in Chapter 2.

It is imperative to closely represent with a small sample the character of the entire stockpile or face of a natural deposit. Care should be taken to avoid segregated areas of the stockpile, or selection of specific soil materials. Sampling criteria should follow ASTM D 75. Stockpile samples are obtained with the assistance of a front end loader to dig in at least two to three buckets into the face of the pile. Once the pile has been "opened," one loader bucket scoop is obtained to represent the bottom, middle, and top of the pile. A square shovel is used to obtain three to four scoops of the representative sample. Typically a stockpile is sampled in this fashion in three to four locations. This procedure yields a large field sample. The large sample is reduced by quartering or dividing to produce a smaller sample for testing (see Figure 2–29). A large or small riffle sample may also be used to split the sample into smaller composite specimens. In natural deposits random test pits are typically utilized to obtain representative specimens of the soil materials. Sampling is discussed in detail in the section "Sampling" in Chapter 2.

Procedure*

Step 1: Oven-dry sample at a temperature of 110 ± 5°C (230 ± 9°F). The test sample size will depend on the nominal maximum aggregate size of the product being tested. Minimum sample size requirements from ASTM C136 are as given below:

Fine aggregate: 300 ± grams minimum
Coarse Aggregate:

Nominal maximum size	Minimum mass of test sample
9.5 mm (3/8″)	1 kg (2 lb)
12.5 mm (1/2″)	2 kg (4 lb)

*Please note that the procedure is for educational purposes to demonstrate gradation of coarse or fine aggregate following general ASTM requirements; for noneducational applications, see the ASTM standards referenced.

19.0 mm (3/4″)	5 kg (11 lb)
25.0 mm (1″)	10 kg (22 lb)
37.5 mm (1 1/2″)	15 kg (33 lb)

Step 2: Select sieves to obtain the required information. Nest (stack) the sieves in decreasing size from top to bottom remembering to place the pan at the bottom.

Step 3: Record initial sample weight.

Step 4: Pour the sample into the nest of sieves. Place sieves in the mechanical shaking device and shake for approximately 10 minutes. Excessive shaking may damage the sample; however, incomplete shaking will result in materials not passing through the sieve openings to their true size leading to erroneous test results.

Step 5: Determine the mass of the material retained on each sieve individually to the nearest 0.1 percent or 0.1 g and record. Do not forget to obtain the mass of the material in the pan. Make sure to remove all materials from the sieve screen openings and include in the mass retained—use of a wire brush on sieve sizes No. 30 and larger and a soft hair bristle for smaller sieves will assist in particle removal.

The total mass of the material after sieving should check closely with the original sample size. ASTM requires that the difference between the sieved product and the original sample size must be less than or equal to 0.3 percent. If the results fall outside this range the test is invalid and must be redone.

Calculations

1. Sum the mass retained on each sieve.
2. Calculate the loss–gain ratio (L/G):

$$L/G = \left(\frac{\text{total sample mass after} - \text{initial sample mass}}{\text{initial sample mass}} \right) \times 100$$

L/G percent less than or equal to 0.3 percent (If the sample is within 0.3 percent of original weight then it is valid and analysis can continue).

3. First calculate the percent retained on each sieve to the nearest 0.1 percent

$$\frac{\text{mass on sieve}}{\text{initial sample mass}} \times 100$$

   (a) Then calculate percent passing = {100 − cumulative percent retained on preceding (larger) sieves} to the nearest 0.1 percent
   (b) Cumulative percent retained = sum of each percent retained.
4. Plot data on 0.45 power chart (grain size chart)
5. Calculate the fineness modulus (FM):
   (a) FM = the sum of the cumulative percentages retained on the following when evaluating fine aggregate for use in concrete:
      i. Fine aggregate: No. 4, No. 8, No. 16, No. 30, No. 50, No. 100

Report

In preparing the laboratory report plot the grain size versus the percent passing for each aggregate type. Discuss the results of your grain size distribution curves. Report the total percentage of material passing each sieve. Report the percentage data for coarse and fine aggregate to the nearest percent. Report the fineness modulus and discuss its uses. Discuss the effects of the procedure on the loss–gain ratio. Compare the results obtained to ASTM C 136 Section 11: Precision and Bias.

TABLE 1 Grading Requirements for Lightweight Aggregate for Structural Concrete

Nominal Size Designation	Percentages (Mass) Passing Sieves Having Square Openings								
	25.0 mm (1 in.)	19.0 mm (¾ in.)	12.5 mm (½ in.)	9.5 mm (⅜ in.)	4.75 mm (No. 4)	2.36 mm (No. 8)	1.18 mm (No. 16)	300 μm (No. 50)	150 μm (No. 100)
Fine aggregate:									
4.75 mm to 0	...	...	...	100	85–100	...	40–80	10–35	5–25
Coarse aggregate:									
25.0 m to 4.75 mm	95–100	...	25–60	...	0–10	...	...	...	...
19.0 mm to 4.75 mm	100	90–100	...	10–50	0–15	...	...	...	...
12.5 mm to 4.75 mm	...	100	90–100	40–80	0–20	0–10	...	...	...
9.5 mm to 2.36 mm	...	...	100	80–100	5–40	0–20	0–10	...	...
Combined fine and coarse aggregate:									
12.5 mm to 0	...	100	95–100	...	50–80	...	...	5–20	2–15
9.5 mm to 0	...	...	100	90–100	65–90	35–65	...	10–25	5–15

Source: ASTM C 33

TABLE 2 Grading Requirements for Coarse Aggregates

Size Number	Nominal Size (Sieves with Square Openings)	Amounts Finer than Each Laboratory Sieves (Square-Openings), Mass Percent													
		100 mm (4 in.)	90 mm (3½ in.)	75 mm (3 in.)	63 mm (2½ in.)	50 mm (2 in.)	37.5 mm (1½ in.)	25.0 mm (1 in.)	19.0 mm (¾ in.)	12.5 mm (½ in.)	9.5 mm (⅜ in.)	4.75 mm (No. 4)	2.36 mm (No. 8)	1.18 mm (No. 16)	300 μm (No. 50)
1	90 to 37.5 mm (3½ to 1½ in.)	100	90 to 100	...	25 to 60	...	0 to 15	...	0 to 5	...	...	...	...	...	...
2	63 to 37.5 mm (2½ to 1½ in.)	...	...	100	90 to 100	35 to 70	0 to 15	...	0 to 5	...	...	...	...	...	...
3	50 to 25.0 mm (2 to 1 in.)	...	...	...	100	90 to 100	35 to 70	0 to 15	...	0 to 5	...	...	...	...	...
357	50 to 4.75 mm (2 in. to No. 4)	...	...	...	100	95 to 100	...	35 to 70	...	10 to 30	...	0 to 5	...	...	...
4	37.5 to 19.0 mm (1½ to ¾ in.)	...	...	...	...	100	90 to 100	20 to 55	0 to 15	...	0 to 5	...	...	...	...
467	37.5 to 4.75 mm (1½ in. to No. 4)	...	...	...	...	100	95 to 100	...	35 to 70	...	10 to 30	0 to 5	...	...	...
5	25.0 to 12.5 mm (1 to ½ in.)	...	...	...	...	...	100	90 to 100	20 to 55	0 to 10	0 to 5	...	...	...	...
56	25.0 to 9.5 mm (1 to ⅜ in.)	...	...	...	...	...	100	90 to 100	40 to 85	10 to 40	0 to 15	0 to 5	...	...	...
57	25.0 to 4.75 mm (1 in. to No. 4)	...	...	...	...	...	100	95 to 100	...	25 to 60	...	0 to 10	0 to 5	...	...
6	19.0 to 9.5 mm (¾ to ⅜ in.)	...	...	...	...	...	...	100	90 to 100	20 to 55	0 to 15	0 to 5	...	...	...
67	19.0 to 4.75 mm (¾ in. to No. 4)	...	...	...	...	...	...	100	90 to 100	20 to 55	20 to 55	0 to 10	0 to 5	...	...
7	12.5 to 4.75 mm (½ in. to No. 4)	...	...	...	...	...	...	100	90 to 100	90 to 100	40 to 70	0 to 15	0 to 5	...	...
8	9.5 to 2.36 mm (⅜ in. to No. 8)	...	...	...	...	...	...	...	...	100	85 to 100	10 to 30	0 to 10	0 to 5	...
89	9.5 to 1.18 mm (⅜ in. to No. 16)	...	...	...	...	...	...	...	...	100	90 to 100	20 to 55	5 to 30	0 to 10	0 to 5
9A	4.75 to 1.18 mm (No. 4 to No. 16)	...	...	...	...	...	...	...	...	...	100	85 to 100	10 to 40	0 to 10	0 to 5

A Size number 9 aggregate is defined in Terminology C 125 as a fine aggregate. It is included as a coarse aggregate when it is combined with a size number 8 material to create a size number 89, which is a coarse aggregate as defined by Terminology C 125.

Source: ASTM C 33

FINE AGGREGATE

5. General Characteristics

5.1 Fine aggregate shall consist of natural sand, manufactured sand, or a combination thereof.

6. Grading

6.1 *Sieve Analysis*—Fine aggregate, except as provided in 6.2 and 6.3 shall be graded within the following limits:

Sieve (Specification E 11)	Percent Passing
9.5-mm (⅜-in.)	100
4.75-mm (No. 4)	95 to 100
2.36-mm (No. 8)	80 to 100
1.18-mm (No. 16)	50 to 85
600-μm (No. 30)	25 to 60
300-μm (No. 50)	5 to 30
150-μm (No. 100)	0 to 10

NOTE 2—Concrete with fine aggregate gradings near the minimums for percent passing the 300 μm (No.50) and 150 μm (No.100) sometimes have difficulties with workability, pumping or excessive bleeding. The addition of entrained air, additional cement, or the addition of an approved mineral admixture to supply the deficient fines, are methods used to alleviate such difficulties.

6.2 The fine aggregate shall have not more than 45 % passing any sieve and retained on the next consecutive sieve of those shown in 6.1, and its fineness modulus shall be not less than 2.3 nor more than 3.1.

6.3 Fine aggregate failing to meet these grading requirements shall meet the requirements of this section provided that the supplier can demonstrate to the purchaser or specifier that concrete of the class specified, made with fine aggregate under consideration, will have relevant properties (see Note 4) at least equal to those of concrete made with the same ingredients, with the exception that the reference fine aggregate shall be selected from a source having an acceptable performance record in similar concrete construction.

NOTE 3—Fine aggregate that conforms to the grading requirements of a specification, prepared by another organization such as a state transportation agency, which is in general use in the area, should be considered as having a satisfactory service record with regard to those concrete properties affected by grading.

Source: ASTM C 33

EXPERIMENT NO. 3   Specific Gravity and Absorption of Fine Aggregates (ASTM C 128)

References

ASTM C 128: Standard Test Method for Density, Relative Density (Specific Gravity), and Absorption of Fine Aggregates

ASTM D 75: Practice for Sampling Aggregates

AASHTO T 84: Specific Gravity and Absorption of Fine Aggregates

Objective

To determine the ratio of the weight of fine aggregate to the weight of an equal volume of water.

Equipment

Balance or scale accurate to 0.1 percent of the sample mass or sensitive to 0.1 g with a minimum capacity of 1 kg, pycnometer, a flask or suitable container for placement of the fine aggregate. The volume of the flask must be equal to or greater than 50 percent above the space needed for the test sample of aggregate (i.e., volume of flask must be two times the volume of aggregate at a minimum).

Mold and tamper meeting the requirements of ASTM C 128—mold in the form of a frustum of a cone.

Suitable drying equipment—oven maintaining a uniform temperature of $110 \pm 5°C$ ($230 \pm 9°F$)

Containers and utensils

Background

Specific gravity is the ratio of the weight of a given volume of aggregate to the weight of an equal volume of water. Water at a temperature of 23°C (73.4°F) has a specific gravity of 1.0. For example, an aggregate with a specific gravity of 2.0 would be two times as heavy as water for the same volume. Specific gravity of materials is important because it is used in the calculation of the solid volume. The solid volume of the aggregate does not change and is a constant parameter in mix designs. Absorption is the increase in mass due to water in the pores of the material. How much water an aggregate will absorb is extremely important in the development of mix design parameters and in quality control in the production of materials.

Specific gravity is used to calculate air voids, voids in mineral aggregate, and voids filled by asphalt in hot mix design. All these properties are critical to a durable asphalt mix. Water absorption can also be an indicator of asphalt absorption. A highly absorptive aggregate could lead to a low durability asphalt mix.

In concrete mix design the specific gravity of the aggregate is needed to calculate the percentage of voids and the solid volume for computations of yield. The absorption is important in determining the total water required in the mix design process.

Procedure*

Step 1:   Obtain a sample of fine aggregate of at least 1000 g with 100 percent passing the No. 4 sieve. Dry sample to a constant weight.

*Please note that the procedure is for educational purposes to demonstrate the measurement of specific gravity of fine aggregate following general ASTM requirements; for noneducational applications, see the ASTM standards referenced.

Step 2:   Immerse sample in water for a period of 24 hours.

Step 3:   The wet aggregate is dried until it reaches a saturated, surface-dry condition (SSD). To determine if the sample has reached this condition requires constant monitoring of the drying process with frequent testing, utilizing the cone method throughout the process. Ensure uniform drying of the sample by stirring the sample occasionally and/or applying a gentle warm-flowing stream of air.

    Cone method: Fill cone to overflowing with the drying aggregate. Lightly tamp the material 25 times into the mold. Make sure that each drop of the tamper starts at about 1/5 in. above the top surface of the fine aggregate. Remove the loose soil from the base and carefully lift the mold vertically. If surface moisture is present the fine aggregate retains the mold shape. This indicates that there is sufficient moisture on the particle surfaces to cause cohesion. When the fine aggregate achieves SSD condition the cone shape slumps upon removal.

Step 4:   Record as B, the weight of the pycnometer filled to its calibrated capacity with water at 23 ± 1.7°C (73.4 ± 3°F) to the nearest 0.1 g.

Step 5:   Place a 500 g ± sample of fine aggregate (SSD condition) into the pycnometer. Note that it is easier to add a small amount of water in pycnometer then tare the scale prior to adding the aggregate to ensure precise weight of SSD material. Once aggregate is added record this weight as S—SSD weight. Be sure to rezero scale once SSD sample weight transferred into the pycnometer is determined.

Step 6:   Fill pycnometer with water at 23 ± 1.7°C (73.4 ± 3°F) to 90 percent of pycnometer.

Step 7:   Tap and roll the pycnometer, holding the device at a slight angle, to allow for the removal of all air bubbles—approximately 15 to 20 minutes.

Step 8:   Once all air is removed, add water to bring the pycnometer to its calibrated capacity (to the meniscus). If bubbles are present adding a few drops of isopropyl alcohol is recommended to disperse the foam.

Step 9:   Record as C, the total weight of the pycnometer, specimen and water to the nearest 0.1 g.

Step 10:   Remove entire contents from the pycnometer into a separate container suitable for oven drying. Additional rinsing may be required to remove all of the sample from the pycnometer.

Step 11:   Place sample in the oven. Dry to a constant weight at a temperature of 110 ± 5°C (230 ± 9°F). Record this as A—oven-dry weight.

Calculations

$$\text{Bulk specific gravity (dry)} = \frac{A}{(B + S - C)}$$

$$\text{Bulk specific gravity (SSD)} = \frac{S}{(B + S - C)}$$

$$\text{Apparent specific gravity} = \frac{A}{(B + A - C)}$$

$$\text{Absorption} = \left[\frac{(S - A)}{A} \right] \times 100$$

where

    A = oven dry weight

    B = weight of pycnometer filled with water to its calibrated capacity

    C = weight of pycnometer and SSD sample filled to its calibrated capacity

    S = SSD sample weight

Report

In preparing the laboratory report indicate the specific gravity results to the nearest 0.01 and the percent absorption results to the nearest 0.1 percent. Compare the results obtained to data published by the Department of Transportation or local supplier for the same material. Check the precision of the results to Table 1 in ASTM C 128–01. Comment on the laboratory procedures used and any difficulties that may have occurred during sample preparations.

EXPERIMENT NO. 4 Specific Gravity and Absorption of Coarse Aggregates

References

ASTM C 127: Standard Test Method for Density, Relative Density (Specific Gravity), and Absorption of Coarse
Aggregates

AASHTO T 85: Specific Gravity and Absorption of Coarse Aggregates

Objective

To determine the bulk specific gravity (dry and SSD), apparent specific gravity, and absorption of coarse aggregates per
ASTM C127.

Sample

Of sufficient quantity for nominal maximum size of aggregate to be tested in accordance with requirements
of ASTM C 127. Dry Sample to constant weight. Remove minuS #4 material

Equipment

Balance or scale accurate to 0.05 percent of the sample mass or 0.5 g whichever is greater with a capacity of 5 kg.
The balance shall be equipped with suitable apparatus for suspending the sample in water.

Wire basket of 3.35 mm (No. 6) or smaller mesh, with a capacity of 4 to 7 L (1 to 2 gal) to hold a maximum size
aggregate of 37.5 mm ($1\frac{1}{2}$ in.).

Water tank—large enough to fully submerge the sample and basket, equipped with an overflow valve to keep water
level constant and watertight. For educational purposes a 5 gal pail works well.

Suspension apparatus—smallest practical diameter wire.

Sieves—No. 4 sieve

Containers and Utensils

Background

Specific gravity is the ratio of the weight of a given volume of aggregate to the weight of an equal volume of water.
Water at a temperature of 23°C (73.4°F) has a specific gravity of 1.0. For example, an aggregate with a specific gravity
of 2.0 would be two times as heavy as water for the same volume. Specific gravity of materials is important because it
is used in the calculation of the solid volume. The solid volume of the aggregate does not change and is a constant
parameter in mix designs. Absorption is the increase in mass due to water in the pores of the material. How much
water an aggregate will absorb is extremely important in the development of mix design parameters and in the quality
control and analysis of materials.

Specific gravity is used to calculate air voids, voids in mineral aggregate, and voids filled by asphalt in hot mix
design. All these properties are critical to a durable asphalt mix. Water absorption can also be an indicator of asphalt
absorption.

In concrete mix design the specific gravity of the aggregate is needed to calculate the percentage of voids and the
solid volume for computations of yield. The absorption is important in determining the total water required in the mix
design process.

Procedure*

Step 1: Attach wire basket to scale suspended over water bath. Submerge wire basket attached to scale, and tare scale.
Remove basket from water bath.

Step 2: Remove soaking saturated sample from water and bring to SSD (saturated surface dry). SSD is achieved by
rolling sample in absorbant cloths. No visible fills of water should be evident on the damp aggregate particles.

*Please note that the procedure is for educational purposes to demonstrate the measurement of specific gravity of coarse aggregates following general
ASTM requirements; for noneducational applications, see the ASTM standards referenced.

Step 3: Tare the weight of the basket on second scale. Add SSD sample and record weight as SSD sample weight in air as B.

$$B = \underline{\hspace{3cm}}$$

Step 4: Reattach basket filled with SSD aggregates to scale suspended over water bath. Record as C, SSD sample weight in water.

$$C = \underline{\hspace{3cm}}$$

Step 5: Transfer the aggregate from the wire basket, into a container w/known tare weight and dry in oven to constant weight.

$$A = \underline{\hspace{3cm}}$$

Step 6: Determine the weight of the oven-dry aggregate (A):

Step 7: Compute the specific gravity and absorption of the aggregate as follows:

$$\text{Bulk specific gravity (dry)} = \frac{A}{(B-C)}$$

$$\text{Bulk specific gravity (SSD)} = \frac{B}{(B-C)}$$

$$\text{Apparent specific gravity} = \frac{A}{(A-C)}$$

$$\text{Absorption} = \frac{(B-A)}{A} \times 100\%$$

Record all values to the nearest 0.01

Report

In preparing the laboratory report, indicate the specific gravity results to the nearest 0.01 and the percent absorption results to the nearest 0.1 percent. Compare the results obtained to local data published by the Department of Transportation or history from a local supplier for the same aggregate source. Check the precision of the results to Table 1 in ASTM C 127–01. Comment on the laboratory procedures used and any difficulties that may have occurred during sample preparations.

EXPERIMENT NO. 5   Unit Weight and Voids in Aggregates

References

ASTM C 29: Bulk Density ("Unit Weight") and Voids in Aggregate

ASTM C 127: Standard Test Method for Density, Relative Density (Specific Gravity), and Absorption of Coarse Aggregates

ASTM C 128: Standard Test Method for Density, Relative Density (Specific Gravity), and Absorption of Fine Aggregates

AASHTO T 19: Unit Weight and Voids in Aggregates

Objective

To determine the bulk density ("unit weight") and voids of aggregate in a dry and compacted condition.

Equipment

Balance or scale accurate to 0.05 kg (0.1 lb).

Steel Rod: a round, straight steel rod, 16 mm (5/8 in.) diameter and approximately 600 mm (24 in.) in length. Tamping end must be rounded to a hemispherical tip with a diameter of 5/8 in.

Watertight cylindrical sturdy metal container with 14 L (1/2 cu ft) capacity for $1\frac{1}{2}$ in. or finer size aggregate and 2.8 L (1/10 cu ft) capacity for $\frac{1}{2}$ or finer size aggregate. Inside surface of container must be smooth and continuous.

Glass plate: 6 mm (1/4 in.) thick and 25 mm (1 in.) larger than the diameter of the cylindrical device.

Shovel and scoop

Background

Unit weight (bulk density) is the weight of a material divided by the volume. This quantity is used in many engineering applications. Unit weight is used in load calculations for the design of various structures and foundations. In addition, it is important for making conversions for purchasing the correct amount of materials. In this case the test method is often used to determine unit weight values that are necessary for use for many methods of selecting proportions for concrete mixtures. This value will influence the quantity of aggregate which can be used in the concrete.

Procedure*

Metal Container Calibration

Step 1:   Weigh empty metal container and glass plate. Record as D.

Step 2:   Fill the container with water at room temperature and cover with plate glass to eliminate bubbles and excess water. Use of a small water bottle may be necessary to slowly insert water underneath plate for bubble elimination.

Step 3:   Weigh container filled with the water and glass plate. Record as W.

Step 4:   Measure the temperature of the water and determine the unit weight of water from ASTM C 29/ C 29M Table 3. In absence of the table value use 62.4 lb/cu ft at 77°F.

Step 5:   Calculate the volume of the container: V = volume (cu ft, m³); γ = unit weight of water; W = weight

$$V = \frac{W - D}{\gamma}$$

Compacted Unit Weight (Bulk Density) Procedure

Step 5:   Fill container one-third full with dry aggregate and level with fingers.

Step 6:   Rod the layer of aggregate 25 times equally distributed over the cross section of the sample; do not allow the rod to strike the bottom of the container.

---

*Please note that the procedure is for educational purposes to demonstrate the measurement of following general ASTM requirements; for noneducational applications, see the ASTM standards referenced.

Step 7:   Fill container to two-thirds full and level with fingers.

Step 8:   Repeat Step 6; do not force rod into the previously placed layer.

Step 9:   Fill container to overflowing and rod again as in Step 6. Level surface with a straight edge. Make sure that projections of the larger pieces of the coarse aggregate balance the larger voids in the surface below the top of the measure. Weigh the container filled with rodded aggregate and record to the nearest 0.05 kg (0.1 lb) as G.

Step 10:   Weigh dry bucket empty; record as "T."

Repeat steps 5–10 with a second sample of the same-size aggregate; remember to clean and dry container before reuse.

Calculations

1. Unit weight (bulk density):

$$\gamma = \frac{(G - T)}{V}$$

where,

γ = unit weight (lb/cu ft, kg/m^3);

G = mass of aggregate plus container (lb, kg);

T = mass of the container (lb, kg);

V = volume of container (cu ft, m^3)

2. % Voids

$$\% \text{ voids} = \frac{((S \times \gamma_w) - \gamma)}{(S \times \gamma_w)} \times 100$$

where,

S = bulk specific gravity (dry)

γ = unit weight

γ_w = unit weight of water at 25°C (77°F) which is 62.4 lb/cu ft

Report

In preparing the laboratory report indicate the results for unit weight to the nearest 10 kg/m^3 (1 lb/cu ft^3) for each trial. Average the trials for determining the final unit weight of each sample type. Compare the results of each trial and discuss. Compare the results to ASTM C 29/29M Section 15 Precision and Bias and discuss.

TABLE 3   Density of Water

| Temperature | | lb/ft³ | kg/m³ |
°F	°C		
60	15.6	62.366	999.01
65	18.3	62.336	998.54
70	21.1	62.301	997.97
73.4	23.0	62.274	997.54
75	23.9	62.261	997.32
80	26.7	62.216	996.59
85	29.4	62.166	995.83

Source: ASTM C 29

EXPERIMENT NO. 6   Fine Aggregate Organic Impurities

References

ASTM C 40

Objective

To ascertain the presence of organic impurities in a fine aggregate.

Background

Organic impurities in fine aggregates will have an effect on setting times and compressive strength of mortar and concrete. The value of this test is to indicate whether or not more testing of the fine aggregate may be required before the aggregate is approved for use.

Equipment

12 to 16 oz. Glass bottle with water-tight stopper or cap.

Sodium hydroxide solution (3 percent).

Organic glass plate or a Gardner color glass plate.

Procedure*

Step 1:   Collect a representative test sample of sand weighing about 1 pound by quartering or by the use of sampler.

Step 2:   Fill the bottle with a $4\frac{1}{2}$ oz. volume of the sample of the sand to be treated.

Step 3:   Add the sodium hydroxide, 3 percent, to the bottle until the volume of the sand and liquid, indicated after shaking, is 7 liquid ounces.

Step 4:   Cap the bottle, shake vigorously, and allow to stand for 24 hours.

Step 5:   Compare the color of the supernatant liquid above the sand with that of the organic color plate. Make the color comparison by holding the color plate and the bottle close together and looking through to compare the liquid color to the glass color.

Note: This method of test covers the procedure for an approximate determination of the presence of injurious organic compounds in natural sands which are to be used in mortar, or concrete. The principal value to the test is to furnish a warning that further tests of the sand are necessary before they are approved for use.

Report

By color reference number indicate if the liquid above the sample is darker or lighter than the organic plate number 3 or the Gardner plate number 11. When the liquid is darker than the standard numbers, the aggregate will have to be further tested before use in concrete or mortar.

*Please note that the procedure is for educational purposes to demonstrate the measurement of organic impurities and fine aggregates following general ASTM requirements; for noneducational applications, see the ASTM standards referenced.

EXPERIMENT NO. 7 Proportions for Normal, Heavyweight, and Mass Concrete (Concrete Mix Design)

References

ACI 211.1: Standard Practice for Selecting Proportions for Normal, Heavyweight, and Mass Concrete
ASTM C 143: Standard Test Method for Slump of Hydraulic-Cement Concrete
ASTM C 31: Standard Practice for Making and Curing Concrete Test Specimens in the Field
ASTM C 1064: Standard Test Method for Temperature of Freshly Mixed Hydraulic-Cement Concrete
ASTM C 138: Standard Test Method for Density (Unit Weight) Yield, and Air Content (Gravimetric) of Concrete

Objective

To design and prepare a concrete mixture with a targeted compressive strength between 2000 and 6000 psi. To develop an understanding of the hydration process and how the proportions of the various components of the concrete mix determine the ultimate compressive strength of the cured concrete.

Equipment

Balance or scale accurate to 0.1 percent of the sample mass or sensitive to 0.1 g
Cylindrical sample molds
Tamping rod
Temperature measuring device
Graduated cylinder
Slump cone
Hoe and Pan for mixing of concrete or concrete mixer

Background

ACI 211.1 Proportions for Normal, Heavyweight, and Mass Concrete (Concrete Mix Design) is the standard for designing concrete mixtures and determining the proportions of portland cement, water, and coarse and fine aggregates. This experimental procedure does not use admixtures and assumes an indoor concrete mixture (i.e., not subject to freeze–thaw cycles). Prior results of experiments indicating particle size, specific gravity, and moisture absorption of fine and coarse aggregates should be used. The fineness modulus of the fine aggregate obtained from earlier experiments is also necessary. This information will be used to adjust the water content of the concrete mixture.

Procedure*

Note that initial calculations using this standard are designed for production of 1 cu yd (27 cu ft) of concrete. Since a sample size of 1 cu yd is impractical in a laboratory setting, a reduced sample of 0.5 cu ft is ultimately used as the design sample which allows for the molding of approximately two 6 in. diameter sample cylinders.

Step 1:   List the unit weights, fineness modulus, and specific gravity of the fine and coarse aggregates (based on experiments 1–4) in the table below. Also enter this information in the Mix Table:

	SG	FM	UW	% ABS	% FW
CA					
FA					

Step 2:   Prepare approximately a 500 g sample of FA (fine aggregate) and CA (coarse aggregate) by weighing at a saturated state-weight of measure. Remove moisture from the samples and weigh the dried samples (cooling the pan first). Calculate free water % as follows:

$$\text{Total Water \%} = \frac{\text{Sample Wet} - \text{Sample OD (oven dried)}}{\text{Sample OD}} \times 100$$

$$\text{Free Water \%} = \text{Total Water \%} - \text{Absorbed Moisture \%}$$

*Please note that the procedure is for educational purposes following general ACI and ASTM requirements; for noneducational applications, see the ACI and ASTM standards referenced.

List the free water percentages in the table above and in the Mix Table.

Step 3:   Determine the target compressive strength at 28 days ($F^1_cD_{28}$) for the mixture varying between 2000 and 6000 psi.

$$F^1_cD_28 = \underline{\hspace{2cm}} \text{ psi}$$

Step 4:   Using ACI 211.1 Table 6.3.1 determine the target design slump.

$$\text{Slump} = \underline{\hspace{2cm}} \text{ min} \underline{\hspace{2cm}} \text{ max}$$

Step 5:   Using the results of experiments 1–4 determine the maximum aggregate size.

$$\underline{\hspace{3cm}} \text{ inches}$$

Record results of the following in the Mix table:

Mix Table (record all information developed in the table)							
	INT W	% ABS	LB ABS	% FW	LB FW	ADJ W CY	ADJ W 0.5 CF
W					( )		
C							
CA							
FA							
TOTAL							

Step 6:   Initial Water Weight. Using ACI 211.1 Table 6.3.3 determine the unadjusted weight of water required and the percentage of air.

$$\text{Water} = \underline{\hspace{2cm}} \text{ lb Air} = \underline{\hspace{2cm}} \%$$

Step 7:   Water–Cement Ratio. Using ACI 211.1 Table 6.3.4a determine the water–cement ratio

$$\text{W/C} = \underline{\hspace{2cm}}$$

Note: In general, as the water–cement ratio increases, strength decreases (more water, incomplete hydration) leaving behind microscopic air voids that result in porous cement paste that has less internal and bond (aggregate) strength.

Step 8:   Weight of cement. Determine the weight of portland cement by dividing the weight of water by the water–cement ratio.

$$\text{C} = \underline{\hspace{2cm}} \text{ lb}$$

Step 9:   Weight of Coarse Aggregate. Using ACI 211.1 Table 6.3.6 determine the volume of coarse aggregate per unit volume of concrete

$$\text{Vol/CY} = \underline{\hspace{2cm}}$$

Determine the weight of CA = (Vol/CY) × (Uw of CA) × (27 cf/cy) = $\underline{\hspace{3cm}}$

Step 10: *Estimated* Weight of Fine Aggregate (*based on volume of other mix components*).

10a: Solid Volume of Water in CF = W/62.4 = $\underline{\hspace{3cm}}$ CF

10b: Solid Volume of Air in CF = (% Air/100) × 27 = $\underline{\hspace{3cm}}$ CF

10c: Solid Volume of Cement = C/3.15 × 62.4 = $\underline{\hspace{3cm}}$ CF

Note: 3.15 = SG of Type 1 Portland Cement

10d: Solid Volume CA = CA/SG × 62.4 = $\underline{\hspace{3cm}}$ CF

10e: Solid Volume of Air + Water C + CA = $\underline{\hspace{3cm}}$ CF

10f: Solid Volume FA = 27 − (Air + Water C + CA) = $\underline{\hspace{3cm}}$ CF

10g: Weight of FA = SV of FA × 27 × 62.4 = $\underline{\hspace{3cm}}$ CF

Step 11: Complete the above table by inserting the necessary values.

    11a: List the weights of water, cement, and coarse and fine aggregate.

    11b: List the initial weight of water, cement, coarse and fine aggregate in the Mix Table.

    11c: Calculate the weight of LB ABS for the CA and FA by multiplying the materials initial weight by the percent ABS.

    11d: List the free water % for the CA and FA.

    11e: Calculate the weight of free water for the CA and FA by multiplying the weight of each material by its free water percent.

    11f: Determine the adjusted weight of each component.

    11g: Divide the results of each by 54 = 27CF/0.5CF (0.5 is sample size to be mixed).

Step 12: Prepare concrete mix using the batch design calculations above for a mix of 0.5 cu ft which will be necessary for two test cylinders. Mix concrete in pans after weighing components on scales and subtracting tare weights. Mix concrete placing FA (sand) first, CA (crushed stone) and cement. Add water in increments of 0.5, 0.25, and 0.25.

Step 13: Determine slump (See Experiment No. 8).

Step 14: Sample temperatures for each batch and record results (Experiment No. 9).

Step 15: Determine Unit Weight and Yield of Concrete per Experiment No. 9.

Step 15: Prepare test 2 test cylinders per Experiment No. 10.

Step 16: Cure the concrete cylinders in a curing room, which is most desirable, or by the use of wet coverings such as burlap. Test half of the cylinders for compressive strength at 7 days and the remaining cylinders at 28 days. Label all cylinders.

Calculations

Calculate Unit Weight and Yield of Samples.

Calculate the net mass of the concrete in pounds by subtracting the mass of the measure from the mass of the measure filled with concrete. Calculate the density D in cu ft by dividing the net mass of the concrete by the volume of the measure.

$$D = \frac{(M_c - M_m)}{V_m}$$

Then calculate the yield as follows: $Y(cu\ yd) = M/(D \times 27)$

At 7 and 28 days measure the compressive strength of samples. Record the results below:

Design $F^1_cD_{28}$	Slump	Temp	UW	Strength 7 day	Strength 28 day	% ±from Design
2000 psi						
3000 psi						
4000 psi						
5000 psi						
6000 psi						

Report

Prepare a laboratory report showing the results noted above for the each mixture. Plot the compressive strength at 7 and 28 days and note the percentage variance from the design $F^1_cD_{28}$. How do the results demonstrate the importance of the water–cement ratio and the hydration process? *Note: See Chapter 4 for a specific discussion of concrete mixtures design and compressive strength testing.*

EXPERIMENT NO. 8   Measuring Slump of Freshly Mixed Concrete

References

ASTM C 143: Standard Test Method for Slump of Hydraulic-Cement Concrete, American Concrete Institute ACI Certification, Concrete Field Testing Technician, Grade I

Objective

To measure the slump of freshly mixed concrete.

Background

For many years, the measurement of slump has been interpreted as a verification of water–cement ratio, that is, the higher the slump, the higher the water–cement ratio. However, a slight change in aggregate properties, gradation or air content, and the use of a wide variety of admixtures to enhance properties, has diminished the reliance on slump as an indication of water content, or water–cement ratio. Nonetheless, it is an important, easy to perform quality control tool that indicates that a change worthy of investigation has occurred.

Equipment

1. Slump mold
2. Scoop
3. Rod for tamping, 5/8 in. in diameter, 24 in. long
4. Ruler, readable to $\frac{1}{4}$ in.

Procedure

Step 1: Locate position for performing slump test. Surface must be nonabsorbent, flat and rigid.

Step 2: Moisten "gear" (slump cone and scoop). Slump cone interior must be clean and free from any build-up of previously tested mortar. Place slump cone on testing surface, stand on "foot tabs," and apply constant pressure to handles.

Step 3: Fill the slump cone in three increments, by volume, by using a scoop. After each lift is filled into the cone, rod 25 times, ensuring the entire cross section is rodded, by angling the rod to the dimension of the cone. Take care to only rod the newly placed lift (just penetrate preceding lift). The lift thickness to fill the volume of the cone in thirds should be approximately 2-5/8 in. for the first lift, and 6-1/2 in. the second lift.

Step 4: As rodding is occurring on the final lift of material, the level of the material should not fall below the top of the cone. It is acceptable to add additional material to the cone during the test should the level fall below the top of the cone, as long as the rodding count is maintained.

Step 5: Level final lift with tamping rod. Apply downward pressure on handles, and then step off foot tabs, and lift slump cone, in a purely vertical fashion, for a minimum of 12 in., for a period of time of 3 to 7 seconds. Do not twist the slump cone when removing.

Step 6: Rest the slump cone next to the sample on flat surface at same level. Place rod horizontally on slump cone, and measure the slump, to the displaced center of the sample, in inches, from the bottom of the rod, to the nearest $\frac{1}{4}$ in.

Step 7: The entire test from beginning of charging the slump cone to measuring of displaced center must be completed within an elapsed time of 2.5 minutes.

Step 8: If a marked shearing of one side of the sample occurs, the test is invalid, and the material should be re-sampled.

Report

Report slump in inches, to the nearest 0.25 in.

EXPERIMENT NO. 9 Unit Weight, Yield, and Gravimetric Air Content of Concrete

References

ASTM C 138, Standard Test Method for Density (Unit Weight), Yield and Air Content (Gravimetric) of Concrete, American Concrete Institute, ACI Certification Program Concrete Field Testing Technician- Grade I ASTM C 231 Standard Test Method for Air Content of Freshly Mixed Concrete by the Pressure Method.

Objective

To determine the unit weight (density) of concrete, and to calculate yield, and gravimetric air content of concrete.

Background

The unit weight or density is an important attribute of concrete. The results of the density test, typically performed at the mixing plant or on-site from material discharged from a truck, can indicate whether a change in materials, air content, water content or water–cement ratio has occurred. A change in any one of these factors could ultimately result in a change in compressive strength. Since compressive strength is typically not measured for days to come, the unit weight test is a quality control tool that provides useful information, quickly on the concrete.

The density test can also be used to calculate the *yield* of concrete. The determination of yield is another quality control tool that can provide immediate feedback as to the ultimate compressive strength of the material when it is measured in the future. A lower-than-expected density could indicate that an overyield situation potentially exists, which would indicate that for an established mix design, more material is supplied for a larger volume, and it is assumed the cement content is diluted over that larger volume.

Finally, if the theoretical air free weight of the concrete is known, the density test can be used to determine the air content of concrete.

Equipment

Choose minimum capacity of unit weight measure based on the nominal maximum size of the aggregate in the concrete as indicated in the table below.

NMAS of Aggregate	Capacity
25.0 mm (1.0 in.)	6 L (0.2 cu ft)
37.5 mm (1.5 in.)	11 L (0.4 cu ft)
50.0 mm (2.0 in.)	14 L (0.5 cu ft)
75.0 mm (3.0 in.)	28 L (1.0 cu ft)

In most cases the measuring bowl for Standard Test for Air Content by the Pressure Method (ASTM C 231) will meet the measure capacity requirements for unit weight, and is generally used.

Hemispherical rod, 16 mm (5/8 in.) in diameter, 600 mm (24 in.) in length.

Scoop

Rubber mallet

Balance

Strike-off plate

Personal protective equipment (e.g., gloves and safety glasses)

Procedure

Step 1: Weigh measure empty. Record mass as A.

Step 2: Scoop concrete to fill measure one-third full.

Step 3: Rod throughout the depth, and across the entire cross section of the sample 25 times, taking care not to strike the bottom of the measure. Strike the outside of the measure "smartly" with rubber mallet to level concrete.

Step 4: Repeat steps 2 and 3, at two-thirds full, and again, at overfill for the final lift. Take care to rod only the lift that has been placed. Do not penetrate the previously placed lift.

Step 5: Use metal strike off plate to level the final lift.

Step 6: Remove any excess concrete from the exterior of the measure.

Step 7: Place thermometer in measure filled with concrete. Ensure 75 mm or 3 in. of cover in all directions. Lightly pinch material around stem of thermometer to ensure ambient air temperature does not affect the measure of temperature. Leave thermometer in place a minimum of 2 minutes. If temperature has not stabilized within 2 minutes, leave thermometer in place until temperature has stabilized and an accurate measure of temperature is achieved. Remove thermometer from Unit Weight measure, and finish with strike off if necessary. Record temperature to the nearest 0.5°C or 1°F.

Step 8: Weigh the contents of the measure filled with the concrete. Record mass as B.

Calculations

Unit Weight (Density): $D = (B - A)/V$

where

B = Mass of measure filled with concrete
A = Mass of empty measure
V = Volume of measure

Yield: Y (cu yd) $= M/(D \times 27)$

where

M = Mass of all materials batched
D = Unit weight or density of material

Air Content: $A = [(T - D)/T] \times 100$

where

T = Theoretical air free density[*]
D = Unit weight or density of material

[*]"Theoretical" density is determined in the laboratory during the mix design process and is assumed to be a constant.

$$T = \frac{M}{V}$$

where M = Mass of all materials batched.

V = Total volume of component ingredients in the batch, *NOT INCLUDING AIR.*

Report

Report temperature to the nearest 0.5°C. or 1°F.

Report unit weight to nearest 0.01 pounds per cubic foot.

Report yield to nearest 0.01 cu yd.

Report gravimetric air to nearest 0.1 percent.

Precision:

Unit weight results conducted on the same material by two different operators should be within 2.31 lb/cu ft.

EXPERIMENT NO. 10   Preparing and Testing Cylinder and Beam Specimens

References

ASTM C 39: Standard Test Method for Compressive Strength of Cylindrical Concrete Specimens

ASTM C 78: Standard Test Method for Flexural Strength of Concrete (Using Simple Beam with Third Point Loading)

ASTM C 31: Standard Practice for Making and Curing Concrete Test Specimens in the Field

ASTM C 143: Standard Test Method for Slump of Hydraulic-Cement Concrete

ASTM C 1064: Standard Test Method for Temperature of Freshly Mixed Hydraulic-Cement Concrete

ASTM C 138: Standard Test Method for Density (Unit Weight) Yield, and Air Content (Gravimetric) of Concrete

Objective

To perform standard compressive strength tests of cylindrical concrete specimens and third point loading tests of concrete beams.

Equipment

Testing apparatus meeting the requirements of ASTM C 39 and ASTM C 78

Cylindrical sample molds

Beam sample molds

Balance or scale accurate to 0.1 percent of the sample mass or sensitive to 0.1 g

Mallet and cutting device (for removing plastic cylinders)

Tamping rod

Temperature measuring device

Slump cone

Background

These tests are used to determine the compressive strength of cylindrical concrete specimens and flexural strength of concrete beam specimens which may be reinforced or un-reinforced. Such standard tests allow design engineers to observe small test samples as a means of predicting the performance of structural building components including slabs, beams and columns.

For a further discussion of compressive and flexural strength See chapter 4 of the text.

Procedures*

Cylindrical Samples (6 in. mold)

Step 1:   Prepare cylindrical sample molds

- Place molds on a level rigid horizontal surface free of vibration (label molds prior to filling).
- Place concrete in mold moving the scoop around circumference of the mold as the concrete is discharged into the mold.
- Fill mold in three layers of equal volume. On final lift add extra to account for consolidation, and adjust and strike off at end w/representative concrete.
- Rod each layer 25 times with rounded end (hemispherical tip) of rod.
- Rod bottom layer throughout its depth.
- Rod middle and top layers to a depth of 1 in. into underlying layers.
- Tap the sides of the mold lightly w/open hand (single use mold) 10 to 15 times after rodding each layer.
- Strike off excess with tamping rod with minimal amount of manipulation- produce a flat even surface.

*Please note that the procedure is for educational purposes following general ASTM requirements; for non educational applications, see the ASTM standards referenced.

Step 2:   Sample temperature of test cylinders (see experiment 13)

Step 3:   Curing

Immediately following molding of cylinders, move to curing location. Test half of the cylinders for compressive strength at 7 days and the remaining cylinders at 28 days.

Step 4:   Compressive Strength Testing

Safety Note: Concrete testing apparatus can generate large forces equivalent to thousands of pounds of pressure over a relatively small area. A thorough understanding of the manufacturer's operating instructions and safe use of the equipment is most important. During destructive testing concrete specimens may fail and eject pieces of material. Although equipment typically has safety shields, both operators and witnesses should wear appropriate safety clothing and eye protection when conducting such tests.

For moist cured specimens tests made as soon as practicable, but no later than:

3 days and $+/-$ 2 hours from time specimens were placed in curing room

7 days $+/-$ 6 hours

28 days $+/-$ 20 hours

90 days $+/-$ 48 hours

After removing samples from cylinder forms, specimens should be axially loaded within the testing apparatus. Loading rates shall be at 20 to 50 psi/second. At failure, the testing apparatus will record the total force in pounds.

Calculations

Maximum Load Carried _____ lb

Surface Area of Specimen = πr^2 _____ sq in.

Max Load/Surface Area = _____ psi (record to the nearest 10 psi)

Test cylinders for compressive strength at 7 and 28 days _____

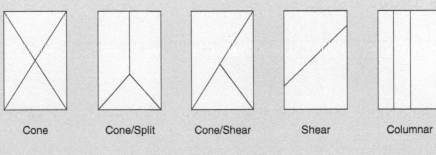

Cone   Cone/Split   Cone/Shear   Shear   Columnar

_____

_____

_____

Beam Specimens

Step 1:   Preparing beam specimens

- Place molds on a level rigid horizontal surface free of vibration (label molds prior to filling)
- Place concrete in mold moving the scoop around circumference of the mold as the concrete is discharged into the mold
- Fill mold in two layers of equal volume. On final lift add extra to account for consolidation, and adjust and strike off at end w/representative concrete.
- Rod each layer 25 times with rounded end (hemispherical tip) of rod.
- Rod bottom layer throughout its depth.
- Strike off excess with tamping rod with minimal amount of manipulation. Produce a flat even surface. After consolidation use a mason's trowel to produce a level top finish to the specimen.

- Prepare one sample without steel reinforcement and one sample with #4 steel reinforcement. Cut the rebar to the length of the mold. Place the reinforcing bar in the center of the sample after the first lift.

Step 2: Rod each layer with the rounded end of the rod using the calculated number of roddings per ASTM C 31.

Step 3: Curing

Cure the concrete cylinders in a curing room, which is most desirable, or by the use of wet coverings such as burlap. Test half of the cylinders for compressive strength at 7 days and the remaining cylinders at 28 days.

Step 4: Immediately following molding of beam specimens move to curing location. Test at specified intervals; typically 7 and 28 days.

Safety Note: Concrete testing apparatus can generate large forces equivalent to thousands of pounds of pressure over a relatively small area. A thorough understanding of the manufacturer's operating instructions and safe use of the equipment is most important. During destructive testing concrete specimens may fail and eject pieces of material. Although equipment typically has safety shields, both operators and witnesses should wear appropriate safety clothing and eye protection when conducting such tests.

After removing samples from forms. Turn the specimen on its side and place in the testing machine using a Third Point Loading method as illustrated below. Loading rates should constantly increase the extreme fiber stress between 125 to 175 psi/minute.

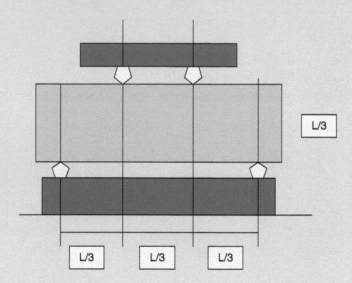

Calculations

Measure the specimen after testing failure by taking dimensions across one of the fractured faces at the center and each edge. Average the width and depth.

Record maximum load applied _____ lb

Span length _____ in.

Average width of specimen _____ in.

Average depth of specimen _____ in.

Calculate modulus of rupture

Fracture in middle third

Calculate Modulus of Rupture

$R = PL/bd^2$

R = modulus of rupture in psi
P = maximum applied load in pounds
L = length of span
b = average width
d = average depth
Modulus of Rupture _____ psi
Observations of Failure of Specimen_____

_____

_____

Fracture in end third

Calculate Modulus of Rupture

$R = 3Pa/bd^2$

Report

Reports of testing of specimens may be made in conjunction with other experiments for concrete design, admixtures, or air content. The report should contain a description of the preparation of specimens, characteristics of failure, and calculations showing compressive strength and modulus of rupture. Each cylinder or beam specimen should have all data noted. *Note: See chapter 4 of the text for a specific discussion of concrete compressive and flexural strength testing.*

EXPERIMENT NO. 11   Air Entrainment and Testing of Air Content

References

ASTM C 231: Standard Test Method for Air Content of Freshly Mixed Concrete by the Pressure Method
ACI 211.1: Standard Practice for Selecting Proportions for Normal, Heavyweight , and Mass Concrete
ASTM C 143: Standard Test Method for Slump of Hydraulic-Cement Concrete
ASTM C 31: Standard Practice for Making and Curing Concrete Test Specimens in the Field
ASTM C 1064: Standard Test Method for Temperature of Freshly Mixed Hydraulic-Cement Concrete
ASTM C 138: Standard Test Method for Density (Unit Weight) Yield, and Air Content (Gravimetric) of Concrete

Objective

To determine the percentage of air entrapped in freshly mixed concrete specimens using an ASTM Type B pressure meter. Further, to compare the relative strength, unit weight, and yield of concrete mixes with varying percentages of entrapped air.

Equipment

Type B pressure meter meeting the requirements of ASTM C 231-08b
Measuring bowl
Brick mason's trowel
Balance or scale accurate to 0.1 percent of the sample mass or sensitive to 0.1 g
Mallet
Strike off bar
Strike off plate
Cylindrical sample molds
Tamping rod
Temperature measuring device
Slump cone

Background

Air entraining admixtures are used to increase the volume of air entrapped in a concrete mix and thereby improve concrete resistance to freeze–thaw cycles and enhance workability. When concrete has been placed and is subjected to water saturation followed by freezing temperatures, the resulting expansion of water as it solidifies into ice can damage concrete. Air entrained concrete creates an air void system in the hardened concrete paste which allows water to freeze and provide room for expansion as the water changes to ice.

The normal mixing of concrete will entrap a certain percentage of air, depending on aggregate size, temperature, and other variables. However, this air is usually in the form of widely spaced bubbles, and under a microscope can be distinguished from entrained air which has a structured appearance resembling a honeycomb. Depending on the design function of the concrete mixture and the potential for exposure to weather and climate, the design engineer will determine the appropriate percentage of entrained air to be introduced into the mixture. Larger percentages may be undesirable, especially in structural elements as each 1 percent increase in entrained air results in a 5 percent decrease in ultimate compressive strength.

See Chapter 4 for a further discussion of air content and air entraining admixtures.

Procedures*

The air content of freshly mixed concrete can be determined from observation of the change in volume of concrete and the resultant change in pressure using a Type B meter. This can be accomplished by equalizing a known volume of air at a known pressure in a sealed air chamber with an unknown volume of air present in the concrete mixture. This method is not valid for lightweight aggregate mixes.

---

*Please note that the procedure is for educational purposes following general ASTM requirements; for noneducational applications see the ASTM standards referenced.

Type B Meter—Pressure Meter

Step 1: Prepare a concrete mixture of the following proportions by volume, sufficient to produce at least eight 6 in. test cylinders to be tested in the Type B pressure meter.

Portland cement = 1 portion

Fine aggregate = 2.6 portions

Coarse aggregate = 2.6 portions

Water = sufficient to produce a slump of $2\frac{1}{2}$ in. to −3 in.

Record the slump in three trials

Slump 1: _____

Slump 2: _____

Slump 3: _____

  Four cylinders should be labeled Non Air Entraining Mixture and four cylinders should be labeled Air Entraining Mixture.

  Split the total mixture into two halves. In the first half no air entraining agent should be added. In the second mixture add a commercial AEA at close of a minimum of 0.3% by weight of cement. Please note that the amount
of air entraining mixture may be increased or decreased depending on the observed results of entrapped air.

Step 2: For *each* sample of non–air entrained and air entrained mixtures proceed as follows:
- Weigh pressure meter container empty bottom portion only and record the weight _____ (lbs).
- Dampen interior of measuring bowl.
- Fill in three equal layers (1/3, 1/3, 1/3- by visual observation).
- Rod each layer 25 times.
- Tap sides w/mallet 10 to 15 times between rodding each layer.
- Last layer should be above the rim ~1/8 in.
- Strike off.
  - Use steel plate two-thirds over top scrape back inward.
  - Strike remaining one-third with plate pushing forward.
- Concrete should be in the pot level and smooth.
- With damp towel clean off the top flange (pot edges) spotless!
- Weigh pot and concrete—for unit weight calculations later _____(lbs).
- Place top on and tighten clamps—insure clamps are tight and fully engaged.
- **Open** both petcocks.
- *Close air* valve between air chamber and the bowl.
- Inject water through petcock until it flows out the other petcock.
- Continue *injecting water into the petcock* while jarring and tapping the meter to insure all air is expelled.
- *Close air bleeder valve* and pump air up to initial pressure line given on the meter as calibration pressure.
- Allow a few seconds for the compressed air to stabilize.
- Adjust the gauge to the initial pressure.
- *Close both petcocks*—turn to 90°.
- *Open air valve* between chamber and bowl—releasing pressure.
- **Rap** sides of bowl smartly with the mallet—watch gauge.
- Read the air percentage after lightly tapping the gauge to stabilize the hand—if the gage needle drops, a leak is present test no good!
- Close air valve cover petcocks with hand and then open to release pressure before removing the cover.
- Record air content.
- Disassemble and clean the meter.

Step 3: Determine slump (Experiment No. 8).

Step 4: Sample temperatures for each batch and record results (Experiment No. 9).

Step 5:   Prepare test 2 test cylinders per batch

- Place molds on a level rigid horizontal surface free of vibration (label molds prior to filling).
- Fill mold in three layers of equal volume—on final lift add extra to account for consolidation, and adjust and strike off at end w/representative concrete.
- Rod bottom layer throughout its depth.
- Tap the sides of the mold lightly w/open hand (single use mold) 10 to 15 times after rodding each layer.

Step 6:   Cure the concrete cylinders in a curing room, which is most desirable, or by the use of wet coverings such as burlap. Test half of the cylinders for compressive strength at 7 days and the remaining cylinders at 28 days. A 3 day test may also be conducted.

Calculations

Note: Calculate unit weight and yield of samples (see Experiment 9).

Calculate the net mass of the concrete in pounds by subtracting the mass of the measure from the mass or the measure filled with concrete. Calculate the density D in cu ft by dividing the net mass of the concrete by the volume of the measure.

$$D = (M_c - M_m)/V_m \quad Note: Vlm = 0.1963 \, cu \, ft^3$$

Non AE	Slump	Slump	UW	% AIR	Strength 7 days	Strength 28 days			Notes
Pressure Meter 1									
Pressure Meter 2									
Pressure Meter 3									
Average readings									
AE	Slump	Slump	UW	% AIR	Strength 7 days	Strength 28 days			Notes
Pressure Meter 1									
Pressure Meter 2									
Pressure Meter 3									
Average readings									

Report

Prepare a laboratory report showing the results noted above for the control batch (i.e., non–air entrained mixture) and each air entrained mixture. Plot the compressive strength of each mixture on a graph and analyze the plot lines to show the effect of the various percentages of air content for each mixture. What conclusions can be drawn about curing times and relative compressive strength, density, and yield of each mixture?

EXPERIMENT NO. 12   Determining the Air Content of Freshly Mixed Concrete using the Volumetric Method or "Rollameter"

References

ASTM C 173: Standard Test Method for Air Content of Freshly Mixed Concrete by the Volumetric Method, American Concrete Institute, Field Testing Technician Grade I, Publication CP-1

Objective

To determine the air content of freshly mixed concrete by using the "rollameter," or volumetric method. This test method is used exclusively for testing mixtures that contain lightweight aggregates although it can also be used for testing normal or heavyweight concrete.

Background

Air content, including entrapped and entrained air voids, is an important measure of concrete quality. While entrained air is desirable, especially in regions where multiple freeze–thaw cycles occur, too much air can lead to a decrease in strength.

Because of the pores in lightweight aggregate, measuring air content by volume is the standard.

Equipment

1. Airmeter consisting of a bowl and top section with graduated neck
2. Funnel
3. Tamping rod
4. Strike-off bar
5. Calibrated cup
6. Rubber mallet
7. Isopropyl alcohol, 70 percent by volume or 65 percent by weight
8. Water syringe

Procedure

Step 1: Moisten the inside of the bowl.

Step 2: Fill bowl in two equal increments. Rod each increment 25 times over the cross section of the sample. Level material by tap side of bowl 10 to 15 times with rubber mallet. Once the bowl is filled, strike off excess from top, and take care to ensure the edge, where top section of the bowl will be placed, is clean and free from any material.

Step 3: Moisten rubber gasket in top section and place and secure to bottom section with clamp provided. Be sure the interface between the top and bottom section is "clean."

Step 4: Place funnel in top of airmeter, and add approximately one pint of water.

Step 5: Add the selected amount of isopropyl alcohol. The selected amount of isopropyl alcohol will be based on experience, and should be the minimum amount required to ensure that the foaming after inversion and rolling does not cause the test to be invalidated because of foam residue greater than 2 percent of air graduations.

Step 6: Add water until the water appears in the graduated neck. Remove the funnel, and carefully fill with water using the syringe, until the bottom of the meniscus reaches the zero mark on the graduated neck.

Step 7: Securely place stopper in top of neck.

Step 8: Loosen concrete from the base of the bowl by inverting and shaking with vigor horizontally for a minimum of 45 seconds. Do not invert the airmeter for more than 5 seconds, so that the material does not escape into the graduated neck of airmeter. Concrete is most effectively removed from the base of the bowl by investing (for no more than 5 secs) 9 times over the 45 second interval.

Step 9: With base of meter on the floor, tilt the neck at a 45° angle, using hand on neck, and hand on the flange to rotate meter, rolling quickly $\frac{1}{4}$ to $\frac{1}{2}$ turns for a period of 1 minute. If during this process water leaks from the airmeter, the test is invalidated.

Step 10: After rolling, stand airmeter upright, and loosen stopper to allow air to rise, and liquid level to stabilize. The stabilization should occur within 6 minutes, and the liquid level is considered stable when it does not change over 0.25 percent of air within a 2-minute period.

Step 11: If the material contains greater than 9 percent air, it will be necessary to add water from the calibrated cup to obtain a reading. Each calibrated cup is equal to 1 percent of the volume of the graduated neck.

Step 12: If stabilization has occurred within 6 minutes, and there is less than 2 percent of air graduations of foam in the neck of airmeter, record air content to nearest 0.25 percent as reading 1.

Step 13: Replace stopper and repeat rolling process for a period of 1 minute. Repeat steps 10 and 11, and report air content as reading 2. If reading 2 is within 0.25 percent air of reading 1, report reading 2 as test result.

Step 14: If reading 2 is greater than 0.25 percent, different than reading 1, repeat steps 10 and 11. Record reading 3. If reading 3 is within 0.25 percent of reading 2, report reading 3 as test result. If reading 3 is greater than 0.25 percent of reading 2, the test is deemed invalid.

Step 15: Remove the airmeter to an area suitable for draining the contents. Remove the top section. Examine the bowl for any remains of consolidated concrete that remains in the sample or in the base of the bowl. If consolidated material remains, the test result is considered invalid.

Calculation

Determine alcohol correction factor per Table 4, if appropriate.

TABLE 4 Correction for the Effect of Isopropyl Alcohol on
C 173/C 173M Air Meter Reading

| | 70 % Isopropyl Alcohol Used | | Correction |
Pints	Ounces	Litres	(Subtract)[A]
0.5	8	0.2	0.0[B]
1.0	16	0.5	0.0[B]
1.5	24	0.7	0.0[B]
2.0	32	0.9	0.0[B]
3.0	48	1.4	0.3
4.0	64	1.9	0.6
5.0	80	2.4	0.9

[A] Subtract from final meter reading.
[B] Corrections less than 0.125 are not significant and are to be applied only when 2.5 pt [1.2 L] or more alcohol is used. The effect occurs when the meter is inverted after being filled with an alcohol-water solution which then becomes further diluted when it is mixed with the water in the concrete. The values given are for air meters that have a bowl volume of 0.075 ft³ [2.1 L] and a top section that is 1.2 times the volume of the bowl.

Source: ASTM C 173

Report

Air Content = Final Meter Reading + Calibrated Cups Water Added − Alcohol Correction Factor

EXPERIMENT NO. 13   CaCl$_2$ Accelerator for Concrete Admixtures

References

ACI 211.1: Standard Practice for Selecting Proportions for Normal, Heavyweight, and Mass Concrete
ASTM C 143: Standard Test Method for Slump of Hydraulic-Cement Concrete
ASTM C 31: Standard Practice for Making and Curing Concrete Test Specimens in the Field
ASTM C 1064: Standard Test Method for Temperature of Freshly Mixed Hydraulic-Cement Concrete
ASTM C 138: Standard Test Method for Density (Unit Weight) Yield, and Air Content (Gravimetric) of Concrete

Objective

To determine the effects of a calcium chloride accelerator when used in preparing and curing concrete test specimens.

Equipment

Balance or scale accurate to 0.1 percent of the sample mass or sensitive to 0.1 g
Cylindrical sample molds
Tamping rod
Temperature measuring device
Graduated cylinder
Slump cone

Background

Calcium chloride (CaCl$_2$) can be used as a concrete admixture which accelerates the rate of hydration producing higher early strength. This is especially useful in cold weather applications and to reduce the time that reusable concrete formwork must be left in place, thereby allowing a more efficient use of such forms.

Because of the presence of the chloride ion, CaCl$_2$ may not be used in prestressed or posttensioned concrete applications or with mixed metals where corrosion of concrete reinforcing will occur.

The purpose of this experiment is to examine the effects of the use of CaCl$_2$ as an admixture and determine the impact on the rate of hydration and effect on compressive strength of concrete samples prepared in the lab.

Procedure*

Four concrete mixes will be prepared with sufficient material to produce two standard test cylinders per batch. The first batch will serve as a control with *no CaCl$_2$ present*. The remaining batches will contain CaCl$_2$ as a percentage of the weight of Type 1 portland cement as follows:

Batch 1 = control = 0 percent CaCl$_2$ W = 2800 ml
Batch 2 = 2 percent CaCl$_2$ C = 9 lb
Batch 3 = 6 percent CaCl$_2$ CA = 28 lb
Batch 4 = 12 percent CaCl$_2$ FA = 28 lb

Step 1:   Determine the amount of CaCl$_2$ to be mixed with water and the water–cement ratio for the trial batches (M = mass C = cement)

    (a)   MC × percent CaCl$_2$ = M CaCl$_2$ × 453.6 (conversion factor lb to g) = M CaCl$_2$ g

        Control = 0

        Batch 2 = _____

        Batch 3 = _____

        Batch 4 = _____

    (b)   Determine the water–cement ratio: 2800 g/453.6 = MW

        lb/MC = W/C = _____

*Please note that the procedure is for educational purposes following general ASTM requirements; for non-educational applications, see the ASTM standards referenced.

Step 2: Determine slump

Record two slump trials and determine average

- Hold mold firmly against the base by standing on the two foot pieces. Do not allow it to move while filling.
- Rod each layer throughout its depth 25 × distributing strokes uniformly over the cross section.
- Strike off concrete level w/top of mold using the tamping rod.
- Lift mold upward 12 in. in one smooth motion without twisting the mold in 5 ± 2 seconds.
- Perform the test from start to finish within 2-1/2 minutes.

Step 3: Place the temperature measuring device in the freshly mixed concrete so that the bulb is submerged a minimum of 3 in. Gently press concrete around the temperature measuring device so that ambient air temperature does not affect the reading.

Leave sensor in concrete for minimum of 2 minutes or until reading stabilizes.

Step 4: Prepare test 2 test cylinders per batch

- Place molds on a level rigid horizontal surface free of vibration (label molds prior to filling).
- Fill mold in three layers of equal volume—on final lift add extra to account for consolidation, and adjust and strike off at end w/representative concrete.
- Rod bottom layer throughout its depth.
- Tap the sides of the mold lightly w/open hand (single use mold) 10 to 15 times after rodding each layer.

Step 5: Cure the concrete cylinders in a curing room, which is most desirable, or by the use of wet coverings such as burlap. Test half of the cylinders for compressive strength at 7 days and the remaining cylinders at 28 days. A 3-day test may also be conducted.

Calculations

Record Results:

Calculate unit weight and yield of samples (see Experiment 9)

At 7 and 28 days measure the compressive strength of samples. Measure the control batch first and then compare results for 2, 6, and 12 percent mixtures. Record the results below:

	Slump	Temp	UW	Strength 7 day	Strength 28 day	% ±
Control						
2 % CaCl$_2$						
6 % CaCl$_2$						
12 % CaCl$_2$						

Report

Prepare a laboratory report showing the results noted above for the control batch and each mixture. Plot the compressive strength of each mixture on a graph and analyze the plot lines to show the effect of the various percentages of CaCl$_2$ for each mixture. What conclusions can be drawn about curing times and relative compressive strength, density, and yield of each mixture?

Note: See Chapter 4 for a specific discussion of concrete admixtures and compressive strength testing.

EXPERIMENT NO. 14   Testing and Evaluation of Concrete Masonry Units (CMUs)

References

ASTM C 140; ASTM C 90

Materials

Three concrete masonry units

Objective

To determine the properties of a concrete masonry unit and evaluate the unit's compliance with specifications.

Background

Concrete masonry units are an important construction material; therefore, their use requires the testing and evaluation of samples at regular intervals. This procedure assures that the materials used are in compliance with current specifications.

Equipment

- Steel Scale
- Caliper
- Oven
- Compression Testing Machine
- Balance
- Immersion Container

Procedure*

Step 1: Measure height (H), length (L), and width (W) of the sample unit. Measure the face shell and web thicknesses $\frac{1}{2}$ in. from the bottom of the unit. Weigh the unit and record as W_r (weight received).

Step 2: Place unit in water at 60 to 80°F (15.6 to 26.7°C) for 24 hours. Weigh specimen suspended in water and record as W_i (weight immersed). Remove unit from water, allow to drain for 1 minute and remove visible water, weigh and record as W_s (saturated weight). Place unit in oven at 212 to 239°F (100 to 115°C) for 24 hours. Remove unit and weigh and record as W_d (oven-dry weight).

Step 3: Calculations

$$\text{Absorption, lb/cu ft} = \frac{W_s - W_d}{W_s - W_i} \times 62.4$$

$$\text{Absorption, \%} = \frac{W_s - W_d}{W_d} \times 100$$

$$\text{Density (D), lb/cu ft} = \frac{W_d}{W_s - W_i} \times 62.4$$

$$\text{Net Volume } (V_n), \text{ cu ft} = \frac{W_d}{D}$$

$$\text{Average Net Area } (A_n), \text{ sq in.} = \frac{V_n \times 1728}{H}$$

*Please note that the procedure is for educational purposes to demonstrate the testing of concrete masonry units following general ASTM requirements; for noneducational applications, see ASTM C 140, C90 standards.

Equivalent web thickness (inches per linear foot): add the measured thicknesses of all webs 0.75 or greater, then multiply by 12 and divide by the length (L) of the unit.

Determine the strength of the unit after capping by applying a load up to $\frac{1}{2}$ the maximum load expected, then set controls so that the load to failure is applied between 1 and 2 minutes.

Note: For this experiment cut pieces of plywood are substituted for actual capping of unit.

Report

Based on data collected from experiment, verify the unit's compliance with Tables 5 and 6.

TABLE 5 Minimum Thickness of Face Shells and Webs

Nominal Width (W) of Units, in. (mm)	Face Shell Thickness (t_{fs}), min, in. (mm)[A]	Web Thickness (t_w)	
		Webs[B] min, in. (mm)	Equivalent Web Thickness, min, in./linear ft[B,C] (mm/linear m)
3 (76.2) and 4 (102)	¾ (19)	¾ (19)	1⅝ (136)
6 (152)	1 (25)[D]	1 (25)	2¼ (188)
8 (203)	1¼ (32)[D]	1 (25)	2¼ (188)
10 (254)	1⅜ (35)[D] 1¼ (32)[D,E]	1⅛ (29)	2½ (209)
12 (305) and greater	1½ (38) 1¼ (32)[D,E]	1⅛ (29)	2½ (209)

[A] Average of measurements on 3 units taken at the thinnest point when measured as described in Test Methods C 140. When this standard is used for split face units, not more than 10 % of a split face shell area shall be less than shown, and the face shell thickness in this area shall be not less than ¾ in. (19.1 mm). When the units are solid grouted the 10 % limit does not apply.

[B] Average of measurements on 3 units taken at the thinnest point when measured as described in Test Methods C 140. The minimum web thickness for units with webs closer than 1 in. (25.4 mm) apart shall be ¾ in. (19.1 mm).

[C] Sum of the measured thicknesses of all webs in the unit, multiplied by 12 and divided by the length of the unit. Equivalent web thickness does not apply to the portion of the unit to be filled with grout. The length of that portion shall be deducted from the overall length of the unit for the calculation of the equivalent web thickness.

[D] For solid grouted masonry construction, minimum face shell thickness shall be not less than ⅝ in. (16 mm).

[E] This face shell thickness (t_{fs}) is applicable where allowable design load is reduced in proportion to the reduction in thickness from basic face shell thicknesses shown, except that allowable design loads on solid grouted units shall not be reduced.

Source: ASTM C 90

TABLE 6 Strength and Absorption Requirements

Compressive Strength,[A] min, psi (MPa)		Water Absorption, max, lb/ft³ (kg/m³) (Average of 3 Units)		
Average Net Area		Weight Classification—Oven-Dry Weight of Concrete, lb/ft³(kg/m³)		
Average of 3 Units	Individual Unit	Lightweight, less than 105 (1680)	Medium Weight, 105 to less than 125 (1680–2000)	Normal Weight, 125 (2000) or more
1900 (13.1)	1700 (11.7)	18 (288)	15 (240)	13 (208)

[A] Higher compressive strengths may be specified where required by design. Consult with local suppliers to determine availability of units of higher compressive strength.

Source: ASTM C 90

EXPERIMENT NO. 15   Preparing and Determining Hot Mix Asphalt (HMA) Bulk Density Specimens with the Superpave Gyratory Compactor

References

AASHTO 312: Preparing and Determining Density of Hot Mix Asphalt (HMA) Specimens by Means of the Superpave Gyratory Compactor.

Objective

To prepare bulk samples using the Superpave gyratory compactor (SGC).

Background

Bulk density samples are prepared for the Superpave design system using the Superpave gyratory compactor. It is recommended that at least two bulk density samples be prepared. After the bulk density samples are prepared, they will be evaluated utilizing AASHTO T 166 to determine bulk specific gravity. Bulk specific gravity test data, combined with AASHTO T 209 Maximum Specific Gravity test data, will yield a determination of air voids of a mixture sampled from a HMA plant, or a prepared mix design from laboratory.

Equipment

Superpave gyratory compactor capable of applying consolidation pressure of 600 kPa (80 psi), at an internal angle of 1.16° and speed of 30 revolutions per minute.

Specimen molds, and top and bottom plates, depending on model of Superpave gyratory compactor (SGC)

Paper protection disks

Thermometers

Balance

Oven

Miscellaneous pans, scoops, metal spatula, and kraft paper for charging Superpave molds.

Insulated gloves and safety glasses

Procedure

Step 1:   Determine mixing and compacting temperatures for HMA from PG Binder supplier. This information is readily available on delivery bill of lading. Mixing temperature is defined at a kinematic viscosity of unaged asphalt at 0.28 ± 0.03 Pa-s, whereas compaction temperature is defined at 0.17 ± 0.02 Pa-s. Typical mixing temperatures range from 300 to 325°F, and compaction temperatures range from 275 to 300°F. These typical ranges are broad, and could result in wide variations of results, especially considering mix design analysis, so it is best to consult supplier for the range, and choose *a* temperature for evaluation. Modified binders will generally have higher mixing and compaction temperatures.

Step 2:   Determine "n design" or gyrations for sample by traffic load (equivalent single axle loads expressed in millions).

ESALs	Ndesign gyrations
< 0.3	50
< 3.0	75
< 30	100
≥ 30	125

Note: Many states have modified gyration levels from those listed here, in an effort to enhance durability of Superpave. Local specifications should be consulted.

Step 3:   When evaluating mixtures in the SGC, the specimen after compaction must have a height of 115 mm $\pm$ 5 mm. This can typically be accomplished for mixtures with combined aggregate specific gravities of 2.55–2.70 with 4500–4700 g of material.

   Age enough mixture to produce desired number of SGC samples at compaction temperature in a forced draft oven for a period of 2 hours. The mixture should be placed no deeper than 1 in. in a flat shallow pan. The mixture should be stirred after 1 hour to ensure uniform aging.

Step 4:   30 to 60 minutes prior to preparing SGC samples, place molds, base and top plates (if applicable), in an oven at compaction temperature.

Step 5:   Following the short-term aging procedure, ensure mixture is at compaction temperature.

Step 6:   Remove mold and base plate from oven. Place paper protective disk in bottom of mold.

Step 7:   Charge, or fill mold with aged materials in one lift. This can be accomplished by lining aging pan with kraft paper, and removing the sample using the kraft paper as a funnel to charge the mold. Level the top of material with spatula. Place two thermometers in sample to ensure the material is at compaction temperature. If material is at compaction temperature, add protective disk to top of sample, place top plate in mold (if applicable, depending on SGC model), and immediately place the mold in compactor for the specified number of gyrations.

Step 8:   Once the SGC has completed compaction, ensure the specimen height is 115 mm $\pm$ 5 mm from machine screen and/or printer output. Record final specimen height. Remove the sample mold from the machine. The sample can be extruded with minimal cooling (ASAP), and with care, move the extruded sample to a rack, near fan for cooling. Remove specimen disks as soon as possible. Label the sample with marking crayon.

Report

Specimen height following compaction at "n design" number of gyrations.

EXPERIMENT NO. 16   Preparation of Hot Mix Asphalt (HMA) Bulk Samples Using the Marshall Hammer

References

ASTM D 1559: Resistance to Plastic Flow of Bituminous Mixtures Using the Marshall Apparatus

Objective

To prepare bulk density samples with a Marshall hammer for determination of HMA air voids, and stability and flow using the Marshall apparatus.

Background

The Marshall mix design method was developed by Bruce Marshall of the Department of Transportation in Mississippi in 1939. It was the most commonly used form of mix design and analysis of HMA mixtures until the 1990s, when it was replaced by Superpave. Some agencies still use the Marshall mix method today, including the Federal Aviation Administration.

Equipment

Specimen mold assembly

Mechanical (or manual) Marshall compaction hammer

Protective paper disks

Scoop and spatula

Balance

Oven

Pans

Thermometers

Marking crayons

Hot plate for maintaining temperature in the base of Marshall hammer

Insulated gloves and safety glasses

Gear for preparing Marshall specimens

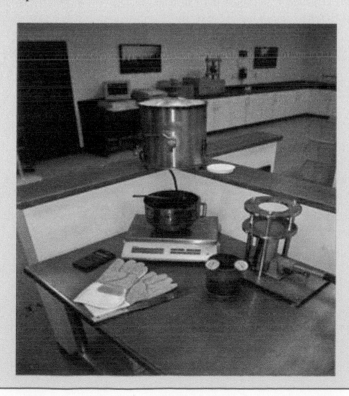

Marshall compactor

Procedure

Step 1:   Obtain sufficient plant or laboratory mixed HMA to fabricate three Marshall specimens of approximately 1200 g. each.

Step 2:   Heat three mold assemblies in the oven at 300°F for approximately 45 minutes.

Step 3:   Remove a mold assembly from the oven. The mold assembly consists of three components: a base plate, an intermediate collar which fits on base plate notch, and a top piece which sits atop the intermediate collar. Care should be taken when removing the assembly from the oven, to support assemble from below the base plate.

Step 4:   Place a protective disk into the bottom of the mold assembly.

Step 5:   Weigh 1200 ± 10 g of HMA into assembly. Spade the material within the mold 15 times around the perimeter, and 10 times in the center.

Step 6:   Slightly mound material at the top after spading, and insert two thermometers. When material is 275°± 5°, place additional protective disk on the top of the sample. Place mold on pedestal, secure clamp, and engage hammer to specified number of blows per side (typically 50 or 75). Once blows are completed on top side of sample, remove clamp, and remove the top section of the mold assembly. Flip the intermediate portion of the assembly, leaving the base on the hammer pedestal. Replace top portion of mold assembly, and engage the hammer to complete the specified number of blows on the bottom portion of the sample.

Step 7:   As soon as possible, remove protective disks, using spatula, taking care not to damage the sample. Cool the sample to room temperature leaving the sample contained in the mold.

Step 8:   Remove the sample using extruder and identify samples with marking crayon.

Report

Mix type, sample identification, and number of blows.

EXPERIMENT NO. 17   Determination of Bulk Specific Gravity of Compacted Hot Mix Asphalt (HMA) Mixtures

References

AASHTO T 166: Standard Method of Test for Bulk Specific Gravity of Compacted Bituminous Mixtures Using Saturated Surface Dry Specimens

Objective

To determine the bulk specific gravity of compacted HMA mixtures.

Background

The determination of air voids is a primary quality characteristic evaluated for hot mix asphalt. Bulk specific gravity is used in the computation of air voids for laboratory compacted specimens, using Marshall compactor or Superpave gyratory compactor, or from samples obtained from in situ pavements for the evaluation of the level of in-place density.

Equipment

Balance
Apparatus for suspending sample from balance in water bath.

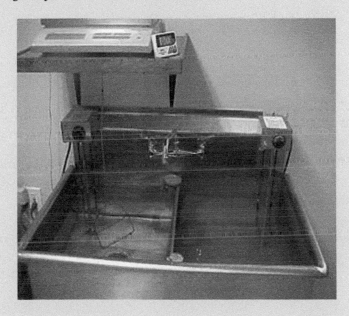

Water bath equipped with overflow valve to ensure constant water level, also capable of maintaining a temperature of 25 ± 1 (77 ± 1.8°F).

Kitchen timer

Damp absorbent cloth

Procedure

Specimen should be dry to a constant mass at room temperature. Specimens molded in laboratory for mix design or quality control of recently produced HMA do not require additional drying. Samples obtained from field for density analysis should be dried to a constant mass at 52 ± 3°C (125 ± 5°F).

Step 1:   Obtain "dry mass" of sample. Denote as A, "dry sample in air."

Step 2:   Ensure scale with suspension apparatus is "zeroed." Ensure water temperature is 25 ± 1°C (77 ± 1.8°F).

Step 3:   Place sample in cradle of suspension apparatus, and suspend sample into water bath. Sample should remain in water bath for 3 ± 1 minute.

Step 4:   Record weight of sample in water as C.

Step 5:   Remove sample from suspension in water and quickly roll and blot with damp towel to remove large fills of water in surface voids, taking care not to dry. Weigh sample after this procedure, and denote weight as B, saturated surface dry (SSD) weight.

Calculation

$$\text{Bulk Specific Gravity (Gmb)} = \frac{A}{(B - C)}$$

where

A = Mass of dry sample in air

B = Mass of SSD sample

C = Mass of sample in water

$(B-C)$ = Volume of sample

Report

Report Gmb to the nearest 0.001. In the case of reporting Marshall bulk specific gravity, one value is reported for the average of three specimens. Typically, only two Superpave gyratory specimens are required for a Gmb value. Should the values of test produced from the same materials within the same lab vary by more than 0.019, the results should be considered suspect, and discarded.

EXPERIMENT NO. 18   Determining the Maximum Theoretical Density (Gmm) or "Rice" of HMA Mixtures, and Using Results with Bulk Specific Gravity (Gmb) to Determine Percent Air Voids in HMA Mixture

References

AASHTO T 209: Theoretical Maximum Specific Gravity and Density of Bituminous Paving Mixtures

AASHTO T 166: Standard Method of Test for Bulk Specific Gravity of Compacted HMA Mixtures Using Saturated Surface Dry Specimens

Objective

To determine the theoretical maximum specific gravity or "voidless" density of an HMA mixture. Once the theoretical maximum density is determined, the results obtained in determining the bulk specific gravity (AASHTO T 166, Experiment No. 17) allow for computation of mixture air voids as a percent of a laboratory or plant prepared mixture, or from roadway sample.

Background

Once the theoretical maximum density is determined, the results obtained in determining the bulk specific gravity (AASHTO T 166) allow for computation of mixture air voids. Air voids can be determined for quality control purposes in the production of HMA, or in the laboratory as part of a mix design, or from an in situ sample from the roadway.

Many agencies also utilize the maximum theoretical density as a reference for their density specifications.

Equipment

Balance—with sufficient capacity and precision for calculations to the nearest 0.001.

Pycnometer—AASHTO T209 describes six acceptable alternatives for pycnometer arrangement. This experiment is designed for use with a "Type E," metal container with 4500 ml capacity.

Vacuum pump

Residual pressure manometer capable of measuring residual pressure to 4.0 or less kPa.

Water bath with overflow capable of maintaining a temperature of 25 ± 1°C or 77 ± 1.8°F

Thermometers

Shaker platform or rubber mallet for agitating specimen in pycnometer

Wire and cradle apparatus for suspending pycnometer from balance in water bath.

Large flat pans for cooling and preparing sample

Flasks under vacuum on shaker table with residual manometer to measure vacuum.

Procedure

Step 1:   Calibration of flask is achieved by suspending flask full with water at 25 ± 1°C or 77 ±1.8° F from scale in the water bath. Record mass as D, "container filled with water at 25°C (77°F)." If multiple flasks are used, each flask must undergo the same procedure, as "D" or calibration is flask dependent.

Step 2:   Obtain sample of HMA. Sample size is based on nominal maximum size of HMA as follows:

NMAS sample size minimum

9.5 mm (3/8 in.) 1000 g.

12.5 mm (1/2 in.) 1500 g.

19.0 mm (3/4 in.) 2000 g.

25.0 mm (1 in.) 2500 g.

37.5 mm (1.5 in.) 4000 g.

Cool sample to room temperature. As sample is cooling stir with large spoon or spatula to prevent agglomeration of the material. When material cools to touch, further break agglomerated pieces to their smallest component by hand.

Step 3: Once sample is prepared, zero pycnometer on scale and obtain initial sample mass by transferring the prepared material into the pycnometer. Record mass as A, "dry sample in air."

Step 4: Add water at 25°C (77°F) to within 2 in. of lip of pycnometer. Attach vacuum lid to pycnometer, and apply vacuum 3.7 ± 0.3 kPa for 15 ± 2 minutes. During this time, trapped air in the sample should be released by either continuous "vibration" mechanically, or by rapping pycnometer with rubber mallet at 2-minute intervals while sample under vacuum.

Step 5: Once vacuum interval is complete, carefully transport pycnometer with deaired sample to water bath, and place sample on cradle suspended on wire from scale, taking care not to introduce air bubbles into the sample while submerging. Sample should remain on cradle for an interval of 10 minutes to ensure the mass of sample, pycnometer, and water is 25°C (77°F). Following a 10-minute temperature conditioning, record mass as E, and "mass" of pycnometer, sample, and water at 25°C (77°F).

Calculations

$$\text{Theoretical Maximum Specific Gravity or Gmm} = \frac{A}{(A + D - E)}$$

where

A = Sample mass in air

D = Mass of pycnometer filled with water at 25°C (77°F).

E = Mass of pycnometer, sample, and water suspended at 25°C (77°F).

Calculation of Air Voids:

$$\frac{(\text{Gmm} - \text{Gmb})}{\text{Gmm}} \times 100$$

where

Gmm = Theoretical Maximum Specific Gravity (AASHTO T 209)

Gmb = Bulk Specific Gravity (AASHTO T 166)

Record

Record Gmm to the nearest 0.001. Two tests completed on the same material should be within 0.011, or 0.019 between laboratories. Results outside of this range should be considered suspect.

Percent air voids should be reported to the nearest 0.1 percent.

EXPERIMENT NO. 19   Measuring HMA Mixture Stability and Resistance to Flow Using Marshall Equipment

References

ASTM D 6927: Standard Test Method for Marshall Stability and Flow of Bituminous Mixtures

AASHTO T 209: Maximum Specific Gravity, Theoretical Maximum Specific Gravity and Density of Bituminous Mixtures

AASHTO T 166 Standard Method of Test for Bulk Specific Gravity of Compacted Hot Mix Asphalt Mixtures Using Saturated Surface Dry Specimens

Objective

To measure Marshall stability and flow for HMA bulk specimens prepared with Marshall hammer.

Background

The Marshall mix design method was developed by Bruce Marshall of the Department of Transportation in Mississippi in 1939. It was the most commonly used form of mix design and analysis of HMA mixtures until the 1990s, when it was replaced by Superpave. Some agencies still use the Marshall mix method today, including the Federal Aviation Administration. Determination of air voids for both Superpave and Marshall is accomplished using AASHTO T 209 (Maximum Theoretical Specific Gravity) and AASHTO T 166 (Bulk Specific Gravity) although the two "systems" can yield different results for the same materials. Marshall stability and flow were intended as a performance predicator for HMA.

Equipment

Compression loading machine capable of applying a vertical load at the rate of 50 ± 5 mm/min (2.00 ± 0.5 inches/min). It should also be equipped with a flowmeter, capable of measuring flow to a sensitivity of 0.25 mm (0.01 in.)

L to R, water bath, compression loading/flowmeter with breaking head and specimen

Breaking head, with upper and lower cylindrical pieces with an internal curvature of 50.8 mm

Water bath, capable of completely submerging samples and maintaining a temperature of 60°C (140°F) or specified temperature.

Tongs for removing samples from water bath

Towel

Manufacturer's graph paper for measuring output

Insulated gloves and safety glasses

Procedure

Step 1: Measure height of specimens (3). Laboratory molded specimens should be 63.5 ± 2.5 mm (2.50 ± 0.10 in.).

Step 2: Condition (soak to temperature) specimens for 30 to 40 minutes in a water bath capable of covering specimens, at a temperature of 60°C (140°), or a specified temperature.

Step 3: Remove sample (1) from the water bath. Lightly towel off excess water, and place in lower section of breaking head. Replace upper section of breaking head, and move the breaking head with sample to loading/flow apparatus for testing. Elapsed time between removing sample from water bath to initiation of test should not exceed 30 seconds.

Step 4: Ensure output marker for measuring stability and flow at zero mark on graph paper.

Step 5: Engage load.

Step 6: Once peak load is reached, remove deformed sample and repeat steps 3 to 5 for remaining two samples.

Calculations

Determine stability from y axis at maximum load. Determine flow at the point where maximum load is also reached. The flow value is taken from the x axis of the graph.

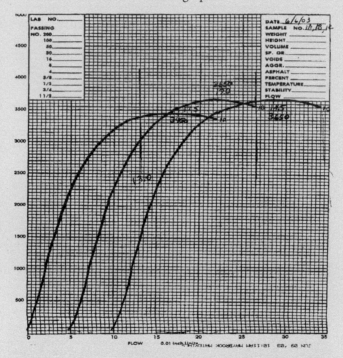

Correct stability by multiplying the measured stability by the correction factor, for specimens at heights other than 63.5 mm (2.5 in.). Additional correction factors are available; however, they are generally outside the deviations expected for laboratory prepared specimens.

Volume	Height (mm)	Height (in.)	Correction Factor
483–495	60.3	2.38	1.09
496–508	61.9	2.44	1.04
509–522	63.5	2.5	1.00
523–535	65.1	2.56	0.93
536–546	68.3	2.6	0.89

Report

Stability (corrected) as an average for three specimens in N (lbs) to nearest 50 N or 10 (lbf). Report flow as an average for three specimens in mm to nearest 0.25 mm (0.01 in.).

Standard Practice for Selecting Proportions for Normal, Heavyweight, and Mass Concrete (ACI 211.1-91)

Reported by ACI Committee 211

Donald E. Dixon,
Chairman

Jack R. Prestrera,
Secretary

George R. U. Burg,*
Chairman, Subcommittee A

Edward A. Abdun-Nur*	David A. Crocker	Mark A. Mearing	George B. Southworth
Stanley G. Barton	Kenneth W. Day	Richard C. Meininger*	Alfred B. Spamer
Leonard W. Bell*	Calvin L. Dodl	Richard W. Narva	Paul R. Stodola
Stanley J. Blas, Jr.	Thomas A. Fox	Leo P. Nicholson	Michael A. Taylor
Ramon L. Carraquillo	Donald A. Graham	James E. Oliverson	Stanely J. Vigalitte
Peggy M. Carraquillo	George W. Hollon	James S. Pierce	William H. Voelker
Alan C. Carter	William W. Hotaling, Jr.	Sandor Popovics*	Jack W. Weber*
Martyn T. Conrey	Robert S. Jenkins	Steven A. Ragan	Dean J. White II
James E. Cook	Paul Klieger	Harry C. Robinson	Milton H. Willis, Jr.
Russel A. Cook*	Frank J. Lahm	Jere H. Rose*	Francis C. Wilson
William A. Cordon	Stanley H. Lee	James A. Scherocman	Robert Yuan
Wayne J. Costa	Gary R. Mass*	James M. Shilstone*	

Committee Members Voting on 1991 Revision

Gary R. Mass†
Chairman

George R. U. Burg†
Chairman, Subcommittee A

Edward A. Abdun-Nur†	David A. Crocker	Richard C. Meininger†	William S. Sypher
William L. Barringer†	Luis H. Diaz	James E. Oliverson	Ava Szypula
Stanley G. Barton	Donald E. Dixon†	James S. Pierce	Jimmie L. Thompson†
Leonard W. Bell†	Calvin L. Dodl	Sandor Popovics	Stanley J. Virgalitte
James E. Bennett, Jr.	Thomas A. Fox	Steven A. Ragan	Woodward L. Vogt
J. Floyd Best	George W. Hollon	Jere H. Rose†	Jack W. Weber
Ramon L. Carrasquillo	Tarif M. Jaber	Donald L. Schlegel	Dean J. White, III
James E. Cook†	Stephen M. Lane	James M. Shilstone, Sr.	Marshall S. Williams
Russell A. Cook	Stanley H. Lee	Paul R. Stodola	John R. Wilson†

Describes, with examples, two methods for selecting and adjusting proportions for normal weight concrete, both with and without chemical admixtures, pozzolanic, and slag materials. One method is based on an estimated weight of the concrete per unit volume; the other is based on calculations of the absolute volume occupied by the concrete ingredients. The procedures take into consideration the requirements for placeability, consistency, strength, and durability. Example calculations are shown for both methods, including adjustments based on the characteristics of the first trial batch.

The proportioning of heavyweight concrete for such purposes as radiation shielding and bridge counterweight structures is described in an appendix. This appendix uses the absolute volume method, which is generally accepted and is more convenient for heavyweight concrete.

There is also an appendix that provides information on the proportioning of mass concrete. The absolute volume method is used because of its general acceptance.

Keywords: absorption; admixtures; aggregates; blast-furnace slag; cementitious materials; concrete durability; concretes; consistency; durability; exposure; fine aggregates; fly ash; heavyweight aggregates; heavyweight concretes; mass concrete; mix proportioning; pozzolans; quality control; radiation shielding; silica fume; slump tests; volume; water-cement ratio; water-cementitious ratio; workability.

CONTENTS

* Members of Subcommittee A who prepared this standard. The committee acknowledges the significant contribution of William L. Barringer to the work of the subcommittee.

† Members of Subcommittee A who prepared the 1991 revision.

This standard supersedes ACI 211.1-89. It was revised by the Expedited Standardization procedure, effective Nov. 1, 1991. This revision incorporates provisions related to the use of the mineral admixture silica fume in concrete. Chapter 4 has been expanded to cover in detail the effects of the use of silica fume on the proportions of concrete mixtures. Editorial changes have also been made in Chapters 2 through 4, and Chapters 6 through 8.

Copyright © 1991, American Concrete Institute.

Chapter 5—Background data, p. 211.1-7

Chapter 6—Procedure, p. 211.1-7

Chapter 7—Sample computations, p. 211.1-13

Chapter 8—References, p. 211.1-18

Appendix 1—Metric system adaptation

Appendix 2—Example problem in metric system

Appendix 3—Laboratory tests

Appendix 4—Heavyweight concrete mix proportioning

Appendix 5—Mass concrete mix proportioning

CHAPTER 1 -- SCOPE

1.1 This Standard Practice describes methods for selecting proportions for hydraulic cement concrete made with and without other cementitious materials and chemical admixtures. This concrete consists of normal and/or high-density aggregates (as distinguished from lightweight aggregates) with a workability suitable for usual cast-in-place construction (as distinguished from special mixtures for concrete products manufacture). Also included is a description of methods used for selecting proportions for mass concrete. Hydraulic cements referred to in this Standard Practice are portland cement (ASTM C 150) and blended cement (ASTM C 595). The Standard does not include proportioning with condensed silica fume.

1.2 The methods provide a first approximation of proportions intended to be checked by trial batches in the laboratory or field and adjusted, as necessary, to produce the desired characteristics of the concrete.

1.3 U.S. customary units are used in the main body of the text. Adaption for the metric system is provided in Appendix 1 and demonstrated in an example problem in Appendix 2.

1.4 Test methods mentioned in the text are listed in Appendix 3.

CHAPTER 2 -- INTRODUCTION

2.1 Concrete is composed principally of aggregates, a portland or blended cement, and water, and may contain other cementitious materials and/or chemical admixtures. It will contain some amount of entrapped air and may also contain purposely entrained air obtained by use of an admixture or air-entraining cement. Chemical admixtures are frequently used to accelerate, retard, improve workability, reduce mixing water requirements, increase strength, or alter other properties of the concrete (see ACI 212.3R). De-

pending upon the type and amount, certain cementitious materials such as fly ash, (see ACI 226.3R) natural pozzolans, ground granulated blast-furnace (GGBF) slag (see ACI 226.1R), and silica fume may be used in conjunction with portland or blended cement for economy or to provide specific properties such as reduced early heat of hydration, improved late-age strength development, or increased resistance to alkali-aggregate reaction and sulfate attack, decreased permeability, and resistance to the intrusion of aggressive solutions (See ACI 225R and ACI 226.1R).

2.2 The selection of concrete proportions involves a balance between economy and requirements for placeability, strength, durability, density, and appearance. The required characteristics are governed by the use to which the concrete will be put and by conditions expected to be encountered at the time of placement. These characteristics should be listed in the job specifications.

2.3 The ability to tailor concrete properties to job needs reflects technological developments that have taken place, for the most part, since the early 1900s. The use of water-cement ratio as a tool for estimating strength was recognized about 1918. The remarkable improvement in durability resulting from the entrainment of air was recognized in the early 1940s. These two significant developments in concrete technology have been augmented by extensive research and development in many related areas, including the use of admixtures to counteract possible deficiencies, develop special properties, or achieve economy (ACI 212.2R). It is beyond the scope of this discussion to review the theories of concrete proportioning that have provided the background and sound technical basis for the relatively simple methods of this Standard Practice. More detailed information can be obtained from the list of references in Chapter 8.

2.4 Proportions calculated by any method must always be considered subject to revision on the basis of experience with trial batches. Depending on the circumstances, the trial mixtures may be prepared in a laboratory, or, perhaps preferably, as full-size field batches. The latter procedure, when feasible, avoids possible pitfalls of assuming that data from small batches mixed in a laboratory environment will predict performance under field conditions. When using maximum-size aggregates larger than 2 in., laboratory trial batches should be verified and adjusted in the field using mixes of the size and type to be used during construction. Trial batch procedures and background testing are described in Appendix 3.

2.5 Frequently, existing concrete proportions not containing chemical admixtures and/or materials other than hydraulic cement are reproportioned to include these materials or a different cement. The performance of the reproportioned concrete should be verified by trial batches in the laboratory or field.

CHAPTER 3 -- BASIC RELATIONSHIP

3.1 Concrete proportions must be selected to provide

necessary placeability, density, strength, and durability for the particular application. In addition, when mass concrete is being proportioned, consideration must be given to generation of heat. Well-established relationships governing these properties are discussed next.

3.2 *Placeability* -- Placeability (including satisfactory finishing properties) encompasses traits loosely accumulated in the terms "workability" and "consistency." For the purpose of this discussion, workability is considered to be that property of concrete that determines its capacity to be placed and consolidated properly and to be finished without harmful segregation. It embodies such concepts as moldability, cohesiveness, and compactability. Workability is affected by: the grading, particle shape, and proportions of aggregate; the amount and qualities of cement and other cementitious materials; the presence of entrained air and chemical admixtures; and the consistency of the mixture. Procedures in this Standard Practice permit these factors to be taken into account to achieve satisfactory placeability economically.

3.3 *Consistency* -- Loosely defined, consistency is the relative mobility of the concrete mixture. It is measured in terms of slump -- the higher the slump the more mobile the mixture -- and it affects the ease with which the concrete will flow during placement. It is related to but not synonymous with workability. In properly proportioned concrete, the unit water content required to produce a given slump will depend on several factors. Water requirement increases as aggregates become more angular and rough textured (but this disadvantage may be offset by improvements in other characteristics such as bond to cement paste). Required mixing water decreases as the maximum size of well-graded aggregate is increased. It also decreases with the entrainment of air. Mixing water requirements usually are reduced significantly by certain chemical water-reducing admixtures.

3.4 *Strength* -- Although strength is an important characteristic of concrete, other characteristics such as durability, permeability, and wear resistance are often equally or more important. Strength at the age of 28 days is frequently used as a parameter for the structural design, concrete proportioning, and evaluation of concrete. These may be related to strength in a general way, but are also affected by factors not significantly associated with strength. In mass concrete, mixtures are generally proportioned to provide the design strength at an age greater than 28 days. However, proportioning of mass concrete should also provide for adequate early strength as may be necessary for form removal and form anchorage.

3.5 *Water-cement or water-cementitious ratio [w/c or w/(c + p)]* -- For a given set of materials and conditions, concrete strength is determined by the net quantity of water used per unit quantity of cement or total cementitious materials. The net water content excludes water absorbed by the aggregates. Differences in strength for a given water-cement ratio w/c or water-cementitious materials ratio $w/(c + p)$ may result from changes in: maximum size of aggregate; grading, surface texture, shape, strength, and stiffness of aggregate particles; differences in cement types and sources; air content; and the use of chemical admixtures that affect the cement hydration process or develop cementitious properties themselves. To the extent that these effects are predictable in the general sense, they are taken into account in this Standard Practice. In view of their number and complexity, it should be obvious that accurate predictions of strength must be based on trial batches or experience with the materials to be used.

3.6 *Durability* -- Concrete must be able to endure those exposures that may deprive it of its serviceability -- freezing and thawing, wetting and drying, heating and cooling, chemicals, deicing agents, and the like. Resistance to some of these may be enhanced by use of special ingredients: low-alkali cement, pozzolans, GGBF slag, silica fume, or aggregate selected to prevent harmful expansion to the alkali-aggregate reaction that occurs in some areas when concrete is exposed in a moist environment; sulfate-resisting cement, GGBF slag, silica fume, or other pozzolans for concrete exposed to seawater or sulfate-bearing soils; or aggregate composed of hard minerals and free of excessive soft particles where resistance to surface abrasion is required. Use of low water-cement or cementitious materials ratio $[w/c$ or $w/(c + p)]$ will prolong the life of concrete by reducing the penetration of aggressive liquids. Resistance to severe weathering, particularly freezing and thawing, and to salts used for ice removal is greatly improved by incorporation of a proper distribution of entrained air. Entrained air should be used in all exposed concrete in climates where freezing occurs. (See ACI 201.2R for further details).

3.7 *Density* -- For certain applications, concrete may be used primarily for its weight characteristic. Examples of applications are counterweights on lift bridges, weights for sinking oil pipelines under water, shielding from radiation, and insulation from sound. By using special aggregates, placeable concrete of densities as high as 350 lb/ft^3 can be obtained--see Appendix 4.

3.8 *Generation of heat* -- A major concern in proportioning mass concrete is the size and shape of the completed structure or portion thereof. Concrete placements large enough to require that measures be taken to control the generation of heat and resultant volume change within the mass will require consideration of temperature control measures. As a rough guide, hydration of cement will generate a concrete temperature rise of 10 to 15 F per 100 lb of portland cement/yd^3 in 18 to 72 hours. If the temperature rise of the concrete mass is not held to a minimum and the heat is allowed to dissipate at a reasonable rate, or if the concrete is subjected to severe temperature differential or thermal gradient, cracking is likely to occur. Temperature control measures can include a relatively low initial placing temperature, reduced quantities of cementitious materials, circulation of chilled water, and, at times, insulation of concrete surfaces as may be required to adjust for these various concrete conditions and exposures. It should be emphasized that mass concrete is not necessarily large-aggregate concrete and that concern about generation of an excessive amount of heat in concrete is not confined to massive dam or foundation structures. Many large structural elements may be massive enough that heat generation should be considered, particularly when the minimum cross-sectional dimensions of a solid concrete member approach or exceed 2 to 3 ft or when cement contents above 600 lb/yd^3 are being used.

CHAPTER 4--EFFECTS OF CHEMICAL ADMIXTURES, POZZOLANIC, AND OTHER MATERIALS ON CONCRETE PROPORTIONS

4.1 *Admixtures* -- By definition (ACI 116R), an admixture is "a material other than water, aggregates, hydraulic cement, and fiber reinforcement used as an ingredient of concrete or mortar and added to the batch immediately before or during its mixing." Consequently, the term embraces an extremely broad field of materials and products, some of which are widely used while others have limited application. Because of this, this Standard Practice is restricted to the effects on concrete proportioning of air-entraining admixtures, chemical admixtures, fly ashes, natural pozzolans, and ground granulated blast-furnace slags (GGBF slag).

4.2 *Air-entraining admixture* -- Air-entrained concrete is almost always achieved through the use of an air-entraining admixture, ASTM C 260, as opposed to the earlier practice in which an air-entraining additive is interground with the cement. The use of an air-entraining admixture gives the concrete producer the flexibility to adjust the entrained air content to compensate for the many conditions affecting the amount of air entrained in concrete, such as: characteristics of aggregates, nature and proportions of constituents of the concrete admixtures, type and duration of mixing, consistency, temperature, cement fineness and chemistry, use of other cementitious materials or chemical admixtures, etc. Because of the lubrication effect of the entrained air bubbles on the mixture and because of the size and grading of the air voids, air-entrained concrete usually contains up to 10 percent less water than non-air-entrained concrete of equal slump. This reduction in the volume of mixing water as well as the volume of entrained and entrapped air must be considered in proportioning.

4.3 *Chemical admixtures* -- Since strength and other important concrete qualities such as durability, shrinkage, and cracking are related to the total water content and the w/c or $w/(c + p)$, water-reducing admixtures are often used to improve concrete quality. Further, since less cement can be used with reduced water content to achieve the same w/c or $w/(c + p)$ or strength, water-reducing and set-controlling admixtures are used widely for reasons of economy (ACI 212.2R).

Chemical admixtures conforming to ASTM C 494, Types A through G, are of many formulations and their purpose or purposes for use in concrete are as follows:

Type A -- Water-reducing
Type B -- Retarding
Type C -- Accelerating
Type D -- Water-reducing and retarding
Type E -- Water-reducing, and accelerating
Type F -- Water-reducing, high-range
Type G -- Water-reducing, high-range, and retarding

The manufacturer or manufacturer's literature should be consulted to determine the required dosage rate for each specific chemical admixture or combination of admixtures. Chemical admixtures have tendencies, when used in large doses, to induce strong side-effects such as excessive retardation and, possibly, increased air entrainment, in accordance with ASTM C 1017. Types A, B, and D, when used by themselves, are generally used in small doses (2 to 7 oz/100 lb of cementitious materials), so the water added to the mixture in the form of the admixture itself can be ignored. Types C, E, F, and G are most often used in large quantities (10 to 90 oz/100 lb of cementitious materials) so their water content should be taken into account when calculating the total unit water content and the w/c or $w/(c + p)$. When Types A, B, and D admixtures are used at higher than normal dosage rates in combination or in an admixture system with an accelerating admixture (Type C or E), their water content should also be taken into account.

Although chemical admixtures are of many formulations, their effect on water demand at recommended dosages is governed by the requirements of ASTM C 494. Recommended dosage rates are normally established by the manufacturer of the admixture or by the user after extensive tests. When used at normal dosage rates, Type A water-reducing, Type D water-reducing and retarding, and Type E water-reducing and accelerating admixtures ordinarily reduce mixing-water requirements 5 to 8 percent, while Type F water-reducing, high-range, and Type G water-reducing, high-range, and retarding admixtures reduce water requirements 12 to 25 percent or more. Types F and G water-reducing, high-range admixtures (HRWR) are often called "superplasticizers."

High-range, water-reducing admixtures are often used to produce flowing concrete with slumps between about 7½ or more with no increase in water demand other than that contained in the admixture itself. Types A, B, or D admixtures at high dosage rates, in combination with Types C or E (for acceleration), may also be used to produce the same effect. When flowing concrete is so produced, it is sometimes possible to increase the amount of coarse aggregate to take advantage of the fluidity of the concrete to flow into place in constricted areas of heavy reinforcement. Flowing concrete has a tendency to segregate; therefore, care must be taken to achieve a proper volume of mortar in the concrete required for cohesion without making the concrete undesirably sticky.

ASTM C 494 lists seven types of chemical admixtures as to their expected performance in concrete. It does not classify chemical admixtures as to their composition. ACI 212.2R lists five general classes of materials used to formulate most water-reducing, set-controlling chemical admixtures. This report, as well as ACI 301 and ACI 318, should be reviewed to determine when restrictions should be placed upon the use of certain admixtures for a given class of concrete. For example, admixtures containing purposely added calcium chloride have been found to accelerate the potential for stress-corrosion of tensioned cables imbedded in concrete when moisture and oxygen are available.

4.4 *Other cementitious materials* -- Cementitious materials other than hydraulic cement are often used in concrete in combination with portland or blended cement for economy, reduction of heat of hydration, improved workability, improved strength and/or improved durability under the anticipated service environment. These materials include fly ash, natural pozzolans (ASTM C 618), GGBF slag (ASTM C 989), and silica fume. Not all of these materials will provide all of the benefits listed.

As defined in ASTM C 618, pozzolans are: "Siliceous or siliceous and aluminous materials which in themselves possess little or no cementitious value, but will, in finely divided form and in the presence of moisture, chemically react with calcium hydroxide at ordinary temperatures to form compounds possessing cementitious properties ... " Fly ash is the "finely divided residue that results from the combustion of ground or powdered coal ... " Fly ash used in concrete is classified into two categories: Class F, which has pozzolanic properties, and Class C, which, in addition to having pozzolanic properties, also has some cementitious properties in that this material may be self-setting when mixed with water. Class C fly ash may contain lime (CaO) amounts higher than 10 percent. The use of fly ash in concrete is more fully described and discussed in ACI 226.3R.

Blast-furnace slag is a by-product of the production of pig iron. When this slag is rapidly quenched and ground, it will possess latent cementitious properties. After processing, the material is known as GGBF slag, whose hydraulic properties may vary and can be separated into grades noted in ASTM C 989. The grade classification gives guidance on the relative strength potential of 50 percent GGBF slag mortars to the reference portland cement at 7 and 28 days. GGBF slag grades are 80, 100, and 120, in order of increasing strength potential.

Silica fume,* as used in concrete, is a by-product resulting from the reduction of high-purity quartz with coal and wood chips in an electric arc furnace during the production of silicon metal or ferrosilicon alloys. The silica fume, which condenses from the gases escaping from the furnaces, has a very high content of amorphous silicon dioxide and consists of very fine spherical particles.

Uses of silica fume in concrete fall into three general categories:

a.   Production of low permeability concrete with enhanced durability.
b.   Production of high-strength concrete.
c.   As a cement replacement (The current economics of cement costs versus silica fume costs do not usually

---
* Other names that have been used include silica dust, condensed or pre-compacted silica fume, and micro silica; the most appropriate is silica fume.

make this a viable use for silica fume in the U.S.).

Silica fume typically has a specific gravity of about 2.2. The lower specific gravity of silica fume compared with that of portland cement means that when replacement is based on weight (mass), a larger volume of silica fume is added than the volume of cement removed. Thus, the volume of cementitious paste increases and there is actually a lowering of the water-cementitious materials ratio on a volume basis.

The particle-size distribution of a typical silica fume shows that most particles are smaller than one micrometer (1 μm with an average diameter of about 0.1 μm, which is approximately one hundred times smaller than the average size cement particle).

The extreme fineness and high silica content of silica fume make it a highly effective pozzolanic material. The silica fume reacts pozzolanically with the calcium hydroxide produced during the hydration of cement to form the stable cementitious compound, calcium silicate hydrate (CSH).

Silica fume has been successfully used to produce very high strength (over 18,000 psi), low permeability, and chemically resistant concretes. Such concretes contain up to 25 percent silica fume by weight (mass) of cement. The use of this high amount of silica fume generally makes the concrete difficult to work. The mixing water demand of a given concrete mixture incorporating silica fume increases with increasing amounts of silica fume.

To maximize the full strength-producing potential of silica fume in concrete, it should always be used with a water-reducing admixture, preferably a high-range, water-reducing (HRWR) admixture. the dosage of the HRWR will depend on the percentages of silica fume and the type of HRWR used.

When proportioning concrete containing silica fume, the following should be considered:

a.   Mixing -- The amount of mixing will depend on the percentage of silica fume used and the mixing conditions. Mixing time may need to be increased to achieve thorough distribution when using large quantities of silica fume with low water content concrete. The use of HRWR assists greatly in achieving uniform dispersion.

b.   Air-entrainment -- The amount of air-entraining admixture to produce a required volume of air in concrete may increase with increasing amounts of silica fume due to the very high surface area of the silica fume and the presence of any carbon within the silica fume. Air entrainment is not usually used in high strength concretes unless they are expected to be exposed to freezing and thawing when saturated with water or to deicing salts.

c.   Workability -- Fresh concrete containing silica fume is generally more cohesive and less prone to segregation than concrete without silica fume. This increase in cohesiveness and reduction to bleeding can provide improved pumping properties. Concrete containing silica fume in excess of 10 percent by weight (mass) of the cementitious materials may become sticky. It may be necessary to increase the slump 2 to 5 in. to maintain the same workability for a given length of time.

   d. Bleeding -- Concrete containing silica fume exhibits reduced bleeding. This reduced bleeding is primarily caused by the high surface area of the silica fume particles, resulting in very little water being left in the mixture for bleeding. As the result of reduced bleeding of concrete containing silica fume, there is a greater tendency for plastic shrinkage cracking to occur.

Typically, the materials listed previously are introduced into the concrete mixer separately. In some cases, however, these same materials may be blended with portland cement in fixed proportions to produce a blended cement, ASTM C 595. Like air-entraining admixtures added to the concrete at the time of batching, the addition of GGBF slag also gives the producer flexibility to achieve desired concrete performance.

When proportioning concrete containing a separately batched, cementitious material such as fly ash, natural pozzolan, GGBF slag, or silica fume, a number of factors must be considered. These include:

   a. Chemical activity of the cementitious material and its effect on concrete strength at various ages.
   b. Effect on the mixing-water demand needed for workability and placeability.
   c. Density (or specific gravity) of the material and its effect on the volume of concrete produced in the batch.
   d. Effect on the dosage rate of chemical admixtures and/or air-entraining admixtures used in the mixture.
   e. Effect of combinations of materials on other critical properties of the concrete, such as time of set under ambient temperature conditions, heat of hydration, rate of strength development, and durability.
   f. Amount of cementitious materials and cement needed to meet the requirements for the particular concrete.

4.4.1 Methods for proportioning and evaluating concrete mixtures containing these supplementary cementitious materials must be based on trial mixtures using a range of ingredient proportions. By evaluating their effect on strength, water requirement, time of set, and other important properties, the optimum amount of cementitious materials can be determined. In the absence of prior information and in the interest of preparing estimated proportions for a first trial batch or a series of trial batches in accordance with ASTM C 192, the following general ranges are given based on the percentage of the ingredients by the total weight of cementitious material used in the batch for structural concrete:

Class F fly ash -- 15 to 25 percent

Class C fly ash -- 15 to 35 percent
Natural pozzolans -- 10 to 20 percent
Ground granulated blast-furnace slag -- 25 to 70 percent
Silica fume -- 5 to 15 percent

For special projects, or to provide certain special required properties, the quantity of the materials used per yd³ of concrete may be different from that shown above.

In cases where high early strengths are required, the total weight of cementitious material may be greater than would be needed if portland cement were the only cementitious material. Where high early strength is not required higher percentages of fly ash are frequently used.

Often, it is found that with the use of fly ash and GGBF slag, the amount of mixing water required to obtain the desired slump and workability of concrete may be lower than that used in a portland cement mixture using only portland cement. When silica fume is used, more mixing water is usually required than when using only portland cement. In calculating the amount of chemical admixtures to dispense for a given batch of concrete, the dosage should generally be applied to the total amount of cementitious material. Under these conditions the reduction in mixing water for conventional water-reducing admixtures (Types A, D, and E) should be at least 5 percent, and for water-reducing, high-range admixtures at least 12 percent. When GGBF slag is used in concrete mixtures containing some high-range water-reducing admixtures, the admixture dosage may be reduced by approximately 25 percent compared to mixtures containing only portland cement.

4.4.2 Due to differences in their specific gravities, a given weight of a supplementary cementitious material will not occupy the same volume as an equal weight of portland cement. The specific gravity of blended cements will be less than that of portland cement. Thus, when using either blended cements or supplementary cementitious materials, the yield of the concrete mixture should be adjusted using the actual specific gravities of the materials used.

4.4.3 Class C fly ash, normally of extremely low carbon content, usually has little or no effect on entrained air or on the air-entraining admixture dosage rate. Many Class F fly ashes may require a higher dosage of air-entraining admixture to obtain specified air contents; if carbon content is high, the dosage rate may be several times that of non-fly ash concrete. The dosage required may also be quite variable. The entrained air content of concrete containing high carbon-content fly ash may be difficult to obtain and maintain. Other cementitious materials may be treated the same as cement in determining the proper quantity of air-entraining admixtures per yd³ of concrete or per 100 lb of cementitious material used.

4.4.4 Concrete containing a proposed blend of cement, other cementitious materials, and admixtures should be tested to determine the time required for setting at various temperatures. The use of most supplementary cementitious materials generally slows the time-of-set of the concrete, and this period may be prolonged by higher percentages of these materials in the cementitious blend, cold weather, and the presence of chemical admixtures not formulated especially for acceleration.

Because of the possible adverse effects on finishing time and consequent labor costs, in some cold climates the proportion of other cementitious materials in the blend may have to be reduced below the optimum amount for strength considerations. Some Class C fly ashes may affect setting time while some other cementitious materials may have little effect on setting time. Any reduction in cement content will reduce heat generation and normally prolong the setting time.

CHAPTER 5 -- BACKGROUND DATA

5.1 To the extent possible, selection of concrete proportions should be based on test data or experience with the materials actually to be used. Where such background is limited or not available, estimates given in this recommended practice may be employed.

5.2 The following information for available materials will be useful:

5.2.1 Sieve analyses of fine and coarse aggregates.

5.2.2 Unit weight of coarse aggregate.

5.2.3 Bulk specific gravities and absorptions of aggregates.

5.2.4 Mixing-water requirements of concrete developed from experience with available aggregates.

5.2.5 Relationships between strength and water-cement ratio or ratio of water-to-cement plus other cementitious materials, for available combinations of cements, other cementitious materials if considered, and aggregates.

5.2.6 Specific gravities of portland cement and other cementitious materials, if used.

5.2.7 Optimum combination of coarse aggregates to meet the maximum density gradings for mass concrete as discussed in Section 5.3.2.1 of Appendix 5.

5.3 Estimates from Tables 6.3.3 and 6.3.4, respectively, may be used when items in Section 5.2.4 and Section 6.3.5 are not available. As will be shown, proportions can be estimated without the knowledge of aggregate-specific gravity and absorption, Section 5.2.3.

CHAPTER 6 -- PROCEDURE

6.1 The procedure for selection of mix proportions given in this section is applicable to normal weight concrete. Although the same basic data and procedures can be used in proportioning heavyweight and mass concretes, additional information and sample computations for these types of concrete are given in Appendixes 4 and 5, respectively.

6.2 Estimating the required batch weights for the concrete involves a sequence of logical, straightforward steps which, in effect, fit the characteristics of the available materials into a mixture suitable for the work. The question of suitability is frequently not left to the individual selecting

the proportions. The job specifications may dictate some or all of the following:

6.2.1 Maximum water-cement or water-cementitious material ratio.

6.2.2 Minimum cement content.

6.2.3 Air content.

6.2.4 Slump.

6.2.5 Maximum size of aggregate.

6.2.6 Strength.

6.2.7 Other requirements relating to such things as strength overdesign, admixtures, and special types of cement, other cementitious materials, or aggregate.

6.3 Regardless of whether the concrete characteristics are prescribed by the specifications or are left to the individual selecting the proportions, establishment of batch weights per yd^3 of concrete can be best accomplished in the following sequence:

6.3.1 *Step 1. Choice of slump* -- If slump is not specified, a value appropriate for the work can be selected from Table 6.3.1. The slump ranges shown apply when vibration is used to consolidate the concrete. Mixes of the stiffest consistency that can be placed efficiently should be used.

Table 6.3.1 — Recommended slumps for various types of construction*

Types of construction	Slump, in.	
	Maximum'	Minimum
Reinforced foundation walls and footings	3	1
Plain footings, caissons, and substructure walls	3	1
Beams and reinforced walls	4	1
Building columns	4	1
Pavements and slabs	3	1
Mass concrete	2	1

*Slump may be increased when chemical admixtures are used, provided that the admixture-treated concrete has the same or lower water-cement or water-cementitious material ratio and does not exhibit segregation potential or excessive bleeding.
'May be increased 1 in. for methods of consolidation other than vibration

6.3.2 *Step 2. Choice of maximum size of aggregate* -- Large nominal maximum sizes of well graded aggregates have less voids than smaller sizes. Hence, concretes with the larger-sized aggregates require less mortar per unit volume of concrete. Generally, the nominal maximum size of aggregate should be the largest that is economically available and consistent with dimensions of the structure. In no event should the nominal maximum size exceed one-fifth of the narrowest dimension between sides of forms, one-third the depth of slabs, nor three-fourths of the minimum clear spacing between individual reinforcing bars, bundles of bars, or pretensioning strands. These limitations are sometimes waived if workability and methods of consolidation are such that the concrete can be placed without honeycomb or void. In areas congested with reinforcing steel, post-tension ducts or conduits, the proportioner should select a nominal maximum size of the aggregate so concrete can be placed without excessive segregation, pockets, or voids. When high strength concrete is desired, best results may be obtained with reduced nominal maximum sizes of aggregate since these produce higher strengths at a given water-cement ratio.

ACI COMMITTEE REPORT

Table 6.3.3 — Approximate mixing water and air content requirements for different slumps and nominal maximum sizes of aggregates

Slump, in.	Water, lb/yd³ of concrete for indicated nominal maximum sizes of aggregate							
	⅜ in.*	½ in.*	¾ in.*	1 in.*	1-½ in.*	2 in.*'	3 in.'ᶻ	6 in.'ᶻ
Non-air-entrained concrete								
1 to 2	350	335	315	300	275	260	220	190
3 to 4	385	365	340	325	300	285	245	210
6 to 7	410	385	360	340	315	300	270	—
More than 7*	—	—	—	—	—	—	—	—
Approximate amount of entrapped air in non-air-entrained concrete, percent	3	2.5	2	1.5	1	0.5	0.3	0.2
Air-entrained concrete								
1 to 2	305	295	280	270	250	240	205	180
3 to 4	340	325	305	295	275	265	225	200
6 to 7	365	345	325	310	290	280	260	—
More than 7*	—	—	—	—	—	—	—	—
Recommended averages¹ total air content, percent for level of exposure:								
Mild exposure	4.5	4.0	3.5	3.0	2.5	2.0	1.5**·''	1.0**·''
Moderate exposure	6.0	5.5	5.0	4.5	4.5	4.0	3.5**·''	3.0**·''
Severe exposure''	7.5	7.0	6.0	6.0	5.5	5.0	4.5**·''	4.0**·''

*The quantities of mixing water given for air-entrained concrete are based on typical total air content requirements as shown for "moderate exposure" in the table above. These quantities of mixing water are for use in computing cement contents for trial batches at 68 to 77 F. They are maximum for reasonably well-shaped angular aggregates graded within limits of accepted specifications. Rounded aggregate will generally require 30 lb less water for non-air-entrained and 25 lb less for air-entrained concretes. The use of water-reducing chemical admixtures, ASTM C 494, may also reduce mixing water by 5 percent or more. The volume of the liquid admixtures is included as part of the total volume of the mixing water. The slump values of more than 7 in. are only obtained through the use of water-reducing chemical admixture; they are for concrete containing nominal maximum size aggregate not larger than 1 in.

'The slump values for concrete containing aggregate larger than 1½ in. are based on slump tests made after removal of particles larger than 1½ in. by wet-screening.

ᶻThese quantities of mixing water are for use in computing cement factors for trail batches when 3 in. or 6 in. nominal maximum size aggregate is used. They are average for reasonably well-shaped coarse aggregates, well-graded from coarse to fine.

¹Additional recommendations for air-content and necessary tolerances on air content for control in the field are given in a number of ACI documents, including ACI 201, 345, 318, 301, and 302. ASTM C 94 for ready-mixed concrete also gives air-content limits. The requirements in other documents may not always agree exactly, so in proportioning concrete consideration must be given to selecting an air content that will meet the needs of the job and also meet the applicable specifications.

**For concrete containing large aggregates that will be wet-screened over the 1½ in. sieve prior to testing for air content, the percentage of air expected in the 1½ in. minus material should be as tabulated in the 1½ in. column. However, initial proportioning calculations should include the air content as a percent of the whole.

''When using large aggregate in low cement factor concrete, air entrainment need not be detrimental to strength. In most cases mixing water requirement is reduced sufficiently to improve the water-cement ratio and to thus compensate for the strength-reducing effect of air-entrained concrete. Generally, therefore, for these large nominal maximum sizes of aggregate, air contents recommended for extreme exposure should be considered even though there may be little or no exposure to moisture and freezing.

''These values are based on the criteria that 9 percent air is needed in the mortar phase of the concrete. If the mortar volume will be substantially different from that determined in this recommended practice, it may be desirable to calculate the needed air content by taking 9 percent of the actual mortar volume.

6.3.3 *Step 3. Estimation of mixing water and air content* -- The quantity of water per unit volume of concrete required to produce a given slump is dependent on: the nominal maximum size, particle shape, and grading of the aggregates; the concrete temperature; the amount of entrained air; and use of chemical admixtures. Slump is not greatly affected by the quantity of cement or cementitious materials within normal use levels (under favorable circumstances the use of some finely divided mineral admixtures may lower water requirements slightly -- see ACI 212.1R). Table 6.3.3 provides estimates of required mixing water for concrete made with various maximum sizes of aggregate, with and without air entrainment. Depending on aggregate texture and shape, mixing water requirements may be somewhat above or below the tabulated values, but they are sufficiently accurate for the first estimate. The differences in water demand are not necessarily reflected in strength since other compensating factors may be involved. A rounded and an angular coarse aggregate, both well and similarly graded and of good quality, can be expected to produce concrete of about the same compressive strength for the same cement factor in spite of differences in w/c or w/(c + p) resulting from the different mixing water requirements.

Particle shape is not necessarily an indicator that an aggregate will be either above or below in its strength-producing capacity.

Chemical admixtures -- Chemical admixtures are used to modify the properties of concrete to make it more workable, durable, and/or economical; increase or decrease the time of set; accelerate strength gain; and/or control temperature gain. Chemical admixtures should be used only after an appropriate evaluation has been conducted to show that the desired effects have been accomplished in the particular concrete under the conditions of intended use. Water-reducing and/or set-controlling admixtures conforming to the requirements of ASTM C 494, when used singularly or in combination with other chemical admixtures, will reduce significantly the quantity of water per unit volume of concrete. The use of some chemical admixtures, even at the same slump, will improve such qualities as workability, finishability, pumpability, durability, and compressive and flexural strength. Significant volume of liquid admixtures should be considered as part of the mixing water. The slumps shown in Table 6.3.1, "Recommended Slumps for Various Types of Construction," may be increased when chemical admixtures are used, providing the admixture-treated concrete has the same or a lower water-cement ratio and does not exhibit segregation potential and excessive bleeding. When only used to increase slump, chemical admixtures may not improve any of the properties of the concrete.

Table 6.3.3 indicates the approximate amount of entrapped air to be expected in non-air-entrained concrete in the upper part of the table and shows the recommended average air content for air-entrained concrete in the lower part of the table. If air entrainment is needed or desired, three levels of air content are given for each aggregate size depending on the purpose of the entrained air and the severity of exposure if entrained air is needed for durability.

Mild exposure -- When air entrainment is desired for a beneficial effect other than durability, such as to improve workability or cohesion or in low cement factor concrete to improve strength, air contents lower than those needed for durability can be used. This exposure includes indoor or outdoor service in a climate where concrete will not be exposed to freezing or to deicing agents.

Moderate exposure -- Service in a climate where freezing is expected but where the concrete will not be continually exposed to moisture or free water for long periods prior to freezing and will not be exposed to deicing agents or other aggressive chemicals. Examples include: exterior beams, columns, walls, girders, or slabs that are not in contact with wet soil and are so located that they will not receive direct applications of deicing salts.

Severe exposure -- Concrete that is exposed to deicing chemicals or other aggressive agents or where the concrete may become highly saturated by continued contact with moisture or free water prior to freezing. Examples include: pavements, bridge decks, curbs, gutters, sidewalks, canal linings, or exterior water tanks or sumps.

The use of normal amounts of air entrainment in concrete with a specified strength near or about 5000 psi may not be possible due to the fact that each added percent of air lowers the maximum strength obtainable with a given combination of materials.[1] In these cases the exposure to water, deicing salts, and freezing temperatures should be carefully evaluated. If a member is not continually wet and will not be exposed to deicing salts, lower air-content values such as those given in Table 6.3.3 for moderate exposure are appropriate even though the concrete is exposed to freezing and thawing temperatures. However, for an exposure condition where the member may be saturated prior to freezing, the use of air entrainment should not be sacrificed for strength. In certain applications, it may be found that the content of entrained air is lower than that specified, despite the use of usually satisfactory levels of air-entraining admixture. This happens occasionally, for example, when very high cement contents are involved. In such cases, the achievement of required durability may be demonstrated by satisfactory results of examination of air-void structure in the paste of the hardened concrete.

When trial batches are used to establish strength relationships or verify strength-producing capability of a mixture, the least favorable combination of mixing water and air content should be used. The air content should be the maximum permitted or likely to occur, and the concrete should be gaged to the highest permissible slump. This will avoid developing an over-optimistic estimate of strength on the assumption that average rather than extreme conditions will prevail in the field. If the concrete obtained in the field has a lower slump and/or air content, the proportions of ingredients should be adjusted to maintain required yield. For additional information on air content recommendations, see ACI 201.2R, 301, and 302.1R.

6.3.4 Step 4. Selection of water-cement or water-cementitious materials ratio -- The required w/c or $w/(c + p)$ is determined not only by strength requirements but also by factors such as durability. Since different aggregates, cements, and cementitious materials generally produce different strengths at the same w/c or $w/(c + p)$, it is highly desirable to have or to develop the relationship between strength and w/c or $w/(c + p)$ for the materials actually to be used. In the absence of such data, approximate and relatively conservative values for concrete containing Type I portland cement can be taken from Table 6.3.4(a). With typical materials, the tabulated w/c or $w/(c + p)$ should produce the strengths shown, based on 28-day tests of specimens cured under standard laboratory conditions. The average strength selected must, of course, exceed the specific strength by a sufficient margin to keep the number of low tests within specific limits -- see ACI 214 and ACI 318.

Table 6.3.4(a) — Relationship between water-cement or water-cementitious materials ratio and compressive strength of concrete

Compressive strength at 28 days, psi*	Water-cement ratio, by weight	
	Non-air-entrained concrete	Air-entrained concrete
6000	0.41	—
5000	0.48	0.40
4000	0.57	0.48
3000	0.68	0.59
2000	0.82	0.74

*Values are estimated average strengths for concrete containing not more than 2 percent air for non-air-entrained concrete and 6 percent total air content for air-entrained concrete. For a constant w/c or $w/(c+p)$, the strength of concrete is reduced as the air content is increased. 28-day strength values may be conservative and may change when various cementitious materials are used. The rate at which the 28 day strength is developed may also change.

Strength is based on 6 x 12 in. cylinders moist-cured for 28 days in accordance with the sections on "Initial Curing" and "Curing of Cylinders for Checking the Adequacy of Laboratory Mixture Proportions for Strength or as the Basis for Acceptance or for Quality Control" of ASTM method C 31 for Making and Curing Concrete Specimens in the Field. These are cylinders cured moist at 73.4 ± 3 F (23 ± 1.7 C) prior to testing.

The relationship in this table assumes a nominal maximum aggregate size of about ¾ to 1 in. For a given source of aggregate, strength produced at a given w/c or $w/(c+p)$ will increase as nominal maximum size of aggregate decreases: see Sections 3.4 and 6.1.2.

For severe conditions of exposure, the w/c or $w/(c + p)$ ratio should be kept low even though strength requirements may be met with a higher value. Table 6.3.4(b) gives limiting values.

When natural pozzolans, fly ash, GGBF slag, and silica fume, hereafter referred to as pozzolanic materials, are used in concrete, a water-to-cement plus pozzolanic materials ratio (or water-to-cement plus other cementitious materials ratio) by weight must be considered in place of the traditional water-cement ratio by weight. There are two ap-

211.1-10   ACI COMMITTEE REPORT

Table 6.3.4(b) — Maximum permissible water-cement or water-cementitious materials ratios for concrete in severe exposures*

Type of structure	Structure wet continuously or frequently and exposed to freezing and thawing'	Structure exposed to sea water or sulfates
Thin sections (railings, curbs, sills, ledges, ornamental work) and sections with less than 1 in. cover over steel	0.45	0.40‡
All other structures	0.50	0.45‡

*Based on report of ACI Committee 201. Cementitious materials other than cement should conform to ASTM C 618 and C 989.
'Concrete should also be air-entrained.
‡If sulfate resisting cement (Type II or Type V of ASTM C 150) is used, permissible water-cement or water-cementitious materials ratio may be increased by 0.05.

proaches normally used in determining the $w/(c + p)$ ratio that will be considered equivalent to the w/c of a mixture containing only portland cement: (1) equivalent weight of pozzolanic materials or (2) equivalent absolute volume of pozzolanic materials in the mixture. For the first approach, the weight equivalency, the total weight of pozzolanic materials remains the same [that is, $w/(c + p) = w/c$ directly]: but the total absolute volume of cement plus pozzolanic materials will normally be slightly greater. With the second approach, using the Eq. (6.3.4.2), a $w/(c + p)$ by weight is calculated that maintains the same absolute volume relationship but that will reduce the total weight of cementitious material since the specific gravities of pozzolanic materials are normally less than that of cement.

The equations for converting a target water-cement ratio w/c to a weight ratio of water to cement plus pozzolanic materials $w/(c + p)$ by (1) weight equivalency or (2) volume equivalency are as follows:

Eq. (6.3.4.1)--Weight equivalency

$$\frac{w}{c+p} \text{ weight ratio, weight equivalency} = \frac{w}{c}$$

where

$$\frac{w}{c+p} = \text{weight of water divided by weight}$$
$$\text{of cement + pozzolanic materials}$$

$$\frac{w}{c} = \text{target water-cement ratio by weight}$$

When the weight equivalency approach is used, the percentage or fraction of pozzolanic materials used in the cementitious material is usually expressed by weight. That is, F_w, the pozzolanic materials percentage by weight of total

cement plus pozzolanic materials, expressed as a decimal factor, is

$$F_w = \frac{p}{c+p}$$

where

F_w = pozzolanic materials percentage by weight, expressed as a decimal factor
p = weight of pozzolanic materials
c = weight of cement

(Note: If only the desired pozzolanic materials percentage factor by absolute volume F_v, is known, it can be converted to F_w as follows

$$F_w = \frac{1}{1 + \left(\dfrac{3.15}{G_p}\right)\left(\dfrac{1}{F_v}-1\right)}$$

where

F_v = pozzolanic materials percentage by absolute volume of the total absolute volume of cement plus pozzolanic materials expressed as a decimal factor
G_p = specific gravity of pozzolanic materials
3.15 = specific gravity of portland cement [use actual value if known to be different])

Example 6.3.4.1 -- Weight equivalency

If a water-cement ratio of 0.60 is required and a fly ash pozzolan is to be used as 20 percent of the cementitous material in the mixture by weight ($F_w = 0.20$), then the required water-to-cement plus pozzolanic material ratio on a weight equivalency basis is

$$\frac{w}{c + p} = \frac{w}{c} = 0.60, \text{ and}$$

$$F_w = \frac{p}{c + p} = 0.20$$

Assuming an estimated mixing-water requirement of 270 lb/yd³, then the required weight of cement + pozzolan is 270 ÷ 0.60 = 450 lb; and the weight of pozzolan is (0.20)(450) = 90 lb. The weight of cement is, therefore, 450 - 90 = 360 lb. If instead of 20 percent fly ash by weight, 20 percent by absolute volume of cement plus pozzolan was specified (F_v = 0.20), the corresponding weight factor is computed as follows for a fly ash with an assumed gravity of 2.40:

$$F_w = \cfrac{1}{1 + \left(\cfrac{3.15}{G_p}\right)\left(\cfrac{1}{F_v} - 1\right)}$$

$$\cfrac{1}{1 + \left(\cfrac{3.15}{2.40}\right)\left(\cfrac{1}{0.2} - 1\right)}$$

$$F_w = \frac{1}{1 + (1.31)(4)} = \frac{1}{1 + 5.24} = \frac{1}{6.24} = 0.16$$

In this case 20 percent by absolute volume is 16 percent by weight, and the weight of pozzolan in the batch would be $(0.16)(450) = 72$ lb, and the weight of cement $450 - 72 = 378$ lb.

Eq. (6.3.4.2) -- Absolute volume equivalency

$$\frac{w}{c + p} \text{ weight ratio, absolute}$$

$$\text{volume equivalency} =$$

$$\frac{3.15 \dfrac{w}{c}}{3.15(1 - F_v) + G_p(F_v)}$$

where

$\dfrac{w}{c + p}$ = weight of water divided by weight of cement + pozzolanic materials

$\dfrac{w}{c}$ = target water-cement ratio by weight

3.15 = specific gravity of portland cement (use actual value if known to be different)

F_v = pozzolan percentage by absolute volume of the total absolute volume of cement plus pozzolan, expressed as a decimal factor

(Note: If only the desired pozzolan percentage by weight F_w is known, it can be converted to F_v as follows

$$F_v = \cfrac{1}{1 + \left(\cfrac{G_p}{3.15}\right)\left(\cfrac{1}{F_w} - 1\right)}$$

where these symbols are the same as defined previously.)

Example 6.3.4.2 -- Absolute volume equivalency
Use the same basic data as Example 6.3.4.1, but it should be specified that the equivalent water-to-cement plus pozzolan ratio be established on the basis of absolute volume, which will maintain, in the mixture, the same ratio of volume of water to volume of cementitious material when changing from cement only to cement plus pozzolan. Again the required water-cement ratio is 0.60, and it is assumed initially that it is desired to use 20 percent by absolute volume of fly ash ($F_v = 0.20$). The specific gravity of the fly ash is assumed to be 2.40 in this example

$$\frac{w}{c + p} = \frac{3.15\left(\dfrac{w}{c}\right)}{3.15(1 - F_v) + G_p(F_v)}$$

$$= \frac{(3.15)(0.60)}{(3.15)(0.80) + (2.40)(0.20)}$$

$$+ \frac{1.89}{2.52 + 0.48} = \frac{1.89}{3.00} = 0.63$$

So the target weight ratio to maintain an absolute volume equivalency is $w/(c + p) = 0.63$. If the mixing water is again 270 lb/yd$_3$, then the required weight of cement + pozzolan is $270 \div 0.63 = 429$ lb; and, since the corresponding weight percentage factor for $F_v = 0.20$ is $F_w = 0.16$ as calculated in Example 6.3.4.1, the weight of fly ash to be used is $(0.16)(429) = 69$ lb and the weight of cement is $429 - 69 = 360$ lb. The volume equivalency procedure provides lower weights of cementitious materials. Checking the absolute volumes

$$\text{fly ash} = \frac{69}{(2.40)(62.4)} = 0.461 ft^3$$

$$\text{cement} = \frac{360}{(3.15)(62.4)} = 1.832 ft^3$$

$$\text{total} = 0.461 + 1.832 = 2.293 ft^3$$

$$\begin{matrix}\text{percent pozzolan} \\ \text{by volume}\end{matrix} = \frac{0.461}{2.293} \times 100 = 20 \text{ percent}$$

If, instead of 20 percent fly ash by volume ($F_v = 0.20$), a weight percentage of 20 percent was specified ($F_w = 0.20$), it could be converted to F_v using $G_p = 2.40$ and the appropriate formula

$$F_v = \cfrac{1}{1 + \left(\cfrac{G_p}{3.15}\right)\left(\cfrac{1}{F_w} - 1\right)}$$

$$F_v = \cfrac{1}{1 + \left(\cfrac{2.40}{3.15}\right)\left(\cfrac{1}{0.2} - 1\right)}$$

$$F_v = \frac{1}{1 + (0.762)(4)} = \frac{1}{4.048} = 0.247$$

In this case 20 percent by weight is almost 25 percent by

absolute volume. The equivalent $w/(c + p)$ ratio by volume will have to be recomputed for this condition since F_v has been changed from that originally assumed in this example

$$\frac{w}{c + p} = \frac{3.15\left(\frac{w}{c}\right)}{3.15(1 - F_v) + G_p(F_v)}$$
$$= \frac{(3.15)(0.60)}{3.15(0.75) + 2.40(0.25)}$$
$$= \frac{1.89}{2.36 + 0.60} = \frac{1.89}{2.96} = 0.64$$

Total cementitious material would be $270 \div 0.64 = 422$ lb. Of this weight 20 percent ($F_w = 0.20$) would be fly ash; $(422)(0.20) = 84$ lb of fly ash and $422 - 84 = 338$ lb of cement.

6.3.5 *Step 5. Calculation of cement content* -- The amount of cement per unit volume of concrete is fixed by the determinations made in Steps 3 and 4 above. The required cement is equal to the estimated mixing-water content (Step 3) divided by the water-cement ratio (Step 4). If, however, the specification includes a separate minimum limit on cement in addition to requirements for strength and durability, the mixture must be based on whichever criterion leads to the larger amount of cement.

The use of pozzolanic or chemical admixtures will affect properties of both the fresh and hardened concrete. See ACI 212.

6.3.6 *Step 6. Estimation of coarse aggregate content* -- Aggregates of essentially the same nominal maximum size and grading will produce concrete of satisfactory workability when a given volume of coarse aggregate, on an oven-dry-rodded basis, is used per unit volume of concrete. Appropriate values for this aggregate volume are given in Table 6.3.6. It can be seen that, for equal workability, the volume of coarse aggregate in a unit volume of concrete is dependent only on its nominal maximum size and the fine-

ness modulus of the fine aggregate. Differences in the amount of mortar required for workability with different aggregates, due to differences in particle shape and grading, are compensated for automatically by differences in oven-dry-rodded void content.

The volume of aggregate in ft³, on an oven-dry-rodded basis, for a yd³ of concrete is equal to the value from Table 6.3.6 multiplied by 27. This volume is converted to dry weight of coarse aggregate required in a yd³ of concrete by multiplying it by the oven-dry-rodded weight per ft³ of the coarse aggregate.

6.3.6.1 For more workable concrete, which is sometimes required when placement is by pump or when concrete must be worked around congested reinforcing steel, it may be desirable to reduce the estimated coarse aggregate content determined using Table 6.3.6 by up to 10 percent. However, caution must be exercised to assure that the resulting slump, water-cement or water-cementitious materials ratio, and strength properties of the concrete are consistent with the recommendations in Sections 6.3.1 and 6.3.4 and meet applicable project specification requirements.

6.3.7 *Step 7. Estimation of fine aggregate content* -- At completion of Step 6, all ingredients of the concrete have been estimated except the fine aggregate. Its quantity is determined by difference. Either of two procedures may be employed: the weight method (Section 6.3.7.1) or the absolute volume method (Section 6.3.7.2).

6.3.7.1 If the weight of the concrete per unit volume is assumed or can be estimated from experience, the required weight of fine aggregate is simply the difference between the weight of fresh concrete and the total weight of the other ingredients. Often the unit weight of concrete is known with reasonable accuracy from previous experience with the materials. In the absence of such information, Table 6.3.7.1 can be used to make a first estimate. Even if the estimate of concrete weight per yd³ is rough, mixture proportions will be sufficiently accurate to permit easy adjustment on the basis of trial batches as will be shown in the examples.

Table 6.3.6 — Volume of coarse aggregate per unit of volume of concrete

Nominal maximum size of aggregate, in.	Volume of oven-dry-rodded coarse aggregate* per unit volume of concrete for different fineness moduli of fine aggregate'			
	2.40	2.60	2.80	3.00
⅜	0.50	0.48	0.46	0.44
½	0.59	0.57	0.55	0.53
¾	0.66	0.64	0.62	0.60
1	0.71	0.69	0.67	0.65
1½	0.75	0.73	0.71	0.69
2	0.78	0.76	0.74	0.72
3	0.82	0.80	0.78	0.76
6	0.87	0.85	0.83	0.81

*Volumes are based on aggregates in oven-dry-rodded condition as described in ASTM C 29.

These volumes are selected from empirical relationships to produce concrete with a degree of workability suitable for usual reinforced construction. For less workable concrete, such as required for concrete pavement construction, they may be increased about 10 percent. For more workable concrete see Section 6.3.6.1.

'See ASTM C 136 for calculation of fineness modulus.

Table 6.3.7.1 — First estimate of weight of fresh concrete

Nominal maximum size of aggregate, in.	First estimate of concrete weight, lb/yd³*	
	Non-air-entrained concrete	Air-entrained concrete
⅜	3840	3710
½	3890	3760
¾	3960	3840
1	4010	3850
1½	4070	3910
2	4120	3950
3	4200	4040
6	4260	4110

*Values calculated by Eq. (6-1) for concrete of medium richness (550 lb of cement per yd³) and medium slump with aggregate specific gravity of 2.7. Water requirements based on values for 3 to 4 in. slump in Table 6.3.3. If desired, the estimated weight may be refined as follows if necessary information is available: for each 10 lb difference in mixing water from the Table 6.3.3 values for 3 to 4 in. slump, correct the weight per yd³ 15 lb in the opposite direction; for each 100 lb difference in cement content from 550 lb, correct the weight per yd³ 15 lb in the same direction; for each 0.1 by which aggregate specific gravity deviates from 2.7, correct the concrete weight 100 lb in the same direction. For air-entrained concrete the air content for severe exposure from Table 6.3.3 was used. The weight can be increased 1 percent for each percent reduction in air content from that amount.

If a theoretically exact calculation of fresh concrete weight per yd³ is desired, the following formula can be used

$$U = 16.85 \, G_a \, (100 - A) + c(1 - G_a/G_c) - w(G_a - 1) \quad \text{(6-1)}$$

where

U = weight in lb of fresh concrete per yd³
G_a = weighted average specific gravity of combined fine and coarse aggregate, bulk SSD*
G_c = specific gravity of cement (generally 3.15)
A = air content, percent
w = mixing water requirement, lb/yd³
c = cement requirement, lb/yd³

6.3.7.2 A more exact procedure for calculating the required amount of fine aggregate involves the use of volumes displaced by the ingredients. In this case, the total volume displaced by the known ingredients--water, air, cementitious materials, and coarse aggregate--is subtracted from the unit volume of concrete to obtain the required volume of fine aggregate. The volume occupied in concrete by any ingredient is equal to its weight divided by the density of that material (the latter being the product of the unit weight of water and the specific gravity of the material).

6.3.8 *Step 8. Adjustments for aggregate moisture* -- The aggregate quantities actually to be weighed out for the concrete must allow for moisture in the aggregates. Generally, the aggregates will be moist and their dry weights should be increased by the percentage of water they contain, both absorbed and surface. The mixing water added to the batch must be reduced by an amount equal to the free moisture contributed by the aggregate -- i.e., total moisture minus absorption.

6.3.8.1 In some cases, it may be necessary to batch an aggregate in a dry condition. If the absorption (normally measured by soaking one day) is higher than approximately one percent, and if the pore structure within the aggregate particles is such that a significant fraction of the absorption occurs during the time prior to initial set, there may be a noticeable increase in the rate of slump loss due to an effective decrease in mixing water. Also, the effective water-cement ratio would be decreased for any water absorbed by the aggregate prior to set; this, of course, assumes that cement particles are not carried into aggregate particle pores.

6.3.8.2 Laboratory trial batch procedures according to ASTM C 192 allow the batching of laboratory air-dried aggregates if their absorption is less than 1.0 percent with an allowance for the amount of water that will be absorbed from the unset concrete. It is suggested by ASTM

---

* SSD indicates saturated-surface-dry basis used in considering aggregate displacement. The aggregate specific gravity used in calculations must be consistent with the moisture condition assumed in the basic aggregate batch weights -- i.e., bulk dry if aggregate weights are stated on a dry basis, and bulk SSD if weights are stated on a saturated-surface-dry basis.

C 192 that the amount absorbed may be assumed to be 80 percent of the difference between the actual amount of water in the pores of the aggregate in their air-dry state and the nominal 24-hr absorption determined by ASTM C 127 or C 128. However, for higher-absorption aggregates, ASTM C 192 requires preconditioning of aggregates to satisfy absorption with adjustments in aggregate weight based on total moisture content and adjustment to include surface moisture as a part of the required amount of mixing water.

6.3.9 *Step 9. Trial batch adjustments* -- The calculated mixture proportions should be checked by means of trial batches prepared and tested in accordance with ASTM C 192 or full-sized field batches. Only sufficient water should be used to produce the required slump regardless of the amount assumed in selecting the trial proportions. The concrete should be checked for unit weight and yield (ASTM C 138) and for air content (ASTM C 138, C 173, or C 231). It should also be carefully observed for proper workability, freedom from segregation, and finishing properties. Appropriate adjustments should be made in the proportions for subsequent batches in accordance with the following procedure.

6.3.9.1 Re-estimate the required mixing water per yd³ of concrete by multiplying the net mixing water content of the trial batch by 27 and dividing the product by the yield of the trial batch in ft³. If the slump of the trial batch was not correct, increase or decrease the re-estimated amount of water by 10 lb for each 1 in. required increase or decrease in slump.

6.3.9.2 If the desired air content (for air-entrained concrete) was not achieved, re-estimate the admixture content required for proper air content and reduce or increase the mixing-water content of Paragraph 6.3.9.1 by 5 lb for each 1 percent by which the air content is to be increased or decreased from that of the previous trial batch.

6.3.9.3 If estimated weight per yd³ of fresh concrete is the basis for proportioning, re-estimate that weight by multiplying the unit weight in lb/ft³ of the trial batch by 27 and reducing or increasing the result by the anticipated percentage increase or decrease in air content of the adjusted batch from the first trial batch.

6.3.9.4 Calculate new batch weights starting with Step 4 (Paragraph 6.3.4), modifying the volume of coarse aggregate from Table 6.3.6 if necessary to provide proper workability.

CHAPTER 7 -- SAMPLE COMPUTATIONS

7.1 Two example problems will be used to illustrate application of the proportioning procedures. The following conditions are assumed:

7.1.1 Type I non-air-entraining cement will be used and its specific gravity is assumed to be 3.15.†

---

† The specific gravity values are not used if proportions are selected to provide a weight of concrete assumed to occupy 1 yd³.

7.1.2 Coarse and fine aggregates in each case are of satisfactory quality and are graded within limits of generally accepted specifications. See ASTM C 33.

7.1.3 The coarse aggregate has a bulk specific gravity of 2.68* and an absorption of 0.5 percent.

7.1.4 The fine aggregate has a bulk specific gravity of 2.64,* an absorption of 0.7 percent, and a fineness modulus of 2.8.

7.2 *Example 1* -- Concrete is required for a portion of a structure that will be below ground level in a location where it will not be exposed to severe weathering or sulfate attack. Structural considerations require it to have an average 28-day compressive strength of 3500 psi.† On the basis of information in Table 6.3.1, as well as previous experience, it is determined that under the conditions of placement to be employed, a slump of 3 to 4 in. should be used and that the available No. 4 to 1½-in. coarse aggregate will be suitable. The dry-rodded weight of coarse aggregate is found to be 100 lb/ft³. Employing the sequence outlined in Section 6, the quantities of ingredients per yd³ of concrete are calculated as follows:

7.2.1 *Step 1* -- As indicated previously, the desired slump is 3 to 4 in.

7.2.2 *Step 2* -- The locally available aggregate, graded from No. 4 to 1½ in., has been indicated as suitable.

7.2.3 *Step 3* -- Since the structure will not be exposed to severe weathering, non-air-entrained concrete will be used. The approximate amount of mixing water to produce 3 to 4-in. slump in non-air-entrained concrete with 1½-in aggregate is found from Table 6.3.3 to be 300 lb/yd³. Estimated entrapped air is shown as 1 percent.

7.2.4 *Step 4* -- From Table 6.3.4(a), the water-cement ratio needed to produce a strength of 3500 psi in non-air-entrained concrete is found to be about 0.62.

7.2.5 *Step 5* -- From the information derived in Steps 3 and 4, the required cement content is found to be 300/0.62 = 484 lb/yd³.

7.2.6 *Step 6* -- The quantity of coarse aggregate is estimated from Table 6.3.6. For a fine aggregate having a fineness modulus of 2.8 and a 1½ in. nominal maximum size of coarse aggregate, the table indicates that 0.71 ft³ of coarse aggregate, on a dry-rodded basis, may be used in each ft³ of concrete. For each yd³, therefore, the coarse aggregate will be 27 x 0.71 = 19.17 ft³. Since it weighs 100 lb per ft³, the dry weight of coarse aggregate is 1917 lb.

7.2.7 *Step 7* -- With the quantities of water, cement, and coarse aggregate established, the remaining material comprising the yd³ of concrete must consist of fine aggregate and whatever air will be entrapped. The required fine aggregate may be determined on the basis of either weight or absolute volume as shown:

7.2.7.1 *Weight basis* -- From Table 6.3.7.1, the weight of a yd³ of non-air-entrained concrete made with ag-

---

* The specific gravity values are not used if proportions are selected to provide a weight of concrete assumed to occupy 1 yd³.

† This is not the specified strength used for structural design but a higher figure expected to be produced on the average. For the method of determining the amount by which average strength should exceed design strength, see ACI 214.

---

gregate having a nominal maximum size of 1½ in. is estimated to be 4070 lb. (For a first trial batch, exact adjustments of this value for usual differences in slump, cement factor, and aggregate specific gravity are not critical.) Weights already known are:

Water, net mixing	300 lb
Cement	484 lb
Coarse aggregate	1917 lb (dry)‡
Total	2701 lb

The weight of fine aggregate, therefore, is estimated to be

$$4070 - 2701 = 1369 \text{ lb (dry)‡}$$

7.2.7.2 *Absolute volume basis* -- With the quantities of cement, water, and coarse aggregate established, and the approximate entrapped air content (as opposed to purposely entrained air) taken from Table 6.3.3, the fine aggregate content can be calculated as follows:

$$\text{Volume of water} = \frac{300}{62.4} = 4.81 \text{ ft}^3$$

$$\text{Solid volume of cement} = \frac{484}{3.15 \times 62.4} = 2.46 \text{ ft}^3$$

$$\text{Solid volume of coarse aggregate} = \frac{1917}{2.68 \times 62.4} = 11.46 \text{ ft}^3$$

$$\text{Volume of entrapped air} = 0.01 \times 27 = 0.27 \text{ ft}^3$$

Total solid volume of ingredients except fine aggregate = 19.00 ft³

$$\text{Solid volume of fine aggregate required} = 27 - 19.00 = 8.00 \text{ ft}^3$$

$$\text{Required weight of dry aggregate} = 8.00 \times 2.64 \times 62.4 = 1318 \text{ lb}$$

7.2.7.3 Batch weights per yd³ of concrete calculated on the two bases are compared as follows:

	Based on estimated concrete weight, lb	Based on absolute volume of ingredients, lb
Water, net mixing	300	300
Cement	484	484
Coarse aggregate, dry	1917	1917
Fine aggregate, dry	1369	1318

---

‡ Aggregate absorption of 0.5 percent is disregarded since its magnitude is unconsequential in relocation to other approximations.

7.2.8 *Step 8* -- Tests indicate total moisture of 2 percent in the coarse aggregate and 6 percent in the fine aggregate. If the trial batch proportions based on assumed concrete weight are used, the adjusted aggregate weights become:

Coarse aggregate, wet	1917 (1.02) = 1955 lb
Fine aggregate, wet	1369 (1.06) = 1451 lb

Absorbed water does not become part of the mixing water and must be excluded from the adjustment in added water. Thus, surface water contributed by the coarse aggregate amounts to 2 - 0.5 = 1.5 percent; that contributed by the fine aggregate to 6 - 0.7 = 5.3 percent. The estimated requirement for added water, therefore, becomes

$$300 - 1917(0.015) - 1369(0.053) = 199 \text{ lb}$$

The estimated batch weights for a yd^3 of concrete are:

Water, to be added	199 lb
Cement	484 lb
Coarse aggregate, wet	1955 lb
Fine aggregate, wet	1451 lb

7.2.9 *Step 9* -- For the laboratory trial batch, it was found convenient to scale the weights down to produce 0.03 yd^3 or 0.81 ft^3 of concrete. Although the calculated quantity of water to be added was 5.97 lb, the amount actually used in an effort to obtain the desired 3 to 4 in. slump is 7.00 lb. The batch as mixed therefore consists of:

Water, to be added	7.00 lb
Cement	14.52 lb
Coarse aggregate, wet	58.65 lb
Fine aggregate, wet	43.53 lb
Total	123.70 lb

The concrete has a measured slump of 2 in. and unit weight of 149.0 lb per ft^3. It is judged to be satisfactory from the standpoint of workability and finishing properties. To provide proper yield and other characteristics for future batches, the following adjustments are made:

7.2.9.1 Since the yield of the trial batch was

$$123.70/149.0 = 0.830 \text{ ft}^3$$

and the mixing water content was 7.00 (added) + 0.86 on coarse aggregate + 2.18 on fine aggregate = 10.04 lb, the mixing water required for a yd^3 of concrete with the same slump as the trial batch should be

$$10.04 \times 27/0.830 = 327 \text{ lb}$$

As indicated in Paragraph 6.3.9.1, this amount must be increased another 15 lb to raise the slump from the measured 2 in. to the desired 3 to 4 in. range, bringing the net mixing water to 342 lb.

7.2.9.2 With the increased mixing water, additional cement will be required to provide the desired water-cement ratio of 0.62. The new cement content becomes

$$342/0.62 = 552 \text{ lb}$$

7.2.9.3 Since workability was found to be satisfactory, the quantity of coarse aggregate per unit volume of concrete will be maintained the same as in the trial batch. The amount of coarse aggregate per yd^3 becomes

$$\frac{58.65}{0.83} \times 27 = 1908 \text{ } lb \text{ } wet$$

which is

$$\frac{1908}{1.02} = 1871 \text{ } lb \text{ } dry$$

and

$$1871 (1.005) = 1880 \text{ SSD*}$$

7.2.9.4 The new estimate for the weight of a yd^3 of concrete is 149.0 x 27 = 4023 lb. The amount of fine aggregate required is therefore

$$4023 - (342 + 552 + 1880) = 1249 \text{ lb SSD}$$

or

$$1249/1.007 = 1240 \text{ lb dry}$$

The adjusted basic batch weights per yd^3 of concrete are:

Water, net mixing	342 lb
Cement	522 lb
Coarse aggregate, dry	1871 lb
Fine aggregate, dry	1240 lb

7.2.10 Adjustments of proportions determined on an absolute volume basis follow a procedure similar to that just outlined. The steps will be given without detailed explanation:

7.2.10.1 Quantities used in nominal 0.81 ft^3 batch are:

Water, added	7.00 lb
Cement	14.52 lb
Coarse aggregate, wet	58.65 lb
Fine aggregate, wet	41.91 lb
Total	122.08 lb

Measured slump 2 in.; unit weight 149.0 lb/ft^3; yield 122.08/149.0 = 0.819 ft^3, workability o.k.

7.2.10.2 Re-estimated water for same slump as

---

* Saturated-surface-dry

trial batch

$$\frac{27(7.00 + 0.86 + 2.09)}{0.819} = 328 \; lb$$

Mixing water required for slump of 3 to 4 in.

$$328 + 15 = 343 \; lb$$

7.2.10.3 Adjusted cement content for increased water

$$343/0.62 = 553 \; lb$$

7.2.10.4 Adjusted coarse aggregate requirement

$$\frac{58.65}{0.819} \; x \; 27 \; = \; 1934 \; lb \; wet$$

or

$$1934/1.02 = 1896 \; lb \; dry$$

7.2.10.5 The volume of ingredients other than air in the original trial batch was

Water	$\frac{9.95}{62.4}$	=	0.159 ft^3
Cement	$\frac{14.52}{3.15 \; x \; 62.4}$	=	0.074 ft^3
Coarse aggregate	$\frac{57.50}{2.68 \; x \; 62.4}$	=	0.344 ft^3
Fine aggregate	$\frac{39.54}{2.64 \; x \; 62.4}$	=	0.240 ft^3
Total		=	0.817 ft^3

Since the yield was 0.819 ft^3, the air content was

$$\frac{0.819 - 0.817}{0.819} = 0.2 \; percent$$

With the proportions of all components except fine aggregate established, the determination of adjusted yd^3 batch quantities can be completed as follows:

Volume of water	=	$\frac{343}{62.4}$	=	5.50 ft^3
Volume of cement	=	$\frac{553}{3.15 \; x \; 62.4}$	=	2.81 ft^3
Volume of air	=	0.002 x 27 =	0.05 ft^3

Volume of coarse aggregate	=	$\frac{1896}{2.68 \; x \; 62.4}$	=	11.34 ft^3
Total volume exclusive of fine aggregate		=	19.70 ft^3
Volume of fine aggregate required	=	27 - 19.70 =	7.30 ft^3
Weight of fine aggregate (dry basis)	=	7.30 x 2.64 x 62.4	=	1203 lb

The adjusted basic batch weights per yd^3 of concrete are then:

Water, net mixing	343 lb
Cement	553 lb
Coarse aggregate, dry	1896 lb
Fine aggregate, dry	1203 lb

These differ only slightly from those given in Paragraph 7.2.9.4 for the method of assumed concrete weight. Further trials or experience might indicate small additional adjustments for either method.

7.3 *Example 2* -- Concrete is required for a heavy bridge pier that will be exposed to fresh water in a severe climate. An average 28-day compressive strength of 3000 psi will be required. Placement conditions permit a slump of 1 to 2 in. and the use of large aggregate, but the only economically available coarse aggregate of satisfactory quality is graded from No. 4 to 1 in. and this will be used. Its dry-rodded weight is found to be 95 lb/ft^3. Other characteristics are as indicated in Section 7.1.

The calculations will be shown in skeleton form only. Note that confusion is avoided if all steps of Section 6 are followed even when they appear repetitive of specified requirements.

7.3.1 *Step 1* -- The desired slump is 1 to 2 in.

7.3.2 *Step 2* -- The locally available aggregate, graded from No. 4 to 1 in., will be used.

7.3.3 *Step 3* -- Since the structure will be exposed to severe weathering, air-entrained concrete will be used. The approximate amount of mixing water to produce a 1 to 2-in. slump in air-entrained concrete with 1-in. aggregate is found from Table 6.3.3 to be 270 lb/yd^3. The recommended air content is 6 percent.

7.3.4 *Step 4* -- From Table 6.3.4(a), the water-cement ratio needed to produce a strength of 3000 psi in air-entrained concrete is estimated to be about 0.59. However, reference to Table 6.3.4(b) reveals that, for the severe weathering exposure anticipated, the water-cement ratio should not exceed 0.50. This lower figure must govern and will be used in the calculations.

7.3.5 *Step 5* -- From the information derived in Steps 3 and 4, the required cement content is found to be 270/0.50

= 540 lb/yd³.

7.3.6 *Step 6* -- The quantity of coarse aggregate is estimated from Table 6.3.6. With a fine aggregate having a fineness modulus of 2.8 and a 1 in. nominal maximum size of coarse aggregate, the table indicates that 0.67 ft³ of coarse aggregate, on a dry-rodded basis, may be used in each ft³ of concrete. For a ft³, therefore, the coarse aggregate will be 27 x 0.67 = 18.09 ft³. Since it weighs 95 lb/ft³, the dry weight of coarse aggregate is 18.09 x 95 = 1719 lb.

7.3.7 *Step 7* -- With the quantities of water, cement, and coarse aggregate established, the remaining material comprising the yd³ of concrete must consist of fine aggregate and air. The required fine aggregate may be determined on the basis of either weight or absolute volume as shown below.

7.3.7.1 *Weight basis* -- From Table 6.3.7.1 the weight of a yd³ of air-entrained concrete made with aggregate of 1 in. maximum size is estimated to be 3850 lb. (For a first trial batch, exact adjustments of this value for differences in slump, cement factor, and aggregate specific gravity are not critical.) Weights already known are:

Water, net mixing	270 lb
Cement	540 lb
Coarse aggregate, dry	1719 lb
Total	2529 lb

The weight of fine aggregate, therefore, is estimated to be

$$3850 - 2529 = 1321 \text{ lb (dry)}$$

7.3.7.2 *Absolute volume basis* -- With the quantities of cement, water, air, and coarse aggregate established, the fine aggregate content can be calculated as follows:

Volume of water	=	$\dfrac{270}{62.4}$	= 4.33 ft³
Solid volume of cement	=	$\dfrac{540}{3.15 \times 62.4}$	= 2.75 ft³
Solid volume of coarse aggregate	=	$\dfrac{1719}{2.68 \times 62.4}$	= 10.28 ft³
Volume of air	=	0.06×27	= 1.62 ft³
Total volume of ingredients except fine aggregate			= 18.98 ft³
Solid volume of fine aggregate required	=	27-18.98	= 8.02 ft³

Required weight of dry fine aggregate = 8.02 x 2.64 x 62.4 = 1321 lb

7.3.7.3 Batch weights per yd³ of concrete calculated on the two bases are compared as follows:

	Based on estimated concrete weight, lb	Based on absolute volume of ingredients, lb
Water, net mixing	270	270
Cement	540	540
Coarse aggregate, dry	1719	1719
Fine aggregate, dry	1321	1321

7.3.8 *Step 8* -- Tests indicate total moisture of 3 percent in the coarse aggregate and 5 percent in the fine aggregate. If the trial batch proportions based on assumed concrete weight are used, the adjusted aggregate weights become:

Coarse aggregate, wet	1719(1.03) = 1771 lb
Fine aggregate, wet	1321(1.05) = 1387 lb

Absorbed water does not become part of the mixing water and must be excluded from the adjustment in added water. Thus, surface water contributed by the coarse aggregate amounts to 3 - 0.5 = 2.5 percent; by the fine aggregate 5 - 0.7 = 4.3 percent. The estimated requirement for added water, therefore, becomes

$$270 - 1719(0.025) - 1321(0.043) = 170 \text{ lb}$$

The estimated batch weights for a yd³ of concrete are:

Water, to be added	170 lb
Cement	540 lb
Coarse aggregate, wet	1771 lb
Fine aggregate, wet	1387 lb
Total	3868 lb

7.3.9 *Step 9* -- For the laboratory trial batch, the weights are scaled down to produce 0.03 yd³ or 0.81 ft³ of concrete. Although the calculated quantity of water to be added was 5.10 lb, the amount actually used in an effort to obtain the desired 1 to 2-in. slump is 4.60 lb. The batch as mixed, therefore, consists of:

Water, added	4.60 lb
Cement	16.20 lb
Coarse aggregate, wet	53.13 lb
Fine aggregate, wet	41.61 lb
Total	115.54 lb

211.1-18   ACI COMMITTEE REPORT

The concrete has a measured slump of 2 in., unit weight of 141.8 lb/ft³ and air content of 6.5 percent. It is judged to be slightly oversanded for the easy placement condition involved. To provide proper yield and other characteristics for future batches, the following adjustments are made.

7.3.9.1 Since the yield of the trial batch was

$$115.543/141.8 = 0.815 \text{ ft}^3$$

and the mixing water content was 4.60 (added) + 1.29 on coarse aggregate + 1.77 on fine aggregate = 7.59 lb, the mixing water required for a yd³ of concrete with the same slump as the trial batch should be

$$\frac{7.59 \times 27}{0.815} = 251 \text{ lb}$$

The slump was satisfactory, but since the air content was too high by 0.5 percent, more water will be needed for proper slump when the air content is corrected. As indicated in Paragraph 6.3.9.2, the mixing water should be increased roughly 5 x 0.5 or about 3 lb, bringing the new estimate to 254 lb/yd³.

7.3.9.2 With the decreased mixing water, less cement will be required to provide the desired water-cement ratio of 0.5. The new cement content becomes

$$254/0.5 = 508 \text{ lb}$$

7.3.9.3 Since the concrete was found to be oversanded, the quantity of coarse aggregate per unit volume will be increased 10 percent to 0.74, in an effort to correct the condition. The amount of coarse aggregate per yd³ becomes

$$0.74 \times 27 \times 95 = 1898 \text{ lb dry}$$

or

$$1898 \times 1.03 = 1955 \text{ wet}$$

and

$$1898 \times 1.005 = 1907 \text{ lb SSD}$$

7.3.9.4 The new estimate for the weight of the concrete with 0.5 percent less air is 141.8/0.995 = 142.50 lb/ft³ or 142.50 x 27 = 3848 lb/yd³. The weight of sand, therefore, is

$$3848 - (254 + 508 + 1907) = 1179 \text{ lb SSD}$$

or

$$1179/1.007 = 1170 \text{ lb dry}$$

The adjusted basic batch weights per yd³ of concrete are:

Water, net mixing	254 lb
Cement	508 lb
Coarse aggregate, dry	1898 lb
Fine aggregate, dry	1170 lb

Admixture dosage must be reduced to provide the desired air content.

7.3.10 Adjustments of proportions determined on an absolute volume basis would follow the procedure outlined in Paragraph 7.2.10, which will not be repeated for this example.

CHAPTER 8 -- REFERENCES

8.1 -- *Recommended references*

The documents of the various standards-producing organizations referred to in this document are listed below with their serial designation, including year of adoption or revision. The documents listed were the latest effort at the time this document was revised. Since some of these documents are revised frequently, generally in minor detail only, the user of this document should check directly with the sponsoring group if it is desired to refer to the latest revision.

American Concrete Institute

116R-90	Cement and Concrete Terminology, SP-19(90)
201.2R-77 (Reapproved 1982)	Guide to Durable Concrete
207.1R-87	Mass Concrete
207.2R-90	Effect of Restraint, Volume Change, and Reinforcement on Cracking of Mass Concrete
207.4R-80(86)	Cooling and Insulating Systems for Mass Concrete
212.3R-89	Chemical Admixtures for Concrete
214-77 (Reapproved 1989)	Recommended Practice for Evaluation of Strength Test Results of Concrete
224R-90	Control of Cracking in Concrete Structures
225R-85	Guide to the Selection and Use of Hydraulic Cements
226.1R-87	Ground Granulated Blast-Furnace Slag as a Cementitious Constituent in Concrete
226.3R-87	Use of Fly Ash in Concrete
301-89	Specifications for Structural Concrete for Buildings
302.1R-89	Guide for Concrete Floor and Slab Construction
304R-89	Guide for Measuring, Mixing, Transporting, and Placing Concrete
304.3R-89	Heavyweight Concrete: Measuring, Mixing, Transporting, and Placing
318-83	Building Code Requirements for Reinforced Concrete

345-82	Standard Practice for Concrete Highway Bridge Deck Construction

ASTM

C 29-78	Standard Test Method for Unit Weight and Voids in Aggregate
31-87a	Standard Method of Making and Curing Concrete Test Specimens in the Field
C 33-86	Standard Specification for Concrete Aggregates
C 39-86	Standard Test Method for Compressive Strength of Cylindrical Concrete Specimens
C 70-79(1985)	Standard Test Method for Surface Moisture in Fine Aggregate
C 78-84	Standard Test Method for Flexural Strength of Concrete (Using Simple Beam with Third-Point Loading)
C 94-86b	Standard Specification for Ready-Mixed Concrete
C 125-86	Standard Definitions of Terms Relating to Concrete and Concrete Aggregates
C 127-84	Standard Test Method for Specific Gravity and Absorption of Coarse Aggregate
C 128-84	Standard Test Method for Specific Gravity and Absorption of Fine Aggregate
C 136-84a	Standard Method for Sieve Analysis of Fine and Coarse Aggregates
C 138-81	Standard Test Method for Unit Weight, Yield, and Air Content (Gravimetric) of Concrete
C 143-78	Standard Test Method for Slump of Portland Cement Concrete
C 150-86	Standard Specification for Portland Cement
C 172-82	Standard Method of Sampling Freshly Mixed Concrete
C 173-78	Standard Test Method for Air Content of Freshly Mixed Concrete by the Volumetric Method
C 192-81	Standard Method of Making and Curing Concrete Test Specimens in the Laboratory
C 231-82	Standard Test Method for Air Content of Freshly Mixed Concrete by the Pressure Method
C 260-86	Standard Specification for Air-Entraining Admixtures for Concrete
C 293-79	Standard Test Method for Flexural Strength of Concrete (Using Simple Beam with Center-Point Loading)
C 494-86	Standard Specification for Chemical Admixtures for Concrete
C 496-86	Standard Test Method for Splitting Tensile Strength of Cylindrical Concrete Specimens

C 566-84	Standard Test Method for Total Moisture Content of Aggregate by Drying
C 595-86	Standard Specification for Blended Hydraulic Cements
C 618-85	Standard Specification for Fly Ash and Raw or Calcined Natural Pozzolan for Use as a Mineral Admixture in Portland Cement Concrete
C 637-84	Standard Specification for Aggregates for Radiation-Shielding Concrete
C 638-84	Standard Descriptive Nomenclature of Constituents of Aggregates for Radiation-Shielding Concrete
C 989-87a	Standard Specification for Granulated Blast-Furnace Slag for Use in Concrete and Mortars
C 1017-85	Standard Specification for Chemical Admixtures for Use in Producing Flowing Concrete
C 1064-86	Standard Test Method for Temperature of Freshly Mixed Portland-Cement Concrete
D 75-82	Standard Practice for Sampling Aggregates
D 3665-82	Standard Practice for Random Sampling of Construction Materials
E 380-84	Standard for Metric Practice

The above publications may be obtained from the following organizations:

American Concrete Institute
P.O. Box 19150
Detroit, MI 48219-0150

ASTM
1916 Race Street
Philadelphia, PA 19103

8.2 -- *Cited references*
1. "Silica Fume in Concrete," ACI Committee 226 Preliminary Report, ACI Materials Journal, *Proceedings* V. 84, Mar.-Apr. 1987, pp. 158-166.

8.3 -- *Additional references*
1. "Standard Practice for Concrete," *Engineer Manual* No. EM 1110-2-2000, Office, Chief of Engineers, U.S. Army Corps of Engineers, Washington, D.C., June 1974.
2. Gaynor, Richard D., "High-Strength Air-Entrained Concrete," *Joint Research Laboratory Publication* No. 17, National Ready Mixed Concrete Association/National Sand and Gravel Association, Silver Spring, 1968, 19 pp.
3. *Proportioning Concrete Mixes*, SP-46, American Concrete Institute, Detroit, 1974, 223 pp.

4. Townsend, Charles L., "Control of Temperature Cracking in Mass Concrete," *Causes, Mechanism, and Control of Cracking in Concrete*, SP-20, American Concrete Institute, Detroit, 1968, pp. 119-139.

5. Townsend, C. L., "Control of Cracking in Mass Concrete Structures," *Engineering Monograph* No. 34, U.S. Bureau of Reclamation, Denver, 1965.

6. Fuller, William B., and Thompson, Sanford E., "The Laws of Proportioning Concrete," *Transactions*, ASCE, V. 59, Dec. 1907, pp. 67-143.

7. Powers, Treval C., *The Properties of Fresh Concrete*, John Wiley & Sons, New York, 1968, pp. 246-256.

8. *Concrete Manual*, 8th Edition, U.S. Bureau of Reclamation, Denver, 1975, 627 pp.

9. Abrams, Duff A., "Design of Concrete Mixtures," *Bulletin* No. 1, Structural Materials Research Laboratory, Lewis Institute, Chicago, 1918, 20 pp.

10. Edwards, L. N., "Proportioning the Materials of Mortars and Concretes by Surface Areas of Aggregates," *Proceedings*, ASTM, V. 18, Part 2, 1918, p. 235.

11. Young, R. B., "Some Theoretical Studies on Proportioning Concrete by the Method of Surface Area Aggregate," *Proceedings*, ASTM, V. 19, Part 2, p. 1919.

12. Talbot, A. N., "A Proposed Method of Estimating the Density and Strength of Concrete and of Proportioning the Materials by Experimental and Analytical Consideration of the Voids in Mortar and Concrete," *Proceedings*, ASTM, V. 21, 1921, p. 940.

13. Weymouth, C. A. G., "A Study of Fine Aggregate in Freshly Mixed Mortars and Concretes," *Proceedings*, ASTM, V. 38, Part 2, 1938, pp. 354-372.

14. Dunagan, W. M., "The Application of Some of the Newer Concepts to the Design of Concrete Mixes," ACI Journal, *Proceedings* V. 36, No. 6, June 1940, pp. 649-684.

15. Goldbeck, A. T., and Gray, J. E., "A Method of Proportioning Concrete for Strength, Workability, and Durability," *Bulletin* No. 11, National Crushed Stone Association, Washington, D.C., Dec. 1942, 30 pp. (Revised 1953 and 1956).

16. Swayze, Myron A., and Gruenwald, Ernst, "Concrete Mix Design--A Modification of Fineness Modulus Method," ACI Journal, *Proceedings* V. 43, No. 7, Mar. 1947, pp. 829-844.

17. Walker, Stanton, and Bartel, Fred F., Discussion of "Concrete Mix Design--A Modification of the Fineness Modulus Method" by Myron A. Swayze and Ernst Gruenwald, ACI Journal, *Proceedings* V. 43, Part 2, Dec. 1947, pp. 844-1-844-17.

18. Henrie, James O., "Properties of Nuclear Shielding Concrete," ACI Journal, *Proceedings* V. 56, No. 1, July 1959, pp. 37-46.

19. Mather, Katharine, "High Strength, High Density Concrete," ACI Journal, *Proceedings* V. 62, No. 8, Aug. 1965, pp. 951-960.

20. Clendenning, T. G.; Kellam, B.; and MacInnis, C., "Hydrogen Evolution from Ferrophosphorous Aggregate in Portland Cement Concrete," ACI Journal, *Proceedings* V. 65, No. 12, Dec. 1968, pp. 1021-1028.

21. Popovics, Sandor, "Estimating Proportions for Structural Concrete Mixtures," ACI Journal, *Proceedings* V. 65, No. 2, Feb. 1968, pp. 143-150.

22. Davis, H. S., "Aggregates for Radiation Shielding Concrete," *Materials Research and Standards*, V. 7, No. 11, Nov. 1967, pp. 494-501.

23. *Concrete for Nuclear Reactors*, SP-34, American Concrete Institute, Detroit, 1972, 1736 pp.

24. Tynes, W. O., "Effect of Fineness of Continuously Graded Coarse Aggregate on Properties of Concrete," *Technical Report* No. 6-819, U.S. Army Engineer Waterways Experiment Station, Vicksburg, Apr. 1968, 28 pp.

25. *Handbook for Concrete and Cement*, CRD-C 3, U.S. Army Engineer Waterways Experiment Station, Vicksburg, 1949 (plus quarterly supplements).

26. Hansen, Kenneth, "Cost of Mass Concrete in Dams," *Publication* No. MS260W, Portland Cement Association, Skokie, 1973, 4 pp.

27. Canon, Robert W., "Proportioning Fly Ash Concrete Mixes for Strength and Economy," ACI Journal, *Proceedings* V. 65, No. 11, Nov. 1968, pp 969-979.

28. Butler, W. B., "Economical Binder Proportioning with Cement Replacement Materials," *Cement, Concrete, and Aggregates*, CCAGDP, V. 10, No. 1, Summer 1988, pp. 45-47.

APPENDIX 1 -- METRIC (SI) SYSTEM ADAPTATION

A1.1 Procedures outlined in this standard practice have been presented using inch-pound units of measurement. The principles are equally applicable in SI system with proper adaptation of units. This Appendix provides all of the information necessary to apply the proportioning procedure using SI measurements. Table A1.1 gives relevant conversion factors. A numerical example is presented in Appendix 2.

TABLE A1.1—CONVERSION FACTORS, in.-lb TO SI UNITS*

Quantity	in.-lb unit	SI† unit	Conversion factor (Ratio: in.-lb/SI)
Length	inch (in.)	millimeter (mm)	25.40
Volume	cubic foot (ft³)	cubic meter (m³)	0.02832
	cubic yard (yd³)	cubic meter (m³)	0.7646
Mass	pound (lb)	kilogram (kg)	0.4536
Stress	pounds per square inch (psi)	megapascal (MPa)	6.895×10^{-2}
Density	pounds per cubic foot (lb/ft³)	kilograms per cubic meter (kg/m³)	16.02
	pounds per cubic yard (lb/yd³)	kilograms per cubic meter (kg/m³)	0.5933
Temperature	degrees Fahrenheit (F)	degrees Celsius (C)	‡

*Gives names (and abbreviations) of measurement units in the inch-pound system as used in the body of this report and in the SI (metric) system, along with multipliers for converting the former to the latter. From ASTM E 380.
†Systéme International d'Unites
‡$C = (F - 32)/1.8$

A1.2 For convenience of reference, numbering of subsequent paragraphs in this Appendix corresponds to the body of the report except that the designation "A1" is prefixed. All tables have been converted and reproduced. Descriptive portions are included only where use of the SI system requires a change in procedure or formula. To the extent practicable, conversions to metric units have been made in such a way that values are realistic in terms of usual practice and significance of numbers. For example, aggregate and sieve sizes in the metric tables are ones commonly used in Europe. Thus, there is not always a precise mathematical correspondence between inch-pound and SI values in corresponding tables.

A1.5.3 *Steps in calculating proportions* -- Except as discussed below, the methods for arriving at quantities of ingredients for a unit volume of concrete are essentially the same when SI units are employed as when inch-pound units are employed. The main difference is that the unit volume of concrete becomes the cubic meter and numerical values must be taken from the proper "A1" table instead of the one referred to in the text.

A1.5.3.1 *Step 1. Choice of slump* -- See Table A1.5.3.1.

TABLE A1.5.3.1 — RECOMMENDED SLUMPS FOR VARIOUS TYPES OF CONSTRUCTION (SI)

Types of construction	Slump, mm	
	Maximum*	Minimum
Reinforced foundation walls and footings	75	25
Plain footings, caissons, and substructure walls	75	25
Beams and reinforced walls	100	25
Building columns	100	25
Pavements and slabs	75	25
Mass concrete	75	25

*May be increased 25 mm for methods of consolidation other than vibration.

A1.5.3.2 *Step 2. Choice of nominal maximum size of aggregate.*

A1.5.3.3 *Step 3. Estimation of mixing water and air content* -- See Table A1.5.3.3.

A1.5.3.4 *Step 4. Selection of water-cement ratio* -- See Table A1.5.3.4.

A1.5.3.5 *Step 5. Calculation of cement content.*

A1.5.3.6 *Step 6. Estimation of coarse aggregate content* -- The dry mass of coarse aggregate required for a cubic meter of concrete is equal to the value from Table A1.5.3.6 multiplied by the dry-rodded unit mass of the aggregate in kilograms per cubic meter.

A1.5.3.7 *Step 7. Estimation of fine aggregate content* -- In the SI, the formula for calculation of fresh concrete mass per cubic meter is:

$$U_M = 10G_a(100 - A) + C_M(1 - G_a/G_c) - W_M(G_a - 1)$$

where

U_M = unit mass of fresh concrete, kg/m³

G_a = weighted average specific gravity of combined fine and coarse aggregate, bulk, SSD

G_c = specific gravity of cement (generally 3.15)

A = air content, percent

W_M = mixing water requirement, kg/m³

C_M = cement requirement, kg/m³

A1.5.3.9 *Step 9. Trial batch adjustments* -- The following "rules of thumb" may be used to arrive at closer approximations of unit batch quantities based on results for a trial batch:

A1.5.3.9.1 The estimated mixing water to produce the same slump as the trial batch will be equal to the net amount of mixing water used divided by the yield of the trial batch in m³. If slump of the trial batch was not correct, increase or decrease the re-estimated water content by 2 kg/m³ of concrete for each increase or decrease of 10 mm in slump desired.

A1.5.3.9.2 To adjust for the effect of

211.1-22   ACI COMMITTEE REPORT

TABLE A1.5.3.3 — APPROXIMATE MIXING WATER AND AIR CONTENT REQUIREMENTS FOR DIFFERENT SLUMPS AND NOMINAL MAXIMUM SIZES OF AGGREGATES (SI)

Slump, mm	Water, Kg/m³ of concrete for indicated nominal maximum sizes of aggregate							
	9.5*	12.5*	19*	25*	37.5*	50†*	75†‡	150†‡
Non-air-entrained concrete								
25 to 50	207	199	190	179	166	154	130	113
75 to 100	228	216	205	193	181	169	145	124
150 to 175	243	228	216	202	190	178	160	—
Approximate amount of entrapped air in non-air-entrained concrete, percent	3	2.5	2	1.5	1	0.5	0.3	0.2
Air-entrained concrete								
25 to 50	181	175	168	160	150	142	122	107
75 to 100	202	193	184	175	165	157	133	119
150 to 175	216	205	197	184	174	166	154	—
Recommended average§ total air content, percent for level of exposure:								
Mild exposure	4.5	4.0	3.5	3.0	2.5	2.0	1.5**††	1.0**††
Moderate exposure	6.0	5.5	5.0	4.5	4.5	4.0	3.5**††	3.0**††
Extreme exposure‡‡	7.5	7.0	6.0	6.0	5.5	5.0	4.5**††	4.0**††

*The quantities of mixing water given for air-entrained concrete are based on typical total air content requirements as shown for "moderate exposure" in the Table above. These quantities of mixing water are for use in computing cement contents for trial batches at 20 to 25 C. They are maximum for reasonably well-shaped angular aggregates graded within limits of accepted specifications. Rounded coarse aggregate will generally require 18 kg less water for non-air-entrained and 15 kg less for air-entrained concretes. The use of water-reducing chemcial admixtures, ASTM C 494, may also reduce mixing water by 5 percent or more. The volume of the liquid admixtures is included as part of the total volume of the mixing water.

†The slump values for concrete containing aggregate larger than 40 mm are based on slump tests made after removal of particles larger than 40 mm by wet-screening.

‡These quantities of mixing water are for use in computing cement factors for trial batches when 75 mm or 150 mm normal maximum size aggregate is used. They are average for reasonably well-shaped coarse aggregates, well-graded from coarse to fine.

§Additional recommendations for air-content and necessary tolerances on air content for control in the field are given in a number of ACI documents, including ACI 201, 345, 318, 301, and 302. ASTM C 94 for ready-mixed concrete also gives air content limits. The requirements in other documents may not always agree exactly so in proportioning concrete consideration must be given to selecting an air content that will meet the needs of the job and also meet the applicable specifications.

**For concrete containing large aggregates which will be wet-screened over the 40 mm sieve prior to testing for air content, the percentage of air expected in the 40 mm minus material should be as tabulated in the 40 mm column. However, initial proportioning calculations should include the air content as a percent of the whole.

††When using large aggregate in low cement factor concrete, air entrainment need not be detrimental to strength. In most cases mixing water requirement is reduced sufficiently to improve the water-cement ratio and to thus compensate for the strength reducing effect of entrained air concrete. Generally, therefore, for these large nominal maximum sizes of aggregate, air contents recommended for extreme exposure should be considered even though there may be little or no exposure to moisture and freezing.

‡‡These values are based on the criteria that 9 percent air is needed in the mortar phase of the concrete. If the mortar volume will be substantially different from that determined in this recommended practice, it may be desirable to calculate the needed air content by taking 9 percent of the actual mortar volume.

incorrect air content in a trial batch of air-entrained concrete on slump, reduce or increase the mixing water content of A1.5.3.9.1 by 3 kg/m³ of concrete for each 1 percent by which the air content is to be increased or decreased from that of the trial batch.

A1.5.3.9.3 The re-estimated unit mass of the fresh concrete for adjustment of trial batch proportions is equal to the unit mass in kg/m³ measured on the trial batch, reduced or increased by the percentage increase or decrease in air content of the adjusted batch from the first trial batch.

TABLE A1.5.3.4(a) — RELATIONSHIPS BETWEEN WATER-CEMENT RATIO AND COMPRESSIVE STRENGTH OF CONCRETE (SI)

Compressive strength at 28 days, MPa*	Water-cement ratio, by mass	
	Non-air-entrained concrete	Air-entrained concrete
40	0.42	—
35	0.47	0.39
30	0.54	0.45
25	0.61	0.52
20	0.69	0.60
15	0.79	0.70

*Values are estimated average strengths for concrete containing not more than 2 percent air for non-air-entrained concrete and 6 percent total air content for air-entrained concrete. For a constant water-cement ratio, the strength of concrete is reduced as the air content is increased.

Strength is based on 152 × 305 mm cylinders moist-cured for 28 days in accordance with the sections on "Initial Curing" and "Curing of Cylinders for Checking the Adequacy of Laboratory Mixture Proportions for Strength or as the Basis for Acceptance or for Quality Control" of ASTM Method C 31 for Making and Curing Concrete Specimens in the Field. These are cylinders cured moist at 23 ± 1.7 C prior to testing.

The relationship in this Table assumes a nominal maximum aggregate size of about 19 to 25 mm. For a given source of aggregate, strength produced at a given water-cement ratio will increase as nominal maximum size of aggregate decreases; see Sections 3.4 and 5.3.2.

TABLE A1.5.3.4(b) — MAXIMUM PERMISSIBLE WATER-CEMENT RATIOS FOR CONCRETE IN SEVERE EXPOSURES (SI)*

Type of structure	Structure wet continuously or frequently and exposed to freezing and thawing†	Structure exposed to sea water or sulfates
Thin sections (railings, curbs, sills, ledges, ornamental work) and sections with less than 5 mm cover over steel	0.45	0.40‡
All other structures	0.50	0.45‡

*Based on ACI 201.2R.

†Concrete should also be air-entrained.

‡If sulfate resisting cement (Type II or Type V of ASTM C 150) is used, permissible water-cement ratio may be increased by 0.05.

TABLE A1.5.3.6 — VOLUME OF COARSE AGGREGATE PER UNIT OF VOLUME OF CONCRETE (SI)

Nominal maximum size of aggregate, mm	Volume of dry-rodded coarse aggregate* per unit volume of concrete for different fineness moduli† of fine aggregate			
	2.40	2.60	2.80	3.00
9.5	0.50	0.48	0.46	0.44
12.5	0.59	0.57	0.55	0.53
19	0.66	0.64	0.62	0.60
25	0.71	0.69	0.67	0.65
37.5	0.75	0.73	0.71	0.69
50	0.78	0.76	0.74	0.72
75	0.82	0.80	0.78	0.76
150	0.87	0.85	0.83	0.81

*Volumes are based on aggregates in dry-rodded condition as described in ASTM C 29.

These volumes are selected from empirical relationships to produce concrete with a degree of workability suitable for usual reinforced construction. For less workable concrete such as required for concrete pavement construction they may be increased about 10 percent. For more workable concrete, such as may sometimes be required when placement is to be by pumping, they may be reduced up to 10 percent.

†See ASTM Method 136 for calculation of fineness modulus.

TABLE A1.5.3.7.1 — FIRST ESTIMATE OF MASS OF FRESH CONCRETE (SI)

Nominal maximum size of aggregate, mm	First estimate of concrete unit mass, kg/m^3*	
	Non-air-entrained concrete	Air-entrained concrete
9.5	2280	2200
12.5	2310	2230
19	2345	2275
25	2380	2290
37.5	2410	2350
50	2445	2345
75	2490	2405
150	2530	2435

*Values calculated by Eq. (A1.5.3.7) for concrete of medium richness (330 kg of cement per m^3) and medium slump with aggregate specific gravity of 2.7. Water requirements based on values for 75 to 100 mm slump in Table A1.5.3.3. If desired, the estimate of unit mass may be refined as follows if necessary information is available: for each 5 kg difference in mixing water from the Table A1.5.3.3 values for 75 to 100 mm slump, correct the mass per m^3 8 kg in the opposite direction; for each 20 kg difference in cement content from 330 kg, correct the mass per m^3 3 kg in the same direction; for each 0.1 by which aggregate specific gravity deviates from 2.7, correct the concrete mass 60 kg in the same direction. For air-entrained concrete the air content for severe exposure from Table A.1.5.3.3 was used. The mass can be increased 1 percent for each percent reduction in air content from that amount.

APPENDIX 2 -- EXAMPLE PROBLEM IN METRIC (SI) SYSTEM

A2.1 *Example 1* -- Example 1 presented in Section 6.2 will be solved here using metric units of measure. Required average strength will be 24 MPa with slump of 75 to 100 mm. The coarse aggregate has a nominal maximum size of 37.5 mm and dry-rodded mass of 1600 kg/m^3. As stated in Section 6.1, other properties of the ingredients are: cement -- Type I with specific gravity of 3.15; coarse aggregate -- bulk specific gravity 2.68 and absorption 0.5 percent; fine aggregate -- bulk specific gravity 2.64, absorption 0.7 percent, and fineness modulus 2.8.

A2.2 All steps of Section 5.3 should be followed in sequence to avoid confusion, even though they sometimes merely restate information already given.

A2.2.1 *Step 1* -- The slump is required to be 75 to 100 mm.

A2.2.2 *Step 2* -- The aggregate to be used has a nominal maximum size of 37.5 mm.

A2.2.3 *Step 3* -- The concrete will be non-air-entrained since the structure is not exposed to severe weathering. From Table A1.5.3.3, the estimated mixing water for a slump of 75 to 100 mm in non-air-entrained concrete made with 37.5 mm aggregate is found to be 181 kg/m^3.

A2.2.4 *Step 4* -- The water-cement ratio for non-air-entrained concrete with a strength of 24 MPa is found from Table A1.5.3.4(a) to be 0.62.

A2.2.5 *Step 5* -- From the information developed in Steps 3 and 4, the required cement content is found to be 181/0.62 = 292 kg/m^3.

A2.2.6 *Step 6* -- The quantity of coarse aggregate is estimated from Table A 1.5.3.6. For a fine aggregate having a fineness modulus of 2.8 and a 37.5 mm nominal maximum size of coarse aggregate, the table indicates that 0.71 m^3 of coarse aggregate, on a dry-rodded basis, may be used in each cubic meter of concrete. The required dry mass is, therefore, 0.71 x 1600 = 1136 kg.

A2.2.7 *Step 7* -- With the quantities of water, cement and coarse aggregate established, the remaining material comprising the cubic meter of concrete must consist of fine aggregate and whatever air will be entrapped. The required fine aggregate may be determined on the basis of either mass or absolute volume as shown below:

A2.2.7.1 *Mass basis* -- From Table A1.5.3.7.1, the mass of a cubic meter of non-air-entrained concrete made with aggregate having a nominal maximum size of 37.5 mm is estimated to be 2410 kg. (For a first trial batch, exact adjustments of this value for usual differences in slump, cement factor, and aggregate specific gravity are not critical.) Masses already known are:

Water (net mixing)	181 kg
Cement	292 kg
Coarse aggregate	1136 kg
Total	1609 kg

The mass of fine aggregate, therefore, is estimated to be

2410 - 1609 = 801 kg

A2.2.7.2 *Absolute volume basis* -- With the quantities of cement, water, and coarse aggregate established, and the approximate entrapped air content (as opposed to purposely entrained air) of 1 percent determined from Table A1.5.3.3, the sand content can be calculated as follows:

$$\text{Volume of water} = \frac{181}{1000} \quad 0.181 \text{ m}^3$$

$$\text{Solid volume of cement} = \frac{292}{3.15 \times 1000} \quad 0.093 \text{ m}^3$$

Solid volume
of coarse   = $\dfrac{1136}{2.68 \times 1000}$ 0.424 m³
aggregate

Volume of entrapped
air   =   0.01 x 1.000   0.010 m³

Total solid volume
of ingredients except
fine aggregate   0.708 m³

Solid volume of
fine aggregate
required   =   1.000 - 0.705   0.292 m³

Required weight
of dry   =   0.292 x 2.64
fine aggregate   x 1000   771 kg

A2.2.7.3 Batch masses per cubic meter of concrete calculated on the two bases are compared below:

	Based on estimated concrete mass, kg	Based on absolute volume of ingredients, kg
Water (net mixing)	181	181
Cement	292	292
Coarse aggregate (dry)	1136	1136
Sand (dry)	801	771

A2.2.8 *Step 8* -- Tests indicate total moisture of 2 percent in the coarse aggregate and 6 percent in the fine aggregate. If the trial batch proportions based on assumed concrete mass are used, the adjusted aggregate masses become

Coarse aggregate (wet) =   1136(1.02) = 1159 kg
Fine aggregates (wet)   =   801(1.06) = 849 kg

Absorbed water does not become part of the mixing water and must be excluded from the adjustment in added water. Thus, surface water contributed by the coarse aggregate amounts to 2 - 0.5 = 1.5 percent; by the fine aggregate 6 - 0.7 = 5.3 percent. The estimated requirement for added water, therefore, becomes

181 - 1136(0.015) - 801(0.053) = 122 kg

The estimated batch masses for a cubic meter of concrete are:

Water (to be added)   122 kg
Cement   292 kg
Coarse aggregate (wet)   1159 kg

Fine aggregate (wet)   849 kg
Total   2422 kg

A2.2.9 *Step 9* -- For the laboratory trial batch, it is found convenient to scale the masses down to produce 0.02 m³ of concrete. Although the calculated quantity of water to be added was 2.44 kg, the amount actually used in an effort to obtain the desired 75 to 100 mm slump is 2.70 kg. The batch as mixed, therefore, consists of

Water (added)   2.70 kg
Cement   5.84 kg
Coarse aggregate (wet)   23.18 kg
Fine aggregate (wet)   16.98 kg
Total   48.70 kg

The concrete has a measured slump of 50 mm and unit mass of 2390 kg/m³. It is judged to be satisfactory from the standpoint of workability and finishing properties. To provide proper yield and other characteristics for future batches, the following adjustments are made:

A2.2.9.1 Since the yield of the trial batch was

$$48.70/2390 = 0.0204 \text{ m}^3$$

and the mixing water content was 2.70 (added) + 0.34 (on coarse aggregate) + 0.84 (on fine aggregate) = 3.88 kg, the mixing water required for a cubic meter of concrete with the same slump as the trial batch should be

$$3.88/0.0204 = 190 \text{ kg}$$

As indicated in A1.5.3.9.1, this amount must be increased another 8 kg to raise the slump from the measured 50 mm to the desired 75 to 100 mm range, bringing the total mixing water to 198 kg.

A2.2.9.2 With the increased mixing water, additional cement will be required to provide the desired water-cement ratio of 0.62. The new cement content becomes

$$198/0.62 = 319 \text{ kg}$$

A2.2.9.3 Since workability was found to be satisfactory, the quantity of coarse aggregate per unit volume of concrete will be maintained the same as in the trial batch. The amount of coarse aggregate per cubic meter becomes

$$\frac{23.18}{0.0204} = 1136 \text{ } kg \text{ } wet$$

which is

$$\frac{1136}{1.02} = 1114 \; kg \; dry$$

and

$$1114 \times 1.005 = 1120 \text{ kg SSD*}$$

A2.2.9.4 The new estimate for the mass of a cubic meter of concrete is the measured unit mass of 2390 kg/m³. The amount of fine aggregate required is, therefore

$$2390 - (198 + 319 + 1120) = 753 \text{ kg SSD*}$$

or

$$753/1.007 = 748 \text{ kg dry}$$

The adjusted basic batch masses per cubic meter of concrete are

Water (net mixing)	198 kg
Cement	319 kg
Coarse aggregate (dry)	1114 kg
Fine aggregate (dry)	748 kg

A2.2.10 Adjustments of proportions determined on an absolute volume basis follow a procedure similar to that just outlined. The steps will be given without detailed explanation:

A2.2.10.1 Quantities used in the nominal 0.02 m³ batch are

Water (added)	2.70 kg
Cement	5.84 kg
Coarse aggregate (wet)	23.18 kg
Fine aggregate (wet)	16.34 kg
Total	48.08 kg

Measured slump 50 mm; unit mass 2390 kg/m³; yield 48.08/2390 = 0.0201 m³; workability o.k.

A2.2.10.2 Re-estimated water for same slump as trial batch:

$$\frac{2.70 + 0.34 + 0.81}{0.0201} = 192 \; kg$$

Mixing water required for slump of 75 to 100 mm:

$$192 + 8 = 200 \text{ kg}$$

A2.2.10.3 Adjusted cement content for increased water:

$$200/0.62 = 323 \text{ kg}$$

A2.2.10.4 Adjusted coarse aggregate requirement:

---
* Saturated-surface-dry.

$$\frac{23.18}{0.0202} = 1153 \; kg \; wet$$

or

$$1163/1.02 = 1130 \text{ kg dry}$$

A2.2.10.5 The volume of ingredients other than air in the original trial batch was

Water	$\dfrac{3.85}{1000}$	$= 0.0039 \; m^3$
Cement	$\dfrac{5.84}{3.15 \times 1000}$	$= 0.0019 \; m^3$
Coarse aggregate	$\dfrac{22.72}{2.68 \times 1000}$	$= 0.0085 \; m^3$
Fine aggregate	$\dfrac{15.42}{2.64 \times 1000}$	$= \underline{0.0058 \; m^3}$

Total	0.0201 m³

Since the yield was also 0.0201 m³, there was no air in the concrete detectable within the precision of the unit mass test and significant figures of the calculations. With the proportions of all components except fine aggregate established, the determination of adjusted cubic meter batch quantities can be completed as follows:

Volume of water	=	$\dfrac{200}{1000}$	=	0.200 m³
Volume of cement	=	$\dfrac{323}{3.15 \times 1000}$	=	0.103 m³
Allowance for volume of cement			=	0.000 m³
Volume of coarse aggregate	=	$\dfrac{1130}{2.68 \times 1000}$	=	0.422 m³
Total volume exclusive of fine aggregate			=	0.725 m³
Volume of fine aggregate required	=	1.000 - 0.725	=	0.275 m³
Mass of fine aggregate (dry basis)	=	0.275 x 2.64 x 1000	=	726 kg

The adjusted basic batch weights per cubic meter of concrete, then, are:

Water (net mixing)	200 kg

Cement	323 kg
Coarse aggregate (dry)	1130 kg
Fine aggregate (dry)	726 kg

These differ only slightly from those given in Paragraph A2.2.9.4 for the method of assumed concrete weight. Further trials or experience might indicate small additional adjustments for either method.

APPENDIX 3 -- LABORATORY TESTS

A.3.1 Selection of concrete mix proportions can be accomplished effectively from results of laboratory tests which determine basic physical properties of materials to be used, establish relationships between water-cement ratio or water to cement and pozzolan ratio, air content, cement content, and strength, and which furnish information on the workability characteristics of various combinations of ingredient materials. The extent of investigation desirable for any given job will depend on its size and importance and on the service conditions involved. Details of the laboratory program will also vary, depending on facilities available and on individual preferences.

A3.2 *Properties of cement*

A3.2.1 Physical and chemical characteristics of cement influence the properties of hardened concrete. However, the only property of cement used directly in computation of concrete mix proportions is specific gravity. The specific gravity of portland cements of the types covered by ASTM C 150 and C 175 may usually be assumed to be 3.15 without introducing appreciable error in mix computations. For other types such as the blended hydraulic cements of ASTM C 595, slag cement in C 989 or pozzolan covered in C 618, the specific gravity for use in volume calculations should be determined by test.

A3.2.2 A sample of cement should be obtained from the mill which will supply the job, or preferably from the concrete supplier. The sample should be ample for tests contemplated with a liberal margin for additional tests that might later be considered desirable. Cement samples should be shipped in airtight containers, or at least in moisture-proof packages. Pozzolans should also be carefully sampled.

A3.3 *Properties of aggregate*

A3.3.1 Sieve analysis, specific gravity, absorption, and moisture content of both fine and coarse aggregate and dry-rodded unit weight of coarse aggregate are physical properties useful for mix computations. Other tests which may be desirable for large or special types of work include petrographic examination and tests for chemical reactivity, soundness, durability, resistance to abrasion, and various deleterious substances. Such tests yield information of value in judging the long-range serviceability of concrete.

A3.3.2 Aggregate gradation as measured by the sieve analysis is a major factor in determining unit water requirement, proportions of coarse aggregate and sand, and cement content for satisfactory workability. Numerous "ideal" aggregate grading curves have been proposed, and these, tempered by practical considerations, have formed the basis for typical sieve analysis requirements in concrete standards. ASTM C 33 provides a selection of sizes and gradings suitable for most concrete. Additional workability realized by

use of air-entrainment permits, to some extent, the use of less restrictive aggregate gradations.

A3.3.3 Samples for concrete mix tests should be representative of aggregate available for use in the work. For laboratory tests, the coarse aggregates should be separated into required size fractions and reconstituted at the time of mixing to assure representative grading for the small test batches. Under some conditions, for work of important magnitude, laboratory investigation may involve efforts to overcome grading deficiencies of the available aggregates. Undesirable sand grading may be corrected by (1) separation of the sand into two or more size fractions and recombining in suitable proportions; (2) increasing or decreasing the quantity of certain sizes to balance the grading; or (3) reducing excess coarse material by grinding or crushing. Undesirable coarse-aggregate gradings may be corrected by: (1) crushing excess coarser fractions; (2) wasting sizes that occur in excess; (3) supplementing deficient sizes from other sources; or (4) a combination of these methods. Whatever grading adjustments are made in the laboratory should be practical and economically justified from the standpoint of job operation. Usually, required aggregate grading should be consistent with that of economically available materials.

A3.4 *Trial batch series*

A3.4.1 The tabulated relationships in the body of this report may be used to make rough estimates of batch quantities for a trial mix. However, they are too generalized to apply with a high degree of accuracy to a specific set of materials. If facilities are available, therefore, it is advisable to make a series of concrete tests to establish quantitative relationships for the materials to be used. An illustration of such a test program is shown in Table A3.4.1.

A3.4.2 First, a batch of medium cement content and usable consistency is proportioned by the described methods. In preparing Mix No. 1, an amount of water is used which will produce the desired slump even if this differs from the estimated requirement. The fresh concrete is tested for slump and unit weight and observed closely for workability and finishing characteristics. In the example, the yield is too high and the concrete is judged to contain an excess of fine aggregate.

A3.4.3 Mix No. 2 is prepared, adjusted to correct the errors in Mix No. 1, and the testing and evaluation repeated. In this case, the desired properties are achieved within close tolerances and cylinders are molded to check the compressive strength. The information derived so far can now be used to select proportions for a series of additional mixes, No. 3 to 6, with cement contents above and below that of Mix No. 2, encompassing the range likely to be needed. Reasonable

TABLE A3.4.1 — TYPICAL TEST PROGRAM TO ESTABLISH CONCRETE-MAKING PROPERTIES OF LOCAL MATERIALS

Mix No.	Cubic yard batch quantities, lb						Concrete characteristics				
	Cement	Sand	Coarse Aggregate	Water Estimated	Water Used	Total used	Slump in.	Unit wt., lb per cu ft	Yield cu ft	28-day Compressive strength, psi	Work-ability
1	500	1375	1810	325	350	4035	4	147.0	27.45	—	Oversanded
2	500	1250	1875	345	340	3965	3	147.0	26.97	3350	o.k.
3	400	1335	1875	345	345	3955	4.5	145.5	27.18	2130	o.k.
4	450	1290	1875	345	345	3960	4	146.2	27.09	2610	o.k.
5	550	1210	1875	345	345	3980	3	147.5	26.98	3800	o.k.
6	600	1165	1875	345	345	3985	3.5	148.3	26.87	4360	o.k.

refinement in these batch weights can be achieved with the help of corrections given in the notes to Table 6.3.7.1.

A3.4.4 Mix No. 2 to 6 provide the background, including the relationship of strength to water-cement ratio for theparticular combination of ingredients, needed to select proportions for a range of specified requirements.

A3.4.5 In laboratory tests, it seldom will be found, even by experienced operators, that desired adjustments will develop as smoothly as indicated in Table A3.4.1. Furthermore, it should not be expected that field results will check exactly with laboratory results. An adjustment of the selected trial mix on the job is usually necessary. Closer agreement between laboratory and field will be assured if machine mixing is employed in the laboratory. This is especially desirable if air-entraining agents are used since the type of mixer influences the amount of air entrained. Before mixing the first batch, the laboratory mixer should be "buttered" or the mix "overmortared" as described in ASTM C 192. Similarly, any processing of materials in the laboratory should simulate as closely as practicable corresponding treatment in the field.

A3.4.6 The series of tests illustrated in Table A3.4.1 may be expanded as the size and special requirements of the work warrant. Variables that may require investigation include: alterative aggregate sources; maximum sizes and gradings; different types and brands of cement; pozzolans; admixtures; and considerations of concrete durability, volume change, temperature rise, and thermal properties.

A3.5 Test methods

A3.5.1 In conducting laboratory tests to provide information for selecting concrete proportions, the latest revisions of the following methods should be used:

A3.5.1.1 For tests of ingredients:
Sampling hydraulic cement--ASTM C 183
Specific gravity of hydraulic cement--ASTM C 188
Sampling stone, slag, gravel, sand, and stone block for use as highway materials--ASTM D 75
Sieve or screen analysis of fine and coarse aggregates--ASTM C 136
Specific gravity and absorption of coarse aggregates--ASTM C 127
Specific gravity and absorption of fine aggregates--ASTM C 128
Surface moisture in fine aggregate--ASTM C 70
Total moisture content of aggregate by drying--ASTM C 566

Unit weight of aggregate--ASTM C 29
Voids in aggregate for concrete--ASTM C 29
Fineness modulus--Terms relating to concrete and concrete aggregates, ASTM C 125

A3.5.1.2 For tests of concrete:
Sampling fresh concrete--ASTM C 172
Air content of freshly mixed concrete by the volumetric method--ASTM C 173
Air content of freshly mixed concrete by the pressure method--ASTM C 231
Slump of portland cement concrete--ASTM C 143
Weight per cubic foot, yield, and air content (gravimetric) of concrete--ASTM C 138
Concrete compression and flexure test specimens, making and curing in the laboratory--ASTM C 192
Compressive strength of molded concrete cylinders--ASTM C 39

TABLE A3.6.1 — CONCRETE MIXES FOR SMALL JOBS

Procedure: Select the proper nominal maximum size of aggregate (see Section 5.3.2). Use Mix B, adding just enough water to produce a workable consistency. If the concrete appears to be undersanded, change to Mix A and, if it appears oversanded, change to Mix C.

Nominal maximum size of aggregate, in.	Mix designation	Approximate weights of solid ingredients per cu ft of concrete, lb				
		Cement	Sand* Air-entrained concrete†	Sand* Concrete without air	Coarse aggregate Gravel or crushed stone	Coarse aggregate Iron blast furnace slag
½	A	25	48	51	54	47
	B	25	46	49	56	49
	C	25	44	47	58	51
¾	A	23	45	49	62	54
	B	23	43	47	64	56
	C	23	41	45	66	58
1	A	22	41	45	70	61
	B	22	39	43	72	63
	C	22	37	41	74	65
1½	A	20	41	45	75	65
	B	20	39	43	77	67
	C	20	37	41	79	69
2	A	19	40	45	79	69
	B	19	38	43	81	71
	C	19	36	41	83	72

*Weights are for dry sand. If damp sand is used, increase tabulated weight of sand 2 lb and, if very wet sand is used, 4 lb

†Air-entrained concrete should be used in all structures which will be exposed to alternate cycles of freezing and thawing. Air-entrainment can be obtained by the use of an air-entraining cement or by adding an air-entraining admixture. If an admixture is used, the amount recommended by the manufacturer will, in most cases, produce the desired air content

Flexural strength of concrete (using simple beam with third-point loading)--ASTM C 78

Flexural strength of concrete (using simple beam with center point loading--ASTM C 293

Splitting tensile strength of molded concrete cylinders--ASTM C 496

A3.6 *Mixes for small jobs*

A3.6.1 For small jobs where time and personnel are not available to determine proportions in accordance with the recommended procedure, mixes in Table A3.6.1 will usually provide concrete that is amply strong and durable if the amount of water added at the mixer is never large enough to make the concrete overwet. These mixes have been predetermined in conformity with the recommended procedure by assuming conditions applicable to the average small job, and for aggregate of medium specific gravity.

Three mixes are given for each nominal maximum size of coarse aggregate. For the selected size of coarse aggregate, Mix B is intended for initial use. If this mix proves to be oversanded, change to Mix C; if it is undersanded, change to Mix A. It should be noted that the mixes listed in the table are based on dry or surface-dry sand. If the fine aggregate is moist or wet, make the corrections in batch weight prescribed in the footnote.

A3.6.2 The approximate cement content per cubic foot of concrete listed in the table will be helpful in estimating cement requirements for the job. These requirements are based on concrete that has just enough water in it to permit ready working into forms without objectionable segregation. Concrete should slide, not run, off a shovel.

APPENDIX 4 -- HEAVYWEIGHT CONCRETE MIX PROPORTIONING

A4.1 Concrete of normal placeability can be proportioned for densities as high as 350 lb per cu ft by using heavy aggregates such as iron ore, iron or steel shot, barite, and iron or steel punchings. Although each of the materials has its own special characteristics, they can be processed to meet the standard requirements for grading, soundness cleanliness, etc. The selection of the aggregate should depend on its intended use. In the case of radiation shielding, determination should be made of trace elements within the material which may become reactive when subjected to radiation. In the selection of materials and proportioning of heavyweight concrete, the data needed and procedures used are similar to those required for normal weight concrete.

Aggregate density and composition for heavyweight concrete should meet requirements of ASTM C 637 and C 638. The following items should be considered.

A4.1.1 Typical materials used as heavy aggregates are listed in Table A4.1.1.

A4.1.2 If the concrete in service is to be exposed to a hot, dry environment resulting in loss of weight, it should

be proportioned so that the fresh unit weight is higher than the required dry unit weight by the amount of the anticipated loss determined by performing an oven dry unit weight on concrete cylinders as follows. Three cylinders are cast and the wet unit weight determined in accordance with ASTM C 138. After 72 hours of standard curing, the cylinders are oven dried to a constant weight at 211 to 230 F and the average unit weight determined. The amount of water lost is determined by subtracting the oven dry unit weight from the wet unit weight. This difference is added to the required dry unit weight when calculating mixture proportions to allow for this loss. Normally, a freshly mixed unit weight is 8 to 10 lb per cu ft higher than the oven dry unit weight[2].

A4.1.3 If entrained air is required to resist conditions of exposure, allowance must be made for the loss in weight due to the space occupied by the air. To compensate for the loss of entrained air as a result of vibration, the concrete mixture should be proportioned with a higher air content to anticipate this loss.

A4.2 Handling of heavyweight aggregates should be in accordance with ACI 304.3R. (See also ASTM C 637 and C 638.) Proportioning of heavyweight concrete to be placed by conventional means can be accomplished in accordance with ACI 211.1 Sections 5.2 through 5.3.7 and the absolute volume method in Section 5.3.7.2. Typical proportions are shown in Table 2 of ACI 304.3R.

A4.3 *Preplaced heavyweight concrete* -- Heavyweight preplaced-aggregate concrete should be proportioned in the same manner as normal weight preplaced-aggregate concrete. (Refer to ACI 304, Table 7.3.2 -- Gradation limits for fine and coarse aggregate for preplaced aggregate concrete.) Example mixture proportions for the preplaced-aggregate method are shown in ACI 304.3R, Table 2 -- Typical proportions for high density concrete, and typical grout proportions can be found in ACI 304.3R, Table 3 -- Typical grout proportions.

A4.4 *Example* -- Concrete is required for counterweights on a lift bridge that will not be subjected to freezing and

TABLE A4.1.1—TYPICAL HEAVYWEIGHT AGGREGATES

Material	Description	Specific gravity	Concrete, unit wt (lb/cu ft)
Limonite Goethite	Hydrous iron ores	3.4-3.8	180-195
Barite	Barium sulfate	4.0-4.4	205-225
Ilmenite Hematite Magnetite	Iron ores	4.2-5.0	215-240
Steel/iron	Shot, pellets, punchings, etc.	6.5-7.5	310-350

Note: Ferrophosphorous and ferrosilicon (heavyweight slags) materials should be used only after thorough investigation. Hydrogen gas evolution in heavyweight concrete containing these aggregates has been known to result from a reaction with the cement.

thawing conditions. An average 28 day compressive strength of 4500 psi will be required. Placement conditions permit a slump of 2 to 3 in. at point of placement and a nominal maximum size aggregate of 1 in. The design of the counter-weight requires* an oven dry unit weight of 225 lb per cu ft. An investigation of economically available materials has indicated the following:

Cement	ASTM C 150 Type I (non-air-entraining)
Fine aggregate	Specular hematite
Coarse aggregate	Ilmenite

Table A4.1.1 indicates that this combination of materials may result in an oven dry unit weight of 215 to 240 lb per cu ft. The following properties of the aggregates have been obtained from laboratory tests.

	Fine aggregate	Coarse aggregate
Fineness modulus	2.30	--
Specific gravity (Bulk SSD)	4.95	4.61
Absorption (percent)	0.05	0.08
Dry rodded weight	--	165 lb per cu ft
Nominal maximum size	--	1 in.

Employing the sequence outlined in Section 5 of this standard practice, the quantities of ingredients per cubic yard of concrete are calculated as follows:

A4.4.1 *Step 1* -- As indicated, the desired slump is 2 to 3 in. at point of placement.

A4.4.2 *Step 2* -- The available aggregate sources have been indicated as suitable, and the course aggregate will be a well-graded and well-shaped crushed ilmenite with a nominal maximum size of 1 in. The fine aggregate will be hematite.

A4.4.3 *Step 3* -- By interpolation in Table 6.3.3, non-air-entrained concrete with a 2 to 3 in. slump and a 1 in. nominal maximum size aggregate requires a water content of approximately 310 lb per cu yd. The estimated entrapped air is 1.5 percent. (Non-air-entrained concrete will be used because (1) the concrete is not to be exposed to severe weather, and (2) a high air content could reduce the dry unit weight of the concrete.)

Note: Values given in Table 6.3.3 for water requirement are based on the use of well-shaped crushed coarse aggregates. Void content of compacted dry fine or coarse aggregate can be used as an indicator of angularity. Void contents of compacted 1 in. coarse aggregate of significantly more than 40 percent indicate angular material that will probably require more water than that listed in Table 5.3.3. Conversely, rounded aggregates with voids below 35 percent will probably need less water.

---

* Oven dry is specified and is considered a more conservative value than that of the air dry.

A4.4.4 *Step 4* -- From Table 6.3.4(a) the water-cement ratio needed to produce a strength of 4500 psi in non-air-entrained concrete is found to be approximately 0.52.

A4.4.5 *Step 5* -- From the information derived in Steps 3 and 4, the required cement content is calculated to be 310/0.52 = 596 lb per cu yd.

A4.4.6 *Step 6* -- The quantity of coarse aggregate is estimated by extrapolation from Table 6.3.6. For a fine aggregate having a fineness modulus of 2.30 and a 1 in. nominal maximum size aggregate, the table indicates that 0.72 cu ft of coarse aggregate, on a dry-rodded basis, may be used in each cubic foot of concrete. For a cubic yard, therefore, the coarse aggregate will be 27 x 0.72 = 19.44 cu ft, and since the dry-rodded unit weight of coarse aggregate is 165 lb per cu ft, the dry weight of coarse aggregate to be used in a cubic yard of concrete will be 19.44 x 165 = 3208 lb. The angularity of the coarse aggregate is compensated for in the ACI proportioning method through the use of the dry-rodded unit weight; however, the use of an extremely angular fine aggregate may require a higher proportion of fine aggregate, an increased cement content, or the use of air entrainment to produce the required workability. The use of entrained air reduces the unit weight of the concrete, but in some instances is necessary for durability.

A4.4.7 *Step 7* -- For heavyweight concrete, the required fine aggregate should be determined on the absolute volume basis. With the quantities of cement, water, air, and coarse aggregate established, the fine aggregate content can be calculated as follows:

$$\text{Volume of water} = \frac{310 \text{ lb}}{62.4 \text{ lb per cu ft}} = 4.97 \text{ cu ft}$$

$$\text{Volume of air} = 0.015 \times 27 \text{ cu ft} = 0.40 \text{ cu ft}$$

$$\text{Solid volume of cement} = \frac{596 \text{ lb}}{3.15 \times 62.4 \text{ lb per cu ft}} = 3.03 \text{ cu ft}$$

$$\text{Solid volume of coarse aggregate} = \frac{3208 \text{ lb}}{4.61 \times 62.4 \text{ lb per cu ft}} = 11.15 \text{ cu ft}$$

Total volume of all ingredients except fine aggregate = 19.55 cu ft

Solid volume of fine aggregate = 27 cu ft - 19.55 cu ft = 7.45 cu ft

Required weight of fine aggregate = 7.45 cu ft x 4.95 x 62.4 lb per cu ft = 2301 lb

The actual test results indicated the concrete possessed the following properties:

Slump	2½ in.
Strength	5000 psi at 28 days

Unit weight (freshly mixed)	235.7 lb per cu ft
Oven dry unit weight	228.2 lb per cu ft
Air content	2.8 percent

Note: Oven dry unit weight of the concrete having a combination of hematite and ilmenite aggregates was 7.5 lb per cu ft less than the freshly mixed unit weight.

APPENDIX 5 -- MASS CONCRETE MIX PROPORTIONING

A5.1 *Introduction* -- Mass concrete is defined as "any volume of concrete with dimensions large enough to require that measures be taken to cope with generation of heat of hydration from the cement and attendant volume change to minimize cracking."[A5.9] The purpose of the mass concrete proportioning procedure is to combine the available cementitious materials, water, fine and coarse aggregate, and admixtures such that the resulting mixture will not exceed some established allowable temperature rise, and yet meet requirements for strength and durability. In some instances, two mixtures may be required -- an interior mass concrete and an exterior concrete for resistance to the various conditions of exposure. Accordingly, concrete technologists and designers during the design stage should consider the effects of temperature on the properties of concrete. A 6-in. wall, for example, will dissipate the generated heat quite readily, but as the thickness and size of the placement increase, a point is reached, whereby, the rate of heat generated far exceeds the rate of heat dissipated. This phenomenon produces a temperature rise within the concrete and may cause sufficient temperature differential between the interior and exterior of the mass or between peak and ultimate stable temperature to induce tensile stresses. The temperature differential between interior and exterior of the concrete generated by decreases in ambient air temperature conditions may cause cracking at exposed surfaces. Furthermore, as the concrete reaches its peak temperature and subsequent cooling takes place, tensile stresses are induced by the cooling if the change in volume is restrained by the foundation or connections to other parts of the structure.

The tensile stress developed by these conditions can be expressed by the equation $S = REeT$; where R is the restraint factor, E is the modulus of elasticity, e is the thermal coefficient of expansion, and T is the temperature difference between the interior and exterior of the concrete or between the concrete at maximum temperature and at ambient air temperature. Detailed discussions on this subject of mass concrete can be found in References A5.1, A5.2, A5.3, A5.5 and A5.14.

Thermal cracking of bridge piers, foundations, floor slabs, beams, columns, and other massive structures (locks and dams) can or may reduce the service life of a structure by promoting early deterioration or excessive maintenance. Furthermore, it should be recognized that the selection of proper mixture proportions is only one means of controlling temperature rise, and that other aspects of the concrete work should be studied and incorporated into the design and construction requirements. For additional information on heat problems and solutions, consult References A5.2 and A5.14.

A5.2 *Mass concrete properties* -- During the design stage of a proposed project, desired specified compressive strength with adequate safety factors for various portions of the structure are normally first established. The engineer will then expand on the other desired properties required of the concrete.

The proportioning of ingredients such that a mass concrete mixture will have the desired properties requires an evaluation of the materials to be used. If adequate data are not available from recent construction projects using the proposed materials, representative samples of all materials proposed for use in the concrete must be tested to determine their properties and conformance with applicable specifications.

A5.3 *Properties of material related to heat generation* --
A5.3.1 *Cementitious materials* -- Cementitious material for mass concrete work may consist of portland cement or blended hydraulic cements as specified in ASTM C 150 and ASTM C 595, respectively, or a combination of portland cement and pozzolan. Pozzolans are specified in ASTM C 618.

A5.3.1.1 *Portland cement* -- The hydration of portland cement is exothermic; that is, heat is generated during the reaction of cement and water. The quantity of heat produced is a function of the chemical composition of the cement as shown in Fig. A5.3 and the initial temperature.

Type II cement is most commonly used in mass concrete, since it is a moderate heat cement and generally has favorable properties for most types of construction. When used with a pozzolanic admixture, which will be discussed later, the heat generated by a combination of Type II and pozzolan is comparative with that of Type IV. In addition, Type II is more readily available than Type IV. Optional heat of hydration requirements may be specified for Type II cement by limitations on the chemical compounds or actual heat of hydration at 7 days.

Low initial concrete placing temperature, commonly used in mass concrete work, will generally decrease the rate of cement hydration and initial heat generated. Correspondingly, strength development in the first few days may also be reduced.

The fineness of the cement also affects the rate of heat of hydration; however, it has little effect on the initial heat

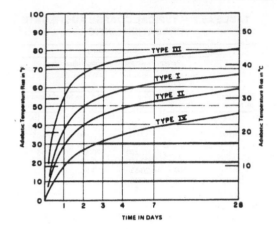

Cement Type	Fineness ASTM C 115 cm²/gm	28-Day Heat of Hydration Calories per gm
I	1790	87
II	1890	76
III	2030	105
IV	1910	60

Fig. A5.3—Temperature rise of mass concrete containing 376 pcy (223 kg/m³) of cement.

generated. Fine-ground cements will produce heat more rapidly during the early ages than a coarse-ground cement, all other cement properties being equal.

A5.3.1.2 *Blended hydraulic cements* -- Blended hydraulic cements conforming to the requirements of ASTM C 595, if available and economical, may be used effectively in mass concrete. These cements are composed of a blend of portland cement and blast-furnace slag or pozzolan. The suffix (MH) or (LH) may be used with the designated type of blended cement to specify moderate heat or low heat requirements where applicable.

A5.3.1.3 *Pozzolans* -- Major economic and temperature rise benefits have been derived from the use of pozzolans. Pozzolan is defined as "a siliceous or siliceous and aluminous material which in itself possesses little or no cementitious value, but will, in finely divided form and in the presence of moisture, chemically react with calcium hydroxide at ordinary temperatures to form compounds possessing cementitious properties."[A5.9] Pozzolans include some diatomaceous earths, opaline cherts and shales, tuffs and volcanic ashes or pumicites, any of which may or may not be processed by calcination, and other various materials requiring calcination to induce satisfactory properties, such as some clays and shales. Fly ash, the finely divided residue that results from the combustion of ground or powdered coal and is transported from the combustion chamber by exhaust gases is also a pozzolan.

Utilization of pozzolans in mass concrete provides a partial replacement of cement with a material which generates considerably less heat at early ages. The early age heat contribution of a pozzolan may conservatively be estimated to range between 15 to 50 percent of that of an equivalent weight of cement.

The effects of pozzolan on the properties of freshly mixed concrete vary with the type and fineness; the chemical, mineralogical and physical characteristics of the pozzolan; the fineness and composition of the cement; the ratio of cement to pozzolan; and the weight of cement plus pozzolan used per unit volume of concrete. For example, it has been reported that some pozzolans may reduce water requirements by as much as 7 percent with a reduction of air-entraining admixture needs by up to about 20 percent. Since certain other pozzolans may require as much as 15 percent additional water and over 60 percent more air-entraining admixture, it is important to evaluate the pozzolan intended for use prior to the start of proportioning.

The proportion of cement to pozzolan depends upon the strength desired at a given age, heat considerations, the chemical and physical characteristics of both cement and pozzolan and the cost of the respective materials. Typical quantities of various types of pozzolan and other materials blended with portland cement to reduce heat generation are shown in Table A5.1.

A5.3.2 *Aggregates* -- The nominal maximum size aggregates recommended for use under various placing conditions are shown in Table A5.2. A nominal maximum size aggregate up to 6 in. (150 mm) should be considered, if large size aggregate is available, economical, and placing conditions permit. Because the larger aggregate provides less surface area to be coated by cement paste, a reduction in the quantity of cement and water can be realized for the same water-cement ratio. This relationship is reflected in Table 6.3.3. Typical gradations for individual size fractions of coarse aggregate are shown in Table A5.3. Gradings and other physical properties of fine aggregate should comply with the requirements of ASTM C 33.

A5.3.2.1 *Coarse aggregate combination* -- Upon determining the nominal maximum size, the individual aggregate size groups available should be combined to produce a gradation approaching maximum density and minimum voids. This results in the maximum amount of mortar available for placeability, workability, and finishability. The dry rodded unit weight method is generally applicable for combining size groups up to a nominal maximum size of 1½ in. (37.5 mm); however, this method is impractical for combining size groups of 3 in. (75 mm) or 6 in. (150 mm) nominal maximum size. Eq. (A5.3) gives an approximate percentage of material passing each sieve size required for a given aggregate type. This equation was developed from work by Fuller and Thompson (Reference A5.13) on the packing characteristics of particulate material. The parabolic curve generated from the equation approximates the ideal gradation for maximum density and minimum voids according to the particle shape of the aggregate. Combining the individual coarse aggregate size groups to approximate the ideal curve is the recommended procedure for use with 6 in. (150 mm) and 3 in. (75 mm) nominal maximum size aggregate mixtures in place of the dry rodded unit weight method.

$$P = \frac{d^x - 0.1875^x}{D^x - 0.1875^x}(100)$$

where

P = cumulative percent passing the d-size sieve
d = sieve opening, in. (mm)
D = nominal maximum size aggregate, in. (mm)
x = exponent (0.5 for rounded and 0.8 for crushed aggregate)

Based on the above equation, the ideal combined gradings for 6 in. and 3 in. (150 and 75 mm) crushed and rounded aggregates are shown in Table A5.4. An acceptable grading for an aggregate that is partially crushed or partially rounded may be interpolated from the gradations in Table A5.4. Using the individual gradation of each size group, 6 in. to 3 in. (150 mm to 75 mm), 3 in. to 1½ in. (75 mm to 37.5 mm)

TABLE A5.1 — TYPICAL QUANTITIES OF POZZOLANS AND OTHER MATERIALS*

Material or class of material	Percent of total cementing material by absolute volume	
	Unexposed concrete†	Exposed concrete‡
Pozzolans (ASTM C 618):		
Class F	35	25
Class N, all types except uncalcined diatomite	30	20
Class N, uncalcined diatomite	20	20
Other materials:		
Slag or natural cement	35	25

*Other quantities of pozzolan or other materials may be used if verified to be acceptable by laboratory mixture evaluations or previous experience. No typical quantities have been established for Class C pozzolan.
†Unexposed concrete for massive structures (i.e., gravity dams, spillways, lock walls, and similar massive structures).
‡Exposed concrete for massive structures (see previous note), and exposed structural concrete (i.e., floodwalls, building foundations, pavements, and similar moderate-size structures).

TABLE A5.2 — NOMINAL MAXIMUM SIZE OF AGGREGATE RECOMMENDED FOR VARIOUS TYPES OF CONSTRUCTION

Features	Nominal maximum size, in. (mm)
Sections over 7½ in. (190 mm) wide, and in which the clear distance between reinforcement bars is at least 2¼ in. (57 mm)	1½(37.5)
Unreinforced sections over 12 in. (300 mm) wide and reinforced sections over 18 in. (457 mm) wide, in which the clear distance between reinforcement bars is over 6 in. (150 mm) and under 10 in. (250 mm)	3(75)
Massive sections in which the clear distance between reinforcement bars is at least 10 in. (250 mm) and for which suitable provision is made for placing concrete containing the larger sizes of aggregate without producing rock pockets or other undesirable conditions.	6(150)

TABLE A5.3 — TYPICAL COARSE AGGREGATE GRADATION LIMITS

Sieve size in. (mm)	Size separation			
	Percent by weight passing individual sieves			
	No. 4 to ¾ in. (4.75 mm to 19 mm)	¾ in. to 1½ in. (19 mm to 37.5 mm)	1½ in. to 3 in. (37.5 mm to 75 mm)	3 in. to 6 in. (75 mm to 150 mm)
7(177)				100
6(150)				90-100
4(100)			100	20-55
3(75)			90-100	0-15
2(50)		100	20-55	0-5
1-1/2(37.5)		90-100	0-10	
1(25)	100	20-55	0-5	
3/4(19)	90-100	0-15		
3/8(9.5)	20-55	0-5		
No. 4(4.75)	0-10			
No. 8(2.36)	0-5			

TABLE A5.4—IDEALIZED COMBINED GRADING FOR 6 IN. (150 mm) AND 3 IN. (75 mm) NOMINAL MAXIMUM SIZE AGGREGATE FROM EQ. (A5.3)

Sieve size — in. (mm)	6 in. (150 mm)		3 in. (75 mm)	
	Percent passing		Percent passing	
	Crushed	Rounded	Crushed	Rounded
6(150)	100	100	—	—
5(125)	85	89	—	—
4(100)	70	78	—	—
3(75)	54	64	100	100
2(50)	38	49	69	75
1-1/2(37.5)	28	39	52	61
1(25)	19	28	34	44
3/4(19)	13	21	25	33
3/8(9.5)	5	9	9	14

1½ in. to ¾ in. (37.5 mm to 19 mm), and ¾ in. to No. 4 (19 mm to 4.75 mm), a trial and error method of selecting the percentage of each size group will be necessary to produce a combined grading of the total coarse aggregate approximating the idealized gradation. Selection of the percentage of each size group can usually be done such that the combined grading is generally within 2 or 3 percent of the ideal grading if the individual size group gradings are within the limits of Table A5.3. Where grading limits other than those of Table A5.3 may be used, more tolerance may be required on certain sieve sizes. Furthermore, natural aggregates in some areas may be deficient of certain sizes and, in such cases, modification of the idealized grading to permit use of this aggregate is recommended.

A5.3.2.2 *Coarse aggregate content* -- The proportion of fine aggregate for mass concrete depends on the final combined grading of coarse aggregate, particle shape, fineness modulus of the fine aggregate, and the quantity of cementitious material. Coarse aggregate amount can be found using the b/b$_o$ method, Table 5.3.6 of ACI 211.1, if the ASTM C 29 bulk unit weight has been determined. For large 3 in. (75 mm) and 6 in. (150 mm) nominal maximum size aggregate Table A5.5 approximates the amount of coarse aggregate as a percent of the total aggregate volume for different moduli of fine aggregate and nominal maximum sizes of coarse aggregate. The table is only applicable for 3 in. (75 mm) and 6 in. (150 mm) nominal maximum size aggregate.

A5.3.3 *Admixtures* -- When proportioning mass concrete use of admixtures should always be considered. The two most commonly used admixtures in mass concrete are air-entraining and water-reducing admixtures.

A5.3.3.1 *Air entrainment* -- Air entrainment in mass concrete is necessary if for no other reason than to increase workability of lean concrete mixtures. The use of air entrainment in mass concrete, as in other concrete, permits a marked improvement in durability, improvement in plasticity and workability, and reduction in segregation and bleeding. The effect of air entrainment on the strength of mass concrete is minimized due to the reduction in the quantity of paste in concrete which contains 3 in. (75 mm) and 6 in. (150 mm) nominal maximum size aggregate. However, such effects should be considered in the design of mass concrete having 1½ in. (37.5 mm) or ¾ in. (19 mm) nominal maximum size aggregate. In lean mixtures strengths are not reduced as much when air entrainment is used; in some

TABLE A5.5 — APPROXIMATE COARSE AGGREGATE CONTENT WHEN USING NATURAL (*N*) OR MANUFACTURE (*M*) FINE AGGREGATE (Percent of total aggregate by absolute volume)

Nominal maximum size and type coarse aggregate	Sand type:	Fineness modulus							
		2.40		2.60		2.80		3.00	
		N	M	N	M	N	M	N	M
6 in. (150 mm) crushed		80	78	79	77	78	76	77	75
6 in. (150 mm) rounded		82	80	81	79	80	78	79	77
3 in. (75 mm) crushed		75	73	74	72	73	71	72	70
3 in. (75 mm) rounded		77	75	76	74	75	73	74	72

Note: For concrete containing 5½ percent air content and a slump of 2 in. (50 mm), both measured on the minus 1½ in. (37.5 mm) portion. The coarse aggregate contents given above may be increased approximately 1 or 2 percent if good control procedures are followed. The coarse aggregate content in the table pertains primarily to the particle shape in the minus 1½ in. (37.5 mm) portion.

TABLE A5.6—APPROXIMATE MORTAR AND AIR CONTENT FOR VARIOUS NOMINAL MAXIMUM SIZE AGGREGATES [1½ in. (37.5 mm) slump and air content of 5 to 6 percent in minus 1½ in. (37.5 mm) portion]

Nominal maximum size and type coarse aggregate	Mortar content cu ft/cu yd ± 0.2 (m 3/m 3 + 0.01)	Air content Total mixture, percent
6 in. (150 mm) crushed	10.5 (0.39)	3.0-4.0
6 in. (150 mm) rounded	10.0 (0.37)	3.0-4.0
3 in. (75 mm) crushed	12.0 (0.44)	3.5-4.5
3 in. (75 mm) rounded	11.5 (0.43)	3.5-4.5

TABLE A5.7 — APPROXIMATE COMPRESSIVE STRENGTHS OF AIR-ENTRAINED CONCRETE FOR VARIOUS WATER-CEMENT RATIOS [Based on the use of 6 × 12-in. (152 × 305-mm) cylinders.]

Water-cement ratio by weight*	Approximate 28-day compressive strength, psi (MPa) (f_c)†	
	Natural aggregate	Crushed aggregate
0.40	4500 (31.0)	5000 (34.5)
0.50	3400 (23.4)	3800 (26.2)
0.60	2700 (18.6)	3100 (21.4)
0.70	2100 (14.5)	2500 (17.2)
0.80	1600 (11.0)	1900 (13.1)

*These W/C ratios may be converted to W/(C + P) ratios by the use of the equation in Section 5.3.4
†90 days when using pozzolan

TABLE A5.8 — MAXIMUM PERMISSIBLE WATER-CEMENT RATIOS FOR MASSIVE SECTIONS

Location of structure	Water-cement ratios, by weight	
	Severe or moderate climate	Mild climate, little snow or frost
At the waterline in hydraulic or waterfront structures where intermittent saturation is possible	0.50	0.55
Unexposed portions of massive structure	No limit*	No limit
Ordinary exposed structures	0.50	0.55
Complete continuous submergence in water	0.58	0.58
Concrete deposited in water	0.45	0.45
Exposure to strong sulfate groundwater or other corrosive liquid, salt or sea water	0.45	0.45
Concrete subjected to high velocity flow of water (>40 f/s) (>12 m/s)	0.45	0.45

Note 1. These W/C ratios may be converted to W/(C + P) ratios by use of equation in Section 5.3.4.
*Limit should be based on the minimum required for workability or Table A5.7 for strength

cases strengths may increase due to the reduction in mixing water requirements with air entrainment. Air contents should be in accordance with those recommended in Table A5.6.

A5.3.3.2 *Water-reducing admixture* -- Water-reducing admixtures meeting the requirements of ASTM C 494 have been found effective in mass concrete mixtures. The water reduction permits a corresponding reduction in the cement content while maintaining a constant water-cement ratio. The amount of water reduction will vary with different concretes; however, 5 to 8 percent is normal. In addition, certain types of water-reducing admixture tend to improve the mobility of concrete and its response to vibration, particularly in large aggregate mixtures.

A5.4 *Strength and durability* -- The procedure for proportioning mass concrete is used primarily for controlling the generation of heat and temperature rise, while satisfying the requirements for strength and durability. The strength and durability properties are primarily governed by the water-cement ratio. The water-cement ratio is the ratio, by

TABLE A5.9 — QUANTITIES OF MATERIALS SUGGESTED
FOR CONCRETE PROPORTIONING TRIAL MIXTURES

| Nominal maximum size aggregate in mixture in. (mm) | Quantities of aggregates, lb (kg) | | | | | Cement, lb. (kg) |
| | Fine aggregate | Coarse aggregates | | | | |
		No. 4 to ¾ in. (4.75 mm to 19 mm)	¾ in. to 1½ in. (19 mm to 37.5 mm)	1½ in. to 3 in. (37.5 to 75 mm)	3 in. to 6 in. (75 mm to 150 mm)	
¾ (19)	1200 (544)	1200 (544)	—	—	—	400 (181)
1½ (37.5)	1000 (454)	1000 (454)	1000 (454)	—	—	400 (181)
3 (75)	2000 (907)	1500 (680)	1000 (454)	2000 (907)	—	500 (227)
6 (150)	3000 (1361)	2000 (907)	1500 (680)	2500 (1134)	3000 (1361)	700 (318)

Note 1. The actual quantity of materials required depends upon the laboratory equipment, availability of materials, and extent of the testing program.
Note 2. If a pozzolan or fly ash is to be used in the concrete, the quantity furnished should be 35 percent of the weight of the cement.
Note 3. One gal. (3.8) of a proposed air-entraining admixture or chemical admixture will be sufficient.

weight, of amount of water, exclusive of that absorbed by theaggregates, to the amount of cement in a concrete or mortar mixture. Unless previous water-cement ratio-compressive strength data are available, the approximate compressive strength of concrete tested in 6 x 12-in. (152 x 305-mm) cylinders for various water-cement ratios can be estimated from Table A5.7. The recommended maximum permissiblewater-cement ratio for concrete subject to various conditions of exposure are shown in Table A5.8. The water-cement ratio determined by calculation should be verified by trial batches to ensure that the specified properties of the concrete are met. Results may show that strength or durability rather than heat generation govern the propor-tions. When this situation occurs alterative measures to control heat will be necessary. For example, in gravity dam construction an exterior-facing mix may be used which con-tains additional cement to provide the required durability. Other measures may include a reduction in the initial temp-erature of concrete at placement or a limitation on the size of the placement. If compressive strengths are given for full mass mixture containing aggregate larger than 1½ in. (75 mm), approximate relationships between strength of the full mass mixture and wet screened 6 x 12-in. (152 x 305-mm) cylinder are available from sources such as Reference A5.6.

A5.5 *Placement and workability* -- Experience has demonstrated that large aggregate mixtures, 3 in. (75 mm) and 6 in. (150 mm) nominal maximum size aggregate, require a minimum mortar content for suitable placing and workability properties. Table A5.6 reflects the total absolute volume of mortar (cement, pozzolan, water, air, and fine aggregate) which is suggested for use in mixtures containing large aggregate sizes. These values should be compared with those determined during the proportioning procedure and appropriate adjustments made by either increasing or decreasing the trial mixture mortar contents for improved workability.

A5.6 *Procedure* -- Upon determining the properties of the materials and knowing the properties of the concrete, the proportioning procedure follows a series of straight-forward steps outlined in A5.6.1 to A5.6.12. Proportions should be determined for the anticipated maximum placing temperature due to the influence on the rate of cement hydration and heat generated. With the use of 3 in. (75 mm)

or 6 in. (150 mm) nominal maximum size aggregate, the pro-cedure may be somewhat different from ACI 211. 1, mainly because of the difficulty in determining the density of the large aggregate by the dry rodded unit weight method. For nominal maximum size aggregate 1½ in. (37.5 mm) or less, proportioning in accordance with ACI 211.1 may be used.

A5.6.1 *Step 1* -- Determine all requirements relating to the properties of the concrete including:

1. Nominal maximum size of aggregates that can be used.
2. Slump range.
3. Water-cement ratio limitations.
4. Expected maximum placing temperature.
5. Air content range.
6. Specified strengths and test ages.
7. Expected exposure conditions.
8. Expected water velocities, when concrete is to be subjected to flowing water.
9. Aggregate quality requirements.
10. Cement and/or pozzolan properties.

A5.6.2 *Step 2* -- Determine the essential properties of materials if sufficient information is not available. Representative samples of all materials to be incorporated in the concrete should be obtained in sufficient quantities to provide verification tests by trial batching. The suggested quantities of materials necessary to complete the required tests are shown in Table A5.9. If pozzolan is economically available, or required by the specification, the percentage as suggested in Table A5.1 should be used as a starting point in the trial mixes.

From the material submitted for the test program, determine the following properties:
1. Sieve analysis of all aggregates.
2. Bulk specific gravity of aggregates.
3. Absorption of aggregates.
4. Particle shape of coarse aggregates.
5. Fineness modulus of fine aggregate.
6. Specific gravity of portland cement, and/or pozzolans and blended cement.
7. Physical and chemical properties of portland cement and/or pozzolans and blended cement including heat of hydration at 7 days.

A complete record of the above properties should be made available for field use; this information will assist in adjusting the mixture should any of the properties of the materials used in the field change from the properties of the materials used in the laboratory trial mix program.

A5.6.3 *Step 3* -- Selection of W/C ratio. If the water-cement ratio is not given in the project document, select from Table A5.8 the maximum permissible water-cement (W/C) ratio for the particular exposure conditions. Compare this W/C ratio with the maximum permissible W/C ratio required in Table A5.7 to obtain the average strength which includes the specified strength plus an allowance for anticipated variation and use the lowest W/C ratio. The W/C ratio should be reduced 0.02 to assure that the maximum permissible W/C ratio is not exceeded during field adjustments. This W/C ratio, if required, can be converted to a water-cement plus pozzolan ratio by the use of Eq. (6.3.4.1).

A5.6.4 *Step 4* -- Estimate of mixing water requirement. Estimate the water requirement from Table 6.3.3 for the specified slump and nominal maximum size aggregate. Initial placing temperature may affect this water requirement; for additional information consult Reference A5.6.

A5.6.5 *Step 5* -- Selection of air content. Select a total air content of the mixture as recommended in Table A5.6. An accurate measure of air content can be made during future adjustment of the mixture by use of Eq. (A5.6).

$$A = \frac{a}{1 + r\left(1 - \frac{a}{100}\right)} \quad (A5.6)$$

where

A = air content of total mixture, expressed as a percent
a = air content of minus 1½ in. (37.5 mm) fraction of mixture, expressed as a percent
r = ratio of the absolute volume of plus 1½ in. (37.5 mm) aggregate to the absolute volume of all other materials in the mixture except air. If 100 percent of the aggregate passes the 1½ in. (37.5 mm) sieve, r = 0, and $A = a$

A5.6.6 *Step 6* -- Compute the required weight of cement from the selected W/C (A5.6.3) and water requirement (A5.6.4).

A5.6.7 *Step 7* -- Determine the absolute volume for the cementitious materials, water content, and air content from information obtained in Steps 4, 5, and 6. Compute individual absolute volumes of cement and pozzolan.

$$V_{c+p} = \frac{C_w}{G_c(62.4)} \text{ cu ft or } \frac{C_w}{G_c(1000)} \text{ m}^3 \quad (A5.6A)$$

$$V_c = V_{c+p}(1 - F_v) \quad (A5.6B)$$

$$V_p = V_{c+p}(F_v) \quad (A5.6C)$$

where

C_w = weight of the equivalent portland cement as determined from Step 6

G_c = specific gravity of portland cement

V_c = volume of cement (cu ft) (m³)

V_p = volume of pozzolan (cu ft) (m³)

V_{c+p} = volume of cement and pozzolan (cu ft) (m³)

F_v = percent pozzolan by absolute volume of the total absolute volume of cement plus pozzolan expressed as a decimal factor

A5.6.8 *Step 8* -- Select percent of coarse aggregate. From Table A5.5, and based on the fineness modulus of the fine aggregate as well as the nominal maximum size and type of coarse aggregate, determine the coarse aggregate percentage of the total volume of aggregate.

A5.6.9 *Step 9* -- Determine the absolute volume of the total aggregate by subtracting from the unit volume the absolute volumes of each material as computed in Step 7. Based on the amount of coarse aggregate selected in Step 8, determine the absolute volume of the coarse aggregate. The remainder of the absolute volume represents the quantity of fine aggregate in the mix.

A5.6.10 *Step 10* -- Establish the desired combination of the separate coarse aggregates size groups. Using the individual coarse aggregates gradings, combine all coarse aggregate to a uniform grading approximating the gradings shown in Table A5.4 for 6 in. (150 mm) nominal maximum size aggregate (NMSA) or 3 in. (75 mm) NMSA. The percentage of each size group should be rounded to the nearest whole percent.

A5.6.11 *Step 11* -- Convert all absolute volumes to weight per unit volume of all ingredients in the mixture.

A5.6.12 *Step 12* -- Check the mortar content. From the absolute volumes computed earlier, compute the mortar content and compare the results with values given in Table A5.6. Values in Table A5.6 will provide an indication of the workability of the mixture as determined by past field performance. Table A5.6 can be used as an aid in making laboratory adjustments of the mixture.

A5.7 *Example problem* -- Concrete is required for a heavy bridge pier that will be exposed to fresh water in a severe climate. The design compressive strength is 3000 psi (20.7 MPa) at 28 days. Placement conditions permit the use of a large nominal maximum size aggregate, and 6 in. (150 mm) nominal maximum size crushed stone is available. Laboratory tests indicate that 6, 3, 1½, and ¾ in. (150, 75, 37.5, and 19 mm) size groups of crushed stone have bulk specific gravities (saturated-surface-dry, S.S.D. basis) of 2.72, 2.70, 2.70, and 2.68, respectively; the natural fine aggregate available has a bulk specific gravity of 2.64 with a fineness modulus of 2.80. A Class F (fly ash) pozzolan is available and should be used to reduce the generation of heat in the concrete. The pozzolan has a specific gravity of 2.45, and Type II portland cement is available.

A5.7.1 *Step 1* -- Determine desired properties. The following properties have been specified upon review of the project documents and consultation with the engineer:

1. A 6 in. (150 mm) nominal maximum size crushed stone aggregate is available and economically feasible to use.
2. Slump range of the concrete will be 1 to 2 in. (25 to 50 mm) as measured in the minus 1½ in. (37.5 mm) portion.
3. Maximum permissible *W/C* ratio by weight required to be 0.50 for durability purposes.
4. Project documents require the concrete to be placed at 65 F (18 C) or below.
5. The concrete is required to be air entrained within a range of 1½ percent of 5 percent when tested on the minus 1½ in. (37.5 mm) material.
6. Assuming a standard deviation of 500 psi (3.45 MPa), considered good overall general construction control, and 80 percent of the tests above design strength, an average compressive strength of no less than 3400 psi (23.4 MPa) at 28 days (90 days with pozzolan) is required in accordance with ACI 214-77.
7. The concrete will be subjected to severe exposure conditions.
8. Water velocities around the concrete will not exceed 40 ft/sec (12 m/s).
9. Aggregates meeting the requirements of the project specifications are available.
10. The project specifications require the use of portland cement Type II and permit the use of pozzolan.

A5.7.2 *Step 2* -- Determine properties of the materials.

1. The coarse aggregates have the following sieve analyses:

	Percent by weight passing individual sieves			
	No. 4 to ¾ in.	¾ in. to 1½ in.	1½ in. to 3 in.	3in. to 6 in.
Sieve size in. (mm)	(4.75 mm to 19 mm)	(19 mm to 37.5 mm)	(37.5 mm to 75 mm)	(75 mm to 150 mm)
7 (175)				100
6 (150)				98
5 (125)				60
4 (100)			100	30
3 (75)			92	10
2 (50)		100	30	2
1½ (37.5)		94	6	
1 (25)	100	36	4	
¾ (19)	92	4		
⅜ (9.5)	30	2		
No. 4 (4.75)	2			

2. The bulk specific gravities (saturated-surface-dry, S.S.D. basis) of the coarse and fine (sand) aggregates are determined to be:

Size group	Specific gravity
6 in. to 3 in. (150 to 75 mm)	2.72
3 in. to 1½ in. (75 to 37.5 mm)	2.70
1½ in. to ¾ in. (37.5 to 19 mm)	2.70
¾ in. to No. 4 (19 to 4.75 mm)	2.68
Fine aggregate	2.64

3. The absorptions of the coarse and fine aggregates are as follows:

Size group	Absorption (percent)
6 in. to 3 in. (150 to 75 mm)	0.5
3 in. to 1½ in. (75 to 37.5 mm)	0.75
1½ in. to ¾ in. (37.5 to 19 mm)	1.0
¾ in. to No. 4 (19 to 4.75 mm)	2.0
Fine aggregate	3.2

4. The coarse and fine aggregate are totally crushed and natural, respectively.
5. The fineness modulus of the fine aggregate is 2.80.
6. Specific gravities of the portland cement and pozzolan are 3.15 and 2.45, respectively.
7. Physical and chemical tests of the portland cement and pozzolan verify compliance with the requirements of the project specifications.

A5.7.3 *Step 3* -- Selection of *W/C* ratio. From Table A5.8, the exposure conditions permit a maximum permissible *W/C* ratio of 0.50 and Table A5.7 recommends a maximum *W/C* ratio of 0.57 to obtain the desired average strength of 3400 psi (23.44 MPa). Since the exposure conditions require the lower *W/C* ratio, the designed *W/C* ratio will be 0.48 or 0.02 less than that permitted to allow for field adjustments.

Since a fly ash pozzolan is available and the quantity of concrete in the project justifies its use economically, 25 percent by volume will be used according to Table A5.1.

A5.7.4 *Step 4* -- Estimate of mixing water requirement. From Table 6.3.3 the estimated water content is 180 lb/cu yd (107 kg/m³) based on the use of a 6 in. (150 mm) crushed stone (NMSA) and a slump of 1 to 2 in. (25 to 50 mm).

A5.7.5 *Step 5* -- Selection of air content. A total air content of 3.2 percent is selected which is within the range recommended in Table A5.6. During later adjustments, after all ingredients are determined, a more accurate total air content can be derived by the use of Eq. (A5.6).

A5.7.6 *Step 6* -- Determine weight of cement from selected *W/C* ratio and water demand.

from Step 3 *W/C* = 0.48

therefore: weight of cement in a total portland cement mixture equals

PROPORTIONS FOR NORMAL, HEAVYWEIGHT, AND MASS CONCRETE   211.1-37

Combined grading computations

Sieve size in. (mm)	Grading of individual size groups percent passing				Trial and error selection Size group percentages and gradings				Combined grading percent passing	Idealize* grading percent passing
					45 percent	25 percent	15 percent	15 percent		
	6 in. to 3 in. (150 mm to 75 mm)	3 in. to 1½ in. (75 mm to 37.5 mm)	1½ in. to ¾ in. (37.5 mm to 19 mm)	¾ in. to No. 4 (19 mm to 4.75 mm)	6 in. to 3 in. (150 mm to 75 mm)	3 in. to 1½ in. (75 mm to 37.5 mm)	1½ in. to ¾ in. (37.5 mm to 19 mm)	¾ in. to No. 4 (19 mm to 4.75 mm)		
7 (175)	100				45	25	15	15	100	
6 (150)	98				44	25	15	15	99	100
4 (100)	30	100			14	25	15	15	69	70
3 (75)	10	92			4	23	15	15	57	54
2 (50)	2	30	100		1	8	15	15	39	38
1½ (37.5)		6	94			2	14	15	31	28
1 (25)		4	36	100		1	5	15	21	21
¾ (19)			4	92			1	14	15	15
⅜ (9.5)			2	30			0	5	5	5
No.4(4.75)				2				0	0	0

*From Table A5.4 for 6 in. (150 mm) nominal maximum size crushed material.

$$\frac{180}{0.48} = 375 \; lb/cu \; yd \; or \; (222 \; kg/m^3)$$

$$V_w = \frac{180}{62.4} = 2.88 \; cu \; ft \; or \left(\frac{107}{1000} = 0.107 \, m^3/m^3\right)$$

A5.7.7 *Step 7* -- Determine absolute volume per cubic yard (cubic meter) for the cementitious materials, water content and air content. As recommended in Table A5.1, 25 percent pozzolan by volume will be used. Using Eq. (A5.6.7A), (B), and (C), the absolute volume of cementitious material can be determined.

$$V_{c+p} = \frac{C_w}{G_c(62.4)} = \frac{375}{3.15(62.4)} = 1.91 \; cu \; ft/cu \; yd \; or$$
$$\left(\frac{222}{3.15(1000)} = 0.070 \, m^3/m^3\right)$$

$$V_c = V_{c+p}(1 - F_v) = 1.91(1 - 0.25) = 1.43 \; cu \; ft/cu \; yd$$
$$or \left(0.070(1 - 0.25) = 0.052 \, m^3/m^3\right)$$

$$V_p = V_{c+p}(F_v) = 1.91(0.25) = 0.48 \; cu \; ft/cu \; yd \; or$$
$$\left(0.070(0.25) = 0.018 \, m^3/m^3\right)$$

$$V_A = 0.032\,(27) = 0.86 \; cu \; ft/cu \; yd \; or$$
$$\left(0.032\,(1.0) = 0.032 \, m^3/m^3\right)$$

A5.7.8 *Step 8* -- For a natural fine aggregate with an F.M. of 2.80 and a 6-in. (152 mm) (NMSA) crushed stone, the volume of coarse aggregate to be used in the trial batch is 78 percent--see Table A5.5.

A5.7.9 *Step 9* -- Determine the absolute volume of fine and coarse aggregates.

$27 - V_w - V_A - V_{c+p}$ = Vol of aggregate/cu yd or
$(1.0 - V_w - V_A - V_{c+p}$ = Vol of aggregate/m³)
$27 - 2.88 - 0.86 - 1.91 = 21.35$ cu ft/cu yd or (0.79m³/m³)

Vol of coarse aggregate	= 21.35 (0.78) cu ft/cu yd
	or [0.79(0.78) m³/m³]
	= 16.65 cu ft/cu yd
	or (0.62 m³/m³)
Vol of fine aggregate	= 21.35(0.22) cu ft/cu yd
	or [0.79(0.22) m³/m³]
	= 4.70 cu ft/cu yd
	or (0.17m³/m³)

A5.7.10 *Step 10* -- Combine the various size groups of coarse aggregate. The existing coarse aggregate gradings were combined by trial-and-error computations, resulting in the following percentages of each size group:

No. 4 to ¾ in.	(4.75 to 19 mm)	15 percent
¾ in. to 1½ in.	(19 to 75 mm)	15 percent
1½ in. to 3 in.	(75 to 150 mm)	25 percent
3 in. to 6 in.	(150 to 300 mm)	45 percent

A5.7.11 *Step 11* -- Convert all absolute volumes to weight per unit volume.

Material	Absolute volume × specific gravity × 62.4	lb/cu yd (kg/m³)
Portland cement	1.43(3.15)62.4	281(167)
Pozzolan	0.48(2.45)62.4	73 (43)
Water	2.88(1.00)62.4	180(107)
Air	0.86	—
Fine aggregate	4.70(2.64)62.4	774(459)S.S.D.*
Coarse aggregate No. 4-¾ in.		
(4.75 –19 mm)	16.65(0.15)(2.68)62.4	418(248)S.S.D.*
¾–1½ in.		
(19mm–75 mm)	16.65(0.15)(2.70)62.4	421(250)S.S.D.*
1½–3 in.		
(75mm–150 mm)	16.65(0.25)(2.70)62.4	701(416)S.S.D.*
3–6 in.		
(150–300 mm)	16.65(0.45)(2.72)62.4	1272(755)S.S.D.*

*Weights based on aggregates in a saturated-surface-dry condition.

A5.7.12 *Step 12* -- Check mortar content and compare with Table A5.6.

$$\text{Mortar content} = V_c + V_p + V_w + V_s + V_A$$
$$= 1.43 + 0.48 + 2.88 + 4.70 + 0.86$$
$$= 10.35 \text{ cu ft/cu yd } (0.383 \text{ m}^3/\text{m}^3)$$

From Table A5.6 the mortar content is estimated to be 10.5 cu ft/cu yd (0.39 m³/m³) which is within the ± 0.2 cu ft (± 0.01 m³) of the actual value.

A5.7.13 *Trial batch* -- From the above information the absolute volume and weight per cubic yard of each ingredient computes as follows:

Material	Absolute volume ft³/yd³(m³/m³)	Weight lb/yd³(kg/m³)
Portland cement	1.43(0.052)	281 (167)
Pozzolan	0.48(0.018)	73 (43)
Water	2.88(0.107)	180 (107)
Air	0.86(0.032)	—
Fine aggregate	4.70(0.174)	774 (459)S.S.D.*
No. 4-¾ in.		
(4.75 –19 mm)	2.50(0.093)	418 (248)S.S.D.*
¾–1½ in.		
(19mm–75 mm)	2.50(0.093)	421 (250)S.S.D.*
1½–3 in.		
(75mm–150 mm)	4.16(0.154)	701 (416)S.S.D.*
3–6 in.		
(150–300 mm)	7.49(0.277)	1272 (755)S.S.D.*
Total	27.00(1.000)	4120(2444)

*Weights are based on aggregate in a saturated-surface-dry condition.

The weights above should be reduced proportionately to facilitate the preparation of trial batches which in turn should be evaluated for proper moisture correction, slump, air content, and general workability. After the necessary adjustments, trial mixtures for strength verification and other desired properties of concrete should be made. Reference 2 will provide guidance in estimating the heat generated by the trial mixture and in determining whether or not other temperature control measures are needed.

A5.8 *References*

A5.1 Townsend, Charles L., "Control of Temperature Cracking in Mass Concrete," *Causes, Mechanism, and Control of Cracking in Concrete*, SP-20, American Concrete Institute, Detroit, 1968, pp. 119-139.

A5.2 ACI Committee 207, "Effect of Restraint, Volume Change, and Reinforcement on Cracking of Massive Concrete," ACI Journal, *Proceedings* V. 70, No. 7, July 1973, pp. 445-470. Also, *ACI Manual of Concrete Practice*, Part 1.

A5.3 Townsend, C. L., "Control of Cracking in Mass Concrete Structures," *Engineering Monograph* No. 34, U.S. Bureau of Reclamation, Denver, 1965.

A5.4 ACI Committee 207, "Mass Concrete for Dams and Other Massive Structures," ACI Journal, *Proceedings* V. 67, No. 4, Apr. 1970, pp. 273-309. Also, ACI Manual of Concrete Practice, Part 1.

A5.5 ACI Committee 224, "Control of Cracking in Concrete Structures," ACI Journal, *Proceedings* V. 69, No. 12, Dec. 1972, pp. 717-753.

A5.6 *Concrete Manual*, 8th Edition, U.S. Bureau of Reclamation, Denver, 1975, 627 pp.

A5.7 *Proportioning Concrete Mixes*, SP-46, American Concrete Institute, Detroit, 1974, 223 pp.

A5.8 Tynes, W. O., "Effect of Fineness of Continuously Graded Coarse Aggregate on Properties of Concrete," *Technical Report* No. 6-819, U.S. Army Engineer Waterways Experiment Station, Vicksburg, Apr. 1968, 28 pp.

A5.9 ACI Committee 116, *Cement and Concrete Terminology*, 2nd Edition, SP-19(78), American Concrete Institute, Detroit, 1978, 50 pp.

A5.10 *Handbook for Concrete and Cement*, CRD-C 3, U.S. Army Engineer Waterways Experiment Station, Vicksburg, 1949 (with quarterly supplements).

A5.11 "Standard Practice for Concrete," EM 1110-2-2000 Office, Chief of Engineers, U.S. Army Corps of Engineers, Washington, D.C., June 1974.

A5.12 Hansen, Kenneth, "Cost of Mass Concrete in Dams," *Publication* No. MS260W, Portland Cement Association, Skokie, 1973, 4 pp.

A5.13 Powers, Treval C., *The Properties of Fresh Concrete*, John Wiley and Sons, New York, 1968, pp. 246-256.

A5.14 ACI Committee 207, "Cooling and Insulating Systems for Mass Concrete," (ACI 207.4R-80) *Concrete International--Design and Construction*, V.2, No. 5, May 1980, pp. 45-64.

GLULAM GARAGE DOOR HEADERS

DESIGN EXAMPLE

The 16-foot 3-inch beam span in Figure 3 supports roof trusses on a 28-foot-wide house with 2-foot overhangs and a 25 psf design snow load. According to the data in Table 1 (Pages 12-13), four different sizes of glulam beams could be selected: 3-1/8 by 13-1/2 inches, 3-1/2 by 12 inches, 5-1/8 by 10-1/2 inches and 5-1/2 by 10-1/2 inches, all suitable to carry the required design loads with the final choice being based on local availability, cost competitiveness and design preference.

FIGURE 3.

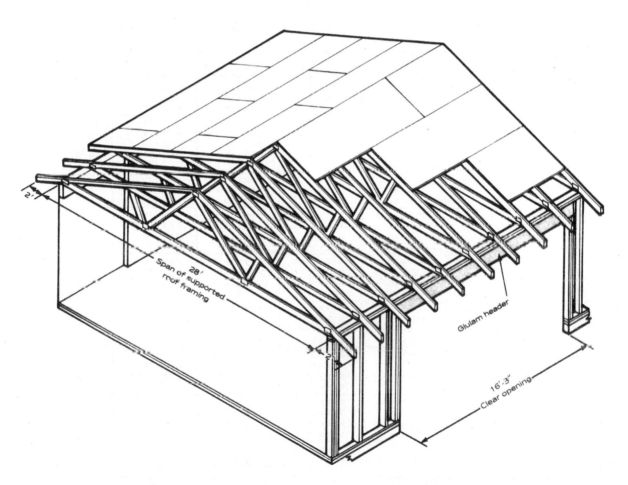

Courtesy of APA—The Engineered Wood Association.

TABLE 1A: 24F *WESTERN SPECIES* GLULAM GARAGE DOOR HEADERS

Load	Clear Door Opening (ft.)	Span of supported roof trusses (ft.) — Glulam Size (in.)				
		24	28	32	36	40
Non-Snow Load (125%) 10 psf Dead 20 psf Live	9'-3"	3-1/8 x 7-1/2 3-1/2 x 7-1/2 5-1/8 x 6 5-1/2 x 6	3-1/8 x 7-1/2 3-1/2 x 7-1/2 5-1/8 x 6 5-1/2 x 6	3-1/8 x 7-1/2 3-1/2 x 7-1/2 5-1/8 x 6 5-1/2 x 6	3-1/8 x 7-1/2 3-1/2 x 7-1/2 5-1/8 x 6 5-1/2 x 6	3-1/8 x 7-1/2 3-1/2 x 7-1/2 5-1/8 x 7-1/2 5-1/2 x 7-1/2
	16'-3"	3-1/8 x 12 3-1/2 x 12 5-1/8 x 10-1/2 5-1/2 x 10-1/2	3-1/8 x 12 3-1/2 x 12 5-1/8 x 10-1/2 5-1/2 x 10-1/2	3-1/8 x 12 3-1/2 x 12 5-1/8 x 10-1/2 5-1/2 x 10-1/2	3-1/8 x 13-1/2 3-1/2 x 12 5-1/8 x 10-1/2 5-1/2 x 10-1/2	3-1/8 x 13-1/2 3-1/2 x 13-1/2 5-1/8 x 12 5-1/2 x 12
	18'-3"	3-1/8 x 13-1/2 3-1/2 x 12 5-1/8 x 10-1/2 5-1/2 x 10-1/2	3-1/8 x 13-1/2 3-1/2 x 13-1/2 5-1/8 x 12 5-1/2 x 12	3-1/8 x 13-1/2 3-1/2 x 13-1/2 5-1/8 x 12 5-1/2 x 12	3-1/8 x 15 3-1/2 x 13-1/2 5-1/8 x 12 5-1/2 x 12	3-1/8 x 15 3-1/2 x 15 5-1/8 x 13-1/2 5-1/2 x 12
Snow Load (115%) 10 psf Dead 25 psf Live	9'-3"	3-1/8 x 7-1/2 3-1/2 x 7-1/2 5-1/8 x 6 5-1/2 x 6	3-1/8 x 7-1/2 3-1/2 x 7-1/2 5-1/8 x 6 5-1/2 x 6	3-1/8 x 9 3-1/2 x 7-1/2 5-1/8 x 7-1/2 5-1/2 x 6	3-1/8 x 9 3-1/2 x 9 5-1/8 x 7-1/2 5-1/2 x 7-1/2	3-1/8 x 9 3-1/2 x 9 5-1/8 x 7-1/2 5-1/2 x 7-1/2
	16'-3"	3-1/8 x 12 3-1/2 x 12 5-1/8 x 10-1/2 5-1/2 x 10-1/2	3-1/8 x 13-1/2 3-1/2 x 12 5-1/8 x 10-1/2 5-1/2 x 10-1/2	3-1/8 x 13-1/2 3-1/2 x 13-1/2 5-1/8 x 12 5-1/2 x 10-1/2	3-1/8 x 15 3-1/2 x 13-1/2 5-1/8 x 12 5-1/2 x 12	3-1/8 x 15 3-1/2 x 15 5-1/8 x 12 5-1/2 x 12
	18'-3"	3-1/8 x 13-1/2 3-1/2 x 13-1/2 5-1/8 x 12 5-1/2 x 12	3-1/8 x 15 3-1/2 x 13-1/2 5-1/8 x 12 5-1/2 x 12	3-1/8 x 15 3-1/2 x 15 5-1/8 x 12 5-1/2 x 12	3-1/8 x 16-1/2 3-1/2 x 15 5-1/8 x 13-1/2 5-1/2 x 13-1/2	3-1/8 x 16-1/2 3-1/2 x 16-1/2 5-1/8 x 13-1/2 5-1/2 x 13-1/2
Snow Load (115%) 10 psf Dead 30 psf Live	9'-3"	3-1/8 x 7-1/2 3-1/2 x 7-1/2 5-1/8 x 6 5-1/2 x 6	3-1/8 x 9 3-1/2 x 7-1/2 5-1/8 x 7-1/2 5-1/2 x 6	3-1/8 x 9 3-1/2 x 9 5-1/8 x 7-1/2 5-1/2 x 7-1/2	3-1/8 x 9 3-1/2 x 9 5-1/8 x 7-1/2 5-1/2 x 7-1/2	3-1/8 x 9 3-1/2 x 9 5-1/8 x 7-1/2 5-1/2 x 7-1/2
	16'-3"	3-1/8 x 13-1/2 3-1/2 x 12 5-1/8 x 10-1/2 5-1/2 x 10-1/2	3-1/8 x 13-1/2 3-1/2 x 13-1/2 5-1/8 x 12 5-1/2 x 10-1/2	3-1/8 x 15 3-1/2 x 13-1/2 5-1/8 x 12 5-1/2 x 12	3-1/8 x 15 3-1/2 x 15 5-1/8 x 12 5-1/2 x 12	3-1/8 x 16-1/2 3-1/2 x 15 5-1/8 x 13-1/2 5-1/2 x 12
	18'-3"	3-1/8 x 15 3-1/2 x 13-1/2 5-1/8 x 12 5-1/2 x 12	3-1/8 x 16-1/2 3-1/2 x 15 5-1/8 x 13-1/2 5-1/2 x 12	3-1/8 x 16-1/2 3-1/2 x 16-1/2 5-1/8 x 13-1/2 5-1/2 x 13-1/2	3-1/8 x 18 3-1/2 x 16-1/2 5-1/8 x 13-1/2 5-1/2 x 13-1/2	3-1/8 x 18 3-1/2 x 18 5-1/8 x 15 5-1/2 x 13-1/2
Snow Load (115%) 10 psf Dead 40 psf Live	9'-3"	3-1/8 x 9 3-1/2 x 9 5-1/8 x 7-1/2 5-1/2 x 7-1/2	3-1/8 x 9 3-1/2 x 9 5-1/8 x 7-1/2 5-1/2 x 7-1/2	3-1/8 x 9 3-1/2 x 9 5-1/8 x 7-1/2 5-1/2 x 7-1/2	3-1/8 x 10-1/2 3-1/2 x 9 5-1/8 x 7-1/2 5-1/2 x 7-1/2	3-1/8 x 12 3-1/2 x 10-1/2 5-1/8 x 9 5-1/2 x 9
	16'-3"	3-1/8 x 15 3-1/2 x 13-1/2 5-1/8 x 12 5-1/2 x 12	3-1/8 x 15 3-1/2 x 15 5-1/8 x 12 5-1/2 x 12	3-1/8 x 16-1/2 3-1/2 x 15 5-1/8 x 13-1/2 5-1/2 x 12	3-1/8 x 18 3-1/2 x 16-1/2 5-1/8 x 13-1/2 5-1/2 x 13-1/2	3-1/8 x 19-1/2 3-1/2 x 18 5-1/8 x 15 5-1/2 x 13-1/2
	18'-3"	3-1/8 x 16-1/2 3-1/2 x 15 5-1/8 x 13-1/2 5-1/2 x 13-1/2	3-1/8 x 18 3-1/2 x 16-1/2 5-1/8 x 13-1/2 5-1/2 x 13-1/2	3-1/8 x 18 3-1/2 x 18 5-1/8 x 15 5-1/2 x 13-1/2	3-1/8 x 19-1/2 3-1/2 x 18 5-1/8 x 15 5-1/2 x 15	3-1/8 x 21 3-1/2 x 19-1/2 5-1/8 x 16-1/2 5-1/2 x 15

Notes:

(1) Service condition = dry.

(2) Maximum deflection under live load = span/240.

(3) Maximum deflection under total load = span/180.

(4) 2-ft roof overhangs.

(5) Beam weight = 35 pcf.

(6) Design properties at normal load duration and dry-use service conditions –
F_b = 2400 psi, f_v = 165 psi, E = 1.8 x 10^6 psi.

TABLE 1B: 24F *SOUTHERN PINE* GLULAM GARAGE DOOR HEADERS

	Clear Door Opening (ft.)	Span of supported roof trusses (ft.)				
		24	28	32	36	40
		Glulam Size (in.)				
Non-Snow Load (125%) 10 psf Dead 20 psf Live	9'-3"	3 x 6-7/8	3 x 6-7/8	3 x 6-7/8	3 x 8-1/4	3 x 8-1/4
		3-1/8 x 6-7/8	3-1/8 x 6-7/8	3-1/8 x 6-7/8	3-1/8 x 8-1/4	3-1/8 x 8-1/4
		5 x 5-1/2	5 x 5-1/2	5 x 6-7/8	5 x 6-7/8	5 x 6-7/8
		5-1/8 x 5-1/2	5-1/8 x 6-7/8	5-1/8 x 6-7/8	5-1/8 x 6-7/8	5-1/8 x 6-7/8
	16'-3"	3 x 12-3/8	3 x 12-3/8	3 x 12-3/8	3 x 13-3/4	3 x 13-3/4
		3-1/8 x 11	3-1/8 x 12-3/8	3-1/8 x 12-3/8	3-1/8 x 13-3/4	3-1/8 x 13-3/4
		5 x 9-5/8	5 x 11	5 x 11	5 x 11	5 x 11
		5-1/8 x 9-5/8	5-1/8 x 11	5-1/8 x 11	5-1/8 x 11	5-1/8 x 11
	18'-3"	3 x 13-3/4	3 x 13-3/4	3 x 13-3/4	3 x 15-1/8	3 x 15-1/8
		3-1/8 x 12-3/8	3-1/8 x 13-3/4	3-1/8 x 13-3/4	3-1/8 x 15-1/8	3-1/8 x 15-1/8
		5 x 11	5 x 12-3/8	5 x 12-3/8	5 x 12-3/8	5 x 12-3/8
		5-1/8 x 11	5-1/8 x 11	5-1/8 x 12-3/8	5-1/8 x 12-3/8	5-1/8 x 12-3/8
Snow Load (115%) 10 psf Dead 25 psf Live	9'-3"	3 x 6-7/8	3 x 8-1/4	3 x 8-1/4	3 x 8-1/4	3 x 9-5/8
		3-1/8 x 6-7/8	3-1/8 x 8-1/4	3-1/8 x 8-1/4	3-1/8 x 8-1/4	3-1/8 x 9-5/8
		5 x 6-7/8	5 x 6-7/8	5 x 6-7/8	5 x 6-7/8	5 x 6-7/8
		5-1/8 x 6-7/8	5-1/8 x 6-7/8	5-1/8 x 6-7/8	5-1/8 x 6-7/8	5-1/8 x 6-7/8
	16'-3"	3 x 12-3/8	3 x 13-3/4	3 x 13-3/4	3 x 15-1/8	3 x 15-1/8
		3-1/8 x 12-3/8	3-1/8 x 13-3/4	3-1/8 x 13-3/4	3-1/8 x 15-1/8	3-1/8 x 15-1/8
		5 x 11	5 x 11	5 x 11	5 x 12-3/8	5 x 12-3/8
		5-1/8 x 11	5-1/8 x 11	5-1/8 x 11	5-1/8 x 12-3/8	5-1/8 x 12-3/8
	18'-3"	3 x 13-3/4	3 x 15-1/8	3 x 16-1/2	3 x 16-1/2	3 x 17-7/8
		3-1/8 x 13-3/4	3-1/8 x 15-1/8	3-1/8 x 15-1/8	3-1/8 x 16-1/2	3-1/8 x 16-1/2
		5 x 12-3/8	5 x 12-3/8	5 x 12-3/8	5 x 13-3/4	5 x 13-3/4
		5-1/8 x 12-3/8	5-1/8 x 12-3/8	5-1/8 x 12-3/8	5-1/8 x 13-3/4	5-1/8 x 13-3/4
Snow Load (115%) 10 psf Dead 30 psf Live	9'-3"	3 x 8-1/4	3 x 8-1/4	3 x 8-1/4	3 x 9-5/8	3 x 9-5/8
		3-1/8 x 8-1/4	3-1/8 x 8-1/4	3-1/8 x 8-1/4	3-1/8 x 9-5/8	3-1/8 x 9-5/8
		5 x 6-7/8	5 x 6-7/8	5 x 6-7/8	5 x 6-7/8	5 x 8-1/4
		5-1/8 x 6-7/8	5-1/8 x 6-7/8	5-1/8 x 6-7/8	5-1/8 x 6-7/8	5-1/8 x 8-1/4
	16'-3"	3 x 13-3/4	3 x 13-3/4	3 x 15-1/8	3 x 16-1/2	3 x 16-1/2
		3-1/8 x 13-3/4	3-1/8 x 13-3/4	3-1/8 x 15-1/8	3-1/8 x 15-1/8	3-1/8 x 16-1/2
		5 x 11	5 x 11	5 x 12-3/8	5 x 12-3/8	5 x 13-3/4
		5-1/8 x 11	5-1/8 x 11	5-1/8 x 12-3/8	5-1/8 x 12-3/8	5-1/8 x 12-3/8
	18'-3"	3 x 15-1/8	3 x 16-1/2	3 x 16-1/2	3 x 17-7/8	3 x 19-1/4
		3-1/8 x 15-1/8	3-1/8 x 15-1/8	3-1/8 x 16-1/2	3-1/8 x 17-7/8	3-1/8 x 17-7/8
		5 x 12-3/8	5 x 12-3/8	5 x 13-3/4	5 x 13-3/4	5 x 15-1/8
		5-1/8 x 12-3/8	5-1/8 x 12-3/8	5-1/8 x 13-3/4	5-1/8 x 13-3/4	5-1/8 x 15-1/8
Snow Load (115%) 10 psf Dead 40 psf Live	9'-3"	3 x 8-1/4	3 x 9-5/8	3 x 9-5/8	3 x 11	3 x 11
		3-1/8 x 8-1/4	3-1/8 x 9-5/8	3-1/8 x 9-5/8	3-1/8 x 9-5/8	3-1/8 x 11
		5 x 6-7/8	5 x 6-7/8	5 x 8-1/4	5 x 8-1/4	5 x 8-1/4
		5-1/8 x 6-7/8	5-1/8 x 6-7/8	5-1/8 x 8-1/4	5-1/8 x 8-1/4	5-1/8 x 8-1/4
	16'-3"	3 x 15-1/8	3 x 16-1/2	3 x 16-1/2	3 x 17-7/8	3 x 19-1/4
		3-1/8 x 15-1/8	3-1/8 x 15-1/8	3-1/8 x 16-1/2	3-1/8 x 17-7/8	3-1/8 x 17-7/8
		5 x 12-3/8	5 x 12-3/8	5 x 13-3/4	5 x 13-3/4	5 x 15-1/8
		5-1/8 x 12-3/8	5-1/8 x 12-3/8	5-1/8 x 13-3/4	5-1/8 x 13-3/4	5-1/8 x 13-3/4
	18'-3"	3 x 16-1/2	3 x 17-7/8	3 x 19-1/4	3 x 19-1/4	3 x 20-5/8
		3-1/8 x 16-1/2	3-1/8 x 17-7/8	3-1/8 x 17-7/8	3-1/8 x 19-1/4	3-1/8 x 20-5/8
		5 x 13-3/4	5 x 13-3/4	5 x 15-1/8	5 x 15-1/8	5 x 16-1/2
		5-1/8 x 13-3/4	5-1/8 x 13-3/4	5-1/8 x 15-1/8	5-1/8 x 15-1/8	5-1/8 x 16-1/2

Notes:

(1) Service condition = dry.

(2) Maximum deflection under live load = span/240.

(3) Maximum deflection under total load = span/180.

(4) 2-ft roof overhangs.

(5) Beam weight = 36 pcf.

(6) Design properties at normal load duration and dry-use service conditions –
F_b = 2400 psi, f_v = 200 psi, E = 1.8 x 10^6 psi.

SUBSTITUTING GLULAMS FOR STEEL OR SOLID-SAWN LUMBER BEAMS

Substitution of *APA EWS* Glulams for solid-sawn beams is simple. Suppose the plans show a solid-sawn 4x14 Select Structural Douglas-fir beam spanning 12 feet and supporting the first floor over a basement in a 24-foot-wide house.

Table 3 shows that either a 3-1/8″ x 12″ or 2-1/2″ x 12″ 24F *APA EWS* Glulam can be substituted.

To check the substitution of a 3-1/8″ x 12″ glulam, calculate:

Load to beam —
  (10 + 40)(12) = 120 plf dead load + 480 plf live load = 600 plf

Table 2 shows that a 3-1/8″ x 12″ *APA EWS* western species glulam will support loads up to a total of 825 plf on a 12-foot span.

Tables 3-4 show glulam equivalents for 4x solid-sawn floor beams. Tables 5-8 show glulam equivalents for 4x solid-sawn roof beams. Tables 9-10 show glulam equivalents for steel floor beams.

TABLE 2
ALLOWABLE LOADS FOR SIMPLE SPAN *WESTERN SPECIES* GLUED LAMINATED FLOOR BEAMS (plf)
(Load Duration Factor = 1.00)

F_b = 2,400 psi, E = 1,800,000 psi, F_v = 165 psi

3-1/8-INCH WIDTH								SPAN (ft)													
Depth (in.)	8	10	12	14	16	18	20	22	24	26	28	30	32	34	36	38	40	42	44	46	48
6	366	187	109	68	—	—	—	—	—	—	—	—	—	—	—	—	—	—	—	—	—
7-1/2	715	366	212	133	89	63	—	—	—	—	—	—	—	—	—	—	—	—	—	—	—
9	952	633	366	231	154	109	79	59	—	—	—	—	—	—	—	—	—	—	—	—	—
10-1/2	1155	875	582	366	245	172	126	94	73	57	—	—	—	—	—	—	—	—	—	—	—
12	1375	1031	825	547	366	257	187	141	109	85	68	56	—	—	—	—	—	—	—	—	—
13-1/2	1614	1198	952	775	521	366	267	201	154	122	97	79	65	54	—	—	—	—	—	—	—
15	1875	1375	1086	897	715	502	366	275	212	167	133	109	89	75	63	53	—	—	—	—	—
16-1/2	2161	1565	1226	1008	856	669	487	366	282	222	178	144	119	99	84	71	61	53	—	—	—
18	2475	1768	1375	1125	952	825	633	475	366	288	231	187	154	129	109	92	79	68	59	52	—
19-1/2	2822	1986	1532	1247	1051	909	792	604	466	366	293	238	196	164	138	117	101	87	76	66	58
21	3208	2221	1699	1375	1155	996	875	751	582	457	366	298	245	205	172	147	126	109	94	83	73
22-1/2	3640	2475	1875	1509	1263	1086	952	848	713	563	450	366	302	252	212	180	154	133	116	102	89
24	4125	2750	2062	1650	1375	1179	1031	917	806	681	547	444	366	305	257	219	187	162	141	123	109
25-1/2	4675	3049	2262	1798	1492	1275	1113	988	888	765	654	533	439	366	309	262	225	194	169	148	130
27	5304	3375	2475	1954	1614	1375	1198	1061	952	852	729	631	521	435	366	311	267	231	201	176	154

TABLE 3
24F GLULAM EQUIVALENTS FOR 4x_ DOUGLAS FIR-LARCH LUMBER FLOOR BEAMS

Designed According to the 1991 NDS
(Load Duration Factor = 1.00)

Span (ft)	Glulam Species	4 x 8 Douglas Fir		4 x 10 Douglas Fir		4 x 12 Douglas Fir		4 x 14 Douglas Fir	
		Select Struc.	No. 1	Select Struc.	No. 1	Select Struc.	No. 1	Select Struc.	No. 1
		24F Glulam Equivalent (in.)							
10	Western Species	2-1/2 x 9 3-1/8 x 7-1/2	2-1/2 x 7-1/2 3-1/8 x 7-1/2	2-1/2 x 9 3-1/8 x 9	2-1/2 x 9 3-1/8 x 9	2-1/2 x 10-1/2 3-1/8 x 9	2-1/2 x 10-1/2 3-1/8 x 9	2-1/2 x 12 3-1/8 x 10-1/2	2-1/2 x 10-1/2 3-1/8 x 10-1/2
	Southern Pine	2-1/2 x 8-1/4 3 x 8-1/4	2-1/2 x 8-1/4 3 x 6-7/8	2-1/2 x 9-5/8 3 x 9-5/8	2-1/2 x 9-5/8 3 x 8-1/4	2-1/2 x 9-5/8 3 x 9-5/8	2-1/2 x 9-5/8 3 x 9-5/8	2-1/2 x 11 3 x 11	2-1/2 x 11 3 x 9-5/8
12	Western Species	2-1/2 x 9 3-1/8 x 9	2-1/2 x 9 3-1/8 x 7-1/2	2-1/2 x 10-1/2 3-1/8 x 10-1/2	2-1/2 x 9 3-1/8 x 9	2-1/2 x 12 3-1/8 x 10-1/2	2-1/2 x 10-1/2 3-1/8 x 10-1/2	**2-1/2 x 12 3-1/8 x 12**	2-1/2 x 12 3-1/8 x 10-1/2
	Southern Pine	2-1/2 x 8-1/4 3 x 8-1/4	2-1/2 x 8-1/4 3 x 8-1/4	2-1/2 x 11 3 x 9-5/8	2-1/2 x 9-5/8 3 x 9-5/8	2-1/2 x 11 3 x 11	2-1/2 x 11 3 x 9-5/8	2-1/2 x 12-3/8 3 x 11	2-1/2 x 11 3 x 11
14	Western Species	2-1/2 x 9 3-1/8 x 9	2-1/2 x 9 3-1/8 x 7-1/2	2-1/2 x 12 3-1/8 x 10-1/2	2-1/2 x 10-1/2 3-1/8 x 9	2-1/2 x 12 3-1/8 x 12	2-1/2 x 10-1/2 3-1/8 x 10-1/2	2-1/2 x 13-1/2 3-1/8 x 12	2-1/2 x 12 3-1/8 x 10-1/2
	Southern Pine	2-1/2 x 8-1/4 3 x 8-1/4	2-1/2 x 8-1/4 3 x 8-1/4	2-1/2 x 11 3 x 11	2-1/2 x 9-5/8 3 x 9-5/8	2-1/2 x 12-3/8 3 x 11	2-1/2 x 11 3 x 11	2-1/2 x 13-3/4 3 x 12-3/8	2-1/2 x 12-3/8 3 x 11
16	Western Species	2-1/2 x 9 3-1/8 x 9	2-1/2 x 9 3-1/8 x 7-1/2	2-1/2 x 12 3-1/8 x 10-1/2	2-1/2 x 10-1/2 3-1/8 x 10-1/2	2-1/2 x 13-1/2 3-1/8 x 12	2-1/2 x 12 3-1/8 x 10-1/2	2-1/2 x 13-1/2 3-1/8 x 13-1/2	2-1/2 x 12 3-1/8 x 12
	Southern Pine	2-1/2 x 8-1/4 3 x 8-1/4	2-1/2 x 8-1/4 3 x 8-1/4	2-1/2 x 11 3 x 11	2-1/2 x 11 3 x 9-5/8	2-1/2 x 12-3/8 3 x 12-3/8	2-1/2 x 11 3 x 11	2-1/2 x 13-3/4 3 x 12-3/8	2-1/2 x 12-3/8 3 x 11
18	Western Species	2-1/2 x 9 3-1/8 x 9	2-1/2 x 9 3-1/8 x 7-1/2	2-1/2 x 10-1/2 3-1/8 x 10-1/2	2-1/2 x 10-1/2 3-1/8 x 10-1/2	2-1/2 x 13-1/2 3-1/8 x 12	2-1/2 x 12 3-1/8 x 10-1/2	2-1/2 x 15 3-1/8 x 13-1/2	2-1/2 x 13-1/2 3-1/8 x 12
	Southern Pine	2-1/2 x 8-1/4 3 x 8-1/4	2-1/2 x 8-1/4 3 x 8-1/4	2-1/2 x 11 3 x 11	2-1/2 x 11 3 x 9-5/8	2-1/2 x 13-3/4 3 x 12-3/8	2-1/2 x 12-3/8 3 x 11	2-1/2 x 13-3/4 3 x 13-3/4	2-1/2 x 12-3/8 3 x 12-3/8
20	Western Species	2-1/2 x 9 3-1/8 x 9	2-1/2 x 9 3-1/8 x 7-1/2	2-1/2 x 10-1/2 3-1/8 x 10-1/2	2-1/2 x 10-1/2 3-1/8 x 10-1/2	2-1/2 x 13-1/2 3-1/8 x 12	2-1/2 x 12 3-1/8 x 12	2-1/2 x 15 3-1/8 x 13-1/2	2-1/2 x 13-1/2 3-1/8 x 12
	Southern Pine	2-1/2 x 8-1/4 3 x 8-1/4	2-1/2 x 8-1/4 3 x 8-1/4	2-1/2 x 11 3 x 11	2-1/2 x 11 3 x 9-5/8	2-1/2 x 13-3/4 3 x 12-3/8	2-1/2 x 12-3/8 3 x 11	2-1/2 x 15-1/8 3 x 13-3/4	2-1/2 x 13-3/4 3 x 12-3/8

Notes:

(1) Span = uniformly loaded simply supported beam.

(2) Maximum deflection = L/360 under live load, based on live/total load = 0.8.

(3) Service condition = dry.

(4) Beam weights for sawn and glulam members are assumed to be the same.

(5) Volume factors for glulam members and size factors for sawn lumber members are in accordance with 1991 NDS.

(6) Minimum glulam sizes considered in the table are: 2-1/2 x 6 and 3 1/8 x 6 (western species), and 2-1/2 x 5-1/2 and 3 x 5-1/2 (southern pine).

(7) Design properties at normal load duration and dry-use service conditions -
Select Structural sawn lumber members: F_b = 1450 psi, F_v = 95 psi, E = 1.9 x 10⁶ psi.
No. 1 sawn lumber members: F_b 1000 psi, F_v = 95 psi, E = 1.7 x 10⁶ psi.
Glulam members: F_b = 2400 psi, F_v = 165 psi (western species) or 200 psi (southern pine), E = 1.8 x 10⁶ psi.

TABLE 4
24F GLULAM EQUIVALENTS FOR 4x_ *SOUTHERN PINE* LUMBER
FLOOR BEAMS

Designed According to the 1991 NDS
(Load Duration Factor = 1.00)

Span (ft)	Glulam Species	4 x 6 Southern Pine		4 x 8 Southern Pine		4 x 10 Southern Pine		4 x 12 Southern Pine	
		Select Struc.	No. 1	Select Struc.	No. 1	Select Struc.	No. 1	Select Struc.	No. 1
		24F Glulam Equivalent (in.)							
10	Western Species	2-1/2 x 7-1/2 3-1/8 x 6	2-1/2 x 7-1/2 3-1/8 x 6	2-1/2 x 9 3-1/8 x 7-1/2	2-1/2 x 9 3-1/8 x 7-1/2	2-1/2 x 9 3-1/8 x 9	2-1/2 x 9 3-1/8 x 9	2-1/2 x 10-1/2 3-1/8 x 9	2-1/2 x 10-1/2 3-1/8 x 9
	Southern Pine	2-1/2 x 6-7/8 3 x 6-7/8	2-1/2 x 6-7/8 3 x 6-7/8	2-1/2 x 8-1/4 3 x 8-1/4	2-1/2 x 8-1/4 3 x 8-1/4	2-1/2 x 9-5/8 3 x 8-1/4	2-1/2 x 9-5/8 3 x 8-1/4	2-1/2 x 9-5/8 3 x 9-5/8	2-1/2 x 9-5/8 3 x 9-5/8
12	Western Species	2-1/2 x 7-1/2 3-1/8 x 6	2-1/2 x 7-1/2 3-1/8 x 6	2-1/2 x 9 3-1/8 x 9	2-1/2 x 9 3-1/8 x 7-1/2	2-1/2 x 10-1/2 3-1/8 x 10-1/2	2-1/2 x 10-1/2 3-1/8 x 9	2-1/2 x 10-1/2 3-1/8 x 10-1/2	2-1/2 x 10-1/2 3-1/8 x 10-1/2
	Southern Pine	2-1/2 x 6-7/8 3 x 6-7/8	2-1/2 x 6-7/8 3 x 6-7/8	2-1/2 x 8-1/4 3 x 8-1/4	2-1/2 x 8-1/4 3 x 8-1/4	2-1/2 x 11 3 x 9-5/8	2-1/2 x 9-5/8 3 x 9-5/8	2-1/2 x 11 3 x 11	2-1/2 x 11 3 x 11
14	Western Species	2-1/2 x 9 3-1/8 x 6	2-1/2 x 6 3-1/8 x 6	2-1/2 x 9 3-1/8 x 9	2-1/2 x 9 3-1/8 x 7-1/2	2-1/2 x 10-1/2 3-1/8 x 10-1/2	2-1/2 x 10-1/2 3-1/8 x 10-1/2	2-1/2 x 12 3-1/8 x 12	2-1/2 x 12 3-1/8 x 10-1/2
	Southern Pine	2-1/2 x 6-7/8 3 x 6-7/8	2-1/2 x 6-7/8 3 x 6-7/8	2-1/2 x 8-1/4 3 x 8-1/4	2-1/2 x 8-1/4 3 x 8-1/4	2-1/2 x 11 3 x 11	2-1/2 x 11 3 x 9-5/8	2-1/2 x 12-3/8 3 x 11	2-1/2 x 12-3/8 3 x 11
16	Western Species	2-1/2 x 7-1/2 3-1/8 x 6	2-1/2 x 6 3-1/8 x 6	2-1/2 x 9 3-1/8 x 7-1/2	2-1/2 x 9 3-1/8 x 7-1/2	2-1/2 x 10-1/2 3-1/8 x 10-1/2	2-1/2 x 10-1/2 3-1/8 x 10-1/2	2-1/2 x 13-1/2 3-1/8 x 12	2-1/2 x 12 3-1/8 x 12
	Southern Pine	2-1/2 x 6-7/8 3 x 6-7/8	2-1/2 x 6-7/8 3 x 6-7/8	2-1/2 x 8-1/4 3 x 8-1/4	2-1/2 x 8-1/4 3 x 8-1/4	2-1/2 x 11 3 x 11	2-1/2 x 11 3 x 9-5/8	2-1/2 x 13-3/4 3 x 12-3/8	2-1/2 x 12-3/8 3 x 11
18	Western Species	2-1/2 x 7-1/2 3-1/8 x 6	2-1/2 x 6 3-1/8 x 6	2-1/2 x 9 3-1/8 x 7-1/2	2-1/2 x 9 3-1/8 x 7-1/2	2-1/2 x 10-1/2 3-1/8 x 10-1/2	2-1/2 x 10-1/2 3-1/8 x 10-1/2	2-1/2 x 13-1/2 3-1/8 x 12	2-1/2 x 13-1/2 3-1/8 x 12
	Southern Pine	2-1/2 x 6-7/8 3 x 6-7/8	2-1/2 x 6-7/8 3 x 6-7/8	2-1/2 x 8-1/4 3 x 8-1/4	2-1/2 x 8-1/4 3 x 8-1/4	2-1/2 x 11 3 x 11	2-1/2 x 11 3 x 9-5/8	2-1/2 x 13-3/4 3 x 12-3/8	2-1/2 x 12-3/8 3 x 12-3/8
20	Western Species	2-1/2 x 7-1/2 3-1/8 x 6	2-1/2 x 6 3-1/8 x 6	2-1/2 x 9 3-1/8 x 7-1/2	2-1/2 x 9 3-1/8 x 7-1/2	2-1/2 x 10-1/2 3-1/8 x 10-1/2	2-1/2 x 10-1/2 3-1/8 x 10-1/2	2-1/2 x 13-1/2 3-1/8 x 12	2-1/2 x 13-1/2 3-1/8 x 12
	Southern Pine	2-1/2 x 6-7/8 3 x 6-7/8	2-1/2 x 6-7/8 3 x 6-7/8	2-1/2 x 8-1/4 3 x 8-1/4	2-1/2 x 8-1/4 3 x 8-1/4	2-1/2 x 11 3 x 11	2-1/2 x 11 3 x 9-5/8	2-1/2 x 13-3/4 3 x 12-3/8	2-1/2 x 12-3/8 3 x 12-3/8

Notes:

(1) Span = uniformly loaded simply supported beam.

(2) Maximum deflection = L/360 under live load, based on live/total load = 0.8.

(3) Service condition = dry.

(4) Beam weights for sawn and glulam members are assumed to be the same.

(5) Volume factors for glulam members and size factors for sawn lumber members are in accordance with 1991 NDS.

(6) Minimum glulam sizes considered in the table are: 2-1/2 x 6 and 3-1/8 x 6 (*western species*), and 2-1/2 x 5-1/2 and 3 x 5-1/2 (*southern pine*).

(7) Design properties at normal load duration and dry-use service conditions -
Select Structural sawn lumber members: F_b = 2300 (8″),1900 (12″) psi, F_v = 90 psi, E = 1.8 x 10⁶ psi.
No. 1 sawn lumber members: F_b 1650 (6″), 1500 (8″), 1300 (10″), 1250 (12″) psi, F_v = 90 psi, E = 1.7 x 10⁶ psi.
Glulam members: F_b = 2400 psi, F_v = 165 psi (*western species*) or 200 psi (*southern pine*), E = 1.8 x 10⁶ psi.

TABLE 5
24F GLULAM EQUIVALENTS FOR 4x_ *DOUGLAS FIR-LARCH* LUMBER
ROOF BEAMS – SNOW LOADS

Designed According to the 1991 NDS
(Load Duration Factor = 1.15)

Span (ft)	Glulam Species	4 x 8 Douglas Fir		4 x 10 Douglas Fir		4 x 12 Douglas Fir		4 x 14 Douglas Fir	
		Select Struc.	No. 1	Select Struc.	No. 1	Select Struc.	No. 1	Select Struc.	No. 1
		24F Glulam Equivalent (in.)							
10	Western Species	2-1/2 x 7-1/2	2-1/2 x 7-1/2	2-1/2 x 9	2-1/2 x 9	2-1/2 x 10-1/2	2-1/2 x 9	2-1/2 x 12	2-1/2 x 10-1/2
		3-1/8 x 7-1/2	3-1/8 x 7-1/2	3-1/8 x 9	3-1/8 x 7-1/2	3-1/8 x 9	3-1/8 x 9	3-1/8 x 10-1/2	3-1/8 x 10-1/2
	Southern Pine	2-1/2 x 8-1/4	2-1/2 x 6-7/8	2-1/2 x 9-5/8	2-1/2 x 8-1/4	2-1/2 x 9-5/8	2-1/2 x 9-5/8	2-1/2 x 11	2-1/2 x 11
		3 x 6-7/8	3 x 6-7/8	3 x 8-1/4	3 x 8-1/4	3 x 9-5/8	3 x 8-1/4	3 x 11	3 x 9-5/8
12	Western Species	2-1/2 x 9	2-1/2 x 7-1/2	2-1/2 x 10-1/2	2-1/2 x 9	2-1/2 x 10-1/2	2-1/2 x 9	2-1/2 x 12	2-1/2 x 10-1/2
		3-1/8 x 7-1/2	3-1/8 x 7-1/2	3-1/8 x 9	3-1/8 x 7-1/2	3-1/8 x 10-1/2	3-1/8 x 9	3-1/8 x 10-1/2	3-1/8 x 10-1/2
	Southern Pine	2-1/2 x 8-1/4	2-1/2 x 8-1/4	2-1/2 x 9-5/8	2-1/2 x 8-1/4	2-1/2 x 11	2-1/2 x 9-5/8	2-1/2 x 12-3/8	2-1/2 x 11
		3 x 8-1/4	3 x 6-7/8	3 x 9-5/8	3 x 8-1/4	3 x 9-5/8	3 x 8-1/4	3 x 11	3 x 9-5/8
14	Western Species	2-1/2 x 9	2-1/2 x 7-1/2	2-1/2 x 10-1/2	2-1/2 x 9	2-1.2 x 12	2-1/2 x 10-1/2	2-1/2 x 13-1/2	2-1/2 x 10-1/2
		3-1/8 x 9	3-1/8 x 7-1/2	3-1/8 x 9	3-1/8 x 9	3-1/8 x 10-1/2	3-1/8 x 9	3-1/8 x 12	3-1/8 x 10-1/2
	Southern Pine	2-1/2 x 8-1/4	2-1/2 x 8-1/4	2-1/2 x 9-5/8	2-1/2 x 9-5/8	2-1/2 x 11	2-1/2 x 9-5/8	2-1/2 x 12-3/8	2-1/2 x 11
		3 x 8-1/4	3 x 6-7/8	3 x 9-5/8	3 x 8-1/4	3 x 11	3 x 9-5/8	3 x 12-3/8	3 x 9-5/8
16	Western Species	2-1/2 x 9	2-1/2 x 9	2-1/2 x 10-1/2	2-1/2 x 9	2-1/2 x 12	2-1/2 x 10-1/2	2-1/2 x 13-1/2	2-1/2 x 10-1/2
		3-1/8 x 9	3-1/8 x 7-1/2	3-1/8 x 10-1/2	3-1/8 x 9	3-1/8 x 10-1/2	3-1/8 x 9	3-1/8 x 12	3-1/8 x 10-1/2
	Southern Pine	2-1/2 x 8-1/4	2-1/2 x 8-1/4	2-1/2 x 11	2-1/2 x 9-5/8	2-1/2 x 11	2-1/2 x 9-5/8	2-1/2 x 12-3/8	2-1/2 x 11
		3 x 8-1/4	3 x 8-1/4	3 x 9-5/8	3 x 8-1/4	3 x 11	3 x 9-5/8	3 x 12-3/8	3 x 11
18	Western Species	2-1/2 x 9	2-1/2 x 9	2-1/2 x 10-1/2	2-1/2 x 10-1/2	2-1/2 x 12	2-1/2 x 10-1/2	2-1/2 x 13-1/2	2-1/2 x 12
		3-1/8 x 9	3-1/8 x 7-1/2	3-1/8 x 10-1/2	3-1/8 x 9	3-1/8 x 12	3-1/8 x 10-1/2	3-1/8 x 12	3-1/8 x 10-1/2
	Southern Pine	2-1/2 x 8-1/4	2-1/2 x 8-1/4	2-1/2 x 11	2-1/2 x 9-5/8	2-1/2 x 12-3/8	2-1/2 x 11	2-1/2 x 12-3/8	2-1/2 x 11
		3 x 8-1/4	3 x 8-1/4	3 x 9-5/8	3 x 9-5/8	3 x 11	3 x 9-5/8	3 x 12-3/8	3 x 11
20	Western Species	2-1/2 x 9	2-1/2 x 9	2-1/2 x 12	2-1/2 x 10-1/2	2-1/2 x 12	2-1/2 x 10-1/2	2-1/2 x 13-1/2	2-1/2 x 12
		3-1/8 x 9	3-1/8 x 7-1/2	3-1/8 x 10-1/2	3-1/8 x 9	3-1/8 x 12	3-1/8 x 10-1/2	3-1/8 x 12	3-1/8 x 10-1/2
	Southern Pine	2-1/2 x 8-1/4	2-1/2 x 8-1/4	2-1/2 x 11	2-1/2 x 9-5/8	2-1/2 x 12-3/8	2-1/2 x 11	2-1/2 x 13-3/4	2-1/2 x 12-3/8
		3 x 8-1/4	3 x 8-1/4	3 x 11	3 x 9-5/8	3 x 12-3/8	3 x 11	3 x 12-3/8	3 x 11

Notes:

(1) Span = uniformly loaded simply supported beam.

(2) Maximum deflection = 1/180 under total load. Deflection under live load must be verified when live/total load > 3/4.

(3) Service condition = dry.

(4) Beam weights for sawn and glulam members are assumed to be the same.

(5) Volume factors for glulam members and size factors for sawn lumber members are in accordance with 1991 NDS.

(6) Minimum glulam sizes considered in the table are: 2-1/2 x 6 and 3-1/8 x 6 (*western species*), and 2-1/2 x 5-1/2 and 3 x 5-1/2 (*southern pine*).

(7) Design properties at normal load duration and dry-use service conditions -
Select Structural sawn lumber members: F_b = 1450 psi, F_v = 95 psi, E = 1.9 x 10^6 psi.
No. 1 sawn lumber members: F_b 1000 psi, F_v = 95 psi, E = 1.7 x 10^6 psi.
Glulam members: F_b = 2400 psi, F_v = 165 psi (*western species*) or 200 psi (*southern pine*), E = 1.8 x 10^6 psi.

TABLE 6
24F GLULAM EQUIVALENTS FOR 4x_ SOUTHERN PINE LUMBER
ROOF BEAMS – SNOW LOADS

Designed According to the 1991 NDS
(Load Duration Factor = 1.15)

Span (ft)	Glulam Species	4 x 6 Southern Pine		4 x 8 Southern Pine		4 x 10 Southern Pine		4 x 12 Southern Pine	
		Select Struc.	No. 1	Select Struc.	No. 1	Select Struc.	No. 1	Select Struc.	No. 1
		24F Glulam Equivalent (in.)							
10	Western Species	2-1/2 x 7-1/2 3-1/8 x 6	2-1/2 x 6 3-1/8 x 6	2-1/2 x 7-1/2 3-1/8 x 7-1/2	2-1/2 x 7-1/2 3-1/8 x 7-1/2	2-1/2 x 9 3-1/8 x 7-1/2	2-1/2 x 9 3-1/8 x 7-1/2	2-1/2 x 10-1/2 3-1/8 x 9	2-1/2 x 10-1/2 3-1/8 x 9
	Southern Pine	2-1/2 x 6-7/8 3 x 6-7/8	2-1/2 x 6-7/8 3 x 5-1/2	2-1/2 x 8-1/4 3 x 6-7/8	2-1/2 x 8-1/4 3 x 6-7/8	2-1/2 x 9-5/8 3 x 8-1/4	2-1/2 x 9-5/8 3 x 8-1/4	2-1/2 x 9-5/8 3 x 9-5/8	2-1/2 x 9-5/8 3 x 9-5/8
12	Western Species	2-1/2 x 7-1/2 3-1/8 x 6	2-1/2 x 7-1/2 3-1/8 x 6	2-1/2 x 9 3-1/8 x 7-1/2	2-1/2 x 7-1/2 3-1/8 x 7-1/2	2-1/2 x 9 3-1/8 x 9	2-1/2 x 9 3-1/8 x 9	2-1/2 x 10-1/2 3-1/8 x 9	2-1/2 x 10-1/2 3-1/8 x 9
	Southern Pine	2-1/2 x 6-7/8 3 x 6-7/8	2-1/2 x 6-7/8 3 x 6-7/8	2-1/2 x 8-1/4 3 x 8-1/4	2-1/2 x 8-1/4 3 x 8-1/4	2-1/2 x 9-5/8 3 x 8-1/4	2-1/2 x 9-5/8 3 x 8-1/4	2-1/2 x 11 3 x 9-5/8	2-1/2 x 11 3 x 9-5/8
14	Western Species	2-1/2 x 7-1/2 3-1/8 x 6	2-1/2 x 6 3-1/8 x 6	2-1/2 x 9 3-1/8 x 9	2-1/2 x 9 3-1/8 x 7-1/2	2-1/2 x 10-1/2 3-1/8 x 9	2-1/2 x 9 3-1/8 x 9	2-1/2 x 12 3-1/8 x 10-1/2	2-1/2 x 10-1/2 3-1/8 x 10-1/2
	Southern Pine	2-1/2 x 6-7/8 3 x 6-7/8	2-1/2 x 6-7/8 3 x 6-7/8	2-1/2 x 8-1/4 3 x 8-1/4	2-1/2 x 8-1/4 3 x 8-1/4	2-1/2 x 9-5/8 3 x 9-5/8	2-1/2 x 9-5/8 3 x 9-5/8	2-1/2 x 11 3 x 11	2-1/2 x 11 3 x 9-5/8
16	Western Species	2-1/2 x 7-1/2 3-1/8 x 6	2-1/2 x 6 3-1/8 x 6	2-1/2 x 9 3-1/8 x 9	2-1/2 x 9 3-1/8 x 7-1/2	2-1/2 x 10-1/2 3-1/8 x 10-1/2	2-1/2 x 10-1/2 3-1/8 x 9	2-1/2 x 12 3-1/8 x 10-1/2	2-1/2 x 10-1/2 3-1/8 x 10-1/2
	Southern Pine	2-1/2 x 6-7/8 3 x 6-7/8	2-1/2 x 6-7/8 3 x 6-7/8	2-1/2 x 8-1/4 3 x 8-1/4	2-1/2 x 8-1/4 3 x 8-1/4	2-1/2 x 11 3 x 11	2-1/2 x 9-5/8 3 x 9-5/8	2-1/2 x 12-3/8 3 x 11	2-1/2 x 11 3 x 11
18	Western Species	2-1/2 x 7-1/2 3-1/8 x 6	2-1/2 x 6 3-1/8 x 6	2-1/2 x 9 3-1/8 x 7-1/2	2-1/2 x 9 3-1/8 x 7-1/2	2-1/2 x 10-1/2 3-1/8 x 10-1/2	2-1/2 x 10-1/2 3-1/8 x 9	2-1/2 x 13-1/2 3-1/8 x 12	2-1/2 x 12 3-1/8 x 10-1/2
	Southern Pine	2-1/2 x 6-7/8 3 x 6-7/8	2-1/2 x 6-7/8 3 x 6-7/8	2-1/2 x 8-1/4 3 x 8-1/4	2-1/2 x 8-1/4 3 x 8-1/4	2-1/2 x 11 3 x 11	2-1/2 x 9-5/8 3 x 9-5/8	2-1/2 x 12-3/8 3 x 12-3/8	2-1/2 x 11 3 x 11
20	Western Species	2-1/2 x 7-1/2 3-1/8 x 6	2-1/2 x 6 3-1/8 x 6	2-1/2 x 9 3-1/8 x 7-1/2	2-1/2 x 9 3-1/8 x 7-1/2	2-1/2 x 10-1/2 3-1/8 x 10-1/2	2-1/2 x 10-1/2 3-1/8 x 10-1/2	2-1/2 x 13-1/2 3-1/8 x 12	2-1/2 x 12 3-1/8 x 10-1/2
	Southern Pine	2-1/2 x 6-7/8 3 x 6-7/8	2-1/2 x 6-7/8 3 x 6-7/8	2-1/2 x 8-1/4 3 x 8-1/4	2-1/2 x 8-1/4 3 x 8-1/4	2-1/2 x 11 3 x 11	2-1/2 x 11 3 x 9-5/8	2-1/2 x 13-3/4 3 x 12-3/8	2-1/2 x 12-3/8 3 x 11

Notes:

(1) Span = uniformly loaded simply supported beam.

(2) Maximum deflection = L/180 under total load. Deflection under live load must be verified when live/total load > 3/4.

(3) Service condition = dry.

(4) Beam weights for sawn and glulam members are assumed to be the same.

(5) Volume factors for glulam members and size factors for sawn lumber members are in accordance with 1991 NDS.

(6) Minimum glulam sizes considered in the table are: 2-1/2 x 6 and 3-1/8 x 6 (western species), and 2-1/2 x 5-1/2 and 3 x 5-1/2 (southern pine).

(7) Design properties at normal load duration and dry-use service conditions -
Select Structural sawn lumber members: F_b = 2300 (8"), 2050 (10"), 1900 (12") psi, F_v = 90 psi, E = 1.8×10^6 psi.
No. 1 sawn lumber members: F_b 1650 (6"), 1500 (8"), 1300 (10"), 1250 (12") psi, F_v = 90 psi, E = 1.7×10^6 psi.
Glulam members: F_b = 2400 psi, F_v = 165 psi (western species) or 200 psi (southern pine), E = 1.8×10^6 psi.

TABLE 7
24F GLULAM EQUIVALENTS FOR 4x_ *DOUGLAS FIR-LARCH* LUMBER
ROOF BEAMS – NON-SNOW LOADS

Designed According to the 1991 NDS
(Load Duration Factor = 1.25)

Span (ft)	Glulam Species	4 x 8 Douglas Fir		4 x 10 Douglas Fir		4 x 12 Douglas Fir		4 x 14 Douglas Fir	
		Select Struc.	No. 1	Select Struc.	No. 1	Select Struc.	No. 1	Select Struc.	No. 1
		24F Glulam Equivalent (in.)							
10	Western Species	2-1/2 x 7-1/2 3-1/8 x 7-1/2	2-1/2 x 7-1/2 3-1/8 x 7-1/2	2-1/2 x 9 3-1/8 x 9	2-1/2 x 9 3-1/8 x 7-1/2	2-1/2 x 10-1/2 3-1/8 x 9	2-1/2 x 9 3-1/8 x 9	2-1/2 x 12 3-1/8 x 10-1/2	2-1/2 x 10-1/2 3-1/8 x 10-1/2
	Southern Pine	2-1/2 x 8-1/4 3 x 8-1/4	2-1/2 x 6-7/8 3 x 6-7/8	2-1/2 x 9-5/8 3 x 8-1/4	2-1/2 x 8-1/4 3 x 8-1/4	2-1/2 x 9-5/8 3 x 9-5/8	2-1/2 x 9-5/8 3 x 8-1/4	2-1/2 x 11 3 x 11	2-1/2 x 11 3 x 9-5/8
12	Western Species	2-1/2 x 9 3-1/8 x 7-1/2	2-1/2 x 7-1/2 3-1/8 x 7-1/2	2-1/2 x 10-1/2 3-1/8 x 9	2-1/2 x 9 3-1/8 x 9	2-1/2 x 10-1/2 3-1/8 x 10-1/2	2-1/2 x 9 3-1/8 x 9	2-1/2 x 12 3-1/8 x 10-1/2	2-1/2 x 10-1/2 3-1/8 x 10-1/2
	Southern Pine	2-1/2 x 8-1/4 3 x 8-1/4	2-1/2 x 8-1/4 3 x 6-7/8	2-1/2 x 9-5/8 3 x 9-5/8	2-1/2 x 8-1/4 3 x 8-1/4	2-1/2 x 11 3 x 9-5/8	2-1/2 x 9-5/8 3 x 9-5/8	2-1/2 x 12-3/8 3 x 11	2-1/2 x 11 3 x 9-5/8
14	Western Species	2-1/2 x 9 3-1/8 x 9	2-1/2 x 7-1/2 3-1/8 x 7-1/2	2-1/2 x 10-1/2 3-1/8 x 9	2-1/2 x 9 3-1/8 x 9	2-1/2 x 12 3-1/8 x 10-1/2	2-1/2 x 10-1/2 3-1/8 x 9	2-1/2 x 13-1/2 3-1/8 x 12	2-1/2 x 10-1/2 3-1/8 x 10-1/2
	Southern Pine	2-1/2 x 8-1/4 3 x 8-1/4	2-1/2 x 8-1/4 3 x 8-1/4	2-1/2 x 11 3 x 9-5/8	2-1/2 x 9-5/8 3 x 8-1/4	2-1/2 x 11 3 x 11	2-1/2 x 9-5/8 3 x 9-5/8	2-1/2 x 12-3/8 3 x 12-3/8	2-1/2 x 11 3 x 9-5/8
16	Western Species	2-1/2 x 9 3-1/8 x 9	2-1/2 x 9 3-1/8 x 7-1/2	2-1/2 x 10-1/2 3-1/8 x 10-1/2	2-1/2 x 9 3-1/8 x 9	2-1/2 x 12 3-1/8 x 10-1/2	2-1/2 x 10-1/2 3-1/8 x 10-1/2	2-1/2 x 13-1/2 3-1/8 x 12	2-1/2 x 12 3-1/8 x 10-1/2
	Southern Pine	2-1/2 x 8-1/4 3 x 8-1/4	2-1/2 x 8-1/4 3 x 8-1/4	2-1/2 x 11 3 x 9-5/8	2-1/2 x 9-5/8 3 x 9-5/8	2-1/2 x 12-3/8 3 x 11	2-1/2 x 11 3 x 9-5/8	2-1/2 x 12-3/8 3 x 12-3/8	2-1/2 x 11 3 x 11
18	Western Species	2-1/2 x 9 3-1/8 x 9	2-1/2 x 9 3-1/8 x 7-1/2	2-1/2 x 12 3-1/8 x 10-1/2	2-1/2 x 10-1/2 3-1/8 x 9	2-1/2 x 12 3-1/8 x 10-1/2	2-1/2 x 10-1/2 3-1/8 x 10-1/2	2-1/2 x 13-1/2 3-1/8 x 12	2-1/2 x 12 3-1/8 x 10-1/2
	Southern Pine	2-1/2 x 8-1/4 3 x 8-1/4	2-1/2 x 8-1/4 3 x 8-1/4	2-1/2 x 11 3 x 11	2-1/2 x 9-5/8 3 x 9-5/8	2-1/2 x 12-3/8 3 x 11	2-1/2 x 11 3 x 11	2-1/2 x 13-3/4 3 x 12-3/8	2-1/2 x 12-3/8 3 x 11
20	Western Species	2-1/2 x 9 3-1/8 x 9	2-1/2 x 9 3-1/8 x 7-1/2	2-1/2 x 12 3-1/8 x 10-1/2	2-1/2 x 10-1/2 3-1/8 x 9	2-1/2 x 13-1/2 3-1/8 x 12	2-1/2 x 12 3-1/8 x 10-1/2	2-1/2 x 13-1/2 3-1/8 x 13-1/2	2-1/2 x 12 3-1/8 x 12
	Southern Pine	2-1/2 x 8-1/4 3 x 8-1/4	2-1/2 x 8-1/4 3 x 8-1/4	2-1/2 x 11 3 x 11	2-1/2 x 9-5/8 3 x 9-5/8	2-1/2 x 12-3/8 3 x 12-3/8	2-1/2 x 11 3 x 11	2-1/2 x 13-3/4 3 x 12-3/8	2-1/2 x 12-3/8 3 x 11

Notes:

(1) Span = uniformly loaded simply supported beam.

(2) Maximum deflection = L/180 under total load. Deflection under live load must be verified when live/total load > 3/4.

(3) Service condition = dry.

(4) Beam weights for sawn and glulam members are assumed to be the same.

(5) Volume factors for glulam members and size factors for sawn lumber members are in accordance with 1991 NDS.

(6) Minimum glulam sizes considered in the table are: 2-1/2 x 6 and 3-1/8 x 6 (western species), and 2-1/2 x 5-1/2 and 3 x 5-1/2 (southern pine).

(7) Design properties at normal load duration and dry-use service conditions -
Select Structural sawn lumber members: F_b = 1450 psi, F_v = 95 psi, E = 1.9 x 10^6 psi.
No. 1 sawn lumber members: F_b 1000 psi, F_v = 95 psi, E = 1.7 x 10^6 psi.
Glulam members: F_b = 2400 psi, F_v = 165 psi (western species) or 200 psi (southern pine), E = 1.8 x 10^6 psi.

TABLE 8
24F GLULAM EQUIVALENTS FOR 4x_ SOUTHERN PINE LUMBER ROOF BEAMS – NON-SNOW LOADS

Designed According to the 1991 NDS
(Load Duration Factor = 1.25)

Span (ft)	Glulam Species	4 x 6 Southern Pine		4 x 8 Southern Pine		4 x 10 Southern Pine		4 x 12 Southern Pine	
		Select Struc.	No. 1	Select Struc.	No. 1	Select Struc.	No. 1	Select Struc.	No. 1
		24F Glulam Equivalent (in.)							
10	Western Species	2-1/2 x 7-1/2 3-1/8 x 6	2-1/2 x 7-1/2 3-1/8 x 6	2-1/2 x 7-1/2 3-1/8 x 7-1/2	2-1/2 x 7-1/2 3-1/8 x 7-1/2	2-1/2 x 9 3-1/8 x 7-1/2	2-1/2 x 9 3-1/8 x 7-1/2	2-1/2 x 10-1/2 3-1/8 x 9	2-1/2 x 10-1/2 3-1/8 x 9
	Southern Pine	2-1/2 x 6-7/8 3 x 6-7/8	2-1/2 x 6-7/8 3 x 6-7/8	2-1/2 x 8-1/4 3 x 6-7/8	2-1/2 x 8-1/4 3 x 6-7/8	2-1/2 x 9-5/8 3 x 8-1/4	2-1/2 x 9-5/8 3 x 8-1/4	2-1/2 x 9-5/8 3 x 9-5/8	2-1/2 x 9-5/8 3 x 9-5/8
12	Western Species	2-1/2 x 7-1/2 3-1/8 x 6	2-1/2 x 7-1/2 3-1/8 x 6	2-1/2 x 9 3-1/8 x 9	2-1/2 x 9 3-1/8 x 7-1/2	2-1/2 x 9 3-1/8 x 9	2-1/2 x 9 3-1/8 x 9	2-1/2 x 10-1/2 3-1/8 x 9	2-1/2 x 10-1/2 3-1/8 x 9
	Southern Pine	2-1/2 x 6-7/8 3 x 6-7/8	2-1/2 x 6-7/8 3 x 6-7/8	2-1/2 x 8-1/4 3 x 8-1/4	2-1/2 x 8-1/4 3 x 8-1/4	2-1/2 x 9-5/8 3 x 9-5/8	2-1/2 x 9-5/8 3 x 8-1/4	2-1/2 x 11 3 x 9-5/8	2-1/2 x 11 3 x 9-5/8
14	Western Species	2-1/2 x 7-1/2 3-1/8 x 6	2-1/2 x 6 3-1/8 x 6	2-1/2 x 9 3-1/8 x 9	2-1/2 x 9 3-1/8 x 9	2-1/2 x 10-1/2 3-1/8 x 10-1/2	2-1/2 x 10-1/2 3-1/8 x 9	2-1/2 x 12 3-1/8 x 10-1/2	2-1/2 x 10-1/2 3-1/8 x 10-1/2
	Southern Pine	2-1/2 x 6-7/8 3 x 6-7/8	2-1/2 x 6-7/8 3 x 6-7/8	2-1/2 x 8-1/4 3 x 8-1/4	2-1/2 x 8-1/4 3 x 8-1/4	2-1/2 x 11 3 x 9-5/8	2-1/2 x 9-5/8 3 x 9-5/8	2-1/2 x 11 3 x 11	2-1/2 x 11 3 x 9-5/8
16	Western Species	2-1/2 x 7-1/2 3-1/8 x 6	2-1/2 x 6 3-1/8 x 6	2-1/2 x 9 3-1/8 x 9	2-1/2 x 9 3-1/8 x 7-1/2	2-1/2 x 10-1/2 3-1/8 x 10-1/2	2-1/2 x 10-1/2 3-1/8 x 9	2-1/2 x 12 3-1/8 x 12	2-1/2 x 12 3-1/8 x 10-1/2
	Southern Pine	2-1/2 x 6-7/8 3 x 6-7/8	2-1/2 x 6-7/8 3 x 6-7/8	2-1/2 x 8-1/4 3 x 8-1/4	2-1/2 x 8-1/4 3 x 8-1/4	2-1/2 x 11 3 x 11	2-1/2 x 9-5/8 3 x 9-5/8	2-1/2 x 12-3/8 3 x 11	2-1/2 x 11 3 x 11
18	Western Species	2-1/2 x 7-1/2 3-1/8 x 6	2-1/2 x 6 3-1/8 x 6	2-1/2 x 9 3-1/8 x 7-1/2	2-1/2 x 9 3-1/8 x 7-1/2	2-1/2 x 10-1/2 3-1/8 x 10-1/2	2-1/2 x 10-1/2 3-1/8 x 10-1/2	2-1/2 x 13-1/2 3-1/8 x 12	2-1/2 x 12 3-1/8 x 10-1/2
	Southern Pine	2-1/2 x 6-7/8 3 x 6-7/8	2-1/2 x 6-7/8 3 x 6-7/8	2-1/2 x 8-1/4 3 x 8-1/4	2-1/2 x 8-1/4 3 x 8-1/4	2-1/2 x 11 3 x 11	2-1/2 x 11 3 x 9-5/8	2-1/2 x 13-3/4 3 x 12-3/8	2-1/2 x 12-3/8 3 x 11
20	Western Species	2-1/2 x 7-1/2 3-1/8 x 6	2-1/2 x 6 3-1/8 x 6	2-1/2 x 9 3-1/8 x 7-1/2	2-1/2 x 9 3-1/8 x 7-1/2	2-1/2 x 10-1/2 3-1/8 x 10-1/2	2-1/2 x 10-1/2 3-1/8 x 10-1/2	2-1/2 x 13-1/2 3-1/8 x 12	2-1/2 x 12 3-1/8 x 12
	Southern Pine	2-1/2 x 6-7/8 3 x 6-7/8	2-1/2 x 6-7/8 3 x 6-7/8	2-1/2 x 8-1/4 3 x 8-1/4	2-1/2 x 8-1/4 3 x 8-1/4	2-1/2 x 11 3 x 11	2-1/2 x 11 3 x 9-5/8	2-1/2 x 13-3/4 3 x 12-3/8	2-1/2 x 12-3/8 3 x 11

Notes

(1) Span = uniformly loaded simply supported beam.

(2) Maximum deflection = L/180 under total load. Deflection under live load must be verified when live/total load > 3/4.

(3) Service condition = dry.

(4) Beam weights for sawn and glulam members are assumed to be the same.

(5) Volume factors for glulam members and size factors for sawn lumber members are in accordance with 1991 NDS.

(6) Minimum glulam sizes considered in the table are: 2-1/2 x 6 and 3-1/8 x 6 (western species), and 2-1/2 x 5-1/2 and 3 x 5-1/2 (southern pine).

(7) Design properties at normal load duration and dry-use service conditions -
Select Structural sawn lumber members: F_b = 2300 (8"), 2050 (10"), 1900 (12") psi, F_v = 90 psi, E = 1.8 x 10^6 psi.
No. 1 sawn lumber members: F_b 1650 (6"), 1500 (8"), 1300 (10"), 1250 (12") psi, F_v = 90 psi, E = 1.7 x 10^6 psi.
Glulam members: F_b = 2400 psi, F_v = 165 psi (western species) or 200 psi (southern pine), E = 1.8 x 10^6 psi.

TABLE 9
24F *WESTERN SPECIES* GLULAM EQUIVALENTS FOR
STEEL FLOOR BEAMS

(Load Duration Factor for Glulam = 1.00)

Span (ft)	W6x9	W8x10	W12x14	W12x16	W12x19	W10x22
	24F *Western Species* Glulam Equivalent (In.)					
12	3-1/8 x 10-1/2 5-1/8 x 9	3-1/8 x 13-1/2 5-1/8 x 10-1/2	3-1/8 x 21 5-1/8 x 15	3-1/8 x 24 5-1/8 x 16-1/2	3-1/8 x 27 5-1/8 x 19-1/2	3-1/8 x 28-1/2 5-1/8 x 21
16	3-1/8 x 10-1/2 5-1/8 x 9	3-1/8 x 13-1/2 5-1/8 x 12	3-1/8 x 18 5-1/8 x 15	3-1/8 x 21 5-1/8 x 15	3-1/8 x 24 5-1/8 x 16-1/2	3-1/8 x 25-1/2 5-1/8 x 18
20	3-1/8 x 10-1/2 5-1/8 x 9	3-1/8 x 13-1/2 5-1/8 x 12	3-1/8 x 18 5-1/8 x 15	3-1/8 x 19-1/2 5-1/8 x 16-1/2	3-1/8 x 21 5-1/8 x 18	3-1/8 x 19-1/2 5-1/8 x 16-1/2
24	3-1/8 x 10-1/2 5-1/8 x 9	3-1/8 x 13-1/2 5-1/8 x 12	3-1/8 x 18 5-1/8 x 16-1/2	3-1/8 x 19-1/2 5-1/8 x 16-1/2	3-1/8 x 21 5-1/8 x 18	3-1/8 x 19-1/2 5-1/8 x 16-1/2
28	3-1/8 x 10-1/2 5-1/8 x 9	3-1/8 x 13-1/2 5-1/8 x 12	3-1/8 x 18 5-1/8 x 16-1/2	3-1/8 x 19-1/2 5-1/8 x 16-1/2	3-1/8 x 21 5-1/8 x 18	3-1/8 x 19-1/2 5-1/8 x 16-1/2
32	3-1/8 x 10-1/2 5-1/8 x 9	3-1/8 x 13-1/2 5-1/8 x 12	3-1/8 x 18 5-1/8 x 16-1/2	3-1/8 x 19-1/2 5-1/8 x 16-1/2	3-1/8 x 21 5-1/8 x 18	3-1/8 x 19-1/2 5-1/8 x 16-1/2
36	3-1/8 x 10-1/2 5-1/8 x 9	3-1/8 x 13-1/2 5-1/8 x 12	3-1/8 x 18 5-1/8 x 16-1/2	3-1/8 x 19-1/2 5-1/8 x 16-1/2	3-1/8 x 21 5-1/8 x 18	3-1/8 x 19-1/2 5-1/8 x 16-1/2
40	3-1/8 x 10-1/2 5-1/8 x 9	3-1/8 x 13-1/2 5-1/8 x 12	3-1/8 x 18 5-1/8 x 16-1/2	3-1/8 x 19-1/2 5-1/8 x 16-1/2	3-1/8 x 21 5-1/8 x 18	3-1/8 x 19-1/2 5-1/8 x 16-1/2

Notes:
(1) Span = uniformly loaded simply supported beam.
(2) Maximum deflection = L/360 under live load, based on live/total load = 0.8.
(3) Service condition for glulam members = dry.
(4) Beam weights for steel and glulam members (assumed 35 pcf) are included.
(5) Volume factors for glulam members are in accordance with 1991 NDS.
(6) Minimum glulam sizes considered in the table are: 3-1/8 x 6 and 5-1/8 x 6.
(7) Design properties for steel members: F_b = 0.66 x 36 ksi, F_v = 0.4 x 36 ksi, E = 29 x 10^6 psi.
(8) Design properties for glulam members at normal load duration and dry-use service conditions: F_b = 2400 psi, F_v = 165 psi, E = 1.8 x 10^6 psi.

TABLE 10
24F *SOUTHERN PINE* GLULAM EQUIVALENTS FOR STEEL FLOOR BEAMS

(Load Duration Factor for Glulam = 1.00)

Span (ft)	W6x9	W8x10	W12x14	W12x16	W12x19	W10x22
	24F *Southern Pine* Glulam Equivalent (in.)					
12	3 x 11	3 x 13-3/4	3 x 19-1/4	3 x 20-5/8	3 x 24-3/4	3 x 26-1/8
	5 x 9-5/8	5 x 11	5 x 13-3/4	5 x 15-1/8	5 x 16-1/2	5 x 17-7/8
16	3 x 11	3 x 13-3/4	3 x 17-7/8	3 x 19-1/4	3 x 22	3 x 23-3/8
	5 x 9-5/8	5 x 11	5 x 15-1/8	5 x 15-1/8	5 x 16-1/2	5 x 17-7/8
20	3 x 11	3 x 13-3/4	3 x 17-7/8	3 x 19-1/4	3 x 20-5/8	3 x 20-5/8
	5 x 9-5/8	5 x 11	5 x 15-1/8	5 x 16-1/2	5 x 17-7/8	5 x 17-7/8
24	3 x 11	3 x 13-3/4	3 x 17-7/8	3 x 19-1/4	3 x 20-5/8	3 x 20-5/8
	5 x 9-5/8	5 x 11	5 x 16-1/2	5 x 16-1/2	5 x 17-7/8	5 x 17-7/8
28	3 x 11	3 x 13-3/4	3 x 17-7/8	3 x 19-1/4	3 x 20-5/8	3 x 20-5/8
	5 x 9-5/8	5 x 11	5 x 16-1/2	5 x 16-1/2	5 x 17-7/8	5 x 17-7/8
32	3 x 11	3 x 13-3/4	3 x 17-7/8	3 x 19-1/4	3 x 20-5/8	3 x 20-5/8
	5 x 9-5/8	5 x 11	5 x 16-1/2	5 x 16-1/2	5 x 17-7/8	5 x 17-7/8
36	3 x 11	3 x 13-3/4	3 x 17-7/8	3 x 19-1/4	3 x 20-5/8	3 x 20-5/8
	5 x 9-5/8	5 x 11	5 x 16-1/2	5 x 16-1/2	5 x 17-7/8	5 x 17-7/8
40	3 x 11	3 x 13-3/4	3 x 17-7/8	3 x 19-1/4	3 x 20-5/8	3 x 19-1/4
	5 x 9-5/8	5 x 12-3/8	5 x 16-1/2	5 x 16-1/2	5 x 17-7/8	5 x 17-7/8

Notes:

(1) Span = uniformly loaded simply supported beam.

(2) Maximum deflection = L/360 under live load, based on live/total load = 0.8.

(3) Service condition for glulam members = dry.

(4) Beam weights for steel and glulam members (assumed 36 pcf) are included.

(5) Volume factors for glulam members are in accordance with 1991 NDS.

(6) Minimum glulam sizes considered in the table are: 3-1/8 x 6 and 5-1/8 x 6.

(7) Design properties for steel members: F_b = 0.66 x 36 ksi, F_v = 0.4 x 36 ksi, E = 29 x 10^6 psi.

(8) Design properties for glulam members at normal load duration and dry-use service conditions: F_b = 2400 psi, F_v = 200 psi, E = 1.8 x 10^6 psi.

INDEX